Comprehensive Approach to Psychiatry

Volume 3

Series Editors
Bernardo Carpiniello, Department of Medical Science and Public Health
University of Cagliari, Cagliari, Italy

Antonio Vita, Department of Clinical and Experimental Science
University of Brescia, Brescia, Italy

Claudio Mencacci, Department of Neuroscience
Luigi Sacco Hospital, Milano, Italy

This book series will cover a broad range of topics with each volume providing an up-to-date overview of subject matters of particular interest for psychiatrists and other professionals in the field of mental health, through a comprehensive approach, including clinical, therapeutical, forensic and psychosocial points of view.

The topics addressed will be selected in collaboration with the Italian Society of Psychiatry.

The book series is aimed at psychiatrists, psychotherapists, psychologists, social workers, nurses, psychiatric rehabilitation technicians, students and researchers.

This book series is published in partnership with the Italian Society of Psychiatry (SIP).

Antonio Vita
Bernardo Carpiniello • Silvana Galderisi
Editors

Negative and Cognitive Symptoms in Schizophrenia

Springer

Editors
Antonio Vita
Department of Clinical & Experimental Sciences
University of Brescia
Brescia, Brescia, Italy

Bernardo Carpiniello
Department of Medical Sciences and Public Health
University of Cagliari
Cagliari, Italy

Silvana Galderisi
Department of Mental and Physical Health
University of Campania "Luigi Vanvitelli"
Naples, Italy

ISSN 2524-8405 ISSN 2524-8413 (electronic)
Comprehensive Approach to Psychiatry
ISBN 978-3-032-06208-6 ISBN 978-3-032-06209-3 (eBook)
https://doi.org/10.1007/978-3-032-06209-3

This Springer imprint is published by the registered company Springer Nature Switzerland AG
The registered company address is: Gewerbestrasse 11, 6330 Cham, Switzerland

Preface

Negative symptoms are common and pervasive in people living with schizophrenia. They can be categorized as primary (core features of schizophrenia) or secondary (attributed to other factors). These symptoms significantly impact an individual's life and are strong predictors of functional outcomes. However, they are often underrecognized and undertreated compared to positive symptoms. Addressing negative symptoms requires a multimodal approach. Enhancing our understanding and developing effective treatment strategies for negative symptoms is essential for improving the well-being of individuals with schizophrenia.

Cognitive impairment represents a core feature of schizophrenia, is present at any stage of the disease and in subjects at high risk for psychosis (UHR), and also has a significant impact on patient's psychosocial functioning. Pharmacological treatment is modestly effective on cognitive dimension while there is considerable evidence of the effectiveness of cognitive remediation interventions on cognition and functioning.

The book, proposed as the third volume of the series within the activities of the Italian Psychiatric Association, has the aim of updating knowledge and discussing available data about the definition of negative and cognitive symptoms, their impact on patients' real-life functioning and quality of life, the validated assessment instruments, also in early intervention settings, and the possible treatments, both pharmacological and non-pharmacological for negative and cognitive symptoms.

Particular attention will be given to discuss the relationships, overlap and independence of these two dimensions in schizophrenia, how to disentangle the relative prevalence and role of such dimensions in everyday patient's life, and the differential effectiveness of treatment options.

To our knowledge, this is the first effort to discuss the complex relationships between negative and cognitive symptoms in a dedicated book, with the contribution of many of the best recognized experts in the field.

We hope that the book will be useful for both researchers in the field of psychoses and for clinicians, very often challenged by these highly prevalent, impactful, and difficult-to-treat symptoms in people living with schizophrenia.

Brescia, Italy — Antonio Vita
Cagliari, Italy — Bernardo Carpiniello
Milano, Italy — Silvana Galderisi

Contents

Part I

Negative Symptoms in Schizophrenia: Clinical and Management Implications

Negative Symptoms in Schizophrenia: Definition, Classification, and Course 1

Pietro Blasio, Giulia M. Giordano, Pasquale Pezzella, Armida Mucci, and Silvana Galderisi

Abstract

Negative symptoms represent a core aspect of schizophrenia, significantly impacting patients' lives and posing challenges to healthcare systems globally. Despite their acknowledged importance, these symptoms have historically been neglected and understudied, also due to classification systems and diagnostic criteria that marginalized their importance. However, recent advancements have brought these symptoms to the forefront of research and clinical practice, recognizing their profound influence on functional outcome and treatment response.

Starting with an exploration of the historical evolution of the negative symptoms concept, this chapter explores the current conceptualization, definition, and longitudinal trajectory of negative symptoms in schizophrenia, describing their complexity and clinical relevance. It offers an updated perspective on these symptoms, underscoring their heterogeneity and the importance of incorporating current conceptualizations on negative symptoms into clinical approaches and research investigations.

Keywords

Blunted affect · Alogia · Anhedonia · Asociality · Avolition · Experiential dimension · Expressive dimension · Primary negative symptoms · Secondary negative symptoms · Persistent negative symptoms · Deficit schizophrenia

P. Blasio · G. M. Giordano · P. Pezzella · A. Mucci · S. Galderisi (✉)
University of Campania "Luigi Vanvitelli", Naples, Italy

A. Vita et al. (eds.), *Negative and Cognitive Symptoms in Schizophrenia*, Comprehensive Approach to Psychiatry 3,
https://doi.org/10.1007/978-3-032-06209-3_1

1.1 Introduction

The expression "negative symptoms" refers to deficits or disruptions of normal functions and behaviors related to motivation, interests, and emotions [1, 2]. Since the first descriptions of schizophrenia, negative symptoms have been acknowledged as a core aspect of the disorder, including two dimensions: loss of motivation and reduction of emotional expression [3, 4]. However, over the years and until the last two decades, negative symptoms have played a marginal role within classification systems and operational criteria for the diagnosis of schizophrenia [5, 6], and only a few dedicated research groups have persisted in highlighting the relevance of negative symptoms for the diagnosis and prognosis of schizophrenia [1, 2, 5]. Recent years have witnessed a significant resurgence of interest in negative symptoms, mainly due to the evidence that they exhibit a high prevalence among individuals with schizophrenia (more than half of the subjects present at least one negative symptom of moderate severity) [7, 8] and are linked to low remission rates, poor real-life functioning, and diminished quality of life [9–15], and a poor response to available treatments [16].

The present chapter will focus on the history of negative symptoms and their current conceptualization. We will also describe the structure of negative symptoms and their classification, as well as the course of these symptoms across various stages of schizophrenia.

1.2 History of Negative Symptoms

The first descriptions of the "negative symptoms" might be dated back to 1809, before the concepts of "Dementia Praecox" and "Schizophrenia" [3, 4]. John Haslam, in his work "Observations on Madness and Melancholy" [17], described a young patient affected by a mental disorder causing him a "thymic deficit" characterized by blunted sensitivity and emotional indifference. In the nineteenth century, J.R. Reynolds, in his monograph on epilepsy, developed the concept of positive symptoms (e.g., delusions, hallucinations, and convulsions) due to a distortion or exaggeration of natural functions, and negative symptoms (e.g., loss of sensation, paralysis, and coma) due to reduction or loss of natural functions [18]. In the same years, J.H. Jackson, in his work "On temporary mental disorders after epileptic paroxysms", hypothesized that negative symptoms were related to a reduction in aspects of higher cognitive and emotional functioning, while positive symptoms were due to excessive functioning [19]. Jackson theorized a hierarchical nervous system model in which a lesion affecting the higher levels of the central nervous system directly produces the negative symptoms while lower-level brain centres, no longer inhibited by the higher ones, produce the positive symptoms. In Jackson's theory, positive symptoms cannot appear without negative ones. In 1889, Emil Kraepelin identified "Dementia Praecox," a mental disorder that mainly affected young people [20]. He classified different forms of this disorder, such as a hebephrenic form (characterized by affective and emotional blunting), a paranoid form (characterized by delusions and hallucinations), and a catatonic form (characterized by a motor deficit syndrome) [4]. In particular, in the hebephrenic form, Kraepelin emphasized the role of "emotional dullness," "failure of mental activities," and "loss of mastery over volition," as the predominant signs and symptoms [4]. Therefore, in the early 1900s, Kraepelin recognized symptoms resembling what we now conceptualize as negative symptoms, placing them in a privileged position as a critical feature of schizophrenia.

In 1908, Bleuler coined the term "Schizophrenia" [3, 21]. He distinguished between fundamental and accessory symptoms of schizophrenia. Fundamental symptoms included weakening of the association process, inadequate or blunted affect, volitional disorders, ambivalence, and autism, which were required for the diagnosis of schizophrenia. Accessory symptoms, on the other hand, included hallucinations and delusions. Probably due to the poorer inter-rater reliability of negative symptoms, compared to positive ones, in the operational diagnostic criteria of schizophrenia, the role of negative symptoms had been de-emphasized [22]. For several years first-rank symptoms described by Kurt Schneider (e.g., thought withdrawal and thought broadcasting) were central element of main diagnostic system criteria, such as PSE-Catego [23], the Diagnostic and Statistical Manual of Mental Disorders (DSM-III and DSM-III-R), the Research Diagnostic Criteria (RDC, [24]) and the International Classification of Diseases (ICD-9, ICD-10). During the 1970s, a renewed interest in negative symptoms was elicited by Strauss and colleagues [25], who categorized three dimensions of schizophrenia, positive symptoms, negative symptoms and relational disorders, and also highlighted the importance of negative symptoms in the diagnosis and prognosis of schizophrenia [25].

In 1988, Carpenter [26] introduced the expression "Deficit Schizophrenia," to describe a specific syndrome characterized by the presence of primary and persistent negative symptoms [26–28]. This syndrome was conceptualized as an illness separate from non-Deficit Schizophrenia, with different signs and symptoms, course of illness, risk factors, biological correlates, and treatment response [28–31].

In the last decades, negative symptoms have gained a renewed interest due to the recognition of their profound impact on the real-life functioning of patients with schizophrenia [11, 14, 15, 32, 33]. Unfortunately, neurobiological mechanisms underlying negative symptoms remain unclear and current therapeutic options have limited effectiveness in treating them [16, 34, 35]. Moreover, although negative symptoms are included in criterion A of the schizophrenia diagnostic criteria in the latest edition of the Diagnostic and Statistical Manual of Mental Disorders (DSM-5), they remain non-essential for the diagnosis of schizophrenia.

1.3 Current Conceptualization of Negative Symptoms

In 2005, Stephen Marder, Wayne Fenton, William T. Carpenter, Jr, and Brian Kirkpatrick organized a consensus development conference, which was held at the NIMH Neuroscience Center in Rockville, Maryland [36]. The objectives outlined in the Consensus Conference were: (i) to review the data concerning the existence of distinct dimensions within negative symptoms, essential for selecting appropriate measures in clinical trials and (ii) to initiate a process to develop or identify universally accepted, evidence-based measures and methodologies necessary to validate the effectiveness of treatments targeting negative symptoms.

One of the consensus points derived from the conference was the inclusion of five constructs in the negative symptom dimension: blunted affect, alogia, anhedonia, asociality, and avolition [36–39]. Blunted affect is traditionally described as a decrease in the observed expression of emotion (facial and vocal expression and body gestures) [2, 36, 40]. The evaluation of blunted affect is based on what is observed during the clinical interview, referring to the spontaneous expression of emotions or the expression of emotion in response to prompts provided by the interviewer, not considering the subjective emotional experience. It has been hypothesized that the pathogenesis of blunted affect includes alterations in emotion identification and discrimination and the ability to perceive nonverbal social signals (facial affect, prosody, and body gestures) [41, 42]. Another hypothesis links the blunted emotional expression to deficits in motor activities [43, 44]. Indeed, individuals with motor abnormalities often experience difficulties in nonverbal communication, such as expressionless faces, reduced expressive gestures, and diminished modulation of the voice [2, 45, 46]. The underlying mechanisms may involve basal ganglia or frontal lobe dysfunctions and may vary across different components of blunted affect [47].

Alogia is a reduction in the quantity of spoken words and spontaneous elaboration of speech. The evaluation of alogia is based on the subjects' language production during the clinical interview. The clinician rates patient's tendency to reply with short answers throughout the interview while not considering the impoverished content of speech.

Multiple studies indicated a link between alogia and impaired performance on verbal fluency tasks, suggesting a possible shared mechanism of alogia and cognitive impairment [48–51]. In support of this hypothesis, Cohen et al. [52] formulated the cognitive resource limitation model, proposing that patients tend to decrease their speech production and general verbal communication in social situations in case of limited cognitive resources. Therefore, this model underscores the vital link between alogia and cognitive impairments.

Anhedonia refers to the diminished capacity to experience (consummatory anhedonia) and/or anticipate pleasure for future activities (anticipatory anhedonia). Consummatory anhedonia is a reduced pleasure experience derived from ongoing enjoyable activities. Anticipatory anhedonia is defined as the reduced ability to anticipate future pleasure. Different studies indicated that individuals diagnosed with schizophrenia might experience similar levels of pleasure in response to emotion-inducing stimuli as healthy individuals [44, 53–56], while they may have impairments in anticipating pleasure [56–58]. The evaluation of anhedonia is based on what the subjects report on the number of pleasant activities/situations they enjoyed or planned, and the intensity of pleasure derived from the same activities/situations and from planning the same activities in the future.

The deficit of pleasure anticipation in schizophrenia is often seen as a component of the broader construct of motivation, as the capacity to anticipate pleasure plays a crucial role in driving behavior toward achieving an expected, albeit presently unavailable, reward [59]. Anomalies in pleasure experience in schizophrenia have also been conceptualized as difficulties in recalling past or predicting future experiences [44]. Consequently, there have been suggestions to avoid the term "anhedonia" in favor of "reduced pleasure-seeking behavior" or "beliefs of low pleasure" [44, 54, 58, 60, 61]. In summary, the current literature suggests that individuals with schizophrenia typically retain the capacity to experience consummatory pleasure. However, they often exhibit a deficiency in anticipating pleasure and engaging in reward-seeking behaviors [2]. The mechanisms responsible for these impairments may pertain to certain facets of motivation, such as anticipation of rewards or assessment of effort, as well as cognitive functioning, such as compromised episodic memory, which hinders the individual's capacity to recollect past enjoyable experiences [2].

Asociality is defined as a reduction in social initiative due to decreased interest in forming close relationships with others; it is often observed before the onset of schizophrenia [2, 62]. It should not be solely assessed considering the observable behavior, such as the number of social interactions and close relationships, but also by exploring the motivation for social interaction (i.e., whether the individual values and seeks out social interactions and close social relationships and attributes importance to social relationships) [39, 63]. Asociality is explored by asking the patients how often they are with others, whether they take the initiative to meet with others, if they desire contact with others, and whether they have close relationships and believe that social relationships are essential. Some authors hypothesized a correlation between asociality and deficits in social cognition [64]. Social cognition refers to the mental activity involved in how individuals perceive, retain, and utilize information regarding others and social contexts [65–67]. The correlation between asociality and social cognition is likely complex: diminished motivation to engage in social activities could lead to isolation and underdeveloped social cognition [64], while deficient social cognition might result in an inability to perceive reward signals during social interactions, potentially leading to social anhedonia, decreased motivation, and ultimately, asocial behavior. Another hypothesis on a possible mechanism of asociality links dysfunctional self-defeating beliefs and asocial behavior: pessimistic expectations regarding future rewards or success in social interactions could diminish motivation to participate in social activities [68].

Avolition is described as a reduced initiation of and persistence in goal-directed activity, due to reduced motivation. Avolition is explored by asking the patients how they spend their time, how much time they spend without doing anything, if they work, study, look for a job, or take care of themselves. At the same time, whether patients are willing to work, study, look for a job, or take care of themselves should be explored. The avolition dimension may be associated with deficits in motivation [69, 70]. These deficits may be related to dysfunctions of the motivation value system, leading to inadequate representation of value (impaired valuation of outcomes and actions), and/or to dysfunctions of the motivational salience system, leading to difficulties in directing attention towards significant stimuli, reduced cognitive engagement, and overall motivation, and/or to poor integration between these two systems [1]. Another hypothesis suggests that deficits in the executive control of behavior might subtend avolition. Indeed, it has been demonstrated that deficits in planning and monitoring might impair goal-directed behavior, with subsequently reduced motivation, which leads to avolition as well as to anhedonia and asociality. However, findings across studies are inconsistent and need further investigation [1].

1.4 Structure of Negative Symptoms

Several studies on negative symptom structure using different assessment instruments, such as the Positive and Negative Syndrome Scale (PANSS), the Scale for the Assessment of Negative Symptoms (SANS), the Brief Negative Symptom Scale (BNSS) or the Clinical Assessment Interview for Negative Symptoms (CAINS), supported a two-factor model comprising the Experiential dimension (including avolition, asociality and anhedonia) and the Expressive dimension (including blunted affect and alogia) (Fig. 1.1) [2, 6, 9, 37, 71]. The two-factor model aligns with the observation that the two dimensions are associated with different clinical and social outcomes and might be subtended by different pathophysiological mechanisms [2, 6, 9, 71, 72].

According to recent studies, the Experiential dimension is associated with dysconnectivity or abnormalities within brain areas belonging to the motivational pathways [1, 9, 73–76]. The Expressive dimension has gained less attention than the Experiential one [1, 9, 77–81]. It might be linked to cognitive impairment (both neurocognition and social cognition) [1, 9, 77–81] and to neurological soft signs,

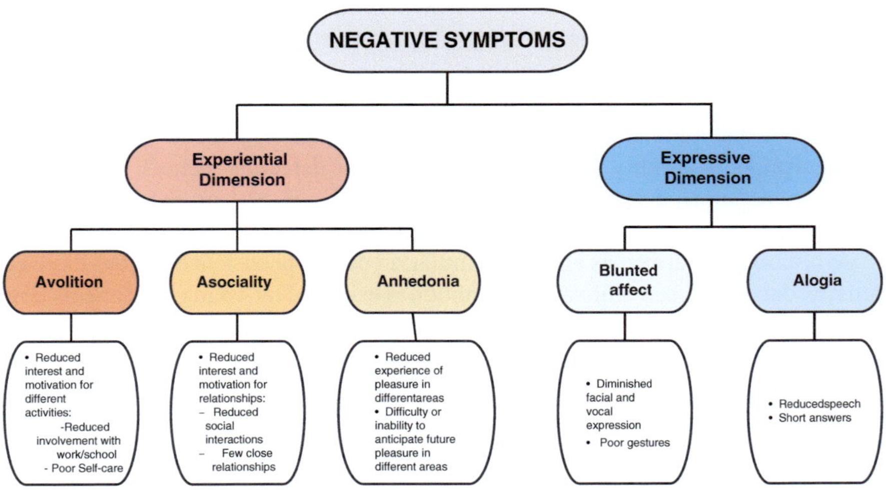

Fig. 1.1 Flow chart illustrating Experiential and Expressive dimensions of negative symptoms

thus suggesting an association with aberrations in neurodevelopmental processes [1, 9, 78, 82].

As regards the relationship with the outcome, the two dimensions show distinct patterns of associations, as we will report in paragraph 6. In particular, the Experimental dimension is a strong predictor of poor functional outcome and has a greater impact on interpersonal relationships and work skills than the Expressive dimension [14, 15].

The two-factor structure of negative symptoms has been supported by several factor analytic studies on both first-generation (such as the PANSS) [83, 84] or SANS [85–87], and second-generation rating scales, (such as CAINS and BNSS) [38, 39, 63, 88–94]. The two-factor model has had a significant impact, shaping the methodology of research exploring the pathophysiological mechanisms of negative symptoms, and of clinical trials aimed at developing new treatments for these symptoms [95–98].

However, more recently, the validity of the two-factor structure has been questioned [99]. Indeed, findings from recent multicenter studies employing confirmatory factor analysis (CFA) or network analyses of negative symptoms assessed with new-generation rating scales, such as the BNSS and the CAINS, have cast doubt on the suitability of the two-factor model [8, 99–104] and a five-factor model has gained prominence. A hierarchical model also emerged, with the five individual negative symptoms serving as first-order factors and the Experiential and Expressive dimensions as second-order factors [100–104]. Recent investigations have demonstrated that the five-factor and hierarchical models strongly fit the data [100, 105, 106]. The good fit of the hierarchical model should not be interpreted as support for the two-factor model but rather as confirming the five-factor model, given that experiential and expressive are considered second-order dimensions in the hierarchical model. Currently, the model of negative symptoms that most accurately explains the latent structure of these symptoms is the five-dimensional model [104, 105, 107–109].

This reinterpretation may have important implications for research on the neurobiological bases of negative symptoms and the development of innovative treatment approaches. However, existing evidence validating distinct neurobiological correlates for each of the five individual negative symptoms is still in its early stages, emphasizing the need for more focused research efforts [2, 110].

1.5 Classification of Negative Symptoms

Different etiologies and trajectories, potentially necessitating diverse treatment approaches, may underlie negative symptoms.

1.5.1 Primary and Secondary Negative Symptoms

The distinction between primary and secondary negative symptoms carries significant diagnostic and therapeutic implications. Primary negative symptoms are intrinsic to the schizophrenia syndrome, typically persist throughout the illness [81, 111–113] and do not respond to available treatments. In contrast, secondary negative symptoms are due to identifiable factors, such as positive symptoms (e.g., social withdrawal due to paranoid delusions), depression, adverse effects of antipsychotic medications (e.g., extrapyramidal side effects), substance use, as well as social deprivation resulting from prolonged hospitalizations or loss of significant relationships [6, 114]. This distinction has therapeutic implications, as some of the factors that cause secondary negative symptoms, such as positive symptoms, depression, or extrapyramidal side effects, can be treated, thus leading to an improvement in negative symptoms and, consequently, in functional outcome.

To identify negative symptoms secondary to positive symptoms, it might be useful to check whether negative symptom severity worsens following the psychotic exacerbations and evaluate whether they could reflect the behavioral consequence of moderate-severity psychotic symptoms (e.g., asociality due to persecutory delusions). Moreover, in these cases, an improvement of negative symptoms is usually observed together with the improvement of positive symptoms, and a worsening occurs during periods of drug discontinuation [1, 114]. In clinical practice, to identify negative symptoms secondary to positive symptoms, it is important to explore patients' internal experiences during psychotic exacerbations, modification of antipsychotic medication, and periods of clinical stability.

Negative symptoms can be due to drug side effects. Expressive deficit symptoms, such as blunted affect and alogia, might often resemble drug-induced extrapyramidal symptoms. To disentangle negative symptoms from extrapyramidal side effects, it is important to identify extrapyramidal signs that do not overlap with the negative ones, such as rigidity or tremor. Furthermore, avolition might be caused by sedation or induced by antipsychotic medications, especially first-generation ones, which cause a blunted response to reward [115, 116]. In these cases, an increase in symptom severity is typically observed following an increase in drug dosage [6, 114].

Negative symptoms secondary to depression. Depressive symptoms resembling negative symptoms, such as pronounced anhedonia, diminished goal-directed behavior, and social withdrawal, frequently occur in individuals with schizophrenia [1, 114, 117–120]. However, recent studies [121–123] suggest that the differential diagnosis (e.g., whether anhedonia is a negative symptom or a depressive symptom) could be enhanced through the utilization of the Calgary Depression Scale for Schizophrenia, regarded as the optimal assessment instrument for evaluating depressive symptoms in this population [6, 124, 125]. To ascertain if depression contributes to secondary negative symptoms in individuals with schizophrenia, this assessment instrument might be employed to examine the patient's subjective experience of depressed mood and thoughts, such as feelings of hopelessness and guilt, which are more related to depression than to negative symptoms.

1.5.2 The Deficit Syndrome

In 1988, Carpenter proposed the concept of Deficit Schizophrenia (DS) to describe patients with schizophrenia characterized by the presence of primary and persistent negative symptoms [26]. Later, Kirkpatrick proposed DS as a diagnostic category, separate from Non-Deficit schizophrenia (NDS) [31]. Other studies have also supported the hypothesis that DS represents a distinct nosological entity [37, 126]. Diagnostic criteria for DS, also summarized in Table 1.1, are the following:

1. Presence of at least two out of the following six negative symptoms: restricted affect (referring to observed behaviors rather than to the patient's subjective experience); diminished emotional range (i.e., reduced range of the patient's subjective emotional experience); poverty of speech; curbing of interests; diminished sense of purpose; diminished social drive.
2. Presence of the above symptoms for at least 12 months, including periods of clinical stability.
3. The above symptoms are primary and not secondary to factors such as anxiety, drug effect, positive symptoms, intellectual disability, and depression.
4. The patient meets DSM (third or later edition) criteria for schizophrenia.

Data from epidemiological studies suggested that DS has a prevalence of 15% at the first episode of psychosis, between 25% and 30% in clinical samples, ad between 14% and 17% in population studies [31, 127], with a higher prevalence in male individuals [26, 128, 129] and subjects with a positive family history of schizophrenia [31, 127, 130, 131]. Different studies have demonstrated that patients with DS differ from those with NDS in terms of risk factors, premorbid functioning, longitudinal course of the disorder, neurobiological correlates, and treatment response [28, 29, 132–134],

Patients with DS, as compared to those with NDS, have a more insidious onset, more pronounced negative symptoms, comparable severity of psychotic symptoms, poorer premorbid functioning, poorer response to treatment, but psychotic symptoms of comparable severity, lower prevalence of dysphoria, hostility, suicidal ideation, depressive symptoms, and substance abuse [28, 29, 31, 135, 136]. Furthermore, DS is more frequently associated with neurological soft signs and spontaneous movement disorders than NDS [28–30, 137, 138], thus suggesting an early deviation from normal neurodevelopmental trajectories [139]. However, it is important to highlight the fact that conflicting results regarding the prevalence of these neurodevelopmental signs have been reported, probably due to the presence of confounding factors (e.g., severity of negative symptoms and pharmacological history) not always controlled for in the analysis [27–29, 137, 138, 140]. Finally, both neuroimaging and neurophysiological data suggested either less or different abnormalities in patients with DS with respect to those with NDS [81], sometimes with evidence of double dissociation [79, 141, 142]. All these findings suggest that DS could be an independent construct with respect to NDS, and not just a more severe form of the disease, as its characteristics and correlates are not just "more of the same" observed in NDS [1, 29, 31, 40, 112, 143].

The gold standard instrument for the diagnosis of Deficit Schizophrenia is the Schedule for the Deficit Syndrome (SDS) [144]. The instrument has good psychometric properties, but also some limitations, such as the need for extensive training, different information sources, and longitudinal assessment to differentiate primary from secondary negative symptoms [29, 31, 81, 112, 143, 145]. To increase the practicality of diagnosing DS, some proxies have been developed using the Brief Psychiatric Rating Scale (BPRS) [146], the Positive and Negative Symptoms Scale (PANSS) [147], or the Scale for the Assessment of Negative Symptoms (SANS) [148]. Although these proxies have shown good sensitivity and specificity, there are concerns about the face validity and the temporal stability of the classification that they provide [81, 133, 149].

1.5.3 Persistent, Predominant, and Prominent Negative Symptoms

The clear distinction between DS and NDS is sometimes difficult, especially at the onset of the psychosis, considering the need for longitudinal observation to distinguish between primary and secondary negative symptoms [145]. To overcome this classification limitation, the concept of "Persistent Negative Symptoms" (PNS) has been introduced [36, 145].

According to the NIMH-MATRICS consensus report, primary or secondary PNS, typically do not respond to available treatments. They significantly impede patients' capacity to fulfil typical role functions and persist even during periods of clinical stability, thus representing a considerable unmet therapeutic challenge [36, 81, 145].

The criteria for the identification of PNS (Table 1.1), according to Buchanan [145], are the following:

1. Presence of at least moderate severity negative symptoms, as assessed using an accepted and validated rating scale;
2. Defined threshold levels of positive symptoms, depression, and extrapyramidal symptoms on accepted and validated rating scales;
3. Persistence for at least 6 months.

Available evidence demonstrates that this concept exhibits construct validity and offers several advantages. Compared to DS, the PNS construct is more easily applicable in clinical trials. Indeed, it allows for the inclusion of some potential sources of secondary negative symptoms while still enabling control of the main confounding factors. It requires less longitudinal observation compared to DS, thus identifying a larger patient population than DS and being suitable for early intervention studies. Nonetheless, there are concerns regarding this construct: it allows the utilization of any validated psychopathological rating scale, including those, like the PANSS and SANS, which contain items unrelated to negative symptoms, and the use of different thresholds for confounding factors (positive, depressive, and extrapyramidal symptoms), thus leading to a lack of consistent definition across studies [6].

Following regulatory agency requirements for assessing drug efficacy in clinical trials targeting negative symptoms, two additional concepts have been developed: "Predominant Negative Symptoms" and "Prominent Negative Symptoms" (Table 1.1). However, neither of these constructs incorporates the assessment of the persistence of negative symptoms over time.

Table 1.1 Criteria for deficit schizophrenia, persistent, predominant, and prominent negative symptoms

	Severity of negative symptoms	Severity of other symptoms	Persistence of negative symptoms
Deficit schizophrenia	At least moderate for at least two negative symptoms [25, 30]	Not specified. Positive, depressive, and anxiety symptoms must be excluded as causes of the negative ones	At least 12 months
Persistent negative symptoms	At least moderate severity of negative symptoms [142]	Positive, depressive, and extrapyramidal symptoms lower than a defined threshold on an accepted and validated rating scale [142]	At least 6 months
Predominant negative symptoms	At least moderate for at least three negative symptoms or at least moderately severe for at least two symptoms [147] or any score on PANSS negative subscale but at least 6 points greater than the PANSS negative subscale score [148] or PANSS negative subscale score of at least 21 and at least 1 point greater than the PANSS positive subscale score [149] or PANSS negative subscale score greater than the PANSS positive subscale score [150]	Positive PANSS subscale score less than 19, depressive, and extrapyramidal symptoms lower than a defined threshold on validated rating scale [147] or not specified [148–150]	Not specified
Prominent negative symptoms	At least moderate for at least three symptoms or at least moderately severe for at least two symptoms [147–150]	Not specified	Not specified

For "Predominant Negative Symptoms," the criteria used in the different studies are heterogeneous and include one criterion from list A and one from list B, as follows:

A

1. Presence of at least moderate for at least three symptoms or at least moderately severe for at least two symptoms [150] or
2. Any score on the PANSS negative subscale but at least 6 points greater than the PANSS positive subscale score [151] or
3. PANSS negative subscale score of at least 21 and at least 1 point greater than the PANSS positive subscale score [152] or
4. PANSS negative subscale score greater than the PANSS positive subscale score [153]

B

1. Positive PANSS subscale score less than 19, depressive and extrapyramidal symptoms lower than a defined threshold on a validated rating scale [150] or
2. Severity of positive, depressive, and extrapyramidal symptoms not specified [151–153]

For "Prominent Negative Symptoms," the criteria used in different studies included at least three symptoms rated "moderate" or at least two symptoms rated "moderately severe" on the PANSS negative [150, 153]. However, limited data are available on whether to consider these constructs (predominant and prominent negative symptoms) in clinical trials [1, 81, 143, 154]. The huge limitations of these two constructs, highlighted by available evidence and expert opinion, are that both encompass a combination of primary and secondary negative symptoms that are likely to fluctuate over time, potentially confounding the results of clinical trials, and that no consensus has been reached to reduce the heterogeneity in the definition of predominant negative symptoms [6].

1.6 Course of Negative Symptoms

Negative symptoms often appear early in the course of the illness, although they typically do not prompt clinical attention [155].

They often precede the first acute psychotic episode and exhibit high stability over time, with a huge impact on functioning, especially in the presence of poor premorbid adjustment and in male patients [1, 111, 155–157]. Additionally, it is important to acknowledge that negative symptoms may also emerge later in the course of the disorder, which is significant for understanding their development and impact. Research indicates that the timing of onset and progression of negative symptoms can be influenced by various factors, including genetics, early brain neurodevelopment, trauma, and responses to psychotic episodes [158, 159]. Further investigation into how these factors interact with the course of negative symptoms, including their relationship with relapse, treatment resistance, and environmental influences, would provide a more comprehensive understanding of their long-term trajectory and implications for treatment.

1.6.1 Negative Symptoms in Subjects at High Risk (HR) of Psychosis

Negative symptoms, especially asociality, appear before attenuated positive symptoms and are more prevalent than the latter ones [160–162]. They have been observed in a significant portion of individuals at HR for psychosis [160, 163, 164], occurring in almost 80% of children and adolescents at HR. More interestingly, in this population, PNS are also identifiable, with a prevalence of almost 6% [165].

The onset of negative symptoms during the prodromal phase has been identified as a risk factor for transitioning to psychosis [162, 164, 166], suggesting that they could be useful as predictors of the transition to psychosis [167]. Furthermore, it has been demonstrated that negative symptoms have a detrimental impact on real-life functioning even in HR subjects and mediate the association between cognition and functioning [165, 168–178] as for the chronic subjects [64, 179, 180].

1.6.2 Negative Symptoms in Subjects with First-Episode Psychosis (FEP)

Negative symptoms are frequent in FEP. In particular, they occur in 60% of children and adolescents and 30–50% of adults experiencing a FEP [158]. The higher prevalence in children and adolescents may be due to different factors. For instance, children and adolescents with FEP frequently present neurodevelopmental alterations [181], show poor or worse premorbid adjustment [182], and experience several cognitive dysfunctions [183].

Negative symptoms in FEP have been associated with impaired premorbid functioning, heightened risk of deliberate self-harm, low treatment adherence, diminished quality of life, and poor functional outcome, the latter persisting for up to 7 years after initial contact with mental health services [184–190].

As regards the PNS, their prevalence in FEP ranges from 4% to 41% [191–203], depending primarily on the different criteria employed to define PNS across studies. While some studies required a single negative symptom of moderate severity [193, 197, 199, 200], others required at least two [192, 196]. Furthermore, some studies excluded [197, 199, 200], while others included symptoms that did not consistently cluster around negative symptom domains [191, 192, 198]. Variability in controlling for secondary negative symptoms also contributed to discrepancies: some studies set thresholds for both depression and Parkinsonism [193, 195, 197, 200], whereas others controlled only for depression severity [198], or positive symptoms [196, 199, 200]. Moreover, differences in follow-up duration further contributed to such heterogeneity, with persistence assessments ranging from 1 month to 3 years in studies involving FEP patients. It is noteworthy that PNS in FEP subjects has been specifically associated with impairment in functional outcome, reduced quality of life, severity of cognitive impairment, poor treatment adherence, and prolonged untreated psychosis [195–197, 201].

1.6.3 Negative Symptoms in Subjects with Chronic Schizophrenia

Several studies indicated that from 54% to 60% of individuals with long-term schizophrenia experience at least one negative symptom of moderate severity [71, 204, 205]. Primary negative symptoms are present in 10–30% of patients with chronic schizophrenia [7, 204]. There is little prevalence data of PNS in chronic schizophrenia [81, 145]. Approximately 35–70% of schizophrenia patients continue to experience clinically significant negative symptoms that persist after treatment of the acute episode [71, 156, 206–208].

In subjects with chronic schizophrenia, negative symptoms have been found to increase the risk for hospital admission and readmission, as well as for longer duration of hospitalization [204]. They directly and indirectly influence the functional outcome of individuals with schizophrenia. Indeed, substantial evidence shows a direct or indirect effect of negative symptoms on social functioning, community activities, work skills, and interpersonal skills [12–14, 180, 209–213]. Moreover, current available data suggested that the two negative symptom dimensions, the Experiential and the Expressive dimensions, showed different relationships with the different aspects of functioning. The Experiential dimension is associated with interpersonal relationships and work skills [32]. In contrast, the Expressive dimension has been found to have a weaker and more indirect relationship with real-life functioning and is associated with everyday life skills [12–15, 32]. Furthermore, in a longitudinal study with a four-year follow-up, only the Experiential dimension predicted interpersonal relationships, while the Expressive dimension did not demonstrate predictive power [15]. Other studies have also supported the differential relationship between the negative symptom dimensions and functioning [214–217].

The long-term trajectory of negative symptoms is still debated, since some studies suggested their relative stability over time, while others indicated fluctuation or significant improvement [111, 218–220]. Therefore, rigorous longitudinal studies are necessary to assess the natural course of negative symptoms, distinguishing primary and secondary symptoms. These studies should implement clear protocols for detecting secondary negative symptoms and addressing potential underlying factors like extrapyramidal side effects, depression, positive symptoms, isolation, and hypostimulation across different stages of the disorder.

1.7 Conclusions

Starting with an exploration of the historical evolution of negative symptoms, this chapter thoroughly explores the current conceptualization, definition, and longitudinal trajectory of negative symptoms in schizophrenia, describing their complex nature and clinical significance.

Negative symptoms represent a core aspect of schizophrenia, significantly impacting patients' functional outcome and quality of life.

However, despite a broad consensus on the definition of negative symptoms, they are not always properly evaluated and are under-recognized by healthcare professionals, caregivers, and patients themselves. Therefore, there is a need to raise awareness among both clinicians and caregivers about the early recognition of these symptoms to allow their management and improve the functioning of individuals affected by schizophrenia.

Furthermore, an accurate assessment of primary and secondary negative symptoms is fundamental to conduct studies aimed at uncovering the pathogenetic mechanisms of these symptoms, promoting research into new treatment options, and advancing knowledge in the management of negative symptoms. Standardized assessment tools are also critical, not only for ensuring consistency in clinical trials but also for improving the reproducibility of research findings. The adoption of such tools would further aid in the study of the neurobiological mechanisms underlying negative symptoms and guide the development of more targeted, effective interventions.

Finally, identifying biomarkers that can define subgroups of individuals with specific negative symptom profiles, considering the type of negative symptoms, the severity, and clinical course, will be critical to better grasp their evolution and variability and for developing targeted therapies.

References

1. Galderisi S, Mucci A, Buchanan RW, Arango C. Negative symptoms of schizophrenia: new developments and unanswered research questions. Lancet Psychiatry. 2018;5(8):664–77.
2. Marder SR, Galderisi S. The current conceptualization of negative symptoms in schizophrenia. World Psychiatry. 2017;16(1):14–24.
3. Bleuler E. Dementia praecox or the group of schizophrenias. Oxford: International Universities Press; 1950. p. 548.
4. Kraepelin E. Dementia praecox and paraphrenia. Chicago: Chicago Medical Book Company; 1919.
5. Dollfus S, Lyne J. Negative symptoms: history of the concept and their position in diagnosis of schizophrenia. Schizophr Res. 2017;186:3–7.
6. Galderisi S, Mucci A, Dollfus S, Nordentoft M, Falkai P, Kaiser S, et al. EPA guidance on assessment of negative symptoms in schizophrenia. Eur Psychiatry. 2021;64(1):e23.
7. Bobes J, Arango C, Garcia-Garcia M, Rejas J. Prevalence of negative symptoms in outpatients with schizophrenia spectrum disorders treated with antipsychotics in routine clinical practice: findings from the CLAMORS study. J Clin Psychiatry. 2010;71(3):280–6.
8. Mucci A, Vignapiano A, Bitter I, Austin SF, Delouche C, Dollfus S, et al. A large European, multicenter, multinational validation study of the brief negative symptom scale. Eur Neuropsychopharmacol. 2019;29(8):947–59.

9. Giordano GM, Caporusso E, Pezzella P, Galderisi S. Updated perspectives on the clinical significance of negative symptoms in patients with schizophrenia. Expert Rev Neurother. 2022;22(7):541–55.
10. Foussias G, Agid O, Fervaha G, Remington G. Negative symptoms of schizophrenia: clinical features, relevance to real world functioning and specificity versus other CNS disorders. Eur Neuropsychopharmacol. 2014;24(5):693–709.
11. García-Fernández L, Romero-Ferreiro V, Sánchez-Pastor L, Dompablo M, Martínez-Gras I, Espejo-Saavedra JM, et al. Impact of negative symptoms on functioning and quality of life in first psychotic episodes of schizophrenia. J Clin Med. 2022;11(4)
12. Galderisi S, Rossi A, Rocca P, Bertolino A, Mucci A, Bucci P, et al. The influence of illness-related variables, personal resources and context-related factors on real-life functioning of people with schizophrenia. World Psychiatry. 2014;13(3):275–87.
13. Galderisi S, Rucci P, Kirkpatrick B, Mucci A, Gibertoni D, Rocca P, et al. Interplay among psychopathologic variables, personal resources, context-related factors, and real-life functioning in individuals with schizophrenia: a network analysis. JAMA Psychiatry. 2018;75(4):396–404.
14. Galderisi S, Rucci P, Mucci A, Rossi A, Rocca P, Bertolino A, et al. The interplay among psychopathology, personal resources, context-related factors and real-life functioning in schizophrenia: stability in relationships after 4 years and differences in network structure between recovered and non-recovered patients. World Psychiatry. 2020;19(1):81–91.
15. Mucci A, Galderisi S, Gibertoni D, Rossi A, Rocca P, Bertolino A, et al. Factors associated with real-life functioning in persons with schizophrenia in a 4-year follow-up study of the Italian network for research on psychoses. JAMA Psychiatry. 2021;78(5):550–9.
16. Galderisi S, Kaiser S, Bitter I, Nordentoft M, Mucci A, Sabé M, et al. EPA guidance on treatment of negative symptoms in schizophrenia. Eur Psychiatry. 2021;64(1):e21.
17. Haslam J. Observations on madness and melancholy. London: J. Callow; 1809.
18. Russell Reynolds J. Lecture on the classification of diseases of the nervous system. Lancet (London, England). 1873;102:405–9.
19. Jackson JH. On temporary mental disorders after epileptic paroxysms. West Riding Lunatic Asylum Med Rep. 1885:105–29.
20. Kraepelin E. Psychiatrie. In: Ein kurzes Lehrbuch für Studierende und Ärzte. 3rd ed. Leipzig: Abel Verlag; 1889.
21. Bleuler E. Die schizophrenen Geistesstörungen im Lichte langjähriger Kranken- und Familiengeschichten. 1908.
22. Feighner JP, Robins E, Guze SB, Woodruff RA Jr, Winokur G, Munoz R. Diagnostic criteria for use in psychiatric research. Arch Gen Psychiatry. 1972;26(1):57–63.
23. Wing JK. At the Research Front*: Standardizing Clinical Diagnostic Judgements the Pse-Catego System: Standardizing Clinical Diagnostic Judgements the Pse-Catego System. Australian & New Zealand Journal of Psychiatry. 1980;14(1):17–20. https://doi.org/10.3109/00048678009159350.
24. Spitzer RL, Endicott J, Robins E. Research diagnostic criteria: rationale and reliability. Arch Gen Psychiatry. 1978;35(6):773–82.
25. Strauss JS, Carpenter WT Jr, Bartko JJ. The diagnosis and understanding of schizophrenia. Part III. Speculations on the processes that underlie schizophrenic symptoms and signs. Schizophr Bull. 1974;11:61–9.
26. Carpenter WT Jr, Heinrichs DW, Wagman AM. Deficit and nondeficit forms of schizophrenia: the concept. Am J Psychiatry. 1988;145(5):578–83.
27. Fenton WS, McGlashan TH. Antecedents, symptom progression, and long-term outcome of the deficit syndrome in schizophrenia. Am J Psychiatry. 1994;151(3):351–6.
28. Kirkpatrick B, Galderisi S. Deficit schizophrenia: an update. World Psychiatry. 2008;7(3):143–7.
29. Galderisi S, Maj M. Deficit schizophrenia: an overview of clinical, biological and treatment aspects. Eur Psychiatry. 2009;24(8):493–500.

30. Galderisi S, Maj M, Mucci A, Cassano GB, Invernizzi G, Rossi A, et al. Historical, psychopathological, neurological, and neuropsychological aspects of deficit schizophrenia: a multicenter study. Am J Psychiatry. 2002;159(6):983–90.
31. Kirkpatrick B, Buchanan RW, Ross DE, Carpenter WT Jr. A separate disease within the syndrome of schizophrenia. Arch Gen Psychiatry. 2001;58(2):165–71.
32. Giuliani L, Giordano GM, Bucci P, Pezzella P, Brando F, Galderisi S. Improving knowledge on pathways to functional outcome in schizophrenia: main results from the Italian network for research on psychoses. Front Psych. 2021;12:791117.
33. Sabe M, Chen C, Perez N, Solmi M, Mucci A, Galderisi S, et al. Thirty years of research on negative symptoms of schizophrenia: a scientometric analysis of hotspots, bursts, and research trends. Neurosci Biobehav Rev. 2023;144:104979.
34. Galderisi S, Kaiser S. The pathophysiology of negative symptoms of schizophrenia: main hypotheses and open challenges. Br J Psychiatry. 2023;223(1):298–300.
35. Collo G, Mucci A, Giordano GM, Merlo Pich E, Galderisi S. Negative symptoms of schizophrenia and dopaminergic transmission: translational models and perspectives opened by iPSC techniques. Front Neurosci. 2020;14:632.
36. Kirkpatrick B, Fenton WS, Carpenter WT Jr, Marder SR. The NIMH-MATRICS consensus statement on negative symptoms. Schizophr Bull. 2006;32(2):214–9.
37. Blanchard JJ, Cohen AS. The structure of negative symptoms within schizophrenia: implications for assessment. Schizophr Bull. 2006;32(2):238–45.
38. Horan WP, Kring AM, Gur RE, Reise SP, Blanchard JJ. Development and psychometric validation of the Clinical Assessment Interview for Negative Symptoms (CAINS). Schizophr Res. 2011;132(2–3):140–5.
39. Kring AM, Gur RE, Blanchard JJ, Horan WP, Reise SP. The Clinical Assessment Interview for Negative Symptoms (CAINS): final development and validation. Am J Psychiatry. 2013;170(2):165–72.
40. Kirkpatrick B. Progress in the study of negative symptoms. Schizophr Bull. 2014;40(Suppl 2):S101–6.
41. Gur RE, Kohler CG, Ragland JD, Siegel SJ, Lesko K, Bilker WB, et al. Flat affect in schizophrenia: relation to emotion processing and neurocognitive measures. Schizophr Bull. 2006;32(2):279–87.
42. Lepage M, Sergerie K, Benoit A, Czechowska Y, Dickie E, Armony JL. Emotional face processing and flat affect in schizophrenia: functional and structural neural correlates. Psychol Med. 2011;41(9):1833–44.
43. Alpert M, Rosenberg SD, Pouget ER, Shaw RJ. Prosody and lexical accuracy in flat affect schizophrenia. Psychiatry Res. 2000;97(2–3):107–18.
44. Kring AM, Moran EK. Emotional response deficits in schizophrenia: insights from affective science. Schizophr Bull. 2008;34(5):819–34.
45. Dworkin RH, Clark SC, Amador XF, Gorman JM. Does affective blunting in schizophrenia reflect affective deficit or neuromotor dysfunction? Schizophr Res. 1996;20(3):301–6.
46. Aghevli MA, Blanchard JJ, Horan WP. The expression and experience of emotion in schizophrenia: a study of social interactions. Psychiatry Res. 2003;119(3):261–70.
47. Walther S, Stegmayer K, Sulzbacher J, Vanbellingen T, Müri R, Strik W, et al. Nonverbal social communication and gesture control in schizophrenia. Schizophr Bull. 2015;41(2):338–45.
48. Fervaha G, Takeuchi H, Foussias G, Agid O, Remington G. Using poverty of speech as a case study to explore the overlap between negative symptoms and cognitive dysfunction. Schizophr Res. 2016;176(2–3):411–6.
49. Stolar N, Berenbaum H, Banich MT, Barch D. Neuropsychological correlates of alogia and affective flattening in schizophrenia. Biol Psychiatry. 1994;35(3):164–72.
50. Joyce EM, Collinson SL, Crichton P. Verbal fluency in schizophrenia: relationship with executive function, semantic memory and clinical alogia. Psychol Med. 1996;26(1):39–49.
51. Bowie CR, Harvey PD, Moriarty PJ, Parrella M, White L, Davis KL. A comprehensive analysis of verbal fluency deficit in geriatric schizophrenia. Arch Clin Neuropsychol. 2004;19(2):289–303.

52. Cohen AS, McGovern JE, Dinzeo TJ, Covington MA. Speech deficits in serious mental illness: a cognitive resource issue? Schizophr Res. 2014;160(1–3):173–9.
53. Cohen AS, Minor KS. Emotional experience in patients with schizophrenia revisited: meta-analysis of laboratory studies. Schizophr Bull. 2010;36(1):143–50.
54. Strauss GP, Gold JM. A new perspective on anhedonia in schizophrenia. Am J Psychiatry. 2012;169(4):364–73.
55. Szczypiński JJ, Gola M. Dopamine dysregulation hypothesis: the common basis for motivational anhedonia in major depressive disorder and schizophrenia? Rev Neurosci. 2018;29(7):727–44.
56. Gard DE, Kring AM, Gard MG, Horan WP, Green MF. Anhedonia in schizophrenia: distinctions between anticipatory and consummatory pleasure. Schizophr Res. 2007;93(1–3):253–60.
57. Kring AM, Caponigro JM. Emotion in schizophrenia: where feeling meets thinking. Curr Dir Psychol Sci. 2010;19(4):255–9.
58. Frost KH, Strauss GP. A review of anticipatory pleasure in schizophrenia. Curr Behav Neurosci Rep. 2016;3(3):232–47.
59. Strauss GP, Waltz JA, Gold JM. A review of reward processing and motivational impairment in schizophrenia. Schizophr Bull. 2014;40(Suppl 2):S107–16.
60. Hu HX, Jiang SY, Shan HD, Chu MY, Lv QY, Yi ZH, et al. Negative belief-updating bias for positive daily life events in individuals with schizophrenia and social anhedonia. Cogn Neuropsychiatry. 2022;27(4):237–54.
61. Visser KF, Chapman HC, Ruiz I, Raugh IM, Strauss GP. A meta-analysis of self-reported anticipatory and consummatory pleasure in the schizophrenia-spectrum. J Psychiatr Res. 2020;121:68–81.
62. Cannon M, Jones P, Gilvarry C, Rifkin L, McKenzie K, Foerster A, et al. Premorbid social functioning in schizophrenia and bipolar disorder: similarities and differences. Am J Psychiatry. 1997;154(11):1544–50.
63. Kirkpatrick B, Strauss GP, Nguyen L, Fischer BA, Daniel DG, Cienfuegos A, et al. The brief negative symptom scale: psychometric properties. Schizophr Bull. 2011;37(2):300–5.
64. Lin CH, Huang CL, Chang YC, Chen PW, Lin CY, Tsai GE, et al. Clinical symptoms, mainly negative symptoms, mediate the influence of neurocognition and social cognition on functional outcome of schizophrenia. Schizophr Res. 2013;146(1–3):231–7.
65. Green MF, Bearden CE, Cannon TD, Fiske AP, Hellemann GS, Horan WP, et al. Social cognition in schizophrenia, part 1: performance across phase of illness. Schizophr Bull. 2012;38(4):854–64.
66. Green MF, Horan WP, Lee J. Social cognition in schizophrenia. Nat Rev Neurosci. 2015;16(10):620–31.
67. Pinkham AE, Penn DL, Green MF, Buck B, Healey K, Harvey PD. The social cognition psychometric evaluation study: results of the expert survey and RAND panel. Schizophr Bull. 2014;40(4):813–23.
68. Beck AT, Grant PM, Huh GA, Perivoliotis D, Chang NA. Dysfunctional attitudes and expectancies in deficit syndrome schizophrenia. Schizophr Bull. 2013;39(1):43–51.
69. Bromberg-Martin ES, Matsumoto M, Hikosaka O. Dopamine in motivational control: rewarding, aversive, and alerting. Neuron. 2010;68(5):815–34.
70. Salamone JD, Yohn SE, López-Cruz L, San Miguel N, Correa M. Activational and effort-related aspects of motivation: neural mechanisms and implications for psychopathology. Brain. 2016;139(Pt 5):1325–47.
71. Correll CU, Schooler NR. Negative symptoms in schizophrenia: a review and clinical guide for recognition, assessment, and treatment. Neuropsychiatr Dis Treat. 2020;16:519–34.
72. Rocca P, Montemagni C, Zappia S, Piterà R, Sigaudo M, Bogetto F. Negative symptoms and everyday functioning in schizophrenia: a cross-sectional study in a real world-setting. Psychiatry Res. 2014;218(3):284–9.
73. Barch DM, Dowd EC. Goal representations and motivational drive in schizophrenia: the role of prefrontal-striatal interactions. Schizophr Bull. 2010;36(5):919–34.

74. Giordano GM, Pezzella P, Giuliani L, Fazio L, Mucci A, Perrottelli A, et al. Resting-state brain activity dysfunctions in schizophrenia and their associations with negative symptom domains: an fMRI study. Brain Sci. 2023;13(1)
75. Giordano GM, Pezzella P, Quarantelli M, Bucci P, Prinster A, Soricelli A, et al. Investigating the relationship between White matter connectivity and motivational circuits in subjects with deficit schizophrenia: a Diffusion Tensor Imaging (DTI) study. J Clin Med. 2021;11(1)
76. Mucci A, Dima D, Soricelli A, Volpe U, Bucci P, Frangou S, et al. Is avolition in schizophrenia associated with a deficit of dorsal caudate activity? A functional magnetic resonance imaging study during reward anticipation and feedback. Psychol Med. 2015;45(8):1765–78.
77. Bègue I, Kaiser S, Kirschner M. Pathophysiology of negative symptom dimensions of schizophrenia - current developments and implications for treatment. Neurosci Biobehav Rev. 2020;116:74–88.
78. Galderisi S, Merlotti E, Mucci A. Neurobiological background of negative symptoms. Eur Arch Psychiatry Clin Neurosci. 2015;265(7):543–58.
79. Giordano GM, Brando F, Perrottelli A, Di Lorenzo G, Siracusano A, Giuliani L, et al. Tracing links between early auditory information processing and negative symptoms in schizophrenia: an ERP study. Front Psych. 2021;12:790745.
80. Kaiser S, Lyne J, Agartz I, Clarke M, Mørch-Johnsen L, Faerden A. Individual negative symptoms and domains - relevance for assessment, pathomechanisms and treatment. Schizophr Res. 2017;186:39–45.
81. Mucci A, Merlotti E, Üçok A, Aleman A, Galderisi S. Primary and persistent negative symptoms: concepts, assessments and neurobiological bases. Schizophr Res. 2017;186:19–28.
82. Peralta V, Moreno-Izco L, Sanchez-Torres A, García de Jalón E, Campos MS, Cuesta MJ. Characterization of the deficit syndrome in drug-naive schizophrenia patients: the role of spontaneous movement disorders and neurological soft signs. Schizophr Bull. 2014;40(1):214–24.
83. Strauss GP, Horan WP, Kirkpatrick B, Fischer BA, Keller WR, Miski P, et al. Deconstructing negative symptoms of schizophrenia: avolition-apathy and diminished expression clusters predict clinical presentation and functional outcome. J Psychiatr Res. 2013;47(6):783–90.
84. Ergül C, Üçok A. Negative symptom subgroups have different effects on the clinical course of schizophrenia after the first episode: a 24-month follow up study. Eur Psychiatry. 2015;30(1):14–9.
85. Liemburg E, Castelein S, Stewart R, van der Gaag M, Aleman A, Knegtering H. Two subdomains of negative symptoms in psychotic disorders: established and confirmed in two large cohorts. J Psychiatr Res. 2013;47(6):718–25.
86. Jang SK, Choi HI, Park S, Jaekal E, Lee GY, Cho YI, et al. A two-factor model better explains heterogeneity in negative symptoms: evidence from the positive and negative syndrome scale. Front Psychol. 2016;7:707.
87. Stiekema AP, Liemburg EJ, van der Meer L, Castelein S, Stewart R, van Weeghel J, et al. Confirmatory factor analysis and differential relationships of the two subdomains of negative symptoms in chronically ill psychotic patients. PLoS One. 2016;11(2):e0149785.
88. Mucci A, Galderisi S, Merlotti E, Rossi A, Rocca P, Bucci P, et al. The Brief Negative Symptom Scale (BNSS): independent validation in a large sample of Italian patients with schizophrenia. Eur Psychiatry. 2015;30(5):641–7.
89. de Medeiros HLV, Vasconcelos SC, Elkis H, Martins DR, de Alexandria Leite RM, de Albuquerque ACL, et al. The brief negative symptom scale: validation in a multicenter Brazilian study. Compr Psychiatry. 2018;85:42–7.
90. Kimhy D, Yale S, Goetz RR, McFarr LM, Malaspina D. The factorial structure of the schedule for the deficit syndrome in schizophrenia. Schizophr Bull. 2006;32(2):274–8.
91. Nakaya M, Ohmori K. A two-factor structure for the schedule for the deficit syndrome in schizophrenia. Psychiatry Res. 2008;158(2):256–9.
92. Strauss GP, Hong LE, Gold JM, Buchanan RW, McMahon RP, Keller WR, et al. Factor structure of the brief negative symptom scale. Schizophr Res. 2012;142(1–3):96–8.

93. Blanchard JJ, Horan WP, Collins LM. Examining the latent structure of negative symptoms: is there a distinct subtype of negative symptom schizophrenia? Schizophr Res. 2005;77(2–3):151–65.
94. Engel M, Fritzsche A, Lincoln TM. Validation of the German version of the Clinical Assessment Interview for Negative Symptoms (CAINS). Psychiatry Res. 2014;220(1–2):659–63.
95. Kring AM, Barch DM. The motivation and pleasure dimension of negative symptoms: neural substrates and behavioral outputs. Eur Neuropsychopharmacol. 2014;24(5):725–36.
96. Marder SR, Kirkpatrick B. Defining and measuring negative symptoms of schizophrenia in clinical trials. Eur Neuropsychopharmacol. 2014;24(5):737–43.
97. Messinger JW, Trémeau F, Antonius D, Mendelsohn E, Prudent V, Stanford AD, et al. Avolition and expressive deficits capture negative symptom phenomenology: implications for DSM-5 and schizophrenia research. Clin Psychol Rev. 2011;31(1):161–8.
98. Watson D, Levin-Aspenson HF, Waszczuk MA, Conway CC, Dalgleish T, Dretsch MN, et al. Validity and utility of hierarchical taxonomy of psychopathology (HiTOP): III. Emotional dysfunction superspectrum. World Psychiatry. 2022;21(1):26–54.
99. Strauss GP, Ahmed AO, Young JW, Kirkpatrick B. Reconsidering the latent structure of negative symptoms in schizophrenia: a review of evidence supporting the 5 consensus domains. Schizophr Bull. 2019;45(4):725–9.
100. Ahmed AO, Kirkpatrick B, Galderisi S, Mucci A, Rossi A, Bertolino A, et al. Cross-cultural validation of the 5-factor structure of negative symptoms in schizophrenia. Schizophr Bull. 2019;45(2):305–14.
101. Ahmed AO, Kirkpatrick B, Granholm E, Rowland LM, Barker PB, Gold JM, et al. Two factors, five factors, or both? external validation studies of negative symptom dimensions in schizophrenia. Schizophr Bull. 2022;48(3):620–30.
102. Chang WC, Strauss GP, Ahmed AO, Wong SCY, Chan JKN, Lee EHM, et al. The latent structure of negative symptoms in individuals with attenuated psychosis syndrome and early psychosis: support for the 5 consensus domains. Schizophr Bull. 2021;47(2):386–94.
103. Li SB, Liu C, Zhang JB, Wang LL, Hu HX, Chu MY, et al. Revisiting the latent structure of negative symptoms in schizophrenia: evidence from two second-generation clinical assessments. Schizophr Res. 2022;248:131–9.
104. Strauss GP, Nuñez A, Ahmed AO, Barchard KA, Granholm E, Kirkpatrick B, et al. The latent structure of negative symptoms in schizophrenia. JAMA Psychiatry. 2018;75(12):1271–9.
105. Rucci P, Caporusso E, Sanmarchi F, Giordano GM, Mucci A, Giuliani L, et al. The structure stability of negative symptoms: longitudinal network analysis of the brief negative symptom scale in people with schizophrenia. BJPsych Open. 2023;
106. Giordano GM, Sanmarchi F, Mucci A, Rucci P, Brando F, Caporusso E, et al. External validation of the five domains of negative symptoms: focus on cognition, functional capacity, and real-world functioning. Eur Psychiatry. 2023;67(1):e3.
107. Haguiara B, Koga G, Diniz E, Fonseca L, Higuchi CH, Kagan S, et al. What is the best latent structure of negative symptoms in schizophrenia? A systematic review. Schizophr Bull Open. 2021;2(1):sgab013.
108. Strauss GP, Esfahlani FZ, Galderisi S, Mucci A, Rossi A, Bucci P, et al. Network analysis reveals the latent structure of negative symptoms in schizophrenia. Schizophr Bull. 2019;45(5):1033–41.
109. Tam MHW, Ling-Ling W, Cheng KM, Wong JOY, Cheung EFC, Lui SSY, et al. Latent structure of self-report negative symptoms in patients with schizophrenia: a preliminary study. Asian J Psychiatr. 2021;61:102680.
110. Shaffer JJ, Peterson MJ, McMahon MA, Bizzell J, Calhoun V, van Erp TG, et al. Neural correlates of schizophrenia negative symptoms: distinct subtypes impact dissociable brain circuits. Mol Neuropsychiatry. 2015;1(4):191–200.
111. Austin SF, Mors O, Budtz-Jørgensen E, Secher RG, Hjorthøj CR, Bertelsen M, et al. Long-term trajectories of positive and negative symptoms in first episode psychosis: a 10year follow-up study in the OPUS cohort. Schizophr Res. 2015;168(1–2):84–91.

112. Kirkpatrick B, Mucci A, Galderisi S. Primary, enduring negative symptoms: an update on research. Schizophr Bull. 2017;43(4):730–6.
113. Kirkpatrick B, Cohen A, Bitter I, Strauss GP. Primary negative symptoms: refining the research target. Schizophr Bull. 2021;47(5):1207–10.
114. Kirschner M, Aleman A, Kaiser S. Secondary negative symptoms - a review of mechanisms, assessment and treatment. Schizophr Res. 2017;186:29–38.
115. Lewander T. Overcoming the neuroleptic-induced deficit syndrome: clinical observations with remoxipride. Acta Psychiatr Scand Suppl. 1994;380:64–7.
116. Lieberman JA, Stroup TS, McEvoy JP, Swartz MS, Rosenheck RA, Perkins DO, et al. Effectiveness of antipsychotic drugs in patients with chronic schizophrenia. N Engl J Med. 2005;353(12):1209–23.
117. Buckley PF, Miller BJ, Lehrer DS, Castle DJ. Psychiatric comorbidities and schizophrenia. Schizophr Bull. 2009;35(2):383–402.
118. Krynicki CR, Upthegrove R, Deakin JFW, Barnes TRE. The relationship between negative symptoms and depression in schizophrenia: a systematic review. Acta Psychiatr Scand. 2018;137(5):380–90.
119. Mulholland CCS. The symptom of depression in schizophrenia and its management. Adv Psychiatr Treat. 2000;6(3):169–77.
120. Rahim T, Rashid R. Comparison of depression symptoms between primary depression and secondary-to-schizophrenia depression. Int J Psychiatry Clin Pract. 2017;21(4):314–7.
121. Grover S, Sahoo S, Dua D, Chakrabarti S, Avasthi A. Scales for assessment of depression in schizophrenia: factor analysis of Calgary depression rating scale and Hamilton depression rating scale. Psychiatry Res. 2017;252:333–9.
122. Lako IM, Wigman JT, Klaassen RM, Slooff CJ, Taxis K, Bartels-Velthuis AA. Psychometric properties of the self-report version of the Quick Inventory of Depressive Symptoms ($QIDS-SR_{16}$) questionnaire in patients with schizophrenia. BMC Psychiatry. 2014;14:247.
123. Schennach R, Riedel M, Obermeier M, Seemüller F, Jäger M, Schmauss M, et al. What are depressive symptoms in acutely ill patients with schizophrenia spectrum disorder? Eur Psychiatry. 2015;30(1):43–50.
124. Addington D, Addington J, Maticka-Tyndale E. Assessing depression in schizophrenia: the Calgary Depression Scale. Br J Psychiatry Suppl. 1993;22:39–44.
125. Müller MJ, Brening H, Gensch C, Klinga J, Kienzle B, Müller KM. The Calgary Depression Rating Scale for schizophrenia in a healthy control group: psychometric properties and reference values. J Affect Disord. 2005;88(1):69–74.
126. Ahmed AO, Strauss GP, Buchanan RW, Kirkpatrick B, Carpenter WT. Are negative symptoms dimensional or categorical? Detection and validation of deficit schizophrenia with taxometric and latent variable mixture models. Schizophr Bull. 2015;41(4):879–91.
127. Kirkpatrick B, Castle D, Murray RM, Carpenter WT Jr. Risk factors for the deficit syndrome of schizophrenia. Schizophr Bull. 2000;26(1):233–42.
128. Bottlender R, Jäger M, Groll C, Strauss A, Möller HJ. Deficit states in schizophrenia and their association with the length of illness and gender. Eur Arch Psychiatry Clin Neurosci. 2001;251(6):272–8.
129. Roy MA, Maziade M, Labbé A, Mérette C. Male gender is associated with deficit schizophrenia: a meta-analysis. Schizophr Res. 2001;47(2–3):141–7.
130. Dollfus S, Germain-Robin S, Chabot B, Brazo P, Delamillieure P, Langlois S, et al. Family history and obstetric complications in deficit and non-deficit schizophrenia: preliminary results. Eur Psychiatry. 1998;13(5):270–2.
131. Ross DE, Kirkpatrick B, Karkowski LM, Straub RE, MacLean CJ, O'Neill FA, et al. Sibling correlation of deficit syndrome in the Irish study of high-density schizophrenia families. Am J Psychiatry. 2000;157(7):1071–6.
132. Cohen AS, Brown LA, Minor KS. The psychiatric symptomatology of deficit schizophrenia: a meta-analysis. Schizophr Res. 2010;118(1–3):122–7.
133. Strauss GP, Harrow M, Grossman LS, Rosen C. Periods of recovery in deficit syndrome schizophrenia: a 20-year multi-follow-up longitudinal study. Schizophr Bull. 2010;36(4):788–99.

134. Galderisi S, Bucci P, Mucci A, Kirkpatrick B, Pini S, Rossi A, et al. Categorical and dimensional approaches to negative symptoms of schizophrenia: focus on long-term stability and functional outcome. Schizophr Res. 2013;147(1):157–62.
135. Kirkpatrick B. Recognizing primary vs secondary negative symptoms and apathy vs expression domains. J Clin Psychiatry. 2014;75(4):e09.
136. Bucci P, Mucci A, Piegari G, Nobile M, Pini S, Rossi A, et al. Characterization of premorbid functioning during childhood in patients with deficit vs. non-deficit schizophrenia and in their healthy siblings. Schizophr Res. 2016;174(1–3):172–6.
137. Cimmer C, Szendi I, Csifcsák G, Szekeres G, Ambrus Kovács Z, Somogyi I, et al. Abnormal neurological signs, visual contrast sensitivity, and the deficit syndrome of schizophrenia. Prog Neuro-Psychopharmacol Biol Psychiatry. 2006;30(7):1225–30.
138. Arango C, Kirkpatrick B, Buchanan RW. Neurological signs and the heterogeneity of schizophrenia. Am J Psychiatry. 2000;157(4):560–5.
139. Melle I. Cognition in schizophrenia: a marker of underlying neurodevelopmental problems? World Psychiatry. 2019;18(2):164–5.
140. Telfer S, Shivashankar S, Krishnadas R, McCreadie RG, Kirkpatrick B. Tardive dyskinesia and deficit schizophrenia. Acta Psychiatr Scand. 2011;124(5):357–62.
141. Giordano GM, Perrottelli A, Mucci A, Di Lorenzo G, Altamura M, Bellomo A, et al. Investigating the relationships of P3b with negative symptoms and neurocognition in subjects with chronic schizophrenia. Brain Sci. 2021;11(12)
142. Giuliani L, Koutsouleris N, Giordano GM, Koenig T, Mucci A, Perrottelli A, et al. A multivariate approach to investigate the associations of electrophysiological indices with schizophrenia clinical and functional outcome. Eur Psychiatry. 2023;66(1):e46.
143. Bucci P, Galderisi S. Categorizing and assessing negative symptoms. Curr Opin Psychiatry. 2017;30(3):201–8.
144. Kirkpatrick B, Buchanan RW, McKenney PD, Alphs LD, Carpenter WT Jr. The Schedule for the Deficit syndrome: an instrument for research in schizophrenia. Psychiatry Res. 1989;30(2):119–23.
145. Buchanan RW. Persistent negative symptoms in schizophrenia: an overview. Schizophr Bull. 2006;33(4):1013–22.
146. Overall JE, Gorham DR. The brief psychiatric rating scale. Psychol Rep. 1962;10:799–812.
147. Kay SR, Fiszbein A, Opler LA. The positive and negative syndrome scale (PANSS) for schizophrenia. Schizophr Bull. 1987;13(2):261–76.
148. Andreasen NC. The Scale for the Assessment of Negative Symptoms (SANS): conceptual and theoretical foundations. Br J Psychiatry Suppl. 1989;7:49–58.
149. Subotnik KL, Nuechterlein KH, Ventura J, Green MF, Hwang SS. Prediction of the deficit syndrome from initial deficit symptoms in the early course of schizophrenia. Psychiatry Res. 1998;80(1):53–9.
150. Stauffer VL, Song G, Kinon BJ, Ascher-Svanum H, Chen L, Feldman PD, et al. Responses to antipsychotic therapy among patients with schizophrenia or schizoaffective disorder and either predominant or prominent negative symptoms. Schizophr Res. 2012;134(2–3):195–201.
151. Olié JP, Spina E, Murray S, Yang R. Ziprasidone and amisulpride effectively treat negative symptoms of schizophrenia: results of a 12-week, double-blind study. Int Clin Psychopharmacol. 2006;21(3):143–51.
152. Riedel M, Müller N, Strassnig M, Spellmann I, Engel RR, Musil R, et al. Quetiapine has equivalent efficacy and superior tolerability to risperidone in the treatment of schizophrenia with predominantly negative symptoms. Eur Arch Psychiatry Clin Neurosci. 2005;255(6):432–7.
153. Rabinowitz J, Werbeloff N, Caers I, Mandel FS, Stauffer V, Menard F, et al. Negative symptoms in schizophrenia--the remarkable impact of inclusion definitions in clinical trials and their consequences. Schizophr Res. 2013;150(2–3):334–8.
154. Marder SR, Alphs L, Anghelescu IG, Arango C, Barnes TR, Caers I, et al. Issues and perspectives in designing clinical trials for negative symptoms in schizophrenia. Schizophr Res. 2013;150(2–3):328–33.
155. der Heiden W, Häfner H. The epidemiology of onset and course of schizophrenia. Eur Arch Psychiatry Clin Neurosci. 2000;250(6):292–303.

156. An der Heiden W, Leber A, Häfner H. Negative symptoms and their association with depressive symptoms in the long-term course of schizophrenia. Eur Arch Psychiatry Clin Neurosci. 2016;266(5):387–96.
157. Lyne J, O'Donoghue B, Roche E, Renwick L, Cannon M, Clarke M. Negative symptoms of psychosis: a life course approach and implications for prevention and treatment. Early Interv Psychiatry. 2018;12(4):561–71.
158. Salazar de Pablo G, Catalan A, Vaquerizo Serrano J, Pedruzo B, Alameda L, Sandroni V, et al. Negative symptoms in children and adolescents with early-onset psychosis and at clinical high-risk for psychosis: systematic review and meta-analysis. Br J Psychiatry. 2023;223(1):282–94.
159. Chen EYH, Wong SMY, Hui CLM, Suen YN, Chan SKW. The emergence of primary negative symptoms: relevance of timing? Br J Psychiatry. 2023;223(1):280–1.
160. Lencz T, Smith CW, Auther A, Correll CU, Cornblatt B. Nonspecific and attenuated negative symptoms in patients at clinical high-risk for schizophrenia. Schizophr Res. 2004;68(1):37–48.
161. Yung AR, Phillips LJ, Yuen HP, Francey SM, McFarlane CA, Hallgren M, et al. Psychosis prediction: 12-month follow up of a high-risk ("prodromal") group. Schizophr Res. 2003;60(1):21–32.
162. Velthorst E, Nieman DH, Becker HE, van de Fliert R, Dingemans PM, Klaassen R, et al. Baseline differences in clinical symptomatology between ultra high risk subjects with and without a transition to psychosis. Schizophr Res. 2009;109(1–3):60–5.
163. Iyer SN, Boekestyn L, Cassidy CM, King S, Joober R, Malla AK. Signs and symptoms in the pre-psychotic phase: description and implications for diagnostic trajectories. Psychol Med. 2008;38(8):1147–56.
164. Piskulic D, Addington J, Cadenhead KS, Cannon TD, Cornblatt BA, Heinssen R, et al. Negative symptoms in individuals at clinical high risk of psychosis. Psychiatry Res. 2012;196(2–3):220–4.
165. Yung AR, Nelson B, McGorry PD, Wood SJ, Lin A. Persistent negative symptoms in individuals at ultra high risk for psychosis. Schizophr Res. 2019;206:355–61.
166. Spada G, Molteni S, Pistone C, Chiappedi M, McGuire P, Fusar-Poli P, et al. Identifying children and adolescents at ultra high risk of psychosis in Italian neuropsychiatry services: a feasibility study. Eur Child Adolesc Psychiatry. 2016;25(1):91–106.
167. Schmidt A, Cappucciati M, Radua J, Rutigliano G, Rocchetti M, Dell'Osso L, et al. Improving prognostic accuracy in subjects at clinical high risk for psychosis: systematic review of predictive models and meta-analytical sequential testing simulation. Schizophr Bull. 2017;43(2):375–88.
168. Lee SJ, Kim KR, Lee SY, An SK. Impaired social and role function in ultra-high risk for psychosis and first-episode schizophrenia: its relations with negative symptoms. Psychiatry Investig. 2017;14(5):539–45.
169. Devoe DJ, Braun A, Seredynski T, Addington J. Negative symptoms and functioning in youth at risk of psychosis: a systematic review and meta-analysis. Harv Rev Psychiatry. 2020;28(6):341–55.
170. Üçok A, Direk N, Kaya H, Çağlar N, Çıkrıkçılı U, Noyan H, et al. Relationship of negative symptom severity with cognitive symptoms and functioning in subjects at ultra-high risk for psychosis. Early Interv Psychiatry. 2021;15(4):966–74.
171. Strauss GP, Pelletier-Baldelli A, Visser KF, Walker EF, Mittal VA. A review of negative symptom assessment strategies in youth at clinical high-risk for psychosis. Schizophr Res. 2020;222:104–12.
172. Carrión RE, Demmin D, Auther AM, McLaughlin D, Olsen R, Lencz T, et al. Duration of attenuated positive and negative symptoms in individuals at clinical high risk: associations with risk of conversion to psychosis and functional outcome. J Psychiatr Res. 2016;81:95–101.
173. Schlosser DA, Campellone TR, Biagianti B, Delucchi KL, Gard DE, Fulford D, et al. Modeling the role of negative symptoms in determining social functioning in individuals at clinical high risk of psychosis. Schizophr Res. 2015;169(1–3):204–8.

174. Demjaha A, Valmaggia L, Stahl D, Byrne M, McGuire P. Disorganization/cognitive and negative symptom dimensions in the at-risk mental state predict subsequent transition to psychosis. Schizophr Bull. 2012;38(2):351–9.
175. Glenthøj LB, Jepsen JR, Hjorthøj C, Bak N, Kristensen TD, Wenneberg C, et al. Negative symptoms mediate the relationship between neurocognition and function in individuals at ultrahigh risk for psychosis. Acta Psychiatr Scand. 2017;135(3):250–8.
176. Gerritsen C, Maheandiran M, Lepock J, Ahmed S, Kiang M, Bagby RM, et al. Negative symptoms in the clinical high-risk state for psychosis: connection with cognition and primacy in impacting functioning. Early Interv Psychiatry. 2020;14(2):188–95.
177. Glenthøj LB, Kristensen TD, Wenneberg C, Hjorthøj C, Nordentoft M. Experiential negative symptoms are more predictive of real-life functional outcome than expressive negative symptoms in clinical high-risk states. Schizophr Res. 2020;218:151–6.
178. Kim KR, Song YY, Park JY, Lee EH, Lee M, Lee SY, et al. The relationship between psychosocial functioning and resilience and negative symptoms in individuals at ultra-high risk for psychosis. Aust N Z J Psychiatry. 2013;47(8):762–71.
179. Giordano GM, Pezzella P, Mucci A, Austin SF, Erfurth A, Glenthøj B, et al. Negative symptoms and social cognition as mediators of the relationship between neurocognition and functional outcome in schizophrenia. Front Psych. 2024;15:1333711.
180. Ventura J, Hellemann GS, Thames AD, Koellner V, Nuechterlein KH. Symptoms as mediators of the relationship between neurocognition and functional outcome in schizophrenia: a meta-analysis. Schizophr Res. 2009;113(2–3):189–99.
181. Owen MJ, O'Donovan MC. Schizophrenia and the neurodevelopmental continuum: evidence from genomics. World Psychiatry. 2017;16(3):227–35.
182. Díaz-Caneja CM, Pina-Camacho L, Rodríguez-Quiroga A, Fraguas D, Parellada M, Arango C. Predictors of outcome in early-onset psychosis: a systematic review. NPJ Schizophr. 2015;1:14005.
183. Kao YC, Liu YP. Effects of age of onset on clinical characteristics in schizophrenia spectrum disorders. BMC Psychiatry. 2010;10:63.
184. Challis S, Nielssen O, Harris A, Large M. Systematic meta-analysis of the risk factors for deliberate self-harm before and after treatment for first-episode psychosis. Acta Psychiatr Scand. 2013;127(6):442–54.
185. Barajas A, Ochoa S, Obiols JE, Lalucat-Jo L. Gender differences in individuals at high-risk of psychosis: a comprehensive literature review. TheScientificWorldJOURNAL. 2015;2015:430735.
186. Thorup A, Petersen L, Jeppesen P, Ohlenschlaeger J, Christensen T, Krarup G, et al. Gender differences in young adults with first-episode schizophrenia spectrum disorders at baseline in the Danish OPUS study. J Nerv Ment Dis. 2007;195(5):396–405.
187. Quattrone D, Di Forti M, Gayer-Anderson C, Ferraro L, Jongsma HE, Tripoli G, et al. Transdiagnostic dimensions of psychopathology at first episode psychosis: findings from the multinational EU-GEI study. Psychol Med. 2019;49(8):1378–91.
188. Santesteban-Echarri O, Paino M, Rice S, González-Blanch C, McGorry P, Gleeson J, et al. Predictors of functional recovery in first-episode psychosis: a systematic review and meta-analysis of longitudinal studies. Clin Psychol Rev. 2017;58:59–75.
189. Watson P, Zhang JP, Rizvi A, Tamaiev J, Birnbaum ML, Kane J. A meta-analysis of factors associated with quality of life in first episode psychosis. Schizophr Res. 2018;202:26–36.
190. Leclerc E, Noto C, Bressan RA, Brietzke E. Determinants of adherence to treatment in first-episode psychosis: a comprehensive review. Revista Brasileira de psiquiatria (Sao Paulo, Brazil : 1999). 2015;37(2):168–76.
191. Mayerhoff DI, Loebel AD, Alvir JM, Szymanski SR, Geisler SH, Borenstein M, et al. The deficit state in first-episode schizophrenia. Am J Psychiatry. 1994;151(10):1417–22.
192. Edwards J, McGorry PD, Waddell FM, Harrigan SM. Enduring negative symptoms in first-episode psychosis: comparison of six methods using follow-up data. Schizophr Res. 1999;40(2):147–58.
193. Malla AK, Norman RM, Takhar J, Manchanda R, Townsend L, Scholten D, et al. Can patients at risk for persistent negative symptoms be identified during their first episode of psychosis? J Nerv Ment Dis. 2004;192(7):455–63.

194. Rapado-Castro M, Soutullo C, Fraguas D, Arango C, Payá B, Castro-Fornieles J, et al. Predominance of symptoms over time in early-onset psychosis: a principal component factor analysis of the positive and negative syndrome scale. J Clin Psychiatry. 2010;71(3):327–37.
195. Chang WC, Hui CL, Tang JY, Wong GH, Lam MM, Chan SK, et al. Persistent negative symptoms in first-episode schizophrenia: a prospective three-year follow-up study. Schizophr Res. 2011;133(1–3):22–8.
196. Hovington CL, Bodnar M, Joober R, Malla AK, Lepage M. Identifying persistent negative symptoms in first episode psychosis. BMC Psychiatry. 2012;12:224.
197. Galderisi S, Mucci A, Bitter I, Libiger J, Bucci P, Fleischhacker WW, et al. Persistent negative symptoms in first episode patients with schizophrenia: results from the European first episode schizophrenia trial. Eur Neuropsychopharmacol. 2013;23(3):196–204.
198. Üçok A, Ergül C. Persistent negative symptoms after first episode schizophrenia: a 2-year follow-up study. Schizophr Res. 2014;158(1–3):241–6.
199. Unrau J, Percie du Sert O, Joober R, Malla A, Lepage M, Raucher-Chéné D. Subtyping negative symptoms in first-episode psychosis: contrasting persistent negative symptoms with a data-driven approach. Schizophr Res. 2022;248:219–27.
200. Raucher-Chéné D, Bodnar M, Lavigne KM, Malla A, Joober R, Lepage M. Dynamic interplay between insight and persistent negative symptoms in first episode of psychosis: a longitudinal study. Schizophr Bull. 2022;48(1):211–9.
201. Puig O, Baeza I, de la Serna E, Cabrera B, Mezquida G, Bioque M, et al. Persistent negative symptoms in first-episode psychosis: early cognitive and social functioning correlates and differences between early and adult onset. J Clin Psychiatry. 2017;78(9):1414–22.
202. Bucci P, Mucci A, van Rossum IW, Aiello C, Arango C, Baandrup L, et al. Persistent negative symptoms in recent-onset psychosis: relationship to treatment response and psychosocial functioning. Eur Neuropsychopharmacol. 2020;34:76–86.
203. Pelizza L, Leuci E, Maestri D, Quattrone E, Azzali S, Paulillo G, et al. Longitudinal persistence of negative symptoms in young individuals with first episode schizophrenia: a 24-month multi-modal program follow-up. Nord J Psychiatry. 2022;76(7):530–8.
204. Patel R, Jayatilleke N, Broadbent M, Chang CK, Foskett N, Gorrell G, et al. Negative symptoms in schizophrenia: a study in a large clinical sample of patients using a novel automated method. BMJ Open. 2015;5(9):e007619.
205. Szkultecka-Dębek M, Walczak J, Augustyńska J, Miernik K, Stelmachowski J, Pieniążek I, et al. Epidemiology and treatment guidelines of negative symptoms in Schizo-phrenia in central and Eastern Europe: a literature review. Clin Pract Epidemiol Mental Health: CP & EMH. 2015;11:158–65.
206. Mäkinen J, Miettunen J, Isohanni M, Koponen H. Negative symptoms in schizophrenia: a review. Nord J Psychiatry. 2008;62(5):334–41.
207. Herbener ES, Harrow M. Longitudinal assessment of negative symptoms in schizophrenia/schizoaffective patients, other psychotic patients, and depressed patients. Schizophr Bull. 2001;27(3):527–37.
208. Sauvé G, Brodeur MB, Shah JL, Lepage M. The prevalence of negative symptoms across the stages of the psychosis continuum. Harv Rev Psychiatry. 2019;27(1):15–32.
209. Strassnig M, Bowie C, Pinkham AE, Penn D, Twamley EW, Patterson TL, et al. Which levels of cognitive impairments and negative symptoms are related to functional deficits in schizophrenia? J Psychiatr Res. 2018;104:124–9.
210. Kiang M, Christensen BK, Remington G, Kapur S. Apathy in schizophrenia: clinical correlates and association with functional outcome. Schizophr Res. 2003;63(1–2):79–88.
211. Bowie CR, Reichenberg A, Patterson TL, Heaton RK, Harvey PD. Determinants of real-world functional performance in schizophrenia subjects: correlations with cognition, functional capacity, and symptoms. Am J Psychiatry. 2006;163(3):418–25.
212. Couture SM, Granholm EL, Fish SC. A path model investigation of neurocognition, theory of mind, social competence, negative symptoms and real-world functioning in schizophrenia. Schizophr Res. 2011;125(2–3):152–60.

213. Moura BM, Isvoranu AM, Kovacs V, Van Rooijen G, Van Amelsvoort T, Simons CJP, et al. The puzzle of functional recovery in schizophrenia-spectrum disorders-replicating a network analysis study. Schizophr Bull. 2022;48(4):871–80.
214. Chang WC, Hui CL, Chan SK, Lee EH, Chen EY. Impact of avolition and cognitive impairment on functional outcome in first-episode schizophrenia-spectrum disorder: a prospective one-year follow-up study. Schizophr Res. 2016;170(2–3):318–21.
215. Foussias G, Mann S, Zakzanis KK, van Reekum R, Agid O, Remington G. Prediction of longitudinal functional outcomes in schizophrenia: the impact of baseline motivational deficits. Schizophr Res. 2011;132(1):24–7.
216. Fulford D, Piskulic D, Addington J, Kane JM, Schooler NR, Mueser KT. Prospective relationships between motivation and functioning in recovery after a first episode of schizophrenia. Schizophr Bull. 2018;44(2):369–77.
217. Ahmed AO, Murphy CF, Latoussakis V, McGovern KE, English J, Bloch A, et al. An examination of neurocognition and symptoms as predictors of post-hospital community tenure in treatment resistant schizophrenia. Psychiatry Res. 2016;236:47–52.
218. Kulhara P, Chandiramani K. Positive and negative subtypes of schizophrenia. A follow-up study from India. Schizophr Res. 1990;3(2):107–16.
219. Lindenmayer JP, Kay SR, Friedman C. Negative and positive schizophrenic syndromes after the acute phase: a prospective follow-up. Compr Psychiatry. 1986;27(4):276–86.
220. Thara R, Henrietta M, Joseph A, Rajkumar S, Eaton WW. Ten-year course of schizophrenia--the Madras longitudinal study. Acta Psychiatr Scand. 1994;90(5):329–36.

The Assessment of Negative Symptoms in Schizophrenia

2

Lucie Métivier and Sonia Dollfus

Abstract

Negative symptoms in schizophrenia significantly impact quality of life and overall functioning. This chapter provides a comprehensive overview of clinical tools for assessing negative symptoms, encompassing hetero-assessments (evaluations based on observer ratings), self-assessments, and digital phenotyping. An examination of clinical trials from 2019 to 2024 reveals a diverse array of tools used to assess these symptoms, with the predominant choice being the negative subscale of the Positive and Negative Syndrome Scale (PANSS), which is not in line with the recent European Psychiatric Association recommendations. The disparity between recommended guidelines and actual clinical practices emphasizes the ongoing need for standardized approaches to ensure consistency and comparability in research outcomes. In essence, the chapter presents an

L. Métivier
Normandie Univ, UNICAEN, PhIND, UMR-S 1237, Neuropresage Team, GIP CYCERON, Caen, France

Normandie Univ, UNICAEN, Fédération Hospitalo-Universitaire (FHU A2M2P), GIP Cyceron, Caen, France

S. Dollfus (✉)
Normandie Univ, UNICAEN, PhIND, UMR-S 1237, Neuropresage Team, GIP CYCERON, Caen, France

Normandie Univ, UNICAEN, Fédération Hospitalo-Universitaire (FHU A2M2P), GIP Cyceron, Caen, France

Normandie Univ, UNICAEN, UFR santé, Caen, France

CHU de CAEN Normandie, Service de Psychiatrie, Centre Esquirol, Caen, France

Service de psychiatrie adulte, Centre Esquirol, Centre Hospitalier Universitaire, Caen, France
e-mail: dollfus-s@chu-caen.fr; dollfus@cyceron.fr; https://sns-dollfus.com/fr/professeur-sonia-dollfus/; https://www.fhu-a2m2p.fr/; https://neuropresage.fr/

A. Vita et al. (eds.), *Negative and Cognitive Symptoms in Schizophrenia*, Comprehensive Approach to Psychiatry 3,
https://doi.org/10.1007/978-3-032-06209-3_2

evolutionary perspective on assessing negative symptoms in schizophrenia, advocating for an integrated approach that combines clinical observations with subjective experiences reported by patients and objective measures through digital phenotyping.

Keywords

Negative symptoms · Schizophrenia · Hetero-evaluation · Self-assessment · Clinical tools · Digital phenotyping

2.1 Introduction

Despite evolving clinical descriptions of schizophrenia over time, negative symptoms, characterized by the reduction of some functions, have been documented since the early nineteenth century. Early authors such as Haslam, Kraepelin, and Bleuler referred to the presence of "blunted sensitivity" and "affective indifference" in patients suffering from schizophrenia [1]. In the early 1970s, Snezhnevsk and Strauss introduced the concept of primary negative symptoms (inherent to the illness) and secondary negative symptoms (linked to other conditions such as depression or the side effects of neuroleptics) [1, 2]. At present, negative symptoms can be categorized into the Experiential dimension, which includes avolition, anhedonia, and asociality, and the Expressive dimension, which includes blunted affect and alogia [3, 4].

The scientific community has become aware of the impact of negative symptoms on schizophrenia outcomes due to their frequency [5, 6], their effect on remission rates, their contribution to disability in daily life, and quality of life [7], and their resistance to pharmacological treatment [8].

Many assessment tools have been developed over the years to measure the presence and intensity of negative symptoms [9]. Hetero-evaluation scales, which require a clinical interview with a trained psychiatrist, are the most widespread. In contrast, self-evaluations, which only involve the patient/person filling out the form themselves and do not require any interview, are often overlooked. Nevertheless, they provide clinical information that may not be detected by the healthcare provider during an interview and offer insights into the personal experiences felt by the individuals themselves.

Moreover, there is an increasing trend towards leveraging objective biobehavioral technologies for digitally phenotyping negative symptoms [10]. Using mobile technology, such as smartphones, allows for the collection of both active and passive data in real-world settings. Active data collection involves tasks such as surveys, while passive approaches gather data automatically. Various active and passive digital phenotyping measures, including geolocation, accelerometry, language processing, and automated analysis of facial expressions and vocal characteristics, have shown promise in capturing negative symptoms [11, 12].

In 2021, the European Psychiatric Association issued "guidance paper" for assessing negative symptoms [13]. This guidance document recommends using evidence-based tools like the Brief Negative Syndrome Scale (BNSS) [14] or the Clinical Assessment Interview for Negative Symptoms (CAINS) [15]. Combining these tools with self-evaluation scales can provide a detailed understanding of the personal experiences of individuals exhibiting negative symptoms.

However, recent clinical trials have not adhered to these guidelines. Between 2019 and 2024, a total of 125 clinical trials had primary or secondary objectives to evaluate negative symptoms in individuals with schizophrenia. Diverse assessment tools were employed, sometimes in combination. The prevailing choice was the negative subscale of the Positive and Negative Syndrome Scale (PANSS-N) [16] (64.8%), followed by the Scale for the Assessment of Negative Symptoms (SANS) [17] (18.4%). Conversely, the BNSS [14] was less frequently used (3.2%), along with the CAINS [15] (1.6%) and the Negative Symptoms Assessment Scale—16 items (NSA-16) [18] (1.6%). Some clinical trials incorporated multiple tools, such as PANSS-N and SANS (4.8%), PANSS-N and BNSS (2.4%), CAINS and SANS (1.6%), and BNSS and CAINS (0.8%). Notably, only one study integrated hetero-evaluation with a self-assessment scale (0.8%) [19]. Finally, very few trials incorporated the recommended tools: the BNSS [20–27], CAINS [19, 24, 28–31] or SNS [19].

In this chapter, we first present the various hetero-evaluation tools, followed by the self-evaluation tools, and finally, the different types of digital phenotyping used for assessing negative symptoms in schizophrenia.

2.2 Hetero-Evaluation Tools

The evolution of assessment tools has led to the development of second-generation scales, which comprehensively cover five domains of negative symptoms: alogia, social withdrawal, anhedonia, blunted affect, and avolition. These scales emerged from the NIMH-Negative Symptom Consensus Development [3], improving upon previous first-generation scales. In this section, we will first introduce the first-generation scales (e.g., Table 2.1), followed by an exploration of second-generation tools (e.g., Table 2.2). Finally, we will delve into scales tailored to evaluate a specific domain of negative symptoms, namely apathy/avolition.

2.2.1 First-Generation Scales

2.2.1.1 Negative Sub-Scale of the Positive and Negative Syndrome Scale (PANSS-N)

Developed in 1987 by Kay et al. [16], the PANSS was developed by incorporating 18 items from the Brief Psychiatric Rating Scale (BPRS) and 12 items from the Psychopathology Rating Schedule (PRS). The administration of this scale involves a 30–40 minutes semi-structured interview conducted by a trained psychiatrist. The 30 items are rated on a 7-point scale (from 1 indicating no symptom to 7 indicating

Table 2.1 First-generation hetero-evaluation scales of negative symptoms

First generation scale	Number of items	Domains of negative symptoms
Negative subscale of the Positive and Negative Syndrome Scale (PANSS)	7	Alogia Social withdrawal Blunted affect
Negative Symptoms Assessment Scale—16 items (NSA-16)	16	Alogia Social withdrawal Blunted affect Avolition *No specific item is designated for anhedonia, which is evaluated alongside emotion: reduced range
Scale for Assessment of Negative Symptoms (SANS)	25	Alogia Social withdrawal Blunted affect Avolition *No specific item is designated for anhedonia, which is evaluated alongside social withdrawal
Schedule for Deficit Syndrome (SDS)	6	Alogia Social withdrawal Blunted affect Avolition *No specific item is designated for anhedonia, which is evaluated alongside diminished emotional range

Table 2.2 Second generation hetero-evaluation scales of negative symptoms

Second-generation scale	Number of items	Domains of negative symptoms
Brief Negative Syndrome Scale (BNSS)	13	Alogia Social withdrawal Blunted affect Avolition Anhedonia
Clinical Assessment Interview for Negative Symptoms (CAINS)	13	Alogia Blunted affect Avolition Anhedonia *No specific item is designated for the evaluation of social withdrawal, which is evaluated alongside avolition and anhedonia

extreme severity), each accompanied by a comprehensive definition and detailed anchoring criteria representing increasing levels of psychopathology. The items are categorized into three subgroups: positive symptoms (7 items), negative symptoms (7 items), and general psychopathology (16 items).

The negative subscale of the PANSS assesses blunted affect, emotional withdrawal, poor rapport, social withdrawal, passivity-apathy, difficulty in abstract thinking, lack of spontaneity and flow of conversation, and stereotyped thinking, with scores ranging from 7 to 49. This subscale has demonstrated good internal consistency (*Cronbach's alpha* = *0.83)* and acceptable test-retest reliability ($r = 0.69$) [16].

However, relying solely on this scale is considered inadequate for a comprehensive assessment of negative symptomatology. There are no specific items assessing anhedonia and avolition. Furthermore, two out of the seven items ("stereotyped thinking" and "difficulty in abstract thinking") seem to be more related to disorganization or cognitive symptoms than to negative symptoms [32].

2.2.1.2 Negative Symptoms Assessment Scale—16 Items (NSA-16)

The NSA was created by Alphs et al. in 1989 [33]. Initially, the scale included 25 items, but shorter versions with 16 items and even 4 items were later developed due to the length of the interview [18, 34]. The NSA-16 is widely used in clinical trials in the United States. Administration of the 16-item version of this scale is conducted during a semi-structured interview lasting 15–30 minutes.

Regarding psychometric properties, this tool has demonstrated good to excellent internal consistency (*Cronbach's alpha* = *0.70–0.90*), as well as convergent and discriminant validities. Factorial analysis revealed five factors: communication, emotion/affect, social involvement, motivation, and retardation [18]. A cutoff point of 31 showed excellent sensitivity and good specificity in distinguishing patients with and without negative symptoms [35]. The uniqueness of this scale lies in its assessment of emotional expression through both mimed emotion and emotional feelings.

2.2.1.3 Scale for Assessment of Negative Symptoms (SANS)

Developed by Andreasen et al. in 1989 [17], the SANS consists of 25 items designed to evaluate five domains: alogia, avolition-apathy, reduced affective range, anhedonia-asociality, and attentional deficit. Each domain includes a global evaluation and a composite score, which sums up the scores of all items in that domain. Items are rated on a scale of 0 to 5, with higher scores indicating more severe symptoms. Numerous studies have examined the psychometric properties of the SANS, which is considered a reliable and internally cohesive scale [17].

However, the scale does have several limitations [15, 36]. The assessment of anhedonia does not consider the feelings of pleasure during activities, and the evaluation of social withdrawal does not account for the patient's environment and intentionality. Additionally, attention deficit is more closely related to cognitive deficits than to negative symptoms. Similarly, items related to inappropriate affect and poverty of speech content reflect symptoms of disorganization rather than negative symptoms [37, 38].

Administration time: 30–40 minutes

2.2.1.4 Schedule for Deficit Syndrome (SDS)

The SDS was developed by Kirkpatrick et al. in 1989 [39]. It involves a semi-structured interview assessing six negative symptoms: diminished emotional range, diminished interest, restricted affect, poverty of speech, diminished social drive, and diminished sense of purpose. Symptoms are rated from 0 (normal) to 4 (severe).

This scale allows for the categorization of schizophrenia patients into deficit and non-deficit subtypes. Deficit schizophrenia is characterized by the presence of at least two moderate (≥2) primary negative symptoms persisting for a minimum of 12 months, including periods without treatment [39]. By primary negative symptoms, the authors mean symptoms directly related to the illness and not secondary to various causes such as depression, delirium, disorganization, antipsychotic side effects, or reduced environmental stimulation.

The scale shows robust inter-rater reliability and convergent validity [40]. Factorial analyses have identified two distinct factors: one related to apathy/avolition (the Experiential dimension, which includes reduced interests, weakened sense of purpose, and diminished social drive), and another representing emotional expression (the Expressive dimension, encompassing restricted affect, diminished emotional range, and poverty of speech) [41].

However, some limitations have been noted with this scale. Each negative symptom is assessed by a single item, and the scale does not fully cover the five required negative domains. For instance, anhedonia is not specifically addressed; the item "diminished emotional range" evaluates the ability to experience pleasure as well as other negative emotional states. Additionally, using this scale in clinical settings is challenging due to the requirement for assessment over 12 months and the difficulty in identifying the primary nature of negative symptoms. This is why the concept of persistent negative symptoms, defined as symptoms persisting for over 6 months, is preferred [42].

Administration time: 5–10 minutes

2.2.2 Second-Generation Scales

2.2.2.1 Brief Negative Syndrome Scale (BNSS)

The BNSS was developed based on the recommendations of the National Institute of Mental Health's Consensus Development Conference on Negative Symptoms [3]. Using this scale involves a semi-structured interview lasting approximately 15 minutes. The BNSS comprises 13 items organized into 6 subscales: anhedonia, avolition, blunted affect, alogia, asociality, and lack of distress. All items are rated on a 7-point (0–6) scale, and the total score ranges from 0 to 78, with higher scores indicating greater impairment. Importantly, the scale has demonstrated robust psychometric properties, showing excellent internal consistency for both the total score and all sub-scores (*Cronbach's alpha = 0.89–0.95*), as well as good test-retest reliability (r = *0.76–0.90*) and convergent validity [14]. Moreover, principal component analysis identified a 2-factor solution corresponding to anhedonia/avolition/asociality and emotional expressivity [14].

There are two strengths of this scale: First, the assessment of anhedonia includes "pleasure experienced during an activity" and "anticipated pleasure." The former can be preserved while the latter is often impaired in schizophrenia patients [43]. Moreover, social withdrawal and avolition are described based on both observed behavior and internal experience of the patient [14].

However, the inclusion of the "lack of distress" item in the BNSS scale is controversial regarding its classification as a negative symptom in schizophrenia. Some researchers argue that it does not align with the traditional definition of negative symptoms [44], while others believe it can predict primary and enduring negative symptoms in the deficit schizophrenia subgroup [45].

Administration time: 15 minutes

2.2.2.2 Clinical Assessment Interview for Negative Symptoms (CAINS)

The CAINS was developed by the Collaboration to Advance Negative Symptoms Assessment in Schizophrenia (CANSAS) [46]. It includes 13 items that assess alogia, pleasure (past-week pleasure and upcoming-week pleasure), motivation, and emotional expression during the past 7 days in all domains of the patient's life. In particular, the scale is organized into two subscales—Motivation and Enjoyment (9 items), which assesses motivation and pleasure in social, occupational, and recreational settings, and Expression (4 items), which assesses facial and vocal expression, expressive gestures, and the amount of speech, which complement each other. The CAINS addresses the concepts of anticipated and experienced pleasures, motivation across social, occupational, and recreational domains. Goal-oriented behaviors are evaluated based on the patient's effort to engage in activities. All items are rated on a scale of 0–4, resulting in a total score ranging from 0 to 52, where a higher score indicates more severe negative symptoms.

The scale has shown strong psychometric properties, including satisfactory internal consistency (*Cronbach's alpha = 0.76*), good test–retest reliability (*r = 0.69*), and convergent validity. Factorial analysis also revealed a two-factor structure, covering expression (4 items) and motivation/pleasure (9 items) [47].

Administration time: 30–40 minutes

2.2.3 Hetero-Evaluation Scales Focused on Avolition

Some hetero-evaluation tools were designed to assess a single domain of negative symptomatology focused exclusively on avolition. Three scales have been developed and validated in the schizophrenia population: The Motor-Affective Social Scale (MASS) [48], the Apathy Evaluation Scale (AES) [49] and the Lille Apathy Rating Scale (LARS) [50, 51] (e.g., Table 2.3). The MASS is ideal for both

Table 2.3 Hetero-evaluation tools focused on one negative dimension

Self-assessment tools that evaluate one domain of negative symptoms	Number of items	Domains of negative symptoms	Validation criteria reported in studies
Motor-Affective Social Scale (MASS) [48]	8	Avolition	Internal consistency Convergent validity
The Apathy Evaluation Scale (AES) [49, 52]	18	Avolition	Convergent validity Discriminant validity [53]
The Lille Apathy Rating Scale (LARS) [50, 51]	33	Avolition	Reliability Convergent validity Discriminant validity

clinical and research settings due to its comprehensive assessment of behavioral, cognitive, emotional, and social dimensions of avolition. The AES offers ease of use for regular clinical follow-ups, with versions available for patients, clinicians, and informants. The LARS provides a detailed evaluation through its specific subscales, making it particularly valuable for in-depth clinical studies. The distinct features of each scale (MASS's multidimensionality, AES's simplicity, and LARS's precision) highlight their respective strengths in evaluating avolition in schizophrenia.

2.3 Self-Assessment Scales

Currently, there are several validated self-assessment scales specifically designed to evaluate negative symptoms in schizophrenia. Two of them assess multiple domains of negative symptoms [54, 55], but the majority focus on a single domain such as apathy or anhedonia.

In this section, we will focus on scales that evaluate various domains of negative symptoms (e.g., Table 2.4), including the two self-assessment tools recommended by the European Psychiatric Association [13]: the Self-evaluation of Negative Symptoms (SNS) [54] and the Motivation and Pleasure Scale—Self-Report (MAP-SR) [56]. Next, we will briefly present scales focused on specific domains of negative symptoms, such as apathy/avolition or anhedonia (e.g., Table 2.5).

2.3.1 Self-Evaluations of Negative Symptoms

2.3.1.1 Self-Evaluation of Negative Symptoms (SNS)

The SNS [54] (available on demand via the website https://sns-dollfus.com) is a brief scale that takes approximately 5 minutes to complete, developed using patients' verbatim. This tool comprises 20 short sentences covering the five domains of

Table 2.4 Self-assessment tools that evaluate various domains of negative symptoms

Self-assessment tools that evaluate various domains of negative symptoms	Number of items	Domains of negative symptoms
Self-evaluation of Negative Symptoms (SNS)	20	Alogia Social withdrawal Reduced emotional range Avolition Anhedonia
Motivation and Pleasure Scale – Self Report (MAP-SR)	15	Anhedonia Social withdrawal Avolition
Negative Symptoms Inventory Self-Report (NSI-SR)	11	Social withdrawal Avolition Anhedonia

Table 2.5 Self-assessment tools that evaluate one domain of negative symptoms

Self-assessment tools that evaluate one domain of negative symptoms	Number of items	Domains of negative symptoms	Validation criteria reported in studies
Trait-Intrinsic and Extrinsic Motivation Self-Report scale [57]	13	Avolition	Test–retest reliability Discriminant validity
Brief Regulation of Motivation Scale (BroMS) [58]	12	Avolition	Internal consistency
BIRT Motivation Questionnaire (BMQ-S) [59]	22	Avolition	Internal consistency Test–retest reliability Convergent validity Factor analysis
Social Anhedonia Scale (SAS) [60]	48	Anhedonia	Internal consistency Convergent validity
Physical Anhedonia Scale (PAS) [60]	40	Anhedonia	Internal consistency
Fawcett-Clark Pleasure Scale (FCPS) [61]	36	Anhedonia	Internal consistency Convergent validity
Snaith Hamilton Pleasure Scale (SHAPS) [62]	14	Anhedonia	Internal consistency Test-retest reliability Convergent validity Discriminant validity
Temporal Experience of Pleasure Scale (TEPS) [63]	18	Anhedonia	Internal consistency Test–retest reliability Discriminant validity Factor analysis
Self-Assessment Anhedonia Scale (SAAS) [64]	27	Anhedonia	Internal consistency Convergent validity Discriminant validity Factor analysis
Anticipatory and Consummatory Interpersonal Pleasure Scale (ACIPS) [65]	17	Anhedonia	Internal consistency Convergent validity Discriminant validity Factor analysis

negative symptoms, with subscores ranging from 0 to 8: social withdrawal, reduced emotional range, alogia, avolition, and anhedonia. The total score ranges from 0 (no negative symptoms) to 40 (severe negative symptoms) [54].

The SNS has demonstrated high internal consistency (*Cronbach's* $\alpha = 0.867$), good convergent and discriminant validity, and high test/retest reliability (*ICC* = *0.94*). Confirming an internal consistency, factorial analysis of the 5 subscores revealed two factors corresponding to the Expressive and Experiential dimensions of negative symptoms [54]. Additionally, factorial analysis of the 20 items highlighted five factors corresponding to the five domains of negative symptoms and the 5 subscales of the scale [66]. Cut-offs have been defined in various populations for screening negative symptoms in schizophrenia (total score ≥ 7) [67], in first-episode psychosis (total score ≥ 11) [68], in adolescents with first psychiatric episode (total score > 13) [69] and in student populations (total score >13) [70].

The strengths of this scale lie in its comprehensive evaluation of negative symptoms in a short account of time, its ease of use, and its assessment of anhedonia, including anticipation and experience during pleasurable activities.

2.3.1.2 Motivation and Pleasure Scale—Self Report (MAP-SR)

The MAP-SR, developed by Llerena et al. [56], is based on the CAINS motivation/pleasure subscale. The initial version consists of 18 items that assess motivation and pleasure related to social relations, work, school, and recreational activities either over the past week or for anticipated activities in the upcoming week. However, three items were removed due to low item-total correlation, resulting in a 15-item version. Items are rated on a 0–4 scale, with a total score range of 0 to 60, where higher scores indicate more severe negative symptoms. Consummatory and anticipatory pleasure related to social, recreational, or work domains is assessed with 6 items. Motivations and feelings towards family members, friends, or romantic partners are evaluated with 3 items. Finally, there are 6 items related to motivation and effort to engage in activities.

The scale shows strong psychometric characteristics, with high internal consistency (*Cronbach's alpha* = *0.90*) and satisfactory convergent and discriminant validities [56].

However, some concerns have been noted regarding this scale: it requires patients to recall and quantify feelings or events from the past, which may be challenging for patients with memory impairment. Furthermore, while anhedonia is comprehensively addressed, blunted affect and alogia are not assessed. Moreover, additional details could enhance the evaluation of asociality and avolition [71].

Administration time: 5–10 minutes

2.3.1.3 Negative Symptoms Inventory—Self-Report (NSI-SR)

In 2023, Raugh et al. developed the NSI-SR [72], an 11-item self-assessment scale designed to evaluate three domains of negative symptoms confirmed by factor analysis: asociality, avolition, and anhedonia. The scale's first section quantifies the number of days (0–7) in the preceding week during which participants engaged in specific activities. The second section assesses internal experiences, rated from 0 to 10 based on the past week, with higher scores indicating more severe negative symptoms.

The scale's psychometric properties were validated across student, ultra-high risk of psychosis, and schizophrenia populations, demonstrating good internal consistency for the total score and each factor (*Cronbach's alpha* = *0.74–0.88*), as well as strong convergent and discriminant validities. Test–retest reliability was robust for the total score ($r = 0.55$), avolition ($r = 0.43$), and anhedonia ($r = 0.50$), but only modest for the asociality factor ($r = 0.26$). The NSI-SR appears to reliably differentiate negative symptoms among schizophrenia, ultra-high risk of psychosis, and community samples [72].

However, several concerns must be addressed. First, the scale assesses only three specific domains of negative symptoms; notably, alogia and blunted affect are not included. Secondly, it relies on patients' ability to recall and quantify past feelings or events, which can be challenging. Thirdly, the assessment of anhedonia could be enhanced with additional details, particularly regarding anticipated pleasure. Lastly, asociality showed low test-retest reliability, raising questions about the scale's utility in tracking clinical changes regarding this symptom.

Administration time: 5 minutes

2.3.2 Self-Evaluation of One Domain of Negative Symptoms

Several self-assessment scales focused on only one domain, such as apathy/avolition or anhedonia, have been developed (e.g., Table 2.5).

In the assessment of anhedonia in patients with schizophrenia, several self-assessment tools capture the range of anhedonia. These tools include the Social Anhedonia Scale (SAS) [60] and the Physical Anhedonia Scale (PAS) [60], which respectively focus on the social and physical aspects of anhedonia. Other scales such as the Fawcett-Clark Pleasure Scale (FCPS) [61], the Snaith Hamilton Pleasure Scale (SHAPS) [62], the Temporal Experience of Pleasure Scale (TEPS) [63], the Self-Assessment Anhedonia Scale (SAAS) [64], and the Anticipatory and Consummatory Interpersonal Pleasure Scale (ACIPS) [65] provide a more detailed assessment of anhedonia. Among these, the TEPS, SAAS, and ACIPS specifically evaluate both anticipatory and consummatory aspects of pleasure. Conversely, the FCPS and SHAPS primarily focus on assessing anhedonia across different life domains without distinguishing between anticipatory and consummatory pleasure.

The self-assessment tools for apathy/avolition in schizophrenia patients include the Trait-Intrinsic and Extrinsic Motivation Self-Report scale (Trait IM-EM) [57], the Brief Regulation of Motivation Scale (BroMS) [58], and the BIRT Motivation Questionnaire (BMQ-S) [59]. Each of these tools focuses on assessing apathy/avolition, but with nuances in measurement and scope of evaluated motivation. The Trait IM-EM emphasizes both intrinsic and extrinsic aspects of motivation [57], while the BroMS focuses on motivation regulation [58]. In contrast, the BMQ-S provides a comprehensive assessment of various aspects of motivation, such as intrinsic motivation, extrinsic motivation, and amotivation [59].

2.4 Digital Phenotyping of Negative Symptoms

Clinical rating scales face challenges due to reliance on patients' memory and willingness to report symptoms, as well as raters' biases in the case of evaluations based on observer ratings. Digital phenotyping, using mobile technologies, offers a solution by providing continuous and objective symptom assessment in real-life contexts. Currently, several measures are being explored, with the most relevant ones being accelerometry, global positional satellite (GPS) sensors in mobile phones, analysis of vocal speech, and facial expressions.

2.4.1 Accelerometry

Accelerometry is a passive digital phenotyping method that involves collecting data on the presence, vigor, and variability of movement. It can be gathered using the internal sensors of smartphones or smartbands. Strauss et al. [73] demonstrate that accelerometry variables show good convergent validity with clinical ratings of negative symptoms (BNSS) and adequate discriminant validity, with low correlations observed with clinical measures of positive, disorganized, and total symptoms assessed using the PANSS total and subscores. Therefore, accelerometry might be a valid objective measure of negative symptoms in schizophrenia, complementing clinical rating scales [73].

2.4.2 Global Positional Satellite (GPS) Sensors

Geolocation using global positioning systems (GPS) on smartphones can provide an objective assessment of a person's movement in time and space through the analysis of "flights" (time spent in transit), and "pauses" (time spent in one location), along with mobility metrics derived from these patterns [74]. Studies on mobility in schizophrenia often analyze movement trajectories to identify potential markers of negative symptoms. Recent studies have shown that individuals with schizophrenia exhibit lower overall mobility compared to those with bipolar disorder and healthy controls [75, 76]. Higher mobility is associated with less severe negative symptoms and better functional outcomes. Specifically, more time spent outside the home and in transit correlates with lower symptom severity measured with BNSS and with higher functioning [76], whereas more time spent at home and a higher likelihood of pausing correlate with higher symptom severity measured with CAINS and with lower functioning [75]. Additionally, deviations from expected mobility patterns can predict relapse rates [77].

2.4.3 Vocal Speech

The analysis of prosody and vocal expression provides objective measures of alogia and blunted affect. A recent meta-analysis highlighted that patients with schizophrenia exhibit significantly lower total talking times and speech rates compared to healthy controls, while their mean pause duration is higher, reflecting poverty of speech and alogia [78]. Additionally, acoustic characteristics assessed through a reading task, which includes texts designed to evoke various emotional responses, differ significantly between patients with schizophrenia and healthy controls. These distinct emotional conditions are correlated with both blunted affect and alogia subscale scores of the BNSS [79].

2.4.4 Facial Expression

Numerous investigations have explored the intricate relationship between facial expressions and clinical manifestations in patients with schizophrenia. Cowan et al. [80] used ambulatory video assessment and machine learning techniques to analyze clinician-rated blunted affect alongside computerized facial metrics. Their findings revealed strong associations between head orientation, eye movement, and aspects of mouth movement, resulting in a noteworthy predictive accuracy of 80–82% for assessing blunted affect. Lotzin et al. [81] investigated dynamics within familial speeches, uncovering a significant reduction in facial expression variability among schizophrenia patients engaged in practical discussions. Additionally, another study examining dynamics in 17 facial muscles demonstrated reduced activity in schizophrenia patients compared to healthy controls, particularly in regions such as the outer eyebrow, upper eyelid, nose, dimple, lower jaw, lip, and blink [82]. Among the lower facial muscle activities, one exhibited correlation with negative symptoms measured with BNSS. Reduced facial muscle activity intensity in patients with schizophrenia may be linked to negative symptoms, indicating a potential defect in the expression of basic social emotions [82].

2.5 Conclusion

In summary, the assessment of negative symptoms in schizophrenia is a dynamic and evolving field. Integrating hetero-evaluation, self-assessment tools, and digital phenotyping provides a comprehensive approach, acknowledging clinical observations, subjective experiences, and objective measures. As research progresses, refining existing tools, adhering to established guidelines, and developing new instruments and measures will be crucial for advancing the understanding of the illness and improving the outcomes of schizophrenia.

Acknowledgments We thank the Carré International of the University of Caen Normandy for proofreading the article.

References

1. Dollfus S. Chapter 1 - Apport des auteurs français dans la nosographie des psychoses. In: Dollfus S, editor. Les schizophrénies. Paris: Lavoisier Médecine Sciences; 2019. p. 3–10.
2. Dollfus S, Lyne J. Negative symptoms: history of the concept and their position in diagnosis of schizophrenia. Schizophr Res. 2017;186:3–7.
3. Kirkpatrick B, Fenton WS, Carpenter WT, Marder SR. The NIMH-MATRICS consensus statement on negative symptoms. Schizophr Bull. 2006;32:214–9.
4. Messinger JW, Trémeau F, Antonius D, Mendelsohn E, Prudent V, Stanford AD, et al. Avolition and expressive deficits capture negative symptom phenomenology: implications for DSM-5 and schizophrenia research. Clin Psychol Rev. 2011;31:161–8.

5. Mäkinen J, Miettunen J, Jääskeläinen E, Veijola J, Isohanni M, Koponen H. Negative symptoms and their predictors in schizophrenia within the Northern Finland 1966 birth cohort. Psychiatry Res. 2010;178:121–5.
6. Sicras-Mainar A, Maurino J, Ruiz-Beato E, Navarro-Artieda R. Impact of negative symptoms on healthcare resource utilization and associated costs in adult outpatients with schizophrenia: a population-based study. BMC Psychiatry. 2014;14:225.
7. Montvidas J, Adomaitienė V, Leskauskas D, Dollfus S. Health-related quality of life prediction using the self-evaluation of negative symptoms scale (SNS) in patients with schizophrenia. Nord J Psychiatry. 2021;75:S21.
8. Stahl SM, Buckley PF. Negative symptoms of schizophrenia: a problem that will not go away. Acta Psychiatr Scand. 2007;115:4–11.
9. Lincoln TM, Dollfus S, Lyne J. Current developments and challenges in the assessment of negative symptoms. Schizophr Res. 2017;186:8–18.
10. Cohen AS, Schwartz E, Le TP, Cowan T, Kirkpatrick B, Raugh IM, et al. Digital phenotyping of negative symptoms: the relationship to clinician ratings. Schizophr Bull. 2021;47:44–53.
11. Harvey PD, Depp CA, Rizzo AA, Strauss GP, Spelber D, Carpenter LL, et al. Technology and mental health: state of the art for assessment and treatment. Am J Psychiatry. 2022;179:897–914.
12. Kirkpatrick B, Luther L, Strauss GP. Negative symptoms in the clinic: we treat what we can describe. Br J Psychiatry. 2023;223:271–2.
13. Galderisi S, Mucci A, Dollfus S, Nordentoft M, Falkai P, Kaiser S, et al. EPA guidance on assessment of negative symptoms in schizophrenia. Eur Psychiatry. 2021;64:e23.
14. Kirkpatrick B, Strauss GP, Nguyen L, Fischer BA, Daniel DG, Cienfuegos A, et al. The brief negative symptom scale: psychometric properties. Schizophr Bull. 2011;37:300–5.
15. Galderisi S, Mucci A, Buchanan RW, Arango C. Negative symptoms of schizophrenia: new developments and unanswered research questions. Lancet Psychiatry. 2018;5:664–77.
16. Kay SR, Fiszbein A, Opler LA. The positive and negative syndrome scale (PANSS) for schizophrenia. Schizophr Bull. 1987;13:261–76.
17. Andreasen NC. The Scale for the Assessment of Negative Symptoms (SANS): conceptual and theoretical foundations. Br J Psychiatry Suppl. 1989:49–58.
18. Axelrod BN, Goldman RS, Alphs LD. Validation of the 16-item negative symptom assessment. J Psychiatr Res. 1993;27:253–8.
19. Cella M, Tomlin P, Robotham D, Green P, Griffiths H, Stahl D, et al. Virtual reality therapy for the negative symptoms of schizophrenia (V-NeST): a pilot randomised feasibility trial. Schizophr Res. 2022;248:50–7.
20. Strauss GP, Zamani Esfahlani F, Sayama H, Kirkpatrick B, Opler MG, Saoud JB, et al. Network analysis indicates that avolition is the most central domain for the successful treatment of negative symptoms: evidence from the Roluperidone randomized clinical trial. Schizophr Bull. 2020;46:964–70.
21. Koblan KS, Kent J, Hopkins SC, Krystal JH, Cheng H, Goldman R, et al. A non-D2-receptor-binding drug for the treatment of schizophrenia. N Engl J Med. 2020;382:1497–506.
22. Gonzalez-Blanco L, Garcia-Portilla MP, Dal Santo F, Garcia-Alvarez L, de la Fuente-Tomas L, Menendez-Miranda I, et al. Predicting real-world functioning in outpatients with schizophrenia: role of inflammation and psychopathology. Psychiatry Res. 2019;280:112509.
23. Pos K, Franke N, Smit F, Wijnen BFM, Staring ABP, Van der Gaag M, et al. Cognitive behavioral therapy for social activation in recent-onset psychosis: randomized controlled trial. J Consult Clin Psychol. 2019;87:151–60.
24. Oh J, Lee E, Cha EJ, Seo H-J, Choi K-H. Community-based multi-site randomized controlled trial of behavioral activation for patients with negative symptoms of schizophrenia. Schizophr Res. 2023;252:118–26.
25. Bryl K, Bradt J, Cechnicki A, Fisher K, Sossin KM, Goodill S. The role of dance/movement therapy in the treatment of negative symptoms in schizophrenia: a mixed methods pilot study. J Ment Health. 2022;31:613–23.
26. Meyer-Lindenberg A, Nielsen J, Such P, Lemming OM, Zambori J, Buller R, et al. A double-blind, randomized, placebo-controlled proof of concept study of the efficacy and

safety of Lu AF11167 for persistent negative symptoms in people with schizophrenia. Eur Neuropsychopharmacol. 2022;61:4–14.
27. Rabinowitz J, Badescu S, Palamarchuk P, Filyk V, Voloshchuk A, Rud V, et al. Personal and social adjustment effects of roluperidone in patients with schizophrenia and negative symptoms: results from an exploratory outcome of a randomized placebo-controlled trial. Schizophr Res. 2019;211:103–4.
28. Reddy LF, Glynn SM, McGovern JE, Sugar CA, Reavis EA, Green MF. A novel psychosocial intervention for motivational negative symptoms in schizophrenia: combined motivational interviewing and CBT. Am J Psychiatry. 2023;180:367–76.
29. Bodén R, Bengtsson J, Thörnblom E, Struckmann W, Persson J. Dorsomedial prefrontal theta burst stimulation to treat anhedonia, avolition, and blunted affect in schizophrenia or depression—a randomized controlled trial. J Affect Disord. 2021;290:308–15.
30. Granholm E, Twamley EW, Mahmood Z, Keller AV, Lykins HC, Parrish EM, et al. Integrated cognitive-behavioral social skills training and compensatory cognitive training for negative symptoms of psychosis: effects in a pilot randomized controlled trial. Schizophr Bull. 2022;48:359–70.
31. Mahmood Z, Van Patten R, Keller AV, Lykins HC, Perivoliotis D, Granholm E, et al. Reducing negative symptoms in schizophrenia: feasibility and acceptability of a combined cognitive-behavioral social skills training and compensatory cognitive training intervention. Psychiatry Res. 2021;295:113620.
32. Garcia-Portilla MP, Garcia-Alvarez L, Saiz PA, Al-Halabi S, Bobes-Bascaran MT, Bascaran MT, et al. Psychometric evaluation of the negative syndrome of schizophrenia. Eur Arch Psychiatry Clin Neurosci. 2015;265:559–66.
33. Alphs LD, Summerfelt A, Lann H, Muller RJ. The negative symptom assessment: a new instrument to assess negative symptoms of schizophrenia. Psychopharmacol Bull. 1989;25:159–63.
34. Alphs L, Morlock R, Coon C, Cazorla P, Szegedi A, Panagides J. Validation of a 4-item Negative Symptom Assessment (NSA-4): a short, practical clinical tool for the assessment of negative symptoms in schizophrenia. Int J Methods Psychiatr Res. 2011;20:e31–7.
35. Garcia-Alvarez L, Garcia-Portilla MP, Saiz PA, Fonseca-Pedrero E, Bobes-Bascaran MT, Gomar J, et al. Spanish validation of the Negative Symptom Assessment-16 (NSA-16) in patients with schizophrenia. Rev Psiquiatr Salud Ment (Engl Ed). 2018;11:169–75.
36. Kumari S, Malik M, Florival C, Manalai P, Sonje S. An assessment of five (PANSS, SAPS, SANS, NSA-16, CGI-SCH) commonly used symptoms rating scales in schizophrenia and comparison to newer scales (CAINS, BNSS). J Addict Res Ther. 2017;8:324.
37. Mueser KT, Sayers SL, Schooler NR, Mance RM, Haas GL. A multisite investigation of the reliability of the scale for the assessment of negative symptoms. Am J Psychiatry. 1994;151:1453–62.
38. Peralta V, Cuesta MJ. Negative symptoms in schizophrenia: a confirmatory factor analysis of competing models. Am J Psychiatry. 1995;152:1450–7.
39. Kirkpatrick B, Buchanan RW, McKenney PD, Alphs LD, Carpenter WT. The schedule for the deficit syndrome: an instrument for research in schizophrenia. Psychiatry Res. 1989;30:119–23.
40. Citak S, Oral ET, Aker AT, Senocak M. Reliability and validity of the schedule for deficit syndrome in schizophrenia. Turk Psikiyatri Derg. 2006;17:115–27.
41. Nakaya M, Ohmori K. A two-factor structure for the schedule for the deficit syndrome in schizophrenia. Psychiatry Res. 2008;158:256–9.
42. Buchanan RW. Persistent negative symptoms in schizophrenia: an overview. Schizophr Bull. 2007;33:1013–22.
43. Kring AM, Elis O. Emotion deficits in people with schizophrenia. Annu Rev Clin Psychol. 2013;9:409–33.
44. Blanchard JJ, Shan L, Andrea A, Savage C, Kring AM, Weittenhiller L. Chapter 7 - Negative symptoms and their assessment in schizophrenia and related disorders. In: Badcock JC, Paulik G, éditeurs. A clinical introduction to psychosis. Academic Press; 2020. p. 153–175.

45. Strauss GP, Keller WR, Buchanan RW, Gold JM, Fischer BA, McMahon RP, et al. Next-generation negative symptom assessment for clinical trials: validation of the brief negative symptom scale. Schizophr Res. 2012;142:88–92.
46. Blanchard JJ, Kring AM, Horan WP, Gur R. Toward the next generation of negative symptom assessments: the collaboration to advance negative symptom assessment in schizophrenia. Schizophr Bull. 2011;37:291–9.
47. Kring AM, Gur RE, Blanchard JJ, Horan WP, Reise SP. The clinical assessment interview for negative symptoms (CAINS): final development and validation. Am J Psychiatry. 2013;170:165–72.
48. Trémeau F, Goggin M, Antonius D, Czobor P, Hill V, Citrome L. A new rating scale for negative symptoms: the motor-affective-social scale. Psychiatry Res. 2008;160:346–55.
49. Marin RS, Biedrzycki RC, Firinciogullari S. Reliability and validity of the apathy evaluation scale. Psychiatry Res. 1991;38:143–62.
50. Sockeel P, Dujardin K, Devos D, Denève C, Destée A, Defebvre L. The Lille apathy rating scale (LARS), a new instrument for detecting and quantifying apathy: validation in Parkinson's disease. J Neurol Neurosurg Psychiatry. 2006;77:579–84.
51. Yazbek H, Norton J, Capdevielle D, Larue A, Boulenger J-P, Gély-Nargeot M-C, et al. The Lille Apathy Rating Scale (LARS): exploring its psychometric properties in schizophrenia. Schizophr Res. 2014;157:278–84.
52. Clarke DE, Van Reekum R, Patel J, Simard M, Gomez E, Streiner DL. An appraisal of the psychometric properties of the clinician version of the Apathy Evaluation Scale (AES-C). Int J Methods Psychiatr Res. 2007;16:97–110.
53. Fervaha G, Foussias G, Takeuchi H, Agid O, Remington G. Measuring motivation in people with schizophrenia. Schizophr Res. 2015;169:423–6.
54. Dollfus S, Mach C, Morello R. Self-evaluation of negative symptoms: a novel tool to assess negative symptoms. Schizophr Bull. 2016;42:571–8.
55. Selten JP, Sijben NE, van den Bosch RJ, Omloo-Visser J, Warmerdam H. The subjective experience of negative symptoms: a self-rating scale. Compr Psychiatry. 1993;34:192–7.
56. Llerena K, Park SG, McCarthy JM, Couture SM, Bennett ME, Blanchard JJ. The Motivation and Pleasure Scale-Self-Report (MAP-SR): reliability and validity of a self-report measure of negative symptoms. Compr Psychiatry. 2013;54:568–74.
57. Didier PR, Moore TM, Calkins ME, Prettyman G, Levinson T, Savage C, et al. Evaluation of a new intrinsic and extrinsic motivation scale in youth with psychosis spectrum symptoms. Compr Psychiatry. 2023;127:152413.
58. Lynch DA, Brown M, Saperstein A, Stefancic A, Medalia A. Assessing metamotivation in schizophrenia: a pilot study of the Brief Regulation of Motivation Scale (BRoMS). Psychiatry Res. 2022;317:114799.
59. Raffard S, Norton J, Van der Linden M, Lançon C, Benoit M, Capdevielle D. Psychometric properties of the BIRT Motivation Questionnaire (BMQ), a self-measure of avolition in individuals with schizophrenia. J Psychiatr Res. 2022;147:274–82.
60. Chapman LJ, Chapman JP, Raulin ML. Scales for physical and social anhedonia. J Abnorm Psychol. 1976;85:374–82.
61. Fawcett J, Clark DC, Scheftner WA, Gibbons RD. Assessing anhedonia in psychiatric patients. Arch Gen Psychiatry. 1983;40:79–84.
62. Snaith RP, Hamilton M, Morley S, Humayan A, Hargreaves D, Trigwell P. A scale for the assessment of hedonic tone the Snaith-Hamilton Pleasure Scale. Br J Psychiatry. 1995;167:99–103.
63. Gard DE, Gard MG, Kring AM, John OP. Anticipatory and consummatory components of the experience of pleasure: a scale development study. J Res Pers. 2006;40:1086–102.
64. Olivares JM, Berrios GE, Bousoño M. Self-assessment anhedonia scale. 2011. https://doi.org/10.1037/t00443-000.
65. Gooding DC, Pflum MJ. The assessment of interpersonal pleasure: introduction of the Anticipatory and Consummatory Interpersonal Pleasure Scale (ACIPS) and preliminary findings. Psychiatry Res. 2014;215:237–43.

66. Dollfus S, Mucci A, Giordano GM, Bitter I, Austin SF, Delouche C, et al. European validation of the Self-Evaluation of Negative Symptoms (SNS): a large multinational and multicenter study. Front Psych. 2022;13:826465.
67. Dollfus S, Delouche C, Hervochon C, Mach C, Bourgeois V, Rotharmel M, et al. Specificity and sensitivity of the Self-assessment of Negative Symptoms (SNS) in patients with schizophrenia. Schizophr Res. 2019;211:51–5.
68. Mallet J, Guessoum SB, Tebeka S, Le Strat Y, Dubertret C. Self-evaluation of negative symptoms in adolescent and young adult first psychiatric episodes. Prog Neuro-Psychopharmacol Biol Psychiatry. 2020;103:109988.
69. Rodríguez-Testal JF, Perona-Garcelán S, Dollfus S, Valdés-Díaz M, García-Martínez J, Ruíz-Veguilla M, et al. Spanish validation of the self-evaluation of negative symptoms scale SNS in an adolescent population. BMC Psychiatry. 2019;19:327.
70. Métivier L, Mauduy M, Beaunieux H, Dollfus S. Revealing the unseen: detecting negative symptoms in students. JCM. 2024;13:1709.
71. Richter J, Hesse K, Eberle M-C, Eckstein KN, Zimmermann L, Schreiber L, et al. Self-assessment of negative symptoms—critical appraisal of the motivation and pleasure—self-report's (MAP-SR) validity and reliability. Compr Psychiatry. 2019;88:22–8.
72. Raugh IM, Luther L, Bartolomeo LA, Gupta T, Ristanovic I, Pelletier-Baldelli A, et al. Negative Symptom Inventory-Self-Report (NSI-SR): initial development and validation. Schizophr Res. 2023;256:79–87.
73. Strauss GP, Raugh IM, Zhang L, Luther L, Chapman HC, Allen DN, et al. Validation of accelerometry as a digital phenotyping measure of negative symptoms in schizophrenia. Schizophrenia (Heidelb). 2022;8:37.
74. Mow JL, Gard DE, Mueser KT, Mote J, Gill K, Leung L, et al. Smartphone-based mobility metrics capture daily social motivation and behavior in schizophrenia. Schizophr Res. 2022;250:13–21.
75. Depp CA, Bashem J, Moore RC, Holden JL, Mikhael T, Swendsen J, et al. GPS mobility as a digital biomarker of negative symptoms in schizophrenia: a case control study. NPJ Digit Med. 2019;2:108.
76. Raugh IM, James SH, Gonzalez CM, Chapman HC, Cohen AS, Kirkpatrick B, et al. Geolocation as a digital phenotyping measure of negative symptoms and functional outcome. Schizophr Bull. 2020;46:1596–607.
77. Henson P, Barnett I, Keshavan M, Torous J. Towards clinically actionable digital phenotyping targets in schizophrenia. NPJ Schizophr. 2020;6:13.
78. Low DM, Bentley KH, Ghosh SS. Automated assessment of psychiatric disorders using speech: a systematic review. Laryngoscope Investig Otolaryngol. 2020;5:96–116.
79. Zhao Q, Wang W-Q, Fan H-Z, Li D, Li Y-J, Zhao Y-L, et al. Vocal acoustic features may be objective biomarkers of negative symptoms in schizophrenia: a cross-sectional study. Schizophr Res. 2022;250:180–5.
80. Cowan T, Rodriguez ZB, Strauss GP, Raugh IM, Cohen AS. Computerized analysis of facial expression reveals objective indices of blunted facial affect. Eur Arch Psychiatry Clin Neurosci. 2023;274(7):1771–5.
81. Lotzin A, Haack-Dees B, Resch F, Romer G, Ramsauer B. Facial emotional expression in schizophrenia adolescents during verbal interaction with a parent. Eur Arch Psychiatry Clin Neurosci. 2013;263:529–36.
82. Du X, Fan HZ, Wang YH, Zhang J, Zhu XL, Zhao YL, et al. Characteristics of facial MuscleActivity intensity in patients with schizophrenia and its relationship to negative symptoms. Front Psych. 2022;13:829363.

Negative Symptoms and Psychosocial Functioning in Schizophrenia

3

Nuria Segarra, Renata Zirilli, Sophie Shatford, Arturo Moreno-Cordon, Noham Wolpe, and Emilio Fernandez-Egea

Abstract

Negative symptoms—such as avolition, anhedonia, alogia, affective flattening, and social withdrawal—are core features of schizophrenia that significantly impact psychosocial functioning. Unlike positive symptoms, which are often more responsive to antipsychotic treatment, negative symptoms tend to persist over time and are strongly associated with poor social, occupational, and interpersonal outcomes. This review explores the relationship between negative symptoms and various domains of psychosocial functioning, including employment, social relationships, independent living, and quality of life. Evidence suggests that primary negative symptoms, rather than secondary effects of medication, depression, or psychosis, are the strongest predictors of long-term disability in individuals with schizophrenia. Additionally, functional outcomes are often mediated by deficits in motivation and emotional engagement, highlighting the importance of targeting these domains in therapeutic interventions. Despite advances in pharmacological and psychosocial treatments, effective

N. Segarra · R. Zirilli · S. Shatford
Cambridgeshire and Peterborough NHS Foundation Trust, Cambridge, UK

A. Moreno-Cordon
Department of Psychiatry, Virgen del Rocio University Hospital, Seville, Spain

N. Wolpe
Department of Physical Therapy, The Stanley Steer School of Health Professions, Faculty of Medical & Health Sciences, Tel Aviv University, Tel Aviv, Israel

Department of Psychiatry, University of Cambridge, Cambridge, UK

E. Fernandez-Egea (✉)
Cambridgeshire and Peterborough NHS Foundation Trust, Cambridge, UK

Department of Psychiatry, University of Cambridge, Cambridge, UK
e-mail: ef280@cam.ac.uk

A. Vita et al. (eds.), *Negative and Cognitive Symptoms in Schizophrenia*, Comprehensive Approach to Psychiatry 3,
https://doi.org/10.1007/978-3-032-06209-3_3

management of negative symptoms remains a significant clinical challenge. Emerging approaches, such as cognitive remediation, social skills training, and novel pharmacological agents, show promise but require further validation. Understanding the complex interplay between negative symptoms and psychosocial functioning is essential for developing comprehensive treatment strategies aimed at improving real-world outcomes in schizophrenia.

Keywords

Negative symptoms · Psychosocial functioning · Schizophrenia · Psychosis · Blunted affect · Apathy · Aanhedonia · Alogia · Asociality · Avolition · Self-efficacy · Motivation · Stigma · Resilience

3.1 Introduction

Negative symptoms are considered core symptoms in schizophrenia. They are included in all diagnostic manuals, such as the Diagnostic and Statistical Manual of Mental Disorders, DSM and International Classification of Diseases, ICD. They refer to the absence or lessening of normal behaviour or function in people with schizophrenia. The term "negative symptoms" is a relatively new addition in the psychiatry literature as it was only first linked to schizophrenia in the 1970s. However, the negative symptom construct was crucial since the early conceptualization of the disorder, sometimes more than the psychotic symptoms of schizophrenia.

Kraepelin described dementia praecox as *'a series of states, the common characteristics of which is a peculiar destruction of the internal connections of the psychic personality. The effects of this injury predominate in the emotional and volitional spheres of mental life'*. The centrality of these symptoms was later acknowledged by Blueler when he coined the term schizophrenia(s). He described schizophrenic patients as "lazy and negligent because they no longer feel the urge to do anything, either on their initiative or at the command of others" [1]. Bleuler's fundamental (or primary) symptoms were the famous four 'As', associations, affect, ambivalence, and autism, with a strong resemblance to what is now conceptualized as negative symptoms, and were the basis of the first scales for measuring negative symptoms.

As described in other book chapters, the last conceptualization of negative symptoms came in the early 2000s. It defined five symptom domains: anhedonia, asociality, avolition, blunted affect, and alogia. These five domains are commonly clustered into two dimensions: motivation and pleasure (or MAP, encompassing anhedonia, asociality, and avolition) and emotional expression (or EE, including blunted affect and alogia) [2], which are reminiscent of Kraepelin's emotional and volitional spheres described a century ago.

In this chapter, we will explore how negative symptoms impact the psychosocial functioning of people with schizophrenia. There is an extensive overlap between negative symptoms and psychosocial functioning. Still, they stem from different conceptual traditions, and psychosocial functioning can also be affected by other psychopathology domains in schizophrenia (e.g. psychosis, affective or cognitive

Table 3.1 Psychosocial functioning and consensus negative symptoms

Category	Definition/description
Psychosocial functioning	Refers to an individual's ability to effectively engage in social relationships, perform daily activities, and fulfil roles within community, work, or educational environments [3]
Consensus negative symptoms [4]	Negative symptoms are divided into two primary subdomains: avolition/apathy and diminished expression
Avolition/apathy	
1. Anhedonia	Reduction in the experience of pleasure for current events or future anticipated activities
2. Asociality	Decrease in social interaction due to reduced drive to engage in relationships
3. Avolition	Reduction in the ability to initiate and persist in goal-directed activities due to a lack of motivation
Emotional Expression EE	
1. Blunted affect	Reduction in emotional expression, including facial expressions, voice tone, and gestures
2. Alogia	Reduction in spoken words and spontaneous elaboration during conversations

symptoms). Here, we will first summarize how psychosocial functioning has been considered and evaluated. Then, we will review the evidence linking schizophrenia and, more specifically, negative symptoms with psychosocial functioning, as we summarize in Table 3.1.

3.2 What Is Psychosocial Functioning?

Psychosocial functioning refers to an individual's capacity to perform and engage in social, occupational, and interpersonal activities, reflecting their overall psychological health and social adaptation. This concept encompasses a broad range of behaviours and processes, including the ability to form and maintain relationships, fulfil roles and responsibilities, and cope with the demands and stresses of everyday life. Psychosocial functioning is integral to an individual's quality of life and well-being, affecting their mental health and ability to interact effectively within society.

By definition, psychosocial functioning is impaired in schizophrenia. Besides the core symptoms of schizophrenia (Criterion A, which includes delusions, hallucinations, disorganization, or negative symptoms), the Diagnostic and Statistical Manual of Mental Disorders, DSM's Criterion B states that for any of these symptoms, the 'level of functioning in one or more major areas, such as work, interpersonal relations, or self-care, is markedly below the level achieved before the onset' [3]. However, it is difficult to capture the whole range of psychosocial impairments that come with having such a chronic illness and to create a framework to which to refer when looking at defining psychosocial functioning and all its dimensions, hence why there is a lack of consensus on how to define and measure psychosocial outcomes [4].

3.2.1 History of the Psychosocial Functioning Concept

The concept of psychosocial functioning has evolved significantly over time, shaped by advances in psychology, psychiatry, and social sciences. Historically, the interplay between an individual's psychological state and social environment has been recognized, but a more structured understanding did not emerge until the twentieth century.

In the early 1900s, Sigmund Freud's psychoanalytic theory emphasized the role of unconscious processes and childhood experiences in shaping behaviour and mental health [5]. Concurrently, the sociological perspectives of Émile Durkheim, particularly his studies on social integration and anomie, highlighted how societal structures and norms impact individual behaviour and mental health. Durkheim's work highlighted the necessity of considering social context in understanding psychological well-being [6].

The mid-twentieth century saw further integration of psychological and social factors [7] with the development of the biopsychosocial model by George Engel in the 1970s [8]. Engel's model challenged the prevailing biomedical model by proposing that health and illness are products of the interplay between biological, psychological, and social factors, and they have to be given the same importance in clinical formulations, as they are all intrinsically connected and function interdependently and reciprocally [8–10]. This holistic approach provided a more comprehensive framework for understanding psychosocial functioning, emphasizing the need to consider the full range of influences on an individual's health and behaviour.

In more recent decades, the concept of psychosocial functioning has been refined and operationalized through various assessment tools and theoretical models. For instance, the Global Assessment of Functioning (GAF) [11] scale, introduced in the DSM-III-R [3], provided a standardized method for assessing an individual's overall level of functioning in psychological, social, and occupational domains.

As described below, there are specific considerations for schizophrenia, and different tools have been developed. Additionally, the rise of positive psychology has further expanded the focus from merely addressing dysfunction to promoting strengths and resilience, emphasizing how positive psychosocial functioning can enhance overall well-being and life satisfaction [12].

3.2.2 The Components of Psychosocial Functioning

Despite the lack of a universally accepted definition of psychosocial functioning, few areas are typically considered when evaluating it in people with schizophrenia. These include:

Employment [13, 14]: This assesses whether the individual is engaged in paid or voluntary work, community-supported employment, or actively seeking employment if currently unemployed. Understanding the nature of employment, including its stability, level of support, and alignment with the individual's abilities and aspirations, is crucial in evaluating psychosocial functioning. The vocational domain is a significant aspect of psychosocial functioning. It includes factors such as job performance, academic achievements, social interactions, the ability to manage work-related stress, and overall work behaviour.

Living status [13, 14]: Evaluation of the individual's living arrangements provides insights into their level of independence, social support network, and environmental stressors. Assessing whether the individual resides alone, with a partner, in supervised housing, or is institutionalized helps to assess their social integration and functional autonomy.

Management of finances and personal belongings [15]: Competence in managing finances and personal belongings reflects the individual's organizational skills, decision-making abilities, and level of independence. Evaluating these aspects helps identify areas of support and intervention to enhance psychosocial functioning from a non-medical but complementary approach.

Self-care [14, 15]: Self-care activities include personal hygiene, meal preparation, medication management, and attendance at medical appointments.

Social mixing [14, 15]: Understanding the individual's engagement in social interactions entails assessing the quality and quantity of their social interactions and the underlying motivations driving their social engagement. Social anxiety and interpersonal communication skills influence the individual's social integration and adaptability.

Leisure time [14, 15]: Consideration of the amount and frequency of leisure activities, their quality, and the degree of participation by others. Leisure activities significantly affect social engagement, emotional well-being, and overall quality of life. Examining the individual's leisure goals, including frequency, variety, and social engagement, offers insights into their social connectedness and recreational preferences.

Social manners [15]: Social manners encompass adherence to social norms, etiquette, and conventions governing interpersonal interactions. Assessing the individual's understanding and application of social manners provides insights into their social awareness, adaptability, and respect for social boundaries.

The inherent complexity in gathering such information exhaustively and reliably, specifically in people with schizophrenia, has prompted the integration of scales into the assessment process.

3.2.3 Measuring Psychosocial Functioning in Schizophrenia

The evaluation of psychosocial functioning has progressed in parallel with its conceptualization. Initially, most instruments were summarized by a single score. However, in some other domains were also considered in this score. Over time, psychosocial functioning has been progressively segmented into different components (work, self-care, social activity) while still including an overall composite score. A summary of the different tools can be found in Table 3.2.

Some functioning scales do not evaluate just the psychosocial component. This is the case with the Global Assessment of Functioning (GAF) [11], which also considers symptoms of schizophrenia [9, 16], and the World Health Organization Disability Assessment Schedule (WHO-DAS), which provides a global assessment of disability in areas such as social interaction, daily activities, and social participation through various items [17]. The Specific Levels of Functioning (SLOF) scale, developed by Schneider et al., evaluates psychosocial functioning (e.g. interpersonal relationships, work skills) but also includes other areas such as physical health [18].

The Quality of Life Scale (QLS), developed by Heinrichs et al. in 1984, is another tool used to assess psychosocial functioning. This scale addresses four domains: intrapsychic foundations, interpersonal relationships, instrumental role functioning, and common objects and activities. However, its final score, a summation of these domains, may not fully capture the entirety of the patient's psychosocial functioning due to the unequal weighting of these domains [19, 20].

With a more focused approach to the psychosocial aspect, the Social and Occupational Functioning Assessment Scale (SOFAS) provides a global score [17]. Building on the SOFAS, the Personal and Social Performance Scale (PSP) [21] was subsequently developed. The PSP evaluates four domains of psychosocial functioning, including socially useful activities, personal and social relationships, self-care, and disturbing and aggressive behaviour [9, 22].

Individuals may be asked to complete surveys that measure various aspects of psychosocial functioning, such as the WHO Quality of Life (WHOQOL) assessment, which focuses on the following domains: physical health, psychological health, social relationships, and environment [23].

Presenting varying approaches yet unified by a focused interest in the psychosocial dimension, other scales such as the Social Functioning Scale (SFS) [24], Global Functioning Scale (GFS), and Social-Adaptive Functioning Evaluation (SAFE) [16, 25, 26].

The UCSD Performance-Based Skills Assessment—Brief Version Functional Capacity (UPSA-B) [27]. It is the only one that evaluates the real performance with some functional tasks during the assessment, including psychosocial functioning as communicating with others and skills such as financial management. The Time Use Survey (TUS; Fowler et al., 2009; Short, 2003), a semi-structured interview designed to assess time spent in structured activities over the previous month, has been used in some studies as well.

These scales have been valuable in assessing psychosocial functioning; however, some have significant limitations, such as the absence of both objective and subjective indicators, which can compromise the consistency and objectivity of the assessment. Additionally, many of these scales evaluate aspects of functioning beyond psychosocial functioning. There is also a lack of consensus regarding the definition and conceptualization of psychosocial functioning and related concepts like well-being. The overlap between these domains can complicate the interpretation of results and the comparison between scales [9]. Familiarity with these scales is

Table 3.2 Tools for assessing psychosocial functioning

Assessment tool	Description	Domains assessed	Scoring and time frame
Global Assessment of Functioning (GAF)	Provides a global measure (0–100) of how patients are functioning, with higher scores indicating better functioning. It evaluates psychological, social, and occupational functioning, scoring the lowest level of functioning within the past week [8].	Psychological, social, occupational	Global Score (0–100); Past week
World Health Organization Disability Assessment Schedule (WHO-DAS 2.0)	Provides a global measure (0–100) of functioning over the last 30 days, with higher scores indicating better functioning [25].	Cognition, mobility, self-care, getting along with others, life activities, participation in society	Global Score (0–100); Past 30 days
WHO Quality of Life (WHOQOL) Assessment	Provides a global measure (0–100), with higher scores indicating better quality of life. It assesses the patient's experiences over the last 2 weeks [17].	Physical health, psychological health, social relationships, environment	Global Score (0–100); Past 2 weeks
Quality of Life Scale (QOLS)	It provides a global score based on five domains, with higher scores indicating a better quality of life. The assessment is based on the patient's perceptions and is not tied to a specific period [16].	Material and physical well-being, relationships with others, social/ community/civic activities, personal development, recreation	Global Score; Not Time-Specific
Social and Occupational Functioning Assessment Scale (SOFAS)	Provides a global measure (0–100) focusing specifically on social and occupational functioning [18].	Social, occupational	Global Score (0–100); No Specific Time Frame Provided
Personal and Social Performance Scale (PSP)	Offers a global score (1–100) assessing four domains over the last 30 days, with higher scores indicating better functioning [18].	Socially useful activities, personal and social relationships, self-care, disturbing and aggressive behaviours	Global Score (1–100); Past 30 days
Social Functioning Scale (SFS)	Provides a global measure (0–100) of social functioning across seven domains, assessing functioning over the past 3 months [21].	Social engagement/ withdrawal, interpersonal communication, independence-performance, independence-competence, recreation, prosocial activities, employment	Global Score (0–100); Past 3 months

(continued)

Table 3.2 (continued)

Assessment tool	Description	Domains assessed	Scoring and time frame
Global Functioning Scale (GFS)	Provides a global measure (0–10), with higher scores indicating better functioning, focusing specifically on social and role functioning [26].	Social functioning, role functioning	Global Score (0–10); Past month
Specific Levels of Functioning (SLOF)	Evaluates psychosocial functioning as well as physical health and personal care. Scores range from 1 to 5, with a higher global score indicating better functioning [27].	Interpersonal relationships, social acceptability, activities of daily living, work skills, physical health, personal care	Global Score; Multiple Items Rated 1–5
Social-Adaptive Functioning Evaluation (SAFE)	Assesses the ability to adapt socially and function in various environments, evaluating functioning over 3 months [28].	Social interactions, daily living skills, occupational/role functioning, community integration	Domain-Specific Scores: Past 3 months
UCSD Performance-Based Skills Assessment—Brief Version (UPSA-B)	Evaluate functional capacity in Financial and Communication Skills, with higher scores reflecting better functioning [24].	Financial skills, communication skills	Global Score (0–100); No Specific Time Frame Provided

essential if we want to compare global scores and determine whether they are assessing the same dimensions.

Psychosocial functioning has typically been assessed using Clinician-Reported Outcome Measures (CROM). However, taking the patient's perspective into account is also essential. We believe that the inclusion of Patient-Reported Outcome Measures (PROM) should be considered alongside the clinician's assessment to evaluate psychosocial functioning. If there is a discrepancy between the two, this provides valuable information, warrants further exploration into the possible reasons for the lack of concordance, and helps the clinician to better understand the patient's perspective.

This table summarizes key tools used to assess psychosocial functioning, highlighting the domains each tool evaluates, the scoring systems, and the relevant time frames for assessment.

3.3 Studies in Psychosocial Functioning in Schizophrenia

Deficits in psychosocial functioning are prevalent features observed in both chronic schizophrenia and first-episode psychosis, to such an extent that some authors have proposed new classifications for schizophrenia that prioritize psychosocial functioning as a more objective and quantifiable measure intended for clinical and research applications [16]. However, assessing psychosocial functioning in schizophrenia is challenging due to a lack of consensus regarding its definition [28].

Additionally, using scales that may distort results, particularly those that synthesize evaluations into overall scores, poses a risk of diluting the impact of negative symptoms (or other factors) and can even suggest misleading trends [26]. These challenges significantly impede the analysis of psychosocial functioning within schizophrenia, affecting its assessment and, consequently, the effectiveness of interventions aimed at improving this aspect [29].

As we discussed above, psychosocial functioning can be considered a core component of schizophrenia, given diagnosis based on the DSM criteria. In this section, we will review the literature, summarizing the factors associated with psychosocial functioning in schizophrenia. Knowledge of this area has the scope to inform more nuanced and targeted treatments to improve the functioning of this population in society.

A recent multicentre study in Italy aimed to disentangle the relative contribution of several factors associated with psychosocial functioning in 921 patients with schizophrenia living in the community [30]. The researchers described three domains: illness-related, personal resources, and context-related factors. Illness-related factors included disorganization, positive symptoms, negative symptoms, depression, and extrapyramidal symptoms. They were assessed using the Positive and Negative Symptoms Scale (PANSS) [31], Brief Negative Symptom Scale (BNSS) [32], Calgary Depression Scale for Schizophrenia [33], and St. Hans Rating Scale (SHRS) [34]. Personal resources considered patients' availability, cooperation, help-seeking, and adherence to treatment, which were measured using the Resilience Scale for Adults [35] and the Service User Engagement Scale SUES) [36]. Context-related factors included socioeconomic status, social networks, and internalized stigma. This was obtained from demographic information and the Internalized Stigma of Mental Illness questionnaire. Finally, real-world functioning was assessed using UPSA-B [27] and the SLOF, summarized in the previous section [37].

The researchers used structural equation modelling. This method enabled a detailed exploration of the relative predictive effects of each variable, thereby highlighting how they each contribute to functioning. In terms of the results, the model found that neurocognition had the strongest effect on real-life functioning. However, this effect was indirect, suggesting it is mediated by other factors in the models, which is consistent with other studies demonstrating a relationship between cognition and functioning in schizophrenia (Shimada et al., 2022; Iwata et al., 2017). Illness-related factors, including positive symptoms, disorganization and avolition, directly and indirectly, predicted real-world functioning, indicating that symptomatology is closely linked with poor functioning. Of note, the relationship between avolition and SLOF scores was mediated by service engagement, resilience, and internalized stigma (through resilience), highlighting the complexity of this relationship through multiple pathways. Interestingly, depression had no significant association and poor emotional expression was only indirectly and weakly linked to real-life functioning.

Several other factors have been linked to psychosocial factors in schizophrenia, which were not included in the aforementioned study. Consideration of these will inform a more holistic formulation of psychosocial functioning in schizophrenia. Here, we will explore insight, trauma, self-stigma, self-efficacy, age of onset, and gender.

Firstly, with regard to insight, studies have found a significant relationship between scores on the Scale to Assess Unawareness of Mental Disorder and PSP, which has been reported in this population, supporting the link between insight and functioning [38]. Consistent with this, recent longitudinal research demonstrated a predictive role of insight (awareness into mental health) on instrumental and global functioning 10 years later, whilst enhancement/maintenance of illness awareness was significantly related to interpersonal functioning [39]. Secondly, research has demonstrated an association between a history of sexual abuse and childhood trauma and poorer role functioning, reduced capacity for intimacy, higher levels of neuroticism, and lower occupational and social functioning [40]. Thirdly, self-stigma is prevalent in individuals with schizophrenia and is considered to be more intense than other mental health difficulties [41, 42]. A review exploring the impact of self-stigma on outcomes in schizophrenia demonstrated a strong negative correlation with quality of life and self-esteem [43]. Several studies consistent with this also highlight associations with hope, impaired recovery, poorer social/vocational function, treatment adherence, greater symptom severity, distress, dysfunctional attitudes, depression, and avoidant coping style [44–50]. Fourthly, self-efficacy is defined as a belief in one's ability to perform tasks (Bandura, 1066) and has been found to be positively related to psychosocial functioning in schizophrenia. Interestingly, this is mediated by negative symptoms [15]. Fifthly, gender differences are more extensively explored in the literature. Studies suggest that women often show better outcomes with fewer negative symptoms, while men experience earlier onset, more social difficulties, and lower functioning [51–55]. However, this is not always reported, perhaps indicative of methodological differences in tools used for psychosocial functioning and/or societal expectations of gender roles across time. Lastly, age is another factor to consider. According to meta-analysis evidence, age onset is related to more severe negative symptoms and poorer premorbid functioning [56].

Taken together, the literature highlights the complexity of the factors linked to psychosocial functioning, as summarized in Table 3.3. Research analysing the relationships between predictors, mediators, and real-life functioning will be a critical direction for the future and strongly support the need for integrated and tailored psychosocial treatments.

3.4 Negative Symptoms and Psychosocial Functioning in Schizophrenia

A recent systematic review and meta-analysis examining the relationship between psychopathology and psychosocial functioning found that impaired social functioning in schizophrenia is primarily driven by negative symptoms [29]. Negative symptoms are recognized as a primary predictor of poor psychosocial functioning,

Table 3.3 Main domains linking to psychosocial functioning in schizophrenia

Domain	Description/findings
Symptomatology [30]	Positive symptoms, disorganization, and avolition have been directly and indirectly linked to real-life functioning. Social cognition, functional capacity, resilience, internalized stigma, and engagement with mental health services served as mediators.
Age at onset [56]	The age at onset is related to more severe negative symptoms and poorer premorbid functioning, such as educational attainment, baseline impaired psychosocial functioning, and poorer work history, as highlighted by systematic reviews and meta-analyses.
Gender [51, 52, 54, 57]	Gender differences are well-documented in schizophrenia, affecting prevalence, age of onset, symptom presentation, and outcomes. Females often show better outcomes with fewer negative symptoms, while males experience earlier onset, more social difficulties, and lower functioning. Methodological differences in measuring psychosocial functioning and societal expectations of gender roles may influence these outcomes.
Cognition [30, 58, 59]	Neurocognition, incorporating processing speed, attention, working memory, verbal memory, visual memory, and problem-solving, is significantly indirectly associated with real-world functioning in structural equation modelling. Tools like BACS have found specific correlations between verbal fluency and aspects of social isolation, interpersonal relationships, and employment/occupation status,
Self-efficacy and insight [38, 39]	Self-efficacy, or belief in one's ability to perform tasks, positively influences psychosocial functioning. Insight into one's condition also plays a crucial role, with better insight predicting superior functioning and interpersonal effectiveness.
Trauma and stigma [40, 43]	A history of trauma, particularly sexual abuse, is linked to poorer role functioning and social difficulties in schizophrenia. Stigma, both public and self-imposed, is associated with negative outcomes like lower self-esteem, quality of life, and social functioning, highlighting the need for multi-level interventions.

depending on the affected domain (MAP versus EE), and different areas of psychosocial functioning can be predicted, which will be discussed further in this section.

A clinical subtype of schizophrenia known as "deficit syndrome" has been suggested, characterized by the persistent and predominant presence of negative symptoms since the onset of the illness. This "deficit syndrome" is associated with significantly poorer psychosocial performance compared to other schizophrenia subtypes, highlighting the critical role negative symptoms play in determining overall patient functioning [60].

Despite the huge clinical impact, there is limited efficacy for current psychopharmacological treatment. Despite the modest efficacy of medications in alleviating negative symptoms [26], the true impact of psychosocial rehabilitation interventions remains inadequately explored and understood [13].

Nevertheless, according to the literature, negative symptoms are a core component of schizophrenia. Given the significant psychosocial impact of these symptoms, accumulated evidence necessitates focused attention on avolition. Avolition is a multidimensional construct that influences various processes, such as hedonic experience, reward prediction, and effort evaluation [1]. As will be further elaborated later, these factors profoundly affect the patient's social prognosis, both directly and by mediating the interaction among numerous other contributing factors.

Avolition is considered a major motivational barrier that, in patients with schizophrenia, can significantly impair psychosocial functioning due to a comorbid disability in 'effort computation', which involves estimating the amount of effort required for goal-directed activities. These patients often feel overwhelmed by the demands of such activities, and their subsequent underperformance tends to reinforce feelings of incapacity and excessive exertion [1].

In more interpersonal domains, asociality interferes with participation in many social activities—whether recreational, work-related, or academic—by diminishing the individual's interest in engaging with others. In a typical scenario, a preserved interest in social interaction could compensate for the increased effort these activities are perceived to require (which, as mentioned earlier, is often overestimated due to inaccuracies in effort computation) [1].

In the following subsections, we will review both cross-sectional and longitudinal studies that provide compelling evidence for the association between negative symptoms and psychosocial functioning. While some studies treat negative symptoms as a unified construct, others distinguish between specific dimensions—such as motivational and expressive deficits—to gain a clearer understanding of their distinct impacts on psychosocial outcomes. This nuanced approach offers deeper insights into how these symptoms differentially influence various domains of functioning in individuals with schizophrenia.

3.4.1 Cross-Sectional Studies: Using Single Scores in Negative Symptoms and Single or Domain-Specific Scores in Psychological Functioning in Schizophrenia

Cross-sectional studies analyse data collected from a population at one specific point in time. Early cross-sectional studies exploring the relationship between negative symptoms in schizophrenia and psychosocial functioning typically categorized negative symptoms into a single score. It could be argued that this is an oversimplification given the breadth of negative symptoms, thought to encompass motivational deficits, a lack of pleasure, and diminished emotional expression. Here, we will summarize cross-sectional studies exploring the relationship between psychosocial functioning and negative symptoms, beginning with those utilizing a single score to categorize negative symptoms.

A study published in 2005 [61] calculated a total negative symptom score from the positive and negative symptoms scale (PANSS) in a sample of patients with a diagnosis of schizophrenia or schizoaffective disorder aged 19–60 years. The authors assessed quality of life and functioning through the Global Assessment of Functioning Scale (GAF) [11]. Total scores on these measures correlated with negative symptoms, suggesting that greater severity of negative symptoms is associated with functioning and quality of life. Negative symptoms were significantly associated with psychological health, social relationships, environment, and a lesser likelihood of living independently.

Another study in 2011 [62] assessed 103 patients with schizophrenia and also calculated a total score for negative symptoms using the PANSS. However, psychosocial functioning was measured using the PSP. In a sample of individuals diagnosed with schizophrenia and schizoaffective disorder aged 21–65 years, the authors report a significant negative correlation between total scores on the PSP and negative symptoms. Looking at the different domains of the PSP, they found a significant association between negative symptoms and activities, self-care, and aggressive behaviour, but not relationships, suggesting that negative symptoms are associated with functioning in these areas.

3.4.2 Cross-Sectional Studies: Using Domain-Specific Scores (Negative Symptoms or/and Psychosocial Functioning)

Several factor analyses have suggested that a unidimensional model to assess negative symptoms is an oversimplification of the complexity of the area [16, 17]. Therefore, more recent studies exploring the relationship between psychosocial functioning and negative symptoms in schizophrenia have used tools that are consistent with a number of negative symptom dimensions/ domains. As an example, the Brief Negative Symptom Scale (BNSS) [63] is a semi-structured interview consisting of 13 items across five domains of negative symptoms: Anhedonia, Alogia, Asociality, Avolition, Blunted affect, and also assessing the reaction to distress. An advantage of this scale is the ability to accurately explore negative symptoms based on scores across different domains as opposed to using a single generalized score. This approach allows for a nuanced understanding of the factors relating to functional difficulties, highlighting targeted areas for treatment.

In terms of psychosocial functioning, one study by Garcia-Portilla et al. in 2021 [64] exploring negative symptoms in outpatients with a diagnosis of schizophrenia receiving antipsychotic monotherapy, reported BNSS subdomains of abulia, asociality, and blunted affect predicted scores on the PSP total score. Therefore, it could be considered that specific deficits in initiating action/speech, poor motivation for social activities, and reduced emotional expression predict psychosocial functioning.

Other studies exploring the relationship between psychosocial functioning and the sub-domains of negative symptoms have used different scales. For example, research done in Ultra-high risk for Psychosis and First-Episode Schizophrenia [65] used the Global Function: Social and Role scale, which separates functioning into social and role domains. Social functioning refers to the quantity and quality of age-appropriate intimate relationships, peer relationships, peer conflict, and family member involvement. Role functioning incorporates the support provided to an individual at school, work, or home and their performance in relation to this. The authors also used the Scale for the Assessment of Negative Symptoms (SANS) [66], obtaining scores on the domain of affective flattening/blunting and avolition-apathy. For social functioning, the authors reported a significant correlation with affective flattening/blunting, whereas avolition-apathy was correlated with role functioning. This may suggest that reduced emotional expression and a lack of motivation/interest impair social contact and role performance, respectively. Of note, participants in this sample were at ultra-high risk of psychosis or first-episode patients with schizophrenia, whereas previous research has used participants with chronic diagnoses of schizophrenia. Therefore, adding to the literature by providing insights into the manifestation of negative symptoms at the early pre-psychotic stage.

Another more recent study, published in 2021 [67], looked at the dimensionality of negative symptoms again using the SANS. Psychosocial functioning was assessed using the PSP. Uniquely, the authors analysed scores relating to the psychosocial functioning of 64 people with schizophrenia using a network analysis. This can be a helpful way of analysing complex relationships between a number of variables, producing a visual map representing the strength of the relationship between different domains. In terms of negative symptoms and psychosocial functioning, findings demonstrated an inverse relationship between asociality, avolition, and blunted affect and scores on the PSP, suggesting deficits in these areas are associated with social functioning. Moreover, these are strongly connected with functioning, underlining the importance of targeting negative symptoms to improve patients' real-world functioning.

In a more recent study [68], the authors examined the relationship between negative symptoms, neurocognition, and social cognition with functioning in schizophrenia. One hundred fifteen participants were assessed using the PANSS and the BNSS to assess negative symptoms, and the PSP was used to evaluate psychosocial functioning. The results of this research study showed that the BNSS motivational deficit domain of negative symptoms strongly predicted both the PSP socially useful activities score and the PSP personal and social relationships score. The implication of these findings is of great importance, demonstrating the need for further research aimed at exploring the impact of these factors on various aspects of functioning in order to identify targeted interventions focusing on enhancing specific skills, thereby improving the lives of people with schizophrenia.

Another study focussed on the Treatment-Resistant Schizophrenia (TRS) population. A research study [17] investigated how schizophrenia symptom domains and neurocognition contribute to predicting psychosocial functioning in a sample of 159 TRS patients. The PANSS was used, and its negative symptom factor was broken down into two dimensions: Diminished Expressivity (DE), measuring deficits in expression, and Social Anhedonia (SA), measuring deficits in motivation/pleasure. The study's findings are consistent with previous research on non-TRS populations, showing a strong correlation between negative symptoms and psychosocial functioning. Negative symptoms, especially the SA dimension, were strong predictors of psychosocial functioning; the lower severity in the SA dimension of negative symptoms was associated with higher levels of functioning.

3.4.3 Longitudinal Studies on Negative Symptoms and Psychosocial Functioning

A meta-analysis focussed exclusively on longitudinal studies [69] to explore how baseline negative symptoms (beyond others) are associated with changes in psychosocial functioning over time. The analysis included 84 studies with a total of 33,456 participants. In this meta-analysis, the authors categorized participants based on the duration of illness at baseline. Results showed that improvements in psychosocial functioning during follow-up were associated with lower baseline levels of negative symptoms, suggesting that negative symptoms may act as a moderator of psychosocial improvement. Specifically, low levels of baseline negative symptoms were associated with better social functioning at follow-up for individuals with a duration of illness of less than 2 years, as well as those with a duration between 5 and 10 years, but not for other subgroups (e.g., those with less than 2 years, between 2 and 5 years, or more than 10 years). Consistent with previous authors, this meta-analysis suggests that the first 5 years of illness are a critical period for recovery, during which patients can achieve significant improvements in social functioning [70]. However, an important limitation of this meta-analysis is that negative symptoms and psychosocial functioning are assessed as a single score, regardless of the type of scale used.

3.4.3.1 Longitudinal Studies. Using Single Scores in Negative Symptoms and Psychosocial Functioning

A study conducted in 1995 explored the relationship between symptoms and psychosocial functioning in 113 patients with schizophrenia at a 1- and 2-year follow-up. Interestingly, negative symptoms, but not positive, were predictive of social functioning. The researchers used the Psychiatric Rating Scale (PRS) to evaluate various domains of social functioning and included a negative symptom as a single score among other clinical outcomes.

Another cohort study published in 2000 [71] followed 85 patients with schizophrenia and 69 with schizoaffective disorder. The negative syndrome was identified using a principal component analysis of the AMDP-System (Association for Methodology and Documentation in Psychiatry) [72].

and was associated with the prediction of lower GAF scores approximately 12.5 years later.

Albert et al. (2011) [73] studied 255 individuals after their first episode of non-affective psychosis and followed them for 5 years. They used the SANS to measure negative symptoms and found that lower global scores on the SANS at baseline predicted higher GAF scores and were associated with increased engagement in productive psychosocial activities such as working or studying.

In a research study carried out by Siegrist et al. (2015) [74], a large European naturalistic sample of individuals with schizophrenia ($N = 1208$) was analysed over 24 months. The authors found a significant association between negative symptoms, psychosocial functioning, and the number of social contacts. Negative symptoms were measured using the PANSS, while psychosocial functioning was assessed using the GAF.

Gee et al. (2016) [70] conducted a 12-month follow-up study on individuals with first-episode psychosis. Negative symptoms were measured using the PANSS, and psychosocial functioning was assessed using the Time Use Survey (TUS), a semi-structured interview designed to evaluate time spent in structured activities over the previous month. The authors found that high levels of negative symptoms at baseline were associated with poor social recovery after 12 months.

3.4.3.2 Longitudinal Studies. Using Domain-Specific Scores (Negative Symptoms or/and Psychosocial Functioning)

Bottlender et al. (2010) [75] conducted a 15-year follow-up study involving 61 patients with schizophrenia and 58 with schizoaffective disorder using the AMDP system. Social function was assessed with the Mannheim Disability Assessment Schedule (DAS-M). The study found that severe social disability was best explained by the presence of an apathetic syndrome in both schizophrenia and schizoaffective patients.

Hwu et al. (2002) [76] conducted a follow-up study involving 163 schizophrenic patients assessed with the PANSS scale at admission and during a 1-year follow-up. The study identified two subgroups: one with marked negative symptoms (GWNEG) and one without (GONEG). The GWNEG group demonstrated a shorter duration of employment and worse social functioning compared to the GONEG group.

Evensen et al. (2012a) [77] studied 301 patients with first-episode psychosis at baseline, with a 10-year follow-up of 178 patients. The Apathy Evaluation Scale (AES-S) was used to assess apathy, and psychosocial functioning was measured using the GAF and the Strauss Carpenter Level of Functioning Scale (SCLFS). The study found that 30% of patients met the threshold for apathy at follow-up, which was associated with impaired functioning.

Chang et al. (2018) [78] followed 160 patients with first-episode psychosis, assessed at baseline and followed up at 1, 2, and 3 years. Negative symptoms were divided into two distinct subdomains: amotivation and diminished emotion, using PANSS ratings. Psychosocial functioning was measured using the SOFAS and the Role Functioning Scale (RFS). Amotivation was significantly related to concurrent and longitudinal (1-year) psychosocial functioning, while diminished emotion was not.

Chang et al. (2011) [79] followed 93 individuals with first-episode schizophrenia. The High Royds Evaluation of Negativity Scale (HEN) [80] was used to measure negative symptoms. At baseline, 25.8% of participants exhibited primary negative symptoms, and 23.7% still had these symptoms after 3 years. The study found that lower premorbid academic and social functioning, as measured by the Premorbid Adjustment Scale (PAS) [81], was associated with persistent primary negative symptoms after 3 years. This study is important as the authors distinguished between primary and secondary negative symptoms.

Galderisi et al. (2013) [82] followed 58 individuals with deficit syndrome and 54 with non-deficit schizophrenia, diagnosed using the Schedule for the Deficit

Syndrome (SDS) scale. At the 5-year follow-up, psychosocial functioning was assessed using the Assessment of Disability (AD). Principal component analysis of the SDS identified two factors explaining 83.2% of the variance: 'poor emotional expression' and 'avolition'. Avolition was associated with poorer psychosocial outcomes and had a higher predictive value for long-term functioning than the poor emotional expression factor.

Ventura et al. (2015) [83] included 149 individuals with recent-onset schizophrenia. Negative symptoms were assessed using the BPRS and SANS, while psychosocial functioning was evaluated using the Social and Occupational Functioning Scale (SCORS). Baseline negative symptoms were significantly associated with poor social functioning. At the 12-month follow-up, negative symptoms were linked to impaired work/school functioning, and at the 8-year follow-up, they remained associated with poor social functioning.

Mucci et al. (2021) [84] conducted a 4-year cohort study involving 618 clinically stable patients with schizophrenia. The study found that avolition, assessed by the BNSS scale, was significantly associated with poorer interpersonal relationships after 4 years of follow-up, as measured by the Specific Level of Functioning Scale (SLOF).

It is vital to consider the influence of third variables, such as premorbid adjustment, duration of untreated psychosis, gender, trauma history, cognitive deficits, self-efficacy, depression, comorbid neurodevelopmental disorders, extrapyramidal symptoms, and other factors that impact both negative symptoms and functioning. These variables must be targeted as part of a holistic intervention strategy to optimize treatment effectiveness.

3.4.4 Conclusions from Cross-Sectional and Longitudinal Studies

Cross-sectional studies provide evidence of the association between negative symptoms and psychosocial functioning. Similarly, longitudinal studies demonstrate the predictive capacity of negative symptoms for future psychosocial outcomes. However, findings regarding the relationship between specific subdomains of negative symptoms (e.g. motivational and pleasure deficits versus emotional expressiveness) and distinct areas of psychosocial functioning (such as socially useful activities, personal and social relationships, self-care, etc.) remain limited. There is no doubt that further research is necessary to better understand these.

It is important to highlight the distinct neuroscience mechanisms underlying these associations, as demonstrated in cross-sectional factor analysis studies [2] and longitudinal network analyses across culturally diverse populations [85, 86]. These studies suggest that different brain processes may be linked to specific negative symptom domains, while varying psychosocial outcomes, such as goal-directed activities and social and personal relationships, may also require distinct neural underpinnings. Understanding the complex interactions between negative symptoms and psychosocial functioning could be enhanced by distinguishing their principal components. If we can provide an individualized profile for each subject, we could more accurately tailor and select appropriate therapeutic interventions, ultimately leading to more targeted and effective treatment strategies.

Nevertheless, it is relevant to emphasize the irreplaceable role of clinical judgment in distinguishing between primary and secondary negative symptoms, as clinician insight remains fundamental in both the assessment process and the interpretation of findings. In schizophrenia, frequent comorbidities such as depression, cognitive deficits, low self-efficacy, extrapyramidal symptoms, trauma, and positive symptoms can either exacerbate or influence the persistence of primary symptoms or contribute to the development of secondary negative symptoms, either individually or through their combined effects. This added complexity further complicates the clinical picture. Therefore, the use of specific tools to assess subdomains within negative symptoms and psychosocial functioning is indispensable in psychiatric practice. Accurate measurement of these subdomains is crucial, as it lays the foundation for effectively targeting clinical interventions.

Consequently, when pharmacological or non-pharmacological treatments lead to improvements in specific dimensions, their positive effects can be more precisely identified and understood. This approach avoids the oversimplification of relying on a single composite score that may obscure the distinct dimensions of negative symptoms and psychosocial functioning, thereby enhancing our capacity to detect and interpret therapeutic outcomes with greater clarity and relevance.

3.5 Psychosocial Intervention Targeting Negative Symptoms

Experts in the field underscore the pressing need to develop targeted interventions for improving negative symptoms, with a dual focus on alleviating symptoms and enhancing functional outcomes. But even just for improving negative symptoms, we observe a lack of clear efficacy for psychological interventions SMD -0.14, –0.29 to 0.01) [87]. This finding is consistent with the observed lack of efficacy in pharmacological intervention [2].

Our team proposes targeted interventions focused on the motivation and pleasure domain (avolition) as it is the domain related to psychosocial functioning as cross-sectional, longitudinal, and network [84, 88] studies have shown. The key components of this intervention will combine individual goal setting, psychoeducation, and specific skills development such as resilience.

Because resilience is strongly associated with avolition and psychosocial functioning [89, 90], this would also be useful to target in supporting individuals' ability to cope with stress. We consider the following aspects of resilience, namely, 'perception of self' and 'perception of future' in our approach and, therefore, propose activities to empower participants to believe in their ability to achieve their goals.

Another component to target is cognition, given the strong links with functional capacity. Problem-solving strategies are one way of helping individuals exercise

Table 3.4 Proposed CBT intervention linking motivation, pleasure, and psychosocial functioning

Component	Description
1. *Individualized goal setting*	
Tailored goals	Each participant identifies a psychosocial goal aligned with their personal interests and intrinsic motivation (e.g. work, leisure, socializing). The goal should be naturally rewarding and appetitive
SMART goals	Goals are framed as Specific, Measurable, Achievable, Relevant, and Time-bound (SMART) to ensure clarity and focus
2. *Psychoeducation*	
Understanding reward processes	Educate participants on the distinction between anticipatory anhedonia (difficulty anticipating pleasure) and intact pleasure at the moment
3. *Group-based learning and practice*	
Skill development and problem-solving	Participants engage in exercises that build problem-solving skills, generate options for overcoming obstacles, and enhance self-efficacy
Attributional style	Discussions and activities help participants differentiate what is within their control and what is not, fostering a realistic sense of agency
Self-efficacy	Practising ways to increment self-efficacy, especially in cases where it is low. Activities aim to empower participants to believe in their ability to achieve their goals

cognitive skills such as option generation, cost-benefit analysis, conflict resolution, and decision-making ([22, 67].

Maladaptive attributional styles can be linked to motivation and maintain and exacerbate low mood states, impacting individuals' locus of control and ability to take action, otherwise referred to as 'learnt helplessness' in the literature. For some individuals, where this is present, this would be useful in addition to the intervention strategy proposed [91]. The key components of the intervention proposed are summarized in Table 3.4.

Focusing on this domain offers a promising strategy for CBT interventions. By tailoring appetitive goals to individual interests and incorporating group-based learning and support, this approach aims to enhance participants' engagement and success in achieving individualized and rewarding psychosocial outcomes.

Rather than discouraging the use of CBT for negative symptoms, we consider this an opportunity to rethink how these symptoms are evaluated, both conceptually and psychometrically [92], and to advocate for designing new tailored interventions that use more appropriate outcome measures, including targeting individualized reward-based goal-directed behaviours.

References

1. Strauss GP, Bartolomeo LA, Harvey PD. Avolition as the core negative symptom in schizophrenia: relevance to pharmacological treatment development. [cited 2024 Feb 8]; Available from: https://doi.org/10.1038/s41537-021-00145-4.

2. Galderisi S, Kaiser S, Bitter I, Nordentoft M, Mucci A, Sabé M, et al. EPA guidance on the treatment of negative symptoms in schizophrenia. Eur Psychiatry. 2021;64(1)
3. American Psychiatric Association. Diagnostic and statistical manual of mental disorders. 3rd ed., Rev. Washington, DC: American Psychiatric Association; 1987.
4. Figueira ML, Brissos S. Measuring psychosocial outcomes in schizophrenia patients. Curr Opin Psychiatry. 2011;24(2)
5. Freud S. Introductory lectures on psycho-analysis (parts I and II). SE. 1916;16
6. Durkheim E. Le Suicide: Etude de sociologie. Livre premier : les facteurs extra-sociaux. Psychol Rev. 1898;5(4)
7. Strupp HH. Childhood and society. By Erik H. Erikson. New York: W. W. Norton & Co., Inc.; 1950. 397 pp. Psychoanal Q. 1951;20(2)
8. Engel GL. The need for a new medical model: a challenge for biomedicine. Science (1979). 1977;196(4286)
9. White S, Dominise C, Naik D, Killaspy H. The reliability of the personal and social performance scale—informing its training and use. Psychiatry Res. 2016;243:312–7.
10. Anczewska M, Switaj P, Roszczynskamichta J, Chrostek A, Charzynska K. Psychosocial difficulties experienced by people diagnosed with schizophrenia—barriers to social inclusion. Bulgarian Comparative Education Society; 2013.
11. Aas IM. Guidelines for rating Global Assessment of Functioning (GAF). Ann General Psychiatry. 2011;10:2.
12. Seligman ME, Csikszentmihalyi. Positive psychology. An introduction. Am Psychol. 2000;55(1)
13. Brekke JS, Levin S, Wolkon GH, Sobel E, Slade E. Psychosocial functioning and subjective experience in schizophrenia. Schizophr Bull [Internet]. 1993 [cited 2024 Feb 5];19(3):599–608. Available from: https://doi.org/10.1093/schbul/19.3.599.
14. Giordano GM, Pezzella P, Mucci A, Austin SF, Erfurth A, Glenthøj B, et al. Negative symptoms and social cognition as mediators of the relationship between neurocognition and functional outcome in schizophrenia. Front Psych. 2024;15:1333711.
15. Pratt SI, Mueser KT, Smith TE, Lu W. Self-efficacy and psychosocial functioning in schizophrenia: a mediational analysis. Schizophr Res. 2005;78(2–3):187–97.
16. Bulletin S, Cowman M, Godfrey E, Walsh T, Frawley E, Fowler D, et al. Measures of social and occupational function in early psychosis: a systematic review and meta-analysis. Schizophr Bull [Internet]. 2023 [cited 2024 Feb 5]; Available from: https://doi.org/10.1093/schbul/sbad062.
17. Li Y, Ang MS, Jie Yin Y, See YM, Lee J. Predictors of functioning in treatment-resistant schizophrenia: the role of negative symptoms and neurocognition. Front Psychiatry. 2024;15:1444843.
18. Schneider LC, Struening EL. SLOF: a behavioural rating scale for assessing the mentally ill. Soc Work Res Abstr. 1983;19(3):9.
19. Abplanalp SJ, Mueser KT, Fulford D. The centrality of motivation in psychosocial functioning: network and bifactor analysis of the Quality of Life Scale in first episode psychosis. Psychol Assess [Internet]. 2022 [cited 2024 Feb 8];34(3):205. Available from: /pmc/articles/PMC10066933/.
20. Heinrichs DW, Hanlon TE, Carpenter WT. The quality of life scale: an instrument for rating the schizophrenic deficit syndrome. Schizophr Bull. 1984;10(3):388.
21. Morosini PL, Magliano L, Brambilla L, Ugolini S, Pioli R. Development, reliability and acceptability of a new version of the DSM- IV Social Occupational Functioning Assessment Scale (SOFAS) to assess routine social functioning. Acta Psychiatr Scand. 2000;101(4):323.
22. Giordano GM, Pezzella P, Mucci A, Austin SF, Erfurth A, Glenthøj B, et al. Negative symptoms and social cognition as mediators of the relationship between neurocognition and functional outcome in schizophrenia. Front Psychiatry. 2024:15.
23. Harper A, Power M, Orley J, Herrman H, Schofield H, Murphy B, et al. Development of the World Health Organization WHOQOL-BREF Quality of Life Assessment. Psychol Med. 1998;28(3)

24. Birchwood M, Smith J, Cochrane R, Wetton S, Copestake S. The Social Functioning Scale. The development and validation of a new scale of social adjustment for use in family intervention programmes with schizophrenic patients. Br J Psychiatry. 1990;157(DEC)
25. Wang Z, Ling Y, Wang Y, Zhu T, Gao J, Tang X, et al. The role of two factors of negative symptoms and cognition on social functioning in male patients with Schizophrenia: a mediator model. Brain Sci. 2023;13(2)
26. Okada H, Hirano D, Taniguchi T. Impact of negative symptom domains and other clinical characteristics on functional outcomes in patients with Schizophrenia. Schizophr Res Treatment [Internet]. 2021 [cited 2024 Feb 8];2021. Available from: /pmc/articles/PMC7914085/.
27. Mausbach BT, Harvey PD, Goldman SR, Jeste DV, Patterson TL. Development of a brief scale of everyday functioning in persons with serious mental illness. Schizophr Bull. 2007;33(6):1364.
28. Suzuki T. Subtyping Schizophrenia by social functioning—a pragmatic proposal for clinics and research. Psychiatr Q [Internet]. 2018 [cited 2024 Feb 8];89(3):533–8. Available from: https://pubmed.ncbi.nlm.nih.gov/29297110/
29. Handest R, Molstrom IM, Henriksen MG, Hjorthøj C, Nordgaard J. A systematic review and meta-analysis of the association between psychopathology and social functioning in Schizophrenia. Schizophr Bull. 2023;49(6):1470–85.
30. Galderisi S, Rossi A, Rocca P, Bertolino A, Mucci A, Bucci P, et al. The influence of illness-related variables, personal resources and context-related factors on real-life functioning of people with schizophrenia. World Psychiatry. 2014;13(3):275.
31. Kay SR, Flszbeln A, Qpjer LA. The positive and negative syndrome scale (PANSS) for schizophrenia [Internet], Vol. 13; 1967. Available from: https://academic.oup.com/schizophreniabulletin/article/13/2/261/1919795
32. Kirkpatrick B. Developing concepts in negative symptoms: primary vs secondary and apathy vs expression. J Clin Psychiatry. 2014;75(1):3.
33. Addington D, Addington J, Maticka-Tyndale E. Assessing depression in schizophrenia: the Calgary depression scale. Br J Psychiatry. 1993;163:39.
34. Gerlach J, Korsgaard S, Clemmesen P, Lauersen A-ML, Magelund G, Noring U, et al. The St. Hans Rating Scale for extrapyramidal syndromes: reliability and validity. Acta Psychiatr Scand. 1993;87(4):244.
35. Friborg O, Hjemdal O, Rosenvinge JH, Martinussen M. A new rating scale for adult resilience: what are the central protective resources behind healthy adjustment? Int J Methods Psychiatr Res. 2003;12(2):65.
36. Tait L, Birchwood M, Trower P. A new scale (SES) to measure engagement with community mental health services. J Ment Health. 2002;11(2):191.
37. Schneider LC, Struening EL. SLOF: a behavioral rating scale for assessing the mentally ill. Soc Work Res Abstr. 1983;19(3):9.
38. Erol A, Delibas H, Bora O, Mete L. The impact of insight on social functioning in patients with schizophrenia. Int J Soc Psychiatry. 2015;61(4):379.
39. Canal-Rivero M, Ayesa-Arriola R, Ruiz-Veguilla M, Ortiz-García de la Foz V, Labad J, Crespo-Facorro B. Insight trajectories and their impact on psychosocial functioning: a 10-year follow-up study in first episode psychosis patients. J Psychopathol Clin Sci. 2022;131(7)
40. Lysaker PH, Meyer PS, Evans JD, Clements CA, Marks KA. Childhood sexual trauma and psychosocial functioning in adults with schizophrenia. Psychiatr Serv. 2001;52(11):1485.
41. Brohan E, Elgie R, Sartorius N, Thornicroft G. Self-stigma, empowerment and perceived discrimination among people with schizophrenia in 14 European countries: the GAMIAN-Europe study. Schizophr Res. 2010;122(1–3)
42. Gerlinger G, Hauser M, De Hert M, Lacluyse K, Wampers M, Correll CU. Personal stigma in schizophrenia spectrum disorders: a systematic review of prevalence rates, correlates, impact and interventions. World Psychiatry. 2013;12:155.
43. Sarraf L, Lepage M, Sauvé G. The clinical and psychosocial correlates of self-stigma among people with schizophrenia spectrum disorders across cultures: a systematic review and meta-analysis. Schizophr Res. 2022;248:64.

44. Corrigan PW, Watson AC. The paradox of self-stigma and mental illness, Clinical psychology: science and practice, vol. 9; 2002.
45. Dubreucq J, Plasse J, Franck N. Self-stigma in serious mental illness: a systematic review of frequency, correlates, and consequences. Schizophr Bull. 2021;47:1261.
46. Livingston JD, Boyd JE. Correlates and consequences of internalized stigma for people living with mental illness: a systematic review and meta-analysis. Soc Sci Med. 2010;71(12):2150.
47. Margetić BA, Jakovljević M, Ivanec D, Margetić B, Tošić G. Relations of internalized stigma with temperament and character in patients with schizophrenia. Compr Psychiatry. 2010;51(6):603.
48. Park K, MinHwa L, Seo M. The impact of self-stigma on self-esteem among persons with different mental disorders. Int J Soc Psychiatry. 2019;65(7–8)
49. Schrank B, Amering M, Hay AG, Weber M, Sibitz I. Insight, positive and negative symptoms, hope, depression and self-stigma: a comprehensive model of mutual influences in schizophrenia spectrum disorders. Epidemiol Psychiatr Sci. 2014;23(3):271.
50. Yanos PT, Roe D, Markus K, Lysaker PH. Pathways between internalized stigma and outcomes related to recovery in schizophrenia spectrum disorders. Psychiatr Serv. 2008;59(12):1437.
51. Castle DJ, Abel K, Noriyoshi T, Murray RM. Gender differences in schizophrenia: hormonal effect or subtypes? Schizophr Bull. 1995;21(1):1.
52. Rotstein A, Shadmi E, Roe D, Gelkopf M, Levine SZ. Gender differences in quality of life and the course of schizophrenia: national study. BJPsych Open. 2022;8(2):e35.
53. Rusakovskaya O, Kharitonova N, Movina L, Papsuev O. Real-life functioning in women with schizophrenia living in residential facilities: gender-based comparison. Front Psychiatry. 2023;14
54. Thorup A, Albert N, Bertelsen M, Petersen L, Jeppesen P, Le Quack P, et al. Gender differences in first-episode psychosis at 5-year follow-up-two different courses of disease? Results from the OPUS study at 5-year follow-up. Eur Psychiatry. 2014;29(1):44.
55. Usall J, Haro JM, Araya S, Moreno B, Muñoz PE, Martínez A, et al. Social functioning in schizophrenia: what is the influence of gender? Eur J Psychiat. 2007;21
56. Immonen J, Jääskeläinen E, Korpela H, Miettunen J. Age at onset and the outcomes of schizophrenia: a systematic review and meta-analysis. Early Interv Psychiatry. 2017;11:453.
57. Grossman LS, Harrow M, Rosen C, Faull R, Strauss GP. Sex differences in schizophrenia and other psychotic disorders: a 20-year longitudinal study of psychosis and recovery. Compr Psychiatry. 2008;49(6):523.
58. Shimada T, Inagaki Y, Shimooka Y, Kawano K, Tanaka S, Kobayashi M. Effect of individualized occupational therapy on social functioning in patients with schizophrenia: a five-year follow-up of a randomized controlled trial. J Psychiatr Res [Internet]. 2022 [cited 2024 Feb 8];156:476–84. Available from: https://pubmed.ncbi.nlm.nih.gov/36347107/
59. Iwata K, Matsuda Y, Sato S, Furukawa S, Watanabe Y, Hatsuse N, et al. Efficacy of cognitive rehabilitation using computer software with individuals living with schizophrenia: a randomized controlled trial in Japan. Psychiatr Rehabil J. 2017;40(1):4.
60. Chan RCK, Wang LL, Lui SSY. Theories and models of negative symptoms in schizophrenia and clinical implications. Nat Rev Psychol. 2022 1:8 [Internet]. 2022 [cited 2024 Feb 8];1(8):454–67. Available from: https://www.nature.com/articles/s44159-022-00065-9
61. Hofer A, Baumgartner S, Edlinger M, Hummer M, Kemmler G, Rettenbacher MA, Schweigkofler H, Schwitzer J, Fleischhacker WW, et al. Patient outcomes in schizophrenia I: correlates with sociodemographic variables, psychopathology, and side effects. Eur Psychiatry. 2005;20
62. Schaub D, Brüne M, Jaspen E, Pajonk FG, Bierhoff HW, Juckel G. The illness and everyday living: close interplay of psychopathological syndromes and psychosocial functioning in chronic schizophrenia. Eur Arch Psychiatry Clin Neurosci. 2011;261(2):85.
63. Kirkpatrick B, Strauss GP, Nguyen L, Fischer BA, Daniel DG, Cienfuegos A, et al. The brief negative symptom scale: psychometric properties. Schizophr Bull. 2011;37(2):300.

64. García-Portilla MP, García-Álvarez L, González-Blanco L, Dal Santo F, Bobes-Bascarán T, Martínez-Cao C, et al. Real-world functioning in patients with schizophrenia: beyond negative and cognitive symptoms. Front Psychiatry. 2021:12.
65. Lee SJ, Kim KR, Lee SY, An SK. Impaired social and role function in ultra-high risk for psychosis and first-episode schizophrenia: its relations with negative symptoms. Psychiatry Investig. 2017;14(5):539.
66. Andreasen NC. Scale for the assessment of negative symptoms (SANS). Br J Psychiatry. 1989;155(Suppl. 7):49.
67. Charernboon T. Interplay among positive and negative symptoms, neurocognition, social cognition, and functioning in clinically stable patients with schizophrenia: a network analysis. F1000Res. 2022:10.
68. Maria Giordano G, Caporusso E, Pezzella P, Galderisi S. Updated perspectives on the clinical significance of negative symptoms in patients with schizophrenia. Expert Rev Neurother 2022 [cited 2024 Feb 8]; Available from: https://www.tandfonline.com/action/journalInformation?journalCode=iern20
69. de Winter L, Couwenbergh C, van Weeghel J, Hasson-Ohayon I, Vermeulen JM, Mulder CL, et al. Changes in social functioning over the course of psychotic disorders–a meta-analysis. Schizophr Res. 2022;239:55–82.
70. Gee B, Hodgekins J, Fowler D, Marshall M, Everard L, Lester H, et al. The course of negative symptom in first episode psychosis and the relationship with social recovery. Schizophr Res. 2016;174(1–3):165–71.
71. Möller HJ, Bottlender R, Wegner U, Wittmann J, Strauß A. Long-term course of schizophrenic, affective and schizoaffective psychosis: focus on negative symptoms and their impact on global indicators of outcome. Acta Psychiatr Scand [Internet]. 2000 [cited 2024 Feb 8];102(407):54–7. Available from: https://onlinelibrary.wiley.com/doi/full/10.1034/j.1600-0447.2000.00010.x
72. Das AMDP System: Manual zur Dokumentation psychiatrischer Befunde (The AMDP System: Manual on the Documentation ofPsychiatric Findings). Third edition. (Pp. 103; $8.00). Springer-Verlag: Berlin. 1979. Psychol Med. 1980;10(2).
73. Albert N, Bertelsen M, Thorup A, Petersen L, Jeppesen P, Le Quack P, et al. Predictors of recovery from psychosis. Analyses of clinical and social factors associated with recovery among patients with first-episode psychosis after 5 years. Schizophr Res. 2011;125(2–3)
74. Siegrist K, Millier A, Amri I, Aballéa S, Toumi M. Association between social contact frequency and negative symptoms, psychosocial functioning and quality of life in patients with schizophrenia. Psychiatry Res. 2015;230(3):860.
75. Bottlender R, Strauß A, Möller HJ. Social disability in schizophrenic, schizoaffective and affective disorders 15 years after first admission. Schizophr Res. 2010;116(1):9–15.
76. Hwu HG, Chen CH, Hwang TJ, Liu CM, Cheng JJ, Lin SK, et al. Symptom patterns and subgrouping of schizophrenic patients: significance of negative symptoms assessed on admission. Schizophr Res. 2002;56(1–2)
77. Evensen J, Røssberg JI, Barder H, Haahr U, Hegelstad WTV, Joa I, et al. Flat affect and social functioning: a 10year follow-up study of first episode psychosis patients. Schizophr Res. 2012;139:1–3.
78. Chang WC, Kwong VWY, Hui CLM, Chan SKW, Lee EHM, Chen EYH. Relationship of amotivation to neurocognition, self-efficacy and functioning in first-episode psychosis: a structural equation modeling approach. Psychol Med. 2017;47(4):755.
79. Chang WC, Hui CLM, Tang JYM, Wong GHY, Lam MML, Chan SKW, et al. Persistent negative symptoms in first-episode schizophrenia: a prospective three-year follow-up study. Schizophr Res. 2011;133(1–3):22–8.
80. Mortimer AM, McKenna PJ, Lund CE, Mannuzza S. Rating of negative symptoms using the High Royds Evaluation of Negativity (HEN) scale. Br J Psychiatry. 1989;155(Suppl. 7):89.
81. Brill N, Reichenberg A, Weiser M, Rabinowitz J. Validity of the premorbid adjustment scale. Schizophr Bull. 2008;34(5):981.

82. Galderisi S, Bucci P, Mucci A, Kirkpatrick B, Pini S, Rossi A, et al. Categorical and dimensional approaches to negative symptoms of schizophrenia: focus on long-term stability and functional outcome. Schizophr Res. 2013;147(1):157.
83. Ventura J, Subotnik KL, Gitlin MJ, Gretchen-Doorly D, Ered A, Villa KF, et al. Negative symptoms and functioning during the first year after a recent onset of schizophrenia and 8years later. Schizophr Res. 2015;161(2–3):407–13.
84. Mucci A, Galderisi S, Rucci P, Rossi A, Rocca P, Bertolino A, et al. Recovery in schizophrenia: a network analysis of inter-relationships among disease-related variables, personal resources, context-related factors and real-life functioning. Eur Psychiatry. 2021;64(S1)
85. Ahmed AO, Kirkpatrick B, Galderisi S, Mucci A, Rossi A, Bertolino A, et al. Cross-cultural validation of the 5-factor structure of negative symptoms in schizophrenia. Schizophr Bull. 2019;45(2):305.
86. Rucci P, Caporusso E, Sanmarchi F, Giordano GM, Mucci A, Giuliani L, et al. The structure stability of negative symptoms: longitudinal network analysis of the Brief Negative Symptom Scale in people with schizophrenia. BJPsych Open [Internet]. 2023;9(5):e168. Available from: https://www.cambridge.org/core/product/identifier/S2056472423005410/type/journal_article
87. Salahuddin NH, Schütz A, Pitschel-Walz G, Mayer SF, Chaimani A, Siafis S, et al. Psychological and psychosocial interventions for treatment-resistant schizophrenia: a systematic review and network meta-analysis. Lancet Psychiatry. 2024;11(7):545–53.
88. Moura BM, Isvoranu AM, Kovacs V, Van Rooijen G, Van Amelsvoort T, Simons CJP, et al. The puzzle of functional recovery in schizophrenia-spectrum disorders—replicating a network analysis study. Schizophr Bull. 2022;48(4):871.
89. Wambua GN, Kilian S, Ntlantsana V, Chiliza B. The association between resilience and psychosocial functioning in schizophrenia: a systematic review and meta-analysis. Psychiatry Res [Internet]. 2020 [cited 2024 Feb 8];293. Available from: https://pubmed.ncbi.nlm.nih.gov/32795771/
90. Galderisi S, Rucci P, Kirkpatrick B, Mucci A, Gibertoni D, Rocca P, et al. Interplay among psychopathologic variables, personal resources, context-related factors, and real-life functioning in individuals with schizophrenia a network analysis. JAMA Psychiatry. 2018;75(4)
91. Lysaker PH, Lancaster RS, Nees MA, Davis LW. Attributional style and symptoms as predictors of social function in schizophrenia. J Rehabil Res Dev. 2004;41(2):225.
92. Fernandez-Egea E, Wolpe N. Is it time to change the term negative symptoms? Eur Neuropsychopharmacol. 2022;59:7–8.

4 Assessment of Negative Symptoms and of Real-Life Functioning: Critical Areas of Overlap

Louise Birkedal Glenthøj, Stephen Austin, Nikolai Albert, and Mette Ødegaard Nielsen

Abstract

Negative symptoms—such as avolition, anhedonia, alogia, affective flattening, and social withdrawal—are central to schizophrenia and have a profound impact on real-life functioning, including social relationships, vocational engagement, and independent living. The assessment of these two domains has traditionally been approached separately, yet there is growing recognition of their substantial overlap. Negative symptoms, particularly diminished motivation and social drive, closely mirror impairments in functional outcomes, making it challenging to distinguish between symptom severity and the consequences of those symptoms on daily life. This overlap raises important methodological and clinical concerns, such as the potential for redundancy in measurement tools and misinterpretation of treatment effects. Current assessment instruments, including the

L. B. Glenthøj (✉)
VIRTU Research Group, Mental Health Center Copenhagen, University of Copenhagen, Copenhagen, Denmark

Department of Psychology, University of Copenhagen, Copenhagen, Denmark
e-mail: louise.birkedal.glenthoej@regionh.dk

S. Austin
Department of Psychology, University of Copenhagen, Copenhagen, Denmark

Mental Health Services, East, Copenhagen University Hospital, Copenhagen, Denmark

N. Albert
Mental Health Center Copenhagen, Research unit, CORE, Copenhagen, Denmark

Mental Health Center Copenhagen, OPUS, Copenhagen, Denmark

M. Ø. Nielsen
Unit for Complicated Schizophrenia, Mental Health Center Glostrup, Mental Health Service Capital Region, Copenhagen, Denmark

Faculty of Health and Medical Sciences, University of Copenhagen, Copenhagen, Denmark

A. Vita et al. (eds.), *Negative and Cognitive Symptoms in Schizophrenia*, Comprehensive Approach to Psychiatry 3,
https://doi.org/10.1007/978-3-032-06209-3_4

Positive and Negative Syndrome Scale, the Brief Negative Symptom Scale, and the Scale for the Assessment of Negative Symptoms for negative symptoms, and tools like the Personal and Social Performance Scale and Specific Levels of Functioning Scale for functioning, often fail to fully disentangle intrinsic motivation deficits from observable performance. Moreover, the subjective experience of patients and contextual factors (e.g., environmental support, stigma) are frequently underrepresented. There is a critical need for integrative, multidimensional assessment strategies that capture both domains while maintaining conceptual clarity. Advances in real-time, performance-based, and ecologically valid assessment tools may enhance our ability to understand the dynamic interplay between negative symptoms and functioning. Improved assessment can lead to more precise interventions, better monitoring of treatment response, and ultimately, enhanced functional recovery for individuals with schizophrenia.

Keywords

Negative symptoms · Schizophrenia · Real-life functioning · Functional assessment · SANS · PANSS · BNSS · CAINS · Functional capacity · PSP · GAF · SOFAS · Social functioning · Avolition · Anhedonia · Asociality · Blunted affect · Alogia · Scale validity · Symptom-function overlap · Second-generation symptom scales · Clinical assessment tools · Functional remission · Psychometric properties

4.1 Negative Symptom Assessment

Scales for the assessment of negative symptoms are commonly divided into first-generation and second-generation. Most widely used among the first-generation scales are the Scale for the Assessment of Negative Symptoms—SANS [1] and Positive and Negative Syndrome Scale—PANSS [2], while the most prominent second-generation scales are Brief Negative Symptom Scale—BNSS [3] and Clinical Assessment Interview for Negative Symptoms—CAINS [4].

The SANS, developed by Andreasen, was the first instrument to provide a comprehensive assessment of negative symptoms in schizophrenia [5]. It is a 25-item instrument that divides negative symptoms into the five subscales of affective blunting, alogia, avolition-apathy, anhedonia-asociality, and attention. Each item is rated on a scale from 0 (no impairment) to 5 (severe impairment), and the five subscales can be rated globally. The duration of the SANS varies depending on individual circumstances but takes on average 20 minutes to complete.

SANS was developed in the early 1980s and was based on prior work from the development of The Scale for the Assessment of Thought, Language, and Communication (TLC) and an affect rating scale [6]. The negative thought disorder elements from the TLC and the affective blunting were combined with new items for measuring Avolition-Apathy, Anhedonia-Asociality, and Attention into the Scale for Assessment of Negative Symptoms. The authors argued that to increase reliability a rating scale should be based on objective observational items, rather than on the "*patient's internal psychological state.*" However, the focus on objective items, in addition to increasing the reliability of the scale, might also increase the risk of overlap with functional rating scales, which is most pronounced for specific domains. The fifth domain, attention, has mostly been omitted in recent studies. The authors themselves have pointed out that the domain does not seem to load specifically into any of the clusters of psychotic, disorganized, or negative symptoms, and it is believed to tap into cognitive functioning for which newer measurements have been developed [7]. (Harvey et al., 2006).

The PANSS scale from 1987 [2] is based on a semi-structured interview of around 40 minutes, after which 30 items are rated 1–7, where 7 defines the most severe presentation of the symptom. To help with reliable rating, each item is described in detail along with what sources the rating should rely on in terms of information from the interview and/or information from others, and each score is defined by anchoring criteria. The PANSS total score will be between 30 and 210. In addition, three subscales were initially defined: The positive and negative subscales each consist of 7 items with sub-scores 7–49, and the remaining 16 items are categorized in a general subscale with subscores 16–112 [2].

The PANSS scale is one of the most widely used rating scales within psychoses, probably because it has repeatedly shown good test-retest reliability and inter-rater reliability. However, the internal consistency of the subcategories has been questioned as factor analyses have repeatedly found a superior characterization of PANSS data using five factors defined as positive, negative, disorganized, excited, and depressed [7–9]. The items N5: "difficulty in abstract thinking" and N7: "stereotyped thinking" repeatedly did not load with the remainder of the negative symptoms, challenging the validity of their inclusion in the negative symptom dimension [10, 11]. After the 2006 NIMH consensus statement on negative symptoms assessment [12] it has further been stressed that the PANSS scale does not cover all five negative symptom domains [13, 14].

While the SANS and PANSS have been widely used in the last 30 years of schizophrenia research, it can be argued that they reflect an obsolete operationalization of negative symptoms. Among the main critiques of these scales are that they include items that are not part of the negative symptom complex (e.g., taps cognitive abilities), and that they do not adequately define pleasure into the renewed understanding of it, comprising both a consummatory and anticipatory aspect. Furthermore, they rely predominantly on the observed behavior of individuals and consequently may miss assessing the core psychological processes that constitute the essential part of negative symptoms [10]. Finally, the SANS and PANSS have items that overlap with negative symptoms and functional assessment, as will be described further below.

Following the evolved knowledge on negative symptoms, the newer scales BNSS and CAINS have been developed to address the limitations of previous scales. The BNSS and CAINS are structured interviews conducted by a trained rater, and both scales have now been translated and validated in multiple other languages. The BNSS was developed by Kirkpatrick, Strauss et al., in 2001 [3], and consists of 13 items grouped into the six negative symptom aspects of blunted affect, alogia, asociality, anhedonia, avolition, and distress in keeping with the definition of negative symptoms in the NIMH-MATRICS consensus statement [12]. Each item is rated on a seven-point scale ranging from 0 (not present) to 6 (severe). While the BNSS has shown strong psychometric properties in several studies [3, 15, 16], it has, however, been argued that it may be suboptimal in separating between primary and secondary negative symptoms, and additionally, that it does not separately report on the quantity and quality of social activities [17]. Furthermore, the BNSS comprises a rather intricate scoring system that relies on experienced and extensive training for accurate administration.

The other second-generation negative symptoms scale, the CAINS, was developed by Blanchard et al., in 2012 [4]. Unlike most previous negative symptom scales that offer one global rating of negative symptoms severity, the CAINS addresses the two primary domains of negative symptoms, the Experiential and the Expressive domains. The Experiential domain evaluates the person's motivation and pleasure experiences across diverse life areas, such as social, vocational, and recreational activities, along with the frequency and intensity of anticipated and experienced pleasure. The Expressive domain assesses the expression of emotion and speech, emphasizing both verbal and non-verbal expressive behavior. The items are rated on a five-point scale, rating from 0 (no impairment) to 4 (severe impairment). A recent systematic review evaluated the psychometric properties of the CAINS and found it was an instrument with the potential to be recommended for use with relatively good results for reliability and convergent and discriminative validity. Further studies were, however, recommended to assess the quality of CAINS in the domains of content validity, consistency, and in particular cross-cultural validity [18].

4.2 Assessment of Functioning in Schizophrenia

It is well-known that schizophrenia can significantly impact the individual's daily functioning. A key outcome for treatment in schizophrenia is return to normal functioning, which is also a central component in clinical recovery. Functional outcomes are also important to the person's quality of life [19] and are one of the main drivers of the global burden of the disorder [20].

While there is a degree of consensus on how to measure symptoms and cognition in schizophrenia, there is little consensus about the gold standard in measuring functioning in schizophrenia [21]. This lack of consensus is reflected in a number of reviews that have identified upwards of 20 different measures used regularly in studies examining functioning within schizophrenia [22, 23].

These measures cover different aspects of functioning, which typically cover work and education activities, social interactions, daily living skills, and leisure activities.

In classifying measures of functioning, they can be broadly divided into activity-based assessments, assessed through interview and self-report measures, and capacity-based measures, assessed through the ability to perform tasks [23].

4.2.1 Activity-Based Assessments of Functioning

Activity-based functioning is commonly assessed by mental health clinicians and physicians in schizophrenia. One of the most widely used interview-rated scales for this purpose is the Global Assessment of Functioning (GAF) [24]. GAF assesses subjectively the social, occupational, and psychological functioning of an individual, i.e., how well one is meeting various problems in living. Scores range from 100 (extremely high functioning) to 1 (severely impaired).

The Social and Occupational Functioning Assessment Scale (SOFAS) [25] is another widely used interview-rated scale for functioning and is a further development of the GAF scale. It also provides a single global score for functioning, thereby having the same advantages and disadvantages as the GAF.

Both the GAF and SOFAS are useful to provide an overall level of impairment of functioning, but as they collapse different aspects of functioning into a single global score, they can lack sensitivity to detect dysfunction in specific domains [26]. Additionally, there have been some questions over the psychometric properties of these measures and their ability to detect changes in functioning [21].

The Personal and Social Performance Scale (PSP) is one of the most widely used interview-based scales of social functioning and has been shown to have good psychometric properties [27]. The PSP has an added advantage of the GAF and SOFAS as it separately scores four specific domains of functioning: Socially useful activities: work/education; Personal and social relationships; Self-care; and disruptive/aggressive behaviour that are combined into a single global score (1–100) [22].

Whilst global scales of functioning are most commonly used in schizophrenia, a recent systematic review noted that specific measures of functioning are the best to detect changes in functioning over time and in response to treatment [21].

The commonly specific measures of functioning that are clinician-rated include:

Social Behaviour Schedule (SBS) [28, 29]. Clinician or observer rated 5–5-point Likert scales based on an interview with the patient or relative, or observation based on the previous month. Global score is based on 21 items 5–5-point anchored scale from 0 (no problem or acceptable behavior) to 4 (serious problem), with a higher score representing lower levels of functioning. Antisocial behavior, depressed behavior, social withdrawal, and thought disturbance.

Socially valued Role Classification Scale (SRCS) is an observer-rated instrument based on semi-structured interview used alongside anchor descriptions. Global score or score per role: Self-care and home duties, caring for others, self-development, formal education, and employment-specific self-report measures of functioning include:

Social Functioning Scale (SFS) [30, 31] is a self-report structured questionnaire including 4- and 5-point Likert scales based on the past 3 months. Global score or per domain comprised 78 items, yielding a raw score which is transformed to a distribution with a mean of 100 and SD of 15. Domains investigated are: pro-social activities; recreation; independence-competence; independence-performance; and employment/occupation.

Functional Remission of General Schizophrenia Scale (FROGS) [32] is a self-report and semi-structured interview used to complete a structured questionnaire with 5-point Likert scales. Global score (19–95) derived from 19 items related to functioning behaviors and abilities over the past month across the domains, each scored from 1 to 5 (1- do not do; 2- do partially; 3- do a significant part; 4- do almost all the activity; 5- do perfectly). Higher scores indicate higher functioning. Domains rated include: daily life; social activities; social functioning; quality of rehabilitation; general health and treatment.

4.2.2 Capacity-Based Measures of Functioning

An alternative to measuring function to "real-world" behavior using activity-based interviews, is to focus on functional capacity. These types of assessments evaluate an individual's ability to complete specific activities under controlled conditions. A number of studies have shown an individual's capacity mediates the relationship between ability and real-world functional outcomes [33]. These measures are used less frequently than activity-based scales, often due to the increased resources required for participants to complete tasks and clinicians to score them.

Two regular used measures of functional capacity used in schizophrenia include: the Performance-Based Skills Assessment (UPSA [34] and the Schizophrenia Cognition Rating Scale (SCoRS) [35].

UPSA is a performance-based functional capacity assessment that requires participants to complete various everyday tasks in simulated situations. The UPSA evaluates up to six domains of daily functioning: (1) household management, (2) communication, (3) financial skills, (4) transportation, (5) comprehension/planning, and (6) medication management [36].

The SCoRS assesses the cognitive domains [35, 37] as identified by the MATRICS initiative, which are considered associated with functional outcomes in patients with schizophrenia [38].

4.2.3 Overlap Between Negative Symptom Assessment and Daily Life Functioning

Assessment of negative symptoms and functioning is to some degree intertwined as they tap overlapping constructs. That is, negative symptom scales often assess social withdrawal and reduced social engagement, and similarly, functioning scales evaluate social functioning, including the quality and quantity of social interactions.

A predominant definition in the field has been to conceptualize negative symptoms as comprising the two independent factors of experience and expression [39, 40]. In this distinction, experiential negative symptoms capture avolition, anhedonia, and asociality, whereas expressive negative symptoms encompass emotional expressivity, alogia, and poverty of thought. Within this distinction, particularly the aspect of experiential negative symptoms, and functioning can be argued to have content overlap, potentially inflating the association between these variables. Later studies have, however, advanced the understanding of the latent structure of negative symptoms, finding it to be best conceptualized as a five-factor structure entailing anhedonia, avolition, asociality, alogia, and blunted affect [41]. Hence, the domains of anhedonia, avolition, and asociality would be expected to have the greatest overlap with functioning.

The most pronounced overlap between negative symptoms and functioning can be found with first-generation negative symptoms scales (i.e. SANS and PANSS), as second-generation scales (CAINS and BNSS) have been designed to minimize this overlap. In fact, second-generation scales focus on more narrowly defined constructs of negative symptoms and distinguish between the core negative symptoms and their behavioral consequences to avoid conflating symptom severity with functional impairment. Additionally, second-generation scales emphasize the assessment of observable behaviors and internal experiences rather than the functional outcomes of these behaviors, which may aid in isolating the symptoms from their impact on functioning. Finally, items in second-generation scales are designed to target specific symptoms rather than general impairments. This specificity may also help in reducing the overlap.

In the following sections, we specify the overlap between the first- and second-generation negative symptom scales and the most commonly used measures of functioning in schizophrenia measures including the GAF, SOFAS, PSP, and SFS.

4.2.4 PANSS and Functional Assessment

For most items on the PANSS scale, anchor points defining a higher rating describe an increased impact on behavior, thereby implicitly indicating how the symptom affects the level of functioning in everyday life. For some items, observations about how behavior and social relations are influenced by the symptom are additionally mentioned in the item definition. By including the functional impact in the rating of symptom severity, there will, of course, be some degree of correlation between the rating of symptoms and functioning, which was recently reported for several of the PANSS subscales [42].

For most items belonging to the original negative symptom domain, the rating of a symptom as mild or moderate will not overlap with the rating of social functions. Only when these symptoms become severe, the anchor points are described in a way that they will inevitably map onto a social functioning rating scale. For instance, item N1; blunted affect: "*unmodulated extreme affective discharges, such as excitement, rage, or inappropriate uncontrolled laughter*" will be rated as a disturbing or aggressive behavior on PSP.

However, the rating of emotional withdrawal (N2) and passive-apathetic social withdrawal (N2) are mainly based on the social behavior and function *reported by family or primary care workers around the patient*. For these items, even anchor points of mild to moderate ratings will tend to overlap with the rating of social function. For instance, patients who are described as "*generally emotionally distanced*" or *"only show occasional interest in social activities"* will also rate as less well-functioning on scales like SBS or SFS. This overlap continues when the symptoms become more severe and anchor points are defined as the patient being *"personally neglectful"* and *"socially isolated."* These definitions indeed overlap with functional scales, in which social isolation will be rated as a marked or severe impairment of personal and social relationships in PSP. Likewise, *"frequent neglect of personal functions for which the patient requires supervision"* will result in a very low score on scales like GAF or SOFAS, where scores below 30 define the patient as being *"unable to function independently"*. Summing up, three of the seven original PANSS negative items or 21/49 points = 43% of the total PANSS negative score, will contribute to a correlation with social function.

4.2.5 SANS and Functional Assessment

Similar to the PANSS, increased severity on the SANS scale is linked to functional impairments in the item description. The Avolition-Apathy subscale in the SANS consists of three items (Grooming and Hygiene, Impersistence at Work or School, and Physical Anergia) and one global rating. The description of Grooming and Hygiene is *"The subject displays less attention to grooming and hygiene than normal. Clothing may appear sloppy, outdated, or soiled. The subject may bathe infrequently and not care for hair, nails, or teeth- leading to such manifestations as greasy or uncombed hair, dirty hands, body odor, or unclean teeth and bad breath. Overall, the appearance is dilapidated and disheveled. In extreme cases, the subject may even have poor toilet habits."* And is rated from 0 (*no evidence of poor grooming and hygiene)* to 5 Severe (*Subject's clothes, body, and environment are dirty and smelly).* On the PSP this overlaps with domain C, Self-Care, where one of the 6 questions regarding the domain is: *"Clothing: Is the patient's clothing neat and clean?"*. Similarly, the SFS includes questions on bathing and washing of clothes in the Independence-performance domain. The vignettes describing rating on the GAF scale only mention hygiene in the most severe cases (interval 11–20, e.g., smears feces), but it is understood that all areas of functioning, including hygiene, should be considered on a continuum and that poor personal hygiene should be considered when rating also less severely impaired individuals.

The final two items within the Avolition-Apathy subscale are also at risk of overlapping with functional ratings. *Impersistence at Work or School's* severity rating is solely linked to a person's ability to attend work or school and *Physical Anergia's* severity rating is related to number of hours per day the individual stays inactive, both descriptions that overlap with measurements of activity on the above-mentioned functional scales.

The Anhedonia—Asociality subscale is less likely to overlap with functional scales as the severity rating of the individual items is more linked to the person's ability to feel and experience joy, feel intimate with friends and family, or enjoy sexual activities, which might not overlap if the individual still maintains functioning in these areas. Only the most severe rating of the item *Ability to Feel Intimacy and Closeness* mentions hostility toward friends and family, which would overlap for instance with the PSP domain of disturbing and aggressive behavior. Further, the item *Relationships with Friends and Peers* from the rating of moderate rates the number of contacts with friends per month in the severity description, which overlaps with functional measurements of social interactions. Also, the general descriptions of the items *Recreational Interests and Activities, Sexual Interest and Activity* and *Ability to Feel Intimacy and Closeness* all describe functional impairment within the item or as part of the item, and the domain will overlap with the *Prosocial* domain on the SFS scale, the *Personal and Social Functioning* domain on the PSP, and the GAF which describes impairment in social functioning as one of the elements to be considered in all the scoring intervals.

Of the original five domains on the SANS scale, only the first four are today widely used as the attention domain overlaps with cognitive functioning and is, in most studies, omitted in favor of newer cognitive measurements. The four remaining domains contain 18 individual items, excluding global domains, where seven (39%), are at risk of overlapping with functional measurements, and similarly, two out of four global domains.

4.2.6 BNSS and Functional Assessment

For the BNSS scale, the subscale focused on lack of distress, blunted affect, and alogia do not tap into elements directly associated with levels of functioning. The anhedonia subscale evaluates the intensity and the frequency of all sources of potential pleasures, which includes the frequency of pleasurable social and job-related activities. Since social/job activities are only a few of several sources for pleasure, the anhedonia score will only be slightly affected in cases where social/job activities are few or absent, but other pleasures are present and frequent. This part of the scale will therefore only have a small overlap with functional scales.

Both the asociality and avolition subscales of BNSS address behavior as well as internal experiences. When rating the behavioral part of asociality, the number, the level of intimacy, and the frequency of social relations are included. This leads to a partial overlap with the functional scales, as both PSP and SFS rate the number and frequency of personal relations. However, by adding the level of intimacy and the subjective need for social relations, the overlap is decreased, since this is not measured in the functional scales. The same is the case for avolition, where there is a high degree of overlap between the behavioral measures of avolition on BNSS and the level of function on GAF, PSP, and SFS. Since the subjective experience of being motivated for job, hobbies, and self-care is also rated as an independent item, the correlation between the total avolition score and scores on the functional scale will be lower. Calculating, the overlap between BNSS and level of function is present in 3/13 items, or 18/78 points, meaning that only 23% of the BNSS score will overlap with the rating on the functional scales.

4.2.7 CAINS and Functional Assessment

The CAINS predominantly focuses on the subjective experiences of engaging in different behaviors (being social with friends/family, recreational interests, and work/educational activities) and the motivation and pleasure associated with these activities rather than the actual levels of functioning in these domains. Therefore, the CAINS has considerably less overlap with functioning than previous scales measuring negative symptoms.

Potentially, the CAINS items that overlap with functioning are: Item 3 (frequency of pleasurable social activities) and Item 8 (frequency of pleasurable recreational activities). As both items include the subjective evaluation of finding the activity "pleasurable," it has less overlap than an item that looks at the frequency of behavior without any subjective evaluation.

Item 3 could also have some overlap with domain 2 in PSP (social activities), while item 8 could have some overlap with domain 1 in PSP (work and educational activities). Items 10–13 focus on expressive aspects of negative symptoms.

One could argue that levels of motivation have an indirect effect on the frequency of work and social behaviors (i.e., functioning) where people with low motivation to engage in work/education and socializing with others would show greater impairment in these functional domains.

Examining the ratio of overlap between CAINS and functional scales (e.g., PSP or GAF) based on an item level, there are 2 items out of 13 items giving an overlap of 15%. This overlap may be even further reduced as these 2 items focus on the frequency of "pleasurable" work or recreational activities.

4.2.8 Recommendations for Assessment of Negative Symptoms and Functioning

As mentioned, negative symptoms and functioning are concepts that are closely linked and consequently, a degree of overlap in the assessment of these constructs cannot be fully avoided. Second-generation negative symptom scales do, however, achieve a clearer separation between negative symptoms and functioning through refined constructs, behavioral specificity, context consistency, structured interviews, and rigorous validation processes. Indeed, these advancements lead to more accurate and focused assessments of negative symptoms.

As first-generation scales (i.e., PANSS and SANS) suffer the abovementioned limitations, such as item overlaps between negative symptoms and functional assessment, it may impede an accurate negative symptom measurement, which may again hinder the potential of optimizing targeted therapeutic approaches for negative symptoms, that currently constitute a pivotal unmet treatment need. Hence, second-generation scales such as the BNSS and CAINS are recommended for use in research and clinical practice, and the use of scales such as the SANS and PANSS should only be a supplement to first-generation scales [43]. With regard to measuring functioning in schizophrenia, it must be acknowledged that functioning is a multifaceted construct that involves various domains of a person's life. Hence, to achieve an optimal functional assessment, multiple sources of information would be preferred, e.g., evaluation of functional *activity/achievements* as captured by scales such as PSP and GAF, and functional *capacity* that addresses the individual's capacity for real-life performance [44]. Functional capacity scales are, however, a more recent development in schizophrenia research, and hence, far less is known about their usefulness in clinical and research contexts. Among the most used scales are the UPSA, SCoRS, and the newer Virtual Reality Functional Capacity Assessment Tool (VRFCAT) [45], with the latter two primarily assessing functioning in the context of neurocognition.

These different aspects of functioning may relate differently to negative symptoms. For example, studies have found that alogia and affective blunting may map closer to social skills functional capacity [46]. It cannot, however, be ruled out that these potential different associations may in part reflect an overlap in the assessment of negative symptoms and functional capacity. For example, expressive negative symptoms (affective expression) may relate directly to social skills capacity, and motivational deficits may be a general underlying factor for poor performance on functional capacity measures.

Functional activity/achievements are the most commonly assessed aspect of functioning in schizophrenia, with the scales PSP, GAF, and SOFAS having the most widespread use. The Functioning Assessment Short Test (FAST) is a newer interview-based measure that is increasingly being used to measure global functioning within schizophrenic populations [47]. Whilst being a promising measure, it does not have as strong psychometric properties as PSP at this stage.

Hence, the PSP may still be preferred due to its robust psychometric properties [22] and its ability to provide the most precise estimate of an individual's daily-life functioning based on four key subdomains. The PSP can therefore be an advantageous measure when aiming to provide a global measure of functioning, but it does, however, collapse different aspects of functioning into a single global score, and it may lack sensitivity to detect changes in dysfunction within specific domains [26].

While the GAF and SOFAS may provide a generalized assessment of functioning, they do lack specific subcategories for detailed scoring of diverse areas of real-life functioning. Additionally, GAF has the disadvantage of combining symptom severity and functional impairments in its scoring system.

Thus, specific measures of functioning such as the Social Behaviour Schedule (SBS), Social Functioning Scale (SFS), Functional Remission of General Schizophrenia Scale (FROGS), or the Daily Activity Report (DAR) [48] that assess specific aspects of social or role functioning [22] may be preferred in studies aiming at detecting functional change [21]. On the other hand, these scales are often more time-consuming to complete and may preclude providing a global functioning score.

In summary, to minimize the overlap between negative symptoms and functional assessment, second-generation negative symptoms scales are recommended for use in research trials and clinical practice in psychosis. The PSP is recommended as a measure of global functioning, whereas more specific measures such as the SBS or SFS may be preferred in longitudinal studies aiming at capturing changes in aspects of functioning over time. Finally, functional capacity measures may provide important additional information on functional deficits in psychosis and could be considered as an adjunct measure of functioning.

References

1. Andreasen NC. Scale for the assessment of negative symptoms. Iowa City, Iowa: Dept. of Psychiatry, College of Medicine, the University of Iowa; 1984. 12 leaves
2. Kay SR, Fiszbein A, Opler LA. The positive and negative syndrome scale (PANSS) for schizophrenia. Schizophr Bull [Internet]. 1987;13(2):261–76. Available from: https://academic.oup.com/schizophreniabulletin/article-lookup/doi/10.1093/schbul/13.2.261
3. Kirkpatrick B, Strauss GP, Nguyen L, Fischer BA, Daniel DG, Cienfuegos A, et al. The brief negative symptom scale: psychometric properties. Schizophr Bull. 2011;37(2):300–5.
4. Kring AM, Gur RE, Blanchard JJ, Horan WP, Reise SP. The clinical assessment interview for negative symptoms (CAINS): final development and validation. Am J Psychiatry [Internet]. 2013;170(2):165–72. Available from: https://pmc.ncbi.nlm.nih.gov/articles/PMC3785242/. https://doi.org/10.1176/appi.ajp.2012.12010109.
5. Andreasen NC. Negative symptoms in schizophrenia. Arch Gen Psychiatry [Internet]. 1982;39(7):784. Available from: http://archpsyc.jamanetwork.com/article.aspx?doi=10.1001/archpsyc.1982.04290070020005
6. Andreasen NC. The scale for the assessment of negative symptoms (SANS): conceptual and theoretical foundations. Br J Psychiatry [Internet]. 1989;155(S7):49–52. Available from: https://www.cambridge.org/core/product/identifier/S0007125000291496/type/journal_article
7. Harvey PD, Koren D, Reichenberg A, Bowie CR. Negative symptoms and cognitive deficits: what is the nature of their relationship? Schizophr Bull. 2006;32(2):250–8. https://doi.org/10.1093/schbul/sbj011. Lindenmayer JP, Bernstein-Hyman R, Grochowski S. A new five factor model of schizophrenia. Psychiatr Q [Internet]. 1994;65(4):299–322. Available from: http://link.springer.com/10.1007/BF02354306
8. Langeveld J, Andreassen OA, Auestad B, Færden A, Hauge LJ, Joa I, et al. Is there an optimal factor structure of the Positive and Negative Syndrome Scale in patients with first-episode psychosis? Scand J Psychol [Internet]. 2013;54(2):160–5. Available from: https://onlinelibrary.wiley.com/doi/10.1111/sjop.12017
9. Wallwork RS, Fortgang R, Hashimoto R, Weinberger DR, Dickinson D. Searching for a consensus five-factor model of the Positive and Negative Syndrome Scale for schizophrenia. Schizophr Res [Internet]. 2012;137(1–3):246–50. Available from: https://linkinghub.elsevier.com/retrieve/pii/S0920996412000564
10. Marder SR, Galderisi S. The current conceptualization of negative symptoms in schizophrenia. World Psychiatry. 2017;16(1):14–24.
11. Baandrup L, Allerup P, Nielsen MØ, Bak N, Düring SW, Leucht S, et al. Rasch analysis of the PANSS negative subscale and exploration of negative symptom trajectories in first-episode schizophrenia—data from the OPTiMiSE trial. Psychiatry Res [Internet]. 2020;289:112970. Available from: https://linkinghub.elsevier.com/retrieve/pii/S0165178119319742
12. Kirkpatrick B, Fenton WS, Carpenter WT, Marder SR. The NIMH-MATRICS consensus statement on negative symptoms. Schizophr Bull [Internet]. 2006;32(2):214–9. Available from: https://academic.oup.com/schizophreniabulletin/article-lookup/doi/10.1093/schbul/sbj053
13. Daniel DG. Issues in selection of instruments to measure negative symptoms. Schizophr Res [Internet]. 2013;150(2–3):343–5. Available from: https://linkinghub.elsevier.com/retrieve/pii/S0920996413003496
14. Marder SR, Kirkpatrick B. Defining and measuring negative symptoms of schizophrenia in clinical trials. Eur Neuropsychopharmacol [Internet]. 2014;24(5):737–43. Available from: https://linkinghub.elsevier.com/retrieve/pii/S0924977X13003258

15. Strauss GP, Keller WR, Buchanan RW, Gold JM, Fischer BA, McMahon RP, et al. Next-generation negative symptom assessment for clinical trials: validation of the brief negative symptom scale. Schizophr Res [Internet]. 2012;142(1–3):88–92. Available from: https://linkinghub.elsevier.com/retrieve/pii/S0920996412005841
16. Kirkpatrick B, Saoud JB, Strauss GP, Ahmed AO, Tatsumi K, Opler M, et al. The brief negative symptom scale (BNSS): sensitivity to treatment effects. Schizophr Res [Internet]. 2018;197:269–73. Available from: https://linkinghub.elsevier.com/retrieve/pii/S0920996417307296
17. Gehr J, Glenthøj B, Ødegaard NM. Validation of the Danish version of the brief negative symptom scale. Nord J Psychiatry [Internet]. 2019;73(7):425–32. https://doi.org/10.1080/08039488.2019.1648549.
18. Wehr S, Weigel L, Davis J, Galderisi S, Mucci A, Leucht S. Clinical Assessment Interview for Negative Symptoms (CAINS): a systematic review of measurement properties. Schizophr Bull [Internet]. 2023. Available from: https://academic.oup.com/schizophreniabulletin/advance-article/doi/10.1093/schbul/sbad137/7408086
19. Cotter J, Bucci S, Drake RJ, Yung AR, Carney R, Edge D. Exploring functional impairment in young people at ultra-high risk for psychosis: a qualitative study. Early Interv Psychiatry [Internet]. 2019;13(4):789–97. Available from: https://onlinelibrary.wiley.com/doi/10.1111/eip.12560
20. Insel TR. Assessing the economic costs of serious mental illness. Am J Psychiatry [Internet]. 2008;165(6):663–5. Available from: http://psychiatryonline.org/doi/abs/10.1176/appi.ajp.2008.08030366
21. Cowman M, Godfrey E, Walsh T, Frawley E, Fowler D, Alvarez-Jimenez M, et al. Measures of social and occupational function in early psychosis: a systematic review and meta-analysis. Schizophr Bull. 2024;50(2):266–85.
22. Long M, Stansfeld JL, Davies N, Crellin NE, Moncrieff J. A systematic review of social functioning outcome measures in schizophrenia with a focus on suitability for intervention research. Schizophr Res [Internet]. 2022;241:275–91. Available from: https://linkinghub.elsevier.com/retrieve/pii/S0920996422000809
23. Searle A, Allen L, Lowther M, Cotter J, Barnett JH. Measuring functional outcomes in schizophrenia in an increasingly digital world. Schizophr Res Cogn [Internet]. 2022;29:100248. Available from: https://linkinghub.elsevier.com/retrieve/pii/S2215001322000130
24. Piersma HL, Boes JL. The GAF and psychiatric outcome: a descriptive report. Community Ment Health J [Internet]. 1997;33(1):35–41. Available from: http://www.ncbi.nlm.nih.gov/pubmed/9061261
25. Goldman HH, Skodol AE, Lave TR. Revising axis V for DSM-IV: a review of measures of social functioning. Am J Psychiatry [Internet]. 1992;149(9):1148–56. Available from: http://psychiatryonline.org/doi/abs/10.1176/ajp.149.9.1148
26. Harvey PD. Assessment of everyday functioning in schizophrenia: implications for treatments aimed at negative symptoms. Schizophr Res [Internet]. 2013;150(2–3):353–5. Available from: https://linkinghub.elsevier.com/retrieve/pii/S0920996413002223
27. Morosini PL, Magliano L, Brambilla L, Ugolini S, Pioli R. Development, reliability and acceptability of a new version of the DSM-IV Social and Occupational Functioning Assessment Scale (SOFAS) to assess routine social functioning. Acta Psychiatr Scand [Internet]. 2000;101(4):323–9. Available from: http://www.ncbi.nlm.nih.gov/pubmed/10782554
28. Cella M, Stratta P, Chahal K, Huddy V, Reeder C, Wykes T. Measuring community functioning in schizophrenia with the Social Behaviour Schedule. Schizophr Res [Internet]. 2014;153(1–3):220–4. Available from: https://linkinghub.elsevier.com/retrieve/pii/S0920996413006944
29. Wykes T, Sturt E. The measurement of social behaviour in psychiatric patients: an assessment of the reliability and validity of the SBS schedule. Br J Psychiatry. 1986;148(JAN):1–11.
30. Chan EHC, Kong SDX, Park SH, Song YJC, Demetriou EA, Pepper KL, et al. Validation of the social functioning scale: comparison and evaluation in early psychosis, autism spectrum

disorder and social anxiety disorder. Psychiatry Res [Internet]. 2019;276:45–55. Available from: https://linkinghub.elsevier.com/retrieve/pii/S0165178118316871
31. Birchwood M, Smith J, Cochrane R, Wetton S, Copestake S. The social functioning scale. The development and validation of a new scale of social adjustment for use in family intervention programmes with schizophrenic patients. Br J Psychiatry. 1990;157(DEC):853–9.
32. Llorca PM, Lançon C, Lancrenon S, Bayle FJ, Caci H, Rouillon F, et al. The "functional remission of general schizophrenia" (FROGS) scale: development and validation of a new questionnaire. Schizophr Res [Internet]. 2009;113(2–3):218–25. Available from: https://linkinghub.elsevier.com/retrieve/pii/S0920996409002175
33. Bowie CR, Harvey PD. Cognitive deficits and functional outcome in schizophrenia profile of cognitive impairments in schizophrenia. Neuropsychiatr Dis Treat. 2006;2(4):531–6.
34. Patterson TL, Goldman S, McKibbin CL, Hughs T, Jeste DV. UCSD performance-based skills assessment: development of a new measure of everyday functioning for severely mentally ill adults. Schizophr Bull [Internet]. 2001;27(2):235–45. Available from: https://academic.oup.com/schizophreniabulletin/article-lookup/doi/10.1093/oxfordjournals.schbul.a006870
35. Keefe RSE, Poe M, Walker TM, Kang JW, Harvey PD. The schizophrenia cognition rating scale: an interview-based assessment and its relationship to cognition, real-world functioning, and functional capacity. Am J Psychiatry [Internet]. 2006;163(March):426–32. Available from: http://www.ncbi.nlm.nih.gov/pubmed/16513863
36. Patterson TL, Moscona S, McKibbin CL, Davidson K, Jeste DV. Social skills performance assessment among older patients with schizophrenia. Schizophr Res. 2001;48(2–3):351–60.
37. Keefe RSE, Davis VG, Spagnola NB, Hilt D, Dgetluck N, Ruse S, et al. Reliability, validity and treatment sensitivity of the Schizophrenia Cognition Rating Scale. Eur Neuropsychopharmacol [Internet]. 2015;25(2):176–84. Available from: https://linkinghub.elsevier.com/retrieve/pii/S0924977X14001771
38. Nuechterlein KH, Green MF, Kern RS, Baade LE, Barch DM, Cohen JD, et al. The MATRICS consensus cognitive battery, part 1: test selection, reliability, and validity. Am J Psychiatry. 2008;165(2):203–13.
39. Blanchard JJ, Cohen AS. The structure of negative symptoms within schizophrenia: implications for assessment. Schizophr Bull. 2006;32(2):238–45.
40. Foussias G, Remington G. Negative symptoms in schizophrenia: avolition and occam's razor. Schizophr Bull. 2010;36(2):359–69.
41. Strauss GP, Nuñez A, Ahmed AO, Barchard KA, Granholm E, Kirkpatrick B, et al. The latent structure of negative symptoms in schizophrenia. JAMA Psychiatry. 2018;75(12):1303.
42. Handest R, Molstrom IM, Gram Henriksen M, Hjorthøj C, Nordgaard J. A systematic review and meta-analysis of the association between psychopathology and social functioning in schizophrenia. Schizophr Bull [Internet]. 2023;49(6):1470–85. Available from: https://academic.oup.com/schizophreniabulletin/article/49/6/1470/7187865
43. Galderisi S, Mucci A, Dollfus S, Nordentoft M, Falkai P, Kaiser S, et al. EPA guidance on assessment of negative symptoms in schizophrenia. Eur Psychiatry. 2021;64(1)
44. Harvey PD, Velligan DI, Bellack AS. Performance-based measures of functional skills: usefulness in clinical treatment studies. Schizophr Bull. 2007;33(5):1138–48.
45. Keefe RSE, Davis VG, Atkins AS, Vaughan A, Patterson T, Narasimhan M, et al. Validation of a computerized test of functional capacity. Schizophr Res [Internet]. 2016;175(1–3):90–6. https://doi.org/10.1016/j.schres.2016.03.038.
46. Glenthøj L, Fagerlund B, Hjorthøj C, Jepsen JR, Bak N, Kristensen TD, et al. Social cognition in patients at ultra-high risk for psychosis: what is the relation to social skills and functioning? Schizophr Res Cogn [Internet]. 2016;5:21–7. https://doi.org/10.1016/j.scog.2016.06.004.
47. González-Ortega I, Rosa A, Alberich S, Barbeito S, Vega P, Echeburúa E, et al. Validation and use of the functioning assessment short test in first psychotic episodes. J Nerv Ment Dis [Internet]. 2010;198(11):836–40. Available from: https://journals.lww.com/00005053-201011000-00010
48. Velligan DI, Mintz J, Sierra C, Martin ML, Fredrick M, Maglinte GA, et al. The daily activity report (DAR) a novel measure of functional outcome for serious mental illness. Schizophr Bull [Internet]. 2016;42(3):579–87. Available from: https://academic.oup.com/schizophreniabulletin/article-lookup/doi/10.1093/schbul/sbv185

Negative Symptoms in Schizophrenia: Neurobiological Bases

5

Konstantinos Pallis, Matthias Kirschner, Stefan Kaiser, and Indrit Bègue

Abstract

Negative symptoms of schizophrenia, including reductions in the Experiential dimension (including anhedonia, asociality, and avolition) and in the Expressive dimension (including alogia and blunted affect), represent a significant burden and a challenge in psychiatric treatment and research. Evidence stemming from behavioral research and studies using structural and functional neuroimaging points to distinct mechanisms underlying negative symptom dimensions. The neurobiology of avolition points to anomalies in reward processing, in particular reward anticipation, neurocognitive and social cognition abilities, and to dysfunctional beliefs about oneself and expectations from interactions with others. Regions such as the ventral striatum, orbitofrontal, and lateral prefrontal cortex have been identified as possible neuro-anatomical components for avolition. The neurobiology of the Expressive dimension points to anomalies in emotion expression/perception, impairments in speech fluency, and cognitive functioning. Neuro-anatomical components in the Expressive dimension consist of alterations in the regions underlying these processes, including motor cortex, superior temporal sulcus, and basal ganglia. Recent findings also suggest a contributory role for the cerebellum in the pathophysiology of negative symptoms (in particular for avolition/experiential dimension) with potential for targeted treatments. However, despite progress, the findings on the pathophysiology of negative

K. Pallis (✉) · I. Bègue
Laboratory for neuroimaging and translational psychiatry, Department of psychiatry, University of Geneva, Geneva, Switzerland

Department of psychiatry, Geneva University Hospitals, Geneva, Switzerland
e-mail: konstantinos.pallis@hug.ch

M. Kirschner · S. Kaiser
Department of psychiatry, Geneva University Hospitals, Geneva, Switzerland

A. Vita et al. (eds.), *Negative and Cognitive Symptoms in Schizophrenia*, Comprehensive Approach to Psychiatry 3,
https://doi.org/10.1007/978-3-032-06209-3_5

symptoms remain heterogeneous. Future research should focus on longitudinal studies and multimodal data integration to better understand the evolution and inter-individual variability of negative symptoms. Integrating more targeted and individualized neurostimulation techniques could enhance treatment efficacy and provide causal insights.

Keywords

Negative symptom heterogeneity · Dimension-specific approach · Reward anticipation · Ventral striatum · Cerebellum

5.1 Introduction

5.1.1 Negative Symptoms of Schizophrenia

Negative symptoms of schizophrenia are a major psychopathological domain in schizophrenia. They include reductions in motivation and pleasure (Experiential dimension) as well as diminished verbal and emotional expression (Expressive dimension) [1, 2]. Negative symptoms are prevalent and manifest across all stages of schizophrenia, from the high-risk state to chronic schizophrenia [3]. In fact, 50–60% of individuals with schizophrenia present at least one negative symptom and 10–30% present two or more negative symptoms of at least moderate intensity [4–6]. The vast majority of patients seeking psychiatric help show negative symptoms at onset, manifesting with significant social withdrawal and emotional disturbances prior to formal schizophrenia diagnosis [7].

Negative symptoms severely impact patients' daily functioning, social interactions, occupational engagement, overall quality of life, and chances for total remission [8–12]. They tend to remain stable or worsen without treatment and respond poorly to traditional antipsychotic medications [7], therefore representing a challenge and an unmet need in psychiatry [3]. Traditionally considered as rather homogenous, compelling evidence has led to the conceptualization of negative symptoms into two dimensions with partly distinct neurobiology: Experiential and Expressive dimensions (Fig. 5.1a) [2, 4, 12–16].

The Experiential dimension includes asociality, anhedonia, and avolition, manifesting as motivational deficits in social interactions, reduction/loss of pleasure, and reduction in goal-directed behavior, whereas the Expressive dimension comprises blunted affect and alogia, relating to emotional and verbal expression impairments [1–5]. It has been suggested that these dimensions have at least partly distinct pathophysiological underpinnings and may therefore show differential treatment responses. Focusing on these dimensions can facilitate the development of targeted interventions, leading to potentially better patient outcomes [4, 16–19]. It is worth highlighting that a more granular five-factor model of negative symptoms has also been recently suggested (see Chap. 1 of the present volume) [20], but the clinical and research application remains to be clarified [21]. The current evidence on

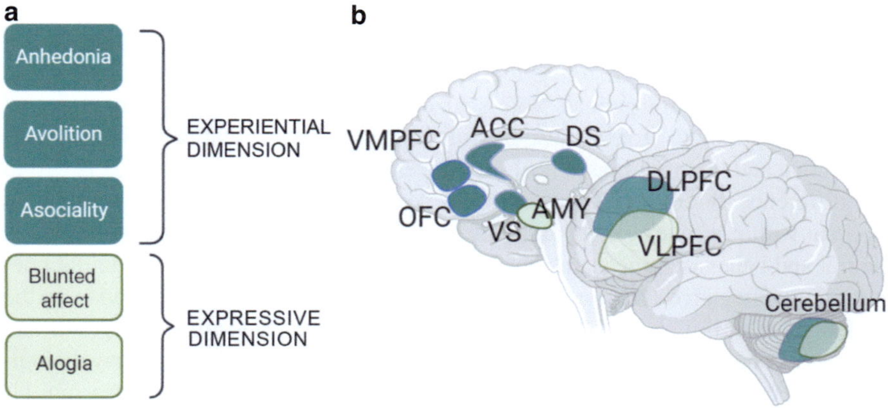

Fig. 5.1 (**a**) Schematic representation of negative symptom dimensions and individual negative symptom domains. (**b**) Schematic summary of brain regions implicated in Experiential dimension (dark green) and Expressive dimension (clear green). Abbreviations: ACC, anterior cingulate cortex; AMY, amygdala; DS, dorsal striatum; OFC, orbito-frontal cortex; DLPFC, dorsolateral prefrontal cortex; VLPFC, ventrolateral prefrontal cortex; VS ventral striatum. (Adapted from Bègue et al. [1])

negative symptom pathophysiology stems predominantly from the two-factor model and will be the focus of the current chapter.

5.1.2 Neurobiology of Negative Symptoms—Preliminary Remarks

When reviewing negative symptoms' pathophysiological mechanisms, some considerations with relevance to negative symptoms emerge. First, the evaluation of negative symptoms is made heterogeneously [22] and without sufficient consideration for their source. To this end, negative symptoms can be classified as primary (characteristic of schizophrenia per se) and secondary negative symptoms (as a consequence of psychosis, depression, medication effects, and so on) [23] and whose mechanisms relate to their primary cause, rather than negative symptoms per se. Second, given that most evidence arises from cross-sectional designs (with few notable exceptions [24]), temporal fluctuations in the clinical manifestation of negative symptoms are not considered [1, 24–27]. Efforts to retrospectively reduce heterogeneity linked to negative symptom evolution can be exemplified by the development of the "deficit syndrome"—a syndrome whereby the presence of primary negative symptoms is required for at least 12 months duration [28]. Another construct in this direction is "persistent negative symptoms," a construct recommended by the EPA guidance group as a potential solution to the heterogeneity problem [22]. A recent study [29] used a latent class analysis to perform a multivariate classification of clinical subtypes based on the demographic, symptomatic, neuropsychological, and functional profiles of negative symptoms of 706 patients with schizophrenia and found three-class models with "deficit," "persistent," and "transient" negative symptom subgroups. Future multivariate research initiatives are promising to explore and refine these constructs to better target treatments and improve outcomes for patients with schizophrenia. Finally, an inherent source of heterogeneity comes from the use of clinical questionnaires that come with several subjective and memory biases [26]. In this respect, another related caveat is tied to the difference in temporality between fast neural processes and questionnaires that operate within a much longer frame, which does not allow capturing fine-grained neural and behavioral mechanisms. A solution in this regard can be the digital phenotyping of negative symptoms [30–32]. Future research will aid in establishing the added value of this approach for better characterization of negative symptoms and their associated neurobiology. In this chapter, we summarize the main psychological and neural mechanisms (Fig. 5.1b) for the two negative symptom dimensions [1, 10, 33].

5.2 Motivation and Pleasure Deficit in Schizophrenia

The experiential dimension in schizophrenia has been conceptualized to arise from impairments in distinct components of goal-directed behavior [34, 35]. Research in the experiential dimension has identified the following impairments: (1) distinct dysfunctions of reward processing, including reward anticipation, reward learning, and effort-based decision-making [36–40], (2) neurocognitive abilities that impact planning and goal generation [41, 42], (3) social cognition [32–35], (4) metacognitive dysfunctional beliefs about oneself and expectations from interacting with others [43–45]. In what follows, we summarize the behavioral, psychological, and neural task-based impairments in reward processing, as well as structural task-free (anatomical, resting state) neural circuitry.

5.2.1 Mechanisms and Neural Bases of the Experiential Dimension

5.2.1.1 Behavioral, Psychological, and Neural Task-Based Impairments in Reward Processing

Motivational deficits, particularly in reward anticipation, are believed to play a crucial role in all negative symptoms included in the Experiential dimension observed in schizophrenia [39, 40, 46–48]. One of the most consistent findings is the blunted activity of the ventral striatum during reward anticipation in patients with schizophrenia vs healthy subjects [47] and its association with negative symptoms related to the experiential dimension [23, 47, 49–51]. However, these results were not homogenous across all studies, see [52–54]. Similar findings have been observed in offspring [55, 56] and first-degree relatives [57] of schizophrenia.

A recent meta-analysis of 17 studies further explored activation patterns in reward anticipation using the monetary incentive delay task [48]. Schizophrenia patients showed significant hypoactivation in brain regions like the striatum and anterior cingulate cortex during reward anticipation. Further, striatal hypoactivation was negatively associated with total negative symptoms and was less pronounced in patients with second-generation antipsychotics, possibly indicating a medication effect on this relationship [48]. Reduced activation in the dorsal striatum during similar tasks has been linked to the Experiential dimension, affecting avolition possibly more than anhedonia [54, 58, 59]. Blunted neural activity in the caudate and putamen, but also anterior cingulate and insula during reward anticipation, was reported to be associated with both clinically assessed deficits in motivation and pleasure (CAINS-MAP), as well as self-reported deficits in an eco-momentary assessment [60]. Given the relative consistency of the striatal hypo-activation during task-based neuro-imaging studies in reward anticipation, one study examined the biomarker status for schizophrenia. During task-based neuroimaging using a reward anticipation task, Carruzzo et al. (2024) [24] replicated the finding of impaired ventral striatal activation during reward anticipation across two centers for patients with schizophrenia, albeit no relation with the Experiential dimension was found. The low test-retest reliability and lack of correlation with symptoms indicate that additional research is necessary before ventral striatal activity can be considered a reliable biomarker for individual patients.

Further studies have explored how the Experiential dimension correlates with dysfunctional activation during reward learning. These include regions such as the striatum, the inferior frontal gyrus, the lateral frontal, the orbitofrontal, and the ventromedial prefrontal cortex [60–65]. Several studies have shown that individuals with schizophrenia may not exhibit generalized deficits in neural prediction error signaling; however, they do show impairments in the neural representation and updating of reward values [61, 64, 66–68]. Schizophrenia patients often show selective impairments in learning from positive outcomes, whereas their ability to learn from negative outcomes remains relatively preserved [69–72]. Such findings seem to point to a disconnection in how positive prediction errors are processed and suggest that decision-making may not be influenced by the potential positive outcomes (reward) associated with an action [61, 69, 71], but it is rather guided by loss avoidance [60]. Some studies have shown that schizophrenia patients tend to rely on habitual responses rather than on actions guided by anticipated positive results [64, 73], again possibly suggestive of a fundamental impairment in reward-based learning and decision-making processes [49]. However, a study [60] involving a small group of patients and controls completing a reinforcement learning task, found no significant differences in performance or neural responses related to reinforcement learning between the two groups. Nonetheless, worse reward learning was associated with more severe clinician ratings of negative symptoms and eco-momentary reports of anticipatory motivation and pleasure negative symptoms.

Effort-based decision-making (EBDM), which involves evaluating the costs and benefits of different actions, is another area that has gained significant attention in the neurobiology of apathy [26, 27]. Converging evidence has highlighted that patients with schizophrenia may overestimate the effort required for tasks to obtain potential rewards, leading them to prefer low-effort, low-reward activities [35, 66, 74–77].

Recent meta-analyses have proven that individuals with schizophrenia show decreased motivation to exert physical effort for rewards, correlating with negative symptom severity [40, 51, 78]. Sub-group analyses from the meta-analysis by Blouzard et al., 2023 [78] have shown a significant difference between patients with schizophrenia and healthy subjects regarding trials with high reward magnitude and probability, but not for the low reward ones. This suggests the existence of a maladaptive effort allocation for increasing magnitudes of reward and probability. Furthermore, both this study and another meta-analysis by Luther et al. (2019) [79] identified significant yet weak correlations between overall reductions in EBDM and negative symptoms in schizophrenia. This implies that deficits in EBDM might play a role in the psychopathology of the Experiential dimension.

Another meta-analysis by Pillny et al., 2024 [80], comparing EBDM in depression and schizophrenia, found that anhedonia and cognitive impairments may vary in severity between the two, being more pronounced in schizophrenia and indicating the presence of disorder-specific pathomechanisms. These mechanisms could stem from impairments in cognitive function and anhedonia, present in schizophrenia, and could account for the more pronounced EBDM deficits observed in schizophrenia [80].

Regarding physical effort, patients with more Experiential dimension-related negative symptoms exhibit less willingness to engage in physically demanding tasks, like using a handgrip for monetary rewards [69, 71, 74]. Cognitive efforts also reflect this trend; studies using tasks like the n-back (where participants identify matching letters across increasing intervals) reveal comparable motivational deficits [81]. Recently, it has been proposed that in schizophrenia, deficits of cognition and reward processing interact, and this interaction is a core feature of the condition [82]. In our research, we assessed the integration of cognition and reward via a classic n-back task with two cognitive load levels modified by reward (reward vs no reward) [83]. We observed that the Experiential dimension, but not the Expressive dimension, correlated with reduced activity in the cerebellar-midbrain-striatal circuitry in unmedicated individuals at risk and in early stages of psychosis. This suggests that the compromised reward-driven allocation of cognitive effort noted in chronic schizophrenia may begin early in the disease's progression, potentially playing a role in the development of avolition/ experiential dimension.

5.2.1.2 Neurocognitive Impairment

The evidence for a direct link between general neurocognitive impairment and the Experiential dimension in schizophrenia is emerging but remains inconclusive to date [78–81]. Certain neurocognitive aspects, particularly those involved in executive functioning and planning, may influence goal-directed behaviors linked to the Experiential dimension [41]. Notably, deficits in generating action options in complex situations correlate with the Experiential dimension [42]. A recent study [82] demonstrated that cognitive performance was uniquely associated with reward sensitivity (defined as the individual's reactivity to increasing rewards). It noted that worse cognitive function was associated with more severe total negative symptom severity. Interestingly, this relationship was primarily driven by blunted expression rather than apathy. Moreover, while some research points to working memory impairment as a potentially contributing factor to the Experiential dimension by affecting the ability to sustain emotional experiences and instrumental learning, the evidence in this area is heterogeneous [83–86]. Finally, disturbances in core aspects of self, described as a diminished sense of self-presence or agency, have been associated with negative symptoms [87, 88]. These results highlight the need for further exploration into how neurocognitive and self-perceptual disturbances relate to specific negative symptom dimensions, including the Experiential dimension.

5.2.1.3 Social Cognition

Social cognition comprises the abilities required to understand, interpret, and respond to the mental states of others and includes processes such as affect sharing, mentalizing, experiencing emotions, and regulating emotions [84, 85]. While social cognition broadly impacts the expression [86, 87], its influence on the Experiential dimension is less established. Research into the relationship between social cognition and overall negative symptoms has yielded mixed results, with some studies identifying significant links [32, 34, 35], while others have not [88, 89].

Regarding the Experiential dimension in particular, impairments in social cognition can lead to reduced social engagement, which may perpetuate a vicious cycle of reduced rewards from social interactions, resulting in impaired social cognitive functions. This cycle could theoretically exacerbate both the symptoms of the Experiential dimension and broader negative symptoms, suggesting a complex interaction that warrants further investigation [90]. In a recent study, individuals with schizophrenia showed normal event-related potential responses to receiving social and nonsocial rewards ("liking"). Still, they exhibited diminished anticipatory responses ("wanting") to social rewards compared to nonsocial ones, in line with a specific deficit in social reward processing related to diminished social motivation [91].

Future research should explore the direct link between apathy and deficits in social cognition. Here, translational research using rodent models could potentially provide valuable insights into these mechanisms; however, refinement of these models and methodologies is necessary to enhance their relevance and applicability to social withdrawal in schizophrenia [92, 93].

5.2.1.4 Dysfunctional Beliefs and Experiential Dimension

Psychological models have identified dysfunctional beliefs as key contributors to the pathophysiology of the Experiential dimension in schizophrenia [44, 45]. These beliefs typically revolve around negative perceptions of oneself and the future, often manifesting as defeatist thoughts like "If I try this, I will surely fail." Such beliefs, combined with internalized stigma, can discourage active engagement and may perpetuate withdrawal from social interactions, ultimately reinforcing negative symptoms related to the Experiential dimension [94–96]. While not all studies align perfectly with this framework [97], the association between motivational, neurocognitive, and psychological factors forms a complex syndrome contributing to the Experiential dimension.

Recent research suggests intricate interactions among these factors [98], underscoring the need for further systematic exploration. Significantly, recent evidence from interventions designed specifically to tackle defeatist thinking, boost reward anticipation, and sustain positive symptoms has shown notable improvements in anhedonia and avolition. These improvements did not affect the Expressive dimension, indicating that such targeted interventions can selectively influence aspects of negative symptoms [99, 100]. Moreover, the benefits of this intervention persisted at a 6-month follow-up [101], suggesting that tailored psychological interventions can have lasting effects on specific negative symptoms in psychiatric conditions.

5.2.2 Neural Circuitry of Avolition/Experiential Dimension: Evidence from Resting State, Connectomic, and Morphometric Imaging Studies

Resting-state functional connectivity studies highlight the critical role of fronto-striatal network dysconnectivity in the manifestation of general negative symptoms in schizophrenia [99, 100, 102], as well as the presence of local striatal blood anomalies [103], although replications in larger cohorts are needed. Recent results indicate that the connectivity between the ventral tegmental area (VTA) and several regions, such as the bilateral insular cortex and the right ventrolateral prefrontal cortex, is inversely related to the Experiential dimension symptoms [104, 105]. Additionally, increased functional connectivity between the left ventral striatum and key default mode network regions (the left ventral anterior insula/putamen and the left inferior frontal gyrus/dorsal anterior insula) during reward anticipation is negatively associated with the Experiential dimension scores (BNSS) and not with the Expressive dimension [25]. Resting-state brain activity in the left orbitofrontal cortex has shown a significant correlation with negative symptoms, particularly within the domains of MAP in this study [106]. Other studies have linked fronto-temporal connectivity abnormalities to both the avolition/experiential dimension and social cognition [107].

Studies have also investigated the role of the cerebellum in the pathogenesis of negative symptoms [108]. Resting-state functional magnetic resonance imaging of first psychotic episode patients showed altered functional connectivity between the cerebellum and cerebral regions responsible for theory of mind processing [109]. It has been proposed that patients with schizophrenia may experience avolition, asociality, or anhedonia due to difficulty understanding facial emotions and processing affect, which is linked to dysfunctional internal models and reduced cerebellar activation [110]. Neuromodulation evidence points to the promising effects of cerebellum stimulation for negative symptoms [111–113] which seem to persist long term [113]. However, only a few pilot studies have been conducted to date and replication of findings is needed.

A landmark study [114] showed that TMS over the cerebellar vermis strengthened cerebellar-prefrontal connectivity in a way that was significantly correlated with improvement in negative symptom severity. In this perspective, cerebellar stimulation, but not sham, increased resting state functional connectivity (RSFC) between the vermian-pallidal areas [115], suggesting a functional linkage with regions in the reward circuitry. Although these studies did not focus specifically on the Experiential dimension, this emerging evidence highlights the cerebellum's role in the neural circuitry underlying negative symptoms in schizophrenia. These findings raise the intriguing possibility that the cerebellum may be implicated in negative symptom pathophysiology via the exertion of a monitoring role on the reward system, analogous to its established role in motor control.

From a connectomic perspective, diffusion-weighted magnetic resonance imaging has shown that schizophrenia patients exhibit altered brain network organization, with increased modularity and reduced connectivity between nodes, including the thalamus, specifically correlating with the severity of total negative symptoms [116]. White matter alterations in the connections between the medial orbitofrontal cortex (mOFC) and the rostral part of the anterior cingulate cortex (rACC) have been associated with more severe negative symptoms, specifically anhedonia-asociality and avolition-apathy, as assessed by the Scale for the Assessment of Negative Symptoms (SANS) [117]. These results fit well with the role of these regions in regulating emotional responses and social behavior in patients with schizophrenia [117]. However, the avolition-apathy domain of the SANS considers the overt behavior in three items ("Grooming and Hygiene," "Impersistence at work or school," "Physical anergia") and does not evaluate the reduced motivation for these goal-directed activities. The behaviors considered by the three items might be due to cognitive impairment, depression, or psychotic symptoms.

Examining the micro-structure alterations of white matter by measuring the fiber density (FD) has also shown promising results. Total negative symptoms in schizophrenia have been linked to orbitofrontal-striatal abnormalities, with variations across the schizophrenia spectrum from schizotypy to chronic stages [118]. In a large-scale coordinated meta-analysis from the ENIGMA consortium, the orbitofrontal cortex thickness was found to negatively associate with total negative symptom severity [119]. Similarly, reduced grey matter in the caudate nucleus and ventral striatum is closely linked to persistent negative symptoms and, in particular, to avolition in schizophrenia [120–122]. Cortical thinning in the temporo-parietal junction and abnormalities in the prefrontal cortex in patients experiencing first-episode psychosis [123, 124] highlight the significant role of brain morphology in the development and maintenance of these symptoms. A negative correlation specifically between the Experiential dimension symptoms and putamen/nucleus accumbens volume was observed in healthy individuals with schizotypy [118]; however, it remains to be clarified whether and how these brain changes precede and evolve with the disorder or antipsychotic medication [118].

5.3 The Expressive Dimension in Schizophrenia

5.3.1 Mechanisms Underlying the Expressive Dimension

Research into diminished expression in schizophrenia has predominantly focused on deficits in emotion expression, emotion functioning, and cognitive resources in particularly during situations of increased social pressure, including speech production and fluency) [125, 126]. Motor aspects of emotional expression have been associated with the Expressive dimension symptoms in schizophrenia [127, 128]. Patients display reductions in the quantity and quality of facial expressions, potentially related to motor cognition [127]. As for emotion functioning, patients present a diminished ability in perceiving emotions, which probably stems from a more general reduction of neurocognitive abilities (abnormal contextual integration) than the impairment of a single function [126, 129]. Recent studies have associated the Expressive dimension, in particular, alogia, to a relative insufficiency of cognitive resources, especially in high cognitive demand conditions, which leads to a reduction of verbal and emotional expression [130]. It is essential to underscore that, in contrast to the Experiential dimension, there is no comprehensive cognitive model for the Expressive dimension. This fact highlights the need for future research in this direction that could lead to a better understanding of the mechanism and potential treatment targets.

5.3.1.1 Behavioral, Psychological, and Neural Task-Based Impairments in Emotion Processing: Expression and Perception

Research has explored the connection between blunted affect and the motor components of emotional expression in schizophrenia patients, noting that patients often exhibit less varied and predominantly neutral facial expressions, potentially due to impairments in motor cognition [127, 128, 131]. These impairments also extend to a reduced ability to mimic facial expressions during tasks of rapid face mimicry [132]. Studies such as those by Lee et al. (2013) [133] have also documented difficulties in imitating facial expressions corresponding to emotional states. While these studies contribute to our understanding of emotional expression in schizophrenia, further research is necessary to definitively link these findings to blunted affect. Task-based fMRI studies examining tasks that require emotional expression or cognitive effort have identified potential regions of interest, including the premotor and motor cortex areas and the inferior parietal lobule, part of the mirror neuron system (Lee et al., 2013). Additionally, findings suggest that diminished expression may be linked to altered activity in the rostral anterior cingulate cortex (ACC) during tasks with cognitive demands, indicating a possible failure to recruit additional cognitive resources [134].

Individuals with schizophrenia have also exhibited significant deficits in face emotion recognition during static/dynamic face emotion recognition tasks. These deficits were linked to reduced activation in the superior temporal sulcus and differing patterns of activation in early visual regions and the pulvinar nucleus of the thalamus, highlighting a disease-specific contribution of visual sensory processing to emotion perception [135]. Further research indicates that the severity of alogia correlates with reduced activity in basal ganglia components during cognitive tasks, suggesting disruptions in both voluntary motor behavior and language processing [136].

The exploration of emotion perception impairment in the Expressive dimension has identified significant challenges since there is little evidence suggesting a direct association between these impairments and negative symptoms (Expressive dimension being more concerned) [137, 138]. Emotional perception studies have shown that blunted affect correlates with decreased activation in the amygdala and related structures during emotional processing tasks [138–140]. Evidence suggests that such abnormalities are closely linked to broader negative symptoms, particularly affecting the Expressive dimension [137]. In the same study, Gur et al. discovered that schizophrenia patients with blunted affect had considerable difficulties differentiating emotion intensity. However, some studies argue that negative symptoms might be associated more with general performance impairment (e.g., abnormal contextual integration of emotional stimuli in prefrontal cortices) than specific emotion perception abnormalities [126, 129]. Thus, while difficulties in processing emotions are evident in schizophrenia, their direct contribution to the Expressive dimension remains unclear. Further, it is not yet established whether emotion-processing impairment contributes directly to the Expressive dimension or indirectly via general cognitive impairment [125, 126, 141, 142].

5.3.1.2 Limited Cognitive Resources Model

The cognitive resource limitation model has been proposed to explain diminished verbal expression, particularly alogia, suggesting that speech relies on various cognitive functions, including controlled retrieval and semantic memory organization [130, 143–146]. Evidence supporting this model shows that deficits in working memory and overall cognitive function are significant contributors to alogia. Cohen et al. [130, 143] have proposed that the Expressive dimension in schizophrenia could stem from insufficient cognitive resources, particularly under high cognitive demand scenarios, like during social interactions. Further studies confirm that the Expressive dimension correlates with cognitive impairments in schizophrenia [86, 87, 147, 148].

Recently, a threefold model has been proposed according to which cognitive symptoms, impaired self-reflectivity, and higher-order language disturbances relate to the Expressive dimension in schizophrenia. Using structural equation modeling on data from 201 individuals, their results showed that self-reflectivity mediated the effects of linguistic and cognitive impairments on diminished expression, suggesting that targeting these areas could improve treatment outcomes [149].

Overall, evidence seems to suggest that high-demand contexts can exacerbate cognitive resource limitations, leading to symptoms like alogia. However, comprehensive cognitive models for the Expressive dimension are lacking, highlighting the need for future research to establish precise treatment targets.

5.3.2 Neural Substrates of Diminished Expression

5.3.2.1 Neural Circuitry of the Expressive Dimension: Evidence from Resting State, Connectomic, and Morphometric Studies

Resting-state fMRI studies have reported abnormalities in connectivity within the default mode network (DMN) associated with total negative symptoms, including the Expressive dimension [141, 142, 150]. A specific pattern of dysconnectivity that correlates with the Expressive dimension(blunted affect and alogia), particularly in the fronto-polar cortex and right inferior temporal gyrus, has been found [151]. It is worth highlighting a longitudinal study that found stable striatal-fronto-parietal-cerebellar patterns of hypoconnectivity in patients with schizophrenia across two independent centers [152]. Notably, weaker sensorimotor striatum connectivity with the superior parietal lobule correlated negatively with the Expressive dimension. It could predict symptom severity 9 months later, suggesting that this pattern may serve as a biomarker for the Expressive dimension in schizophrenia. One explanation is that, in this resting state study, the identified patterns could underlie aberrancies in cerebral flow in regions implicated in affective processing such as cerebellar vermis and fronto-striatal regions broadly [153].

From a connectomic perspective, evidence has indicated reduced connectivity (quantified by the number of connections between brain regions) in the left thalamus and left putamen in patients with schizophrenia compared to the control group. Significantly, a higher level of neural segregation correlated with more severe negative symptoms but not with cognitive impairments, suggesting that this dysconnectivity may be uniquely related to the presence of negative symptoms in schizophrenia [116].

Structural MRI findings have been mixed, with some studies showing correlations between reduced gray matter volume in areas like the anterior cingulate cortex and superior temporal gyrus and symptoms of emotional withdrawal and difficulty in abstract thinking [154]. On the contrary, other studies have found increased gray matter volumes in these regions associated with higher severity of alogia and blunted affect [121]. Alterations in the orbitofrontal cortex and striatum have been observed in patients with schizophrenia, which are linked not only to the Experiential dimension but also to the Expressive dimension [118, 119].

In summary, while significant progress has been made in understanding the neural basis of the Expressive dimension, integrating these findings into a cohesive pathophysiological model remains challenging. Future research should aim to clarify the specific neural mechanisms underlying this dimension of negative symptoms and explore novel treatment targets tailored to the unique needs of individuals exhibiting symptoms of the Expressive dimension.

5.4 Conclusion

Historically, studies have often approached negative symptoms as a uniform group without sufficient consideration for their diverse manifestations and trajectories [27, 155, 156].

Substantial advances have been made in uncovering the behavioral, cognitive, and neural mechanisms contributing to the manifestation of negative symptoms in schizophrenia. Recent research has increasingly recognized the distinct pathological processes associated with specific negative symptom dimensions. Nonetheless, translating these insights into an integrative model that could lead to a more effective understanding of the exact pathophysiological mechanisms remains a significant challenge. For Avolition/Experiential dimension, there is a growing body of evidence linking various motivational deficits and cognitive impairments to its clinical presentation, including challenges in reward anticipation, reward learning, and effort-based decision-making [35, 157, 158]. Neuroimaging studies also suggest that the Experiential dimension is associated with specific regions within a prefrontal-striatal-cerebellar circuitry, contributing to the comprehension of the pathophysiological mechanisms and underscoring the potential for targeted brain stimulation interventions.

In contrast, the mechanisms and neural substrates linked to the Experiential dimension are less well-defined. Current literature points to impairments in emotion expression, emotion perception, and the allocation of cognitive resources for speech production as key areas of dysfunction in individuals with schizophrenia. However, direct links between these deficits and the clinical manifestation of the Expressive dimension are sparse and have rarely been replicated independently. This gap highlights the need for more comprehensive pathophysiological models and targeted treatment approaches.

This chapter focused mostly on neural bases of negative symptoms and the examination of the causal mechanisms underlying these correlates is beyond its immediate scope. It is, however, worth mentioning that work in this direction has focused on how genetic and environmental risk factors contribute to negative symptom emergence through excessive activation of glia-mediated elimination [159]. Further, the role of inflammation is being examined as an important mediator in the pathophysiology of negative symptoms [160]. A need for reproducible methods assessing inflammation-related dysconnectivity is becoming prominent, as well as novel treatment trials targeting inflammation and its effects on the brain [161].

5.5 Future Directions

Moving forward in the investigation of negative symptoms in schizophrenia, it is crucial for future efforts to prioritize longitudinal designs that leverage multimodal data sources—such as neuroimaging, cognitive, behavioral, and digital dimensions—to construct a detailed and nuanced characterization of negative symptoms at the individual level. The combination of multimodal data with causal information obtained from neuromodulation or pharmacotherapy seems a particularly high-yield feat [162] in the spirit of precision psychiatry. Such a combination requires new forms of clinical research designs [163], including translational cross-species designs [164], as well as novel ways of embracing heterogeneity [165]. These comprehensive datasets would be invaluable for identifying patterns that predict changes in symptom severity and the trajectory of negative symptoms. These datasets could also facilitate the development of personalized treatment strategies tailored to the specific negative symptom profiles of individuals. Incorporating neurostimulation techniques, such as transcranial magnetic stimulation, could be essential not only to refine and optimize treatment protocols ("Which treatment for whom?") but also to gain deeper causal insights into the neurobiological mechanisms underpinning negative symptoms.

References

1. Bègue I, Kaiser S, Kirschner M. Pathophysiology of negative symptom dimensions of schizophrenia - current developments and implications for treatment. Neurosci Biobehav Rev. 2020;116:74–88.
2. Kirkpatrick B, Fenton WS, Carpenter WT, Marder SR. The NIMH-MATRICS consensus statement on negative symptoms. Schizophr Bull. 2006;32(2):214–9.
3. Galderisi S, Mucci A, Dollfus S, Nordentoft M, Falkai P, Kaiser S, et al. EPA guidance on assessment of negative symptoms in schizophrenia. Eur Psychiatry. 2021;64(1):e23.
4. Galderisi S, Armida M, Buchanan RW, Celso A. Negative symptoms of schizophrenia: new developments and unanswered research questions. Lancet Psychiatry. 2018;5(8):664–77.
5. Bobes J, Arango C, Garcia-Garcia M, Rejas J, CLAMORS Study Collaborative Group. Prevalence of negative symptoms in outpatients with schizophrenia spectrum disorders treated with antipsychotics in routine clinical practice: findings from the CLAMORS study. J Clin Psychiatry. 2010;71(3):280–6.
6. Szkultecka-Dębek M, Walczak J, Augustyńska J, Miernik K, Stelmachowski J, Pieniążek I, et al. Epidemiology and treatment guidelines of negative symptoms in schizo-phrenia in Central and Eastern Europe: a literature review. Clin Pract Epidemiol Ment Health CP EMH [Internet]. 2015 [cited 2024 Jun 25];11:158–65. Available from: https://www.ncbi.nlm.nih.gov/pmc/articles/PMC4627386/
7. Mosolov SN, Yaltonskaya PA. Primary and secondary negative symptoms in schizophrenia. Front Psychiatry [Internet]. 2022 [cited 2024 Apr 29];12:766692. Available from: https://www.ncbi.nlm.nih.gov/pmc/articles/PMC8761803/
8. Faerden A, Barrett EA, Nesvåg R, Friis S, Finset A, Marder SR, et al. Apathy, poor verbal memory and male gender predict lower psychosocial functioning one year after the first treatment of psychosis. Psychiatry Res [Internet]. 2013 [cited 2024 Apr 29];210(1):55–61. Available from: https://www.sciencedirect.com/science/article/pii/S0165178113000796
9. Fervaha G, George F, Agid O, Remington G. Motivational and neurocognitive deficits are central to the prediction of longitudinal functional outcome in schizophrenia. Acta Psychiatr Scand. 2014;130(4):290–9.
10. Galderisi S, Paola B, Mucci A, Kirkpatrick B, Pini S, Rossi A, Vita A, Maj M. Categorical and dimensional approaches to negative symptoms of schizophrenia: focus on long-term stability and functional outcome. Schizophr Res. 2013;147(1):157–62.
11. Rabinowitz J, Stephen LZ, Garibaldi G, Bugarski Kirola D, Berardo CG, Kapur S. Negative symptoms have greater impact on functioning than positive symptoms in schizophrenia: analysis of CATIE data. Schizophr Res. 2012;137(1):147–50.

12. Strauss GP, William HP, Kirkpatrick B, Fischer BA, Keller WR, Miski P, Buchanan RW, Green MF, Carpenter WT. Deconstructing negative symptoms of schizophrenia: avolition-apathy and diminished expression clusters predict clinical presentation and functional outcome. J Psychiatr Res. 2013;47(6):783–90.
13. Blanchard JJ, Alex CS. The structure of negative symptoms within schizophrenia: implications for assessment. Schizophr Bull. 2005;32(2):238–45.
14. Bucci P, Galderisi S. Categorizing and assessing negative symptoms. Curr Opin Psychiatry. 2017;30(3):201–8.
15. Kimhy D, Delespaul P, Corcoran C, Ahn H, Yale S, Malaspina D. Computerized experience sampling method (ESMc): assessing feasibility and validity among individuals with schizophrenia. J Psychiatr Res. 2006;40(3):221–30.
16. Marder SR, Silvana G. The current conceptualization of negative symptoms in schizophrenia. World Psychiatry Off J World Psychiatr Assoc WPA. 2017;16(1):14–24.
17. Kirkpatrick B. Developing concepts in negative symptoms: primary vs secondary and apathy vs expression. J Clin Psychiatry. 2014;75(suppl 1):3–7.
18. Carpenter WT, Douglas HW, Alphs L. Treatment of negative symptoms. Schizophr Bull. 1985;11(3):440–52.
19. Messinger JW, Fabien T, Antonius D, Mendelsohn E, Prudent V, Stanford A, D; Malaspina, Dolores. Avolition and expressive deficits capture negative symptom phenomenology: implications for DSM-5 and schizophrenia research. Clin Psychol Rev. 2010;31(1):161–8.
20. Ahmed AO, Kirkpatrick B, Granholm E, Rowland LM, Barker PB, Gold JM, et al. Two factors, five factors, or both? External validation studies of negative symptom dimensions in schizophrenia. Schizophr Bull. 2022;48(3):620–30.
21. Kaliuzhna M, Kirschner M, Carruzzo F, Hartmann-Riemer MN, Bischof M, Seifritz E, et al. How far to go in deconstructing negative symptoms? Behavioural and neural level evidence for the amotivation domain. Schizophr Res. 2021;236:41–7.
22. Mucci A, Galderisi S, Gibertoni D, Rossi A, Rocca P, Bertolino A, et al. Factors associated with real-life functioning in persons with schizophrenia in a 4-year follow-up study of the Italian network for research on psychoses. JAMA Psychiatry. 2021;78(5):550–9.
23. Kirschner M, Oliver HM, Bischof M, Hartmann M, Kluge A, Seifritz E, Tobler PN, Kaiser S. Ventral striatal hypoactivation is associated with apathy but not diminished expression in patients with schizophrenia. J Psychiatry Neurosci JPN. 2016;41(3):152–61.
24. Carruzzo F, Kaliuzhna M, Kuenzi N, Geffen T, Katthagen T, Schlagenhauf F, et al. Striatal response to reward anticipation as a biomarker for schizophrenia and negative symptoms: effects, test-retest reliability, and stability across sites. Schizophr Bull [Internet]. 2024 [cited 2024 Apr 29];sbae046. https://doi.org/10.1093/schbul/sbae046.
25. Carruzzo F, Kaiser S, Tobler PN, Kirschner M, Simon JJ. Increased ventral striatal functional connectivity in patients with schizophrenia during reward anticipation. NeuroImage Clin. 2022;33:102944.
26. Lincoln TM, Dollfus S, Lyne J. Current developments and challenges in the assessment of negative symptoms. Schizophr Res. 2016;186:8–18.
27. Savill M, Ciara B, Khanom H, Priebe S. Do negative symptoms of schizophrenia change over time? A meta-analysis of longitudinal data. Psychol Med. 2014;45(8):1613–27.
28. Kirkpatrick B, Armida M, Galderisi S. Primary, enduring negative symptoms: an update on research. Schizophr Bull. 2017;43(4):730–6.
29. Ahmed AO, Strauss GP, Buchanan RW, Kirkpatrick B, Carpenter WT. Schizophrenia heterogeneity revisited: clinical, cognitive, and psychosocial correlates of statistically-derived negative symptoms subgroups. J Psychiatr Res. 2018;97:8–15. https://doi.org/10.1016/j.jpsychires.2017.11.004. Epub 2017 Nov 12
30. Cohen A, Naslund JA, Chang S, Nagendra S, Bhan A, Rozatkar A, et al. Relapse prediction in schizophrenia with smartphone digital phenotyping during COVID-19: a prospective, three-site, two-country, longitudinal study. Schizophrenia [Internet]. 2023 [cited 2024 Apr 30];9(1):1–8. Available from: https://www.nature.com/articles/s41537-023-00332-5

31. Cohen AS, Schwartz E, Le TP, Cowan T, Kirkpatrick B, Raugh IM, et al. Digital phenotyping of negative symptoms: the relationship to clinician ratings. Schizophr Bull. 2021;47(1):44–53.
32. Henson P, Barnett I, Keshavan M, Torous J. Towards clinically actionable digital phenotyping targets in schizophrenia. NPJ Schizophr. 2020;6(1):13.
33. Mørch-Johnsen L, Smelror RE, Andreou D, Barth C, Johannessen C, Wedervang-Resell K, et al. Negative symptom domains are associated with verbal learning in adolescents with early onset psychosis. Front Psych. 2021;12:825681.
34. Husain MR, Jonathan P. Neuroscience of apathy and anhedonia: a transdiagnostic approach. Nat Rev Neurosci. 2018;19(8):470–84.
35. Le Heron CA, Matthew AJ, Husain M. The anatomy of apathy: a neurocognitive framework for amotivated behaviour. Neuropsychologia. 2017;118(Pt B):54–67.
36. Barch DM, Gold JM, Kring AM. Paradigms for assessing hedonic processing and motivation in humans: relevance to understanding negative symptoms in psychopathology. Schizophr Bull [Internet]. 2017 [cited 2024 Apr 29];43(4):701–5. https://doi.org/10.1093/schbul/sbx063.
37. Culbreth AJ, Erin M, K; Deanna, M. Effort-based decision-making in schizophrenia. Behav Curr Opin Sci. 2017;22:1–6.
38. Hartmann-Riemer MN, Matthias K, Kaiser S. Effort-based decision-making paradigms as objective measures of apathy in schizophrenia? Curr Opin Behav Sci. 2018;22:70–5.
39. Kring AM, D M. The motivation and pleasure dimension of negative symptoms: neural substrates and behavioral outputs. Eur Neuropsychopharmacol J Eur Coll Neuropsychopharmacol. 2014;24(5):725–36.
40. Strauss GP, James WA, Gold JM. A review of reward processing and motivational impairment in schizophrenia. Schizophr Bull. 2013;40(Suppl 2):S107–16.
41. Faerden A, Vaskinn A, Finset A, Agartz I, Ann Barrett E, Friis S, et al. Apathy is associated with executive functioning in first episode psychosis. BMC Psychiatry. 2009;9:1.
42. Hartmann M, Agne K, Kalis A, Mojzisch A, Tobler PN, Kaiser S. Apathy in schizophrenia as a deficit in the generation of options for action. J Abnorm Psychol. 2015;124(2):309–18.
43. Beck AT, Robyn H, Bredemeier K, Silverstein SM, Grant PM. What accounts for poor functioning in people with schizophrenia: a re-evaluation of the contributions of neurocognitive v. attitudinal and motivational factors. Psychol Med. 2018;48(16):2776–85.
44. Grant PM, Aaron B, T. Defeatist beliefs as a mediator of cognitive impairment, negative symptoms, and functioning in schizophrenia. Schizophr Bull. 2008;35(4):798–806.
45. Staring ABP, ter Huurne MAB, van der Gaag M. Cognitive behavioral therapy for negative symptoms (CBT-n) in psychotic disorders: a pilot study. J Behav Ther Exp Psychiatry. 2013;44(3):300–6.
46. Leroy A, Amad A, D'Hondt F, Pins D, Jaafari N, Thomas P, et al. Reward anticipation in schizophrenia: a coordinate-based meta-analysis. Schizophr Res [Internet]. 2020 [cited 2024 Jun 25];218:2–6. Available from: https://www.sciencedirect.com/science/article/pii/S0920996419306115
47. Radua J, André S, Borgwardt S, Heinz A, Schlagenhauf F, McGuire P, Fusar Poli P. Ventral striatal activation during reward processing in psychosis: a neurofunctional meta-analysis. JAMA Psychiatry. 2015;72(12):1243–51.
48. Zeng J, Yan J, Cao H, Su Y, Song Y, Luo Y, et al. Neural substrates of reward anticipation and outcome in schizophrenia: a meta-analysis of fMRI findings in the monetary incentive delay task. Transl Psychiatry [Internet]. 2022 [cited 2024 May 14];12:448. Available from: https://www.ncbi.nlm.nih.gov/pmc/articles/PMC9573872/
49. Brandl F, Knolle F, Avram M, Leucht C, Yakushev I, Priller J, et al. Negative symptoms, striatal dopamine and model-free reward decision-making in schizophrenia. Brain J Neurol. 2023;146(2):767–77.
50. Waltz JA, Julie SB, Ross TJ, Kurup PK, Salmeron BJ, Rose EJ, Gold JM, Stein EA. Abnormal responses to monetary outcomes in cortex, but not in the basal ganglia, in schizophrenia. Neuropsychopharmacol Off Publ Am Coll Neuropsychopharmacol. 2010;35(12):2427–39.

51. Wolf DH, Theodore SD, Kantrowitz JJ, Katchmar N, Vandekar L, Elliott MA, Ruparel K. Amotivation in schizophrenia: integrated assessment with behavioral, clinical, and imaging measures. Schizophr Bull. 2014;40(6):1328–37.
52. Esslinger C, Englisch S, Inta D, Rausch F, Schirmbeck F, Mier D, et al. Ventral striatal activation during attribution of stimulus saliency and reward anticipation is correlated in unmedicated first episode schizophrenia patients. Schizophr Res. 2012;140(1–3):114–21.
53. Nielsen MØ, Rostrup E, Wulff S, Bak N, Lublin H, Kapur S, et al. Alterations of the brain reward system in antipsychotic naïve schizophrenia patients. Biol Psychiatry. 2012;71(10):898–905.
54. Mucci A, Danai D, Soricelli A, Volpe U, Bucci P, Frangou S, Prinster A, Salvatore M, Galderisi S, Maj M. Is avolition in schizophrenia associated with a deficit of dorsal caudate activity?: a functional magnetic resonance imaging study during reward anticipation and feedback. Psychol Med. 2015;45(8):1765–78.
55. Vink M, de Leeuw M, Pouwels R, van den Munkhof HE, Kahn RS, Hillegers M. Diminishing striatal activation across adolescent development during reward anticipation in offspring of schizophrenia patients. Schizophr Res [Internet]. 2016 [cited 2024 Jun 25];170(1):73–9. Available from: https://www.sciencedirect.com/science/article/pii/S092099641530061X
56. de Leeuw M, Kahn RS, Vink M. Fronto-striatal dysfunction during reward processing in unaffected siblings of schizophrenia patients. Schizophr Bull [Internet]. 2015 [cited 2024 Jun 25];41(1):94–103. https://doi.org/10.1093/schbul/sbu153.
57. Grimm O, Heinz A, Walter H, Kirsch P, Erk S, Haddad L, et al. Striatal response to reward anticipation: evidence for a systems-level intermediate phenotype for schizophrenia. JAMA Psychiatry [Internet] 2014 [cited 2024 Jun 25];71(5):531–9. https://doi.org/10.1001/jamapsychiatry.2014.9.
58. Morris RW, Stephanie QL, Griffiths K, Green MJ, Balleine BW. Corticostriatal control of goal-directed action is impaired in schizophrenia. Biol Psychiatry. 2014;77(2):187–95.
59. Stepien M, Andrei M, Kubli R, Schneider K, Tobler PN, Seifritz E, Herdener M, Kaiser S, Kirschner M. Investigating the association of ventral and dorsal striatal dysfunction during reward anticipation with negative symptoms in patients with schizophrenia and healthy individuals. PLoS One. 2018;13(6):1–11.
60. Moran EK, Adam CJ, Kandala S, Deanna M. From neuroimaging to daily functioning: a multimethod analysis of reward anticipation in people with schizophrenia. J Abnorm Psychol. 2019;128(7):723–34.
61. Dowd EC, Michael FJ, Collins AGE, Gold JM, Deanna M. Probabilistic reinforcement learning in patients with schizophrenia: relationships to anhedonia and avolition. Biol Psychiatry Cogn Neurosci Neuroimaging. 2016;1(5):460–73.
62. Fuentes-Claramonte P, Garcia-Leon MA, Salgado-Pineda P, Ramiro N, Soler-Vidal J, Torres ML, et al. Do the negative symptoms of schizophrenia reflect reduced responsiveness to reward? Examination using a reward prediction error (RPE) task. Psychol Med. 2023;53(15):7106–15.
63. Kluge A, Matthias K, Hager OM, Bischof M, Habermeyer B, Seifritz E, Walther S, Kaiser S. Combining actigraphy, ecological momentary assessment and neuroimaging to study apathy in patients with schizophrenia. Schizophr Res. 2017;195(195):176–82.
64. Waltz JA, Ziye X, Brown EC, Ruiz R, Frank MJ, Gold JM. Motivational deficits in schizophrenia are associated with reduced differentiation between gain and loss-avoidance feedback in the striatum. Biol Psychiatry Cogn Neurosci Neuroimaging. 2017;3(3):239–47.
65. Katthagen T, Kaminski J, Heinz A, Buchert R, Schlagenhauf F. Striatal dopamine and reward prediction error signaling in unmedicated schizophrenia patients. Schizophr Bull. 2020;46(6):1535–46.
66. Culbreth AJ, Andrew W, Deanna NA. Negative symptoms are associated with an increased subjective cost of cognitive effort. J Abnorm Psychol. 2016;125(4):528–36.
67. Reinen JV, Jared SX, Horga G, Abi Dargham A, Daw ND, Shohamy D. Motivational context modulates prediction error response in schizophrenia. Schizophr Bull. 2016;42(6):1467–75.

68. Spilka MJ, Raugh IM, Berglund AM, Visser KF, Strauss GP. Reinforcement learning profiles and negative symptoms across chronic and clinical high-risk phases of psychotic illness. Eur Arch Psychiatry Clin Neurosci. 2023;273(8):1747–60.
69. Gold JM, James WA, Matveeva TM, Kasanova Z, Strauss GP, Herbener ES, Collins AGE, Frank MJ. Negative symptoms and the failure to represent the expected reward value of actions: behavioral and computational modeling evidence. Arch Gen Psychiatry. 2012;69(2):129–38.
70. Maia TV, Michael FJ. An integrative perspective on the role of dopamine in schizophrenia. Biol Psychiatry. 2016;81(1):52–66.
71. Waltz JA, Michael FJ, Robinson BM, Gold JM. Selective reinforcement learning deficits in schizophrenia support predictions from computational models of striatal-cortical dysfunction. Biol Psychiatry. 2007;62(7):756–64.
72. Yilmaz A, Fatma S, Gonul AS. Reduced reward-related probability learning in schizophrenia patients. Neuropsychiatr Dis Treat. 2012;8:27–34.
73. Hernaus D, Michael FJ, Brown EC, Brown JK, Gold JM, Waltz JA. Impaired expected value computations in schizophrenia are associated with a reduced ability to integrate reward probability and magnitude of recent outcomes. Biol Psychiatry Cogn Neurosci Neuroimaging. 2018;4(3):280–90.
74. Fervaha GG, Ariel G, Zakzanis KK, Foussias G, Agid O, Remington G. Incentive motivation deficits in schizophrenia reflect effort computation impairments during cost-benefit decision-making. J Psychiatr Res. 2013;47(11):1590–6.
75. Hartmann M, Oliver HM, Reimann AV, Chumbley JR, Kirschner M, Seifritz E, Tobler PN, Kaiser S. Apathy but not diminished expression in schizophrenia is associated with discounting of monetary rewards by physical effort. Schizophr Bull. 2014;41(2):503–12.
76. Isıklı S, Bektaş AB, Tamer Ş, Atabay M, Arkalı BD, Bağcı B, et al. Effort-cost decision-making associated with negative symptoms in schizophrenia and bipolar disorder. Behav Brain Res. 2024;467:114996.
77. Gold JM, Gregory SP, Waltz JA, Robinson BM, Brown J, Frank MJ. Negative symptoms of schizophrenia are associated with abnormal effort-cost computations. Biol Psychiatry. 2013;74(2):130–6.
78. Blouzard E, Pouchon A, Polosan M, Bastin J, Dondé C. Effort-cost decision-making among individuals with schizophrenia: a systematic review and meta-analysis. JAMA Psychiatry. 2023;80(6):548–57.
79. Luther L, Fischer MW, Firmin RL, Salyers MP. Clarifying the overlap between motivation and negative symptom measures in schizophrenia research: a meta-analysis. Schizophr Res. 2019;206:27–36.
80. Pillny M, Renz KE, Lincoln TM, Lütje F, Hay A, Fulford D, et al. Effort-based decision-making in psychopathology: a transdiagnostic multilevel meta-analysis [Internet]. OSF; 2024 [cited 2024 Jun 26]. Available from: https://osf.io/j6asc
81. Culbreth AJ, Moran EK, Kandala S, Westbrook A, Barch DM. Effort, avolition and motivational experience in schizophrenia: analysis of behavioral and neuroimaging data with relationships to daily motivational experience. Clin Psychol Sci J Assoc Psychol Sci [Internet] 2020 [cited 2024 Jun 26];8(3):555–68. Available from: https://www.ncbi.nlm.nih.gov/pmc/articles/PMC7983405/
82. Robison AJ, Thakkar KN, Diwadkar VA. Cognition and reward circuits in schizophrenia: synergistic, not separate. Biol Psychiatry [Internet]. 2020 [cited 2024 Jun 26];87(3):204–14. Available from: https://www.sciencedirect.com/science/article/pii/S0006322319317469
83. Bègue I, Brakowski J, Seifritz E, Dagher A, Tobler PN, Kirschner M, et al. Cerebellar and cortico-striatal-midbrain contributions to reward-cognition processes and apathy within the psychosis continuum. Schizophr Res [Internet]. 2022 [cited 2024 Jun 26];246:85–94. Available from: https://www.sciencedirect.com/science/article/pii/S0920996422002444
84. Somogy GSV. Social cognition and psychopathology: a critical overview. World Psychiatry Off J World Psychiatr Assoc WPA. 2015;14(1):5–14.
85. Green MF, William HP, Lee J. Social cognition in schizophrenia. Nat Rev Neurosci. 2015;16(10):620–31.

86. Galderisi S, Alessandro R, Rocca P, Bertolino A, Mucci A, Bucci P, Rucci P, Gibertoni D, Aguglia E, Amore M, Bellomo A, Biondi M, Brugnoli R, Dell'Osso L, De Ronchi D, Di Emidio G, Di Giannantonio M, Fagiolini A, Marchesi C, Monteleone P, Oldani L, Pinna F, Roncone R, Sacchetti E, Santonastaso P, Siracusano A, Vita A, Zeppegno P, Maj M. The influence of illness-related variables, personal resources and context-related factors on real-life functioning of people with schizophrenia. World Psychiatry Off J World Psychiatr Assoc WPA. 2014;13(3):275–87.
87. Hartmann-Riemer MN, Oliver HM, Kirschner M, Bischof M, Kluge A, Seifritz E, Kaiser S. The association of neurocognitive impairment with diminished expression and apathy in schizophrenia. Schizophr Res. 2015;169(1):427–32.
88. Rassovsky Y, William HP, Lee J, Sergi MJ, Green MF. Pathways between early visual processing and functional outcome in schizophrenia. Psychol Med. 2010;41(3):487–97.
89. Mancuso F, William HP, Kern RS, Green MF. Social cognition in psychosis: multidimensional structure, clinical correlates, and relationship with functional outcome. Schizophr Res. 2010;125(2):143–51.
90. Fett AK, Wolfgang V, de Gracia Dominguez M, Penn DL, van Os J, Krabbendam L. The relationship between neurocognition and social cognition with functional outcomes in schizophrenia: a meta-analysis. Neurosci Biobehav Rev. 2010;35(3):573–88.
91. Catalano LT, Wynn JK, Green MF, Gold JM. Reduced neural activity when anticipating social versus nonsocial rewards in schizophrenia: preliminary evidence from an ERP study. Schizophr Res. 2022 Aug;246:7–16.
92. Hazani R, Lavidor M, Weller A. Treatments for social interaction impairment in animal models of schizophrenia: a critical review and meta-analysis. Schizophr Bull [Internet]. 2022 [cited 2024 Jun 26];48(6):1179–93. Available from: https://www.ncbi.nlm.nih.gov/pmc/articles/PMC9673263/
93. Powell SB, Swerdlow NR. The relevance of animal models of social isolation and social motivation for understanding schizophrenia: review and future directions. Schizophr Bull. 2023;49(5):1112–26.
94. Campellone TR, Janelle CM, Kring AM. The power to resist: the relationship between power, stigma, and negative symptoms in schizophrenia. Psychiatry Res. 2013;215(2):280–5.
95. Couture SM, Jack BJ, Bennett ME. Negative expectancy appraisals and defeatist performance beliefs and negative symptoms of schizophrenia. Psychiatry Res. 2011;189(1):43–8.
96. Quinlan T, Scott RC, Granholm E. The role of dysfunctional attitudes in models of negative symptoms and functioning in schizophrenia. Schizophr Res. 2014;157(1):182–9.
97. Pillny ML, Tania M. Predictors of improved functioning in patients with psychosis: the role of amotivation and defeatist performance beliefs. Psychiatry Res. 2016;244:117–22.
98. Reddy LF, William HP, Deanna M, Buchanan RW, Gold JM, Marder SR, Wynn JK, Young JW, Green MF. Understanding the association between negative symptoms and performance on effort-based decision-making tasks: the importance of defeatist performance beliefs. Schizophr Bull. 2017;44(6):1217–26.
99. Shukla D, Joshua C, Sampath H, Kochunov P, Hare SM, Wisner K, Rowland LM, Hong LE. Aberrant frontostriatal connectivity in negative symptoms of schizophrenia. Schizophr Bull. 2018;45(5):1051–9.
100. Tu PC, Chuen HJ, Li CT, Bai YM, Su TP. Cortico-striatal disconnection within the cingulo-opercular network in schizophrenia revealed by intrinsic functional connectivity analysis: a resting fMRI study. NeuroImage. 2011;59(1):238–47.
101. Favrod J, Alexandra N, Chaix J, Pellet J, Frobert L, Fankhauser C, Ismailaj A, Brana A, Tamic G, Suter C, Rexhaj S, Golay P, Bonsack C. Improving pleasure and motivation in schizophrenia: a randomized controlled clinical trial. Psychother Psychosom. 2019;88(2):84–95.
102. Fornito A, Ben JH, Goodby E, Dean AM, Ooi C, Nathan PJ, Lennox BR, Jones PB, Suckling J, Bullmore ET. Functional dysconnectivity of corticostriatal circuitry as a risk phenotype for psychosis. JAMA Psychiatry. 2013;70(11):1143–51.

103. Schneider K, Michels L, Hartmann-Riemer MN, Burrer A, Tobler PN, Stämpfli P, et al. Cerebral blood flow in striatal regions is associated with apathy in patients with schizophrenia. J Psychiatry Neurosci JPN. 2019;44(2):102–10.
104. Brasso C, Stanziano M, Bosco FM, Morese R, Valentini MC, Vercelli A, et al. Alteration of the functional connectivity of the cortical areas characterized by the presence of Von Economo neurons in schizophrenia, a pilot study. J Clin Med. 2023;12(4):1377.
105. Giordano GM, Mario S, Papa M, Mucci A, Prinster A, Soricelli A, Galderisi S. Functional connectivity of the ventral tegmental area and avolition in subjects with schizophrenia: a resting state functional MRI study. Eur Neuropsychopharmacol J Eur Coll Neuropsychopharmacol. 2018;28(5):589–602.
106. Giordano GM, Pezzella P, Giuliani L, Fazio L, Mucci A, Perrottelli A, et al. Resting-state brain activity dysfunctions in schizophrenia and their associations with negative symptom domains: an fMRI study. Brain Sci [Internet]. 2023 [cited 2024 Apr 29];13(1):83. Available from: https://www.ncbi.nlm.nih.gov/pmc/articles/PMC9856573/
107. Abram SV, Krista W, Fox JM, Deanna M, Wang L, Csernansky JG, MacDonald AW, Smith MJ. Fronto-temporal connectivity predicts cognitive empathy deficits and experiential negative symptoms in schizophrenia. Hum Brain Mapp. 2016;38(3):1111–24.
108. Picard H, Amado I, Mouchet-Mages S, Olié JP, Krebs MO. The role of the cerebellum in schizophrenia: an update of clinical, cognitive, and functional evidences. Schizophr Bull [Internet]. 2008 [cited 2024 May 14];34(1):155–72. https://doi.org/10.1093/schbul/sbm049.
109. Choi S, Kim M, Kim T, Choi EJ, Lee J, Moon SY, et al. Fronto-striato-thalamic circuit connectivity and neuromelanin in schizophrenia: an fMRI and neuromelanin-MRI study. Schizophrenia [Internet]. 2023 [cited 2024 Jun 26];9(1):1–7. Available from: https://www.nature.com/articles/s41537-023-00410-8
110. Bernard JA, Mittal VA. Dysfunctional activation of the cerebellum in schizophrenia: a functional neuroimaging meta-analysis. Clin Psychol Sci J Assoc Psychol Sci [Internet]. 2015 [cited 2024 May 14];3(4):545–66. Available from: https://www.ncbi.nlm.nih.gov/pmc/articles/PMC4574495/
111. Demirtas-Tatlidede A, Freitas C, Cromer JR, Safar L, Ongur D, Stone WS, et al. Safety and proof of principle study of cerebellar vermal theta burst stimulation in refractory schizophrenia. Schizophr Res. 2010;124(1–3):91–100.
112. Tikka SK, Garg S, Sinha VK, Nizamie SH, Goyal N. Resting state dense array gamma oscillatory activity as a response marker for cerebellar-repetitive transcranial magnetic stimulation (rTMS) in schizophrenia. J ECT. 2015;31(4):258–62.
113. Han X, Zhu Z, Luan J, Lv P, Xin X, Zhang X, et al. Effects of repetitive transcranial magnetic stimulation and their underlying neural mechanisms evaluated with magnetic resonance imaging-based brain connectivity network analyses. Eur J Radiol Open [Internet]. 2023 [cited 2024 Jun 26];10:100495. Available from: https://www.sciencedirect.com/science/article/pii/S2352047723000217
114. Brady RO, Gonsalvez I, Lee I, Öngür D, Seidman LJ, Schmahmann JD, et al. Cerebellar-prefrontal network connectivity and negative symptoms in schizophrenia. Am J Psychiatry [Internet]. 2019 [cited 2024 Jun 26];176(7):512–20. Available from: https://ajp.psychiatryonline.org/doi/10.1176/appi.ajp.2018.18040429
115. Basavaraju R, Ithal D, Thanki MV, Ramalingaiah AH, Thirthalli J, Reddy RP, et al. Intermittent theta burst stimulation of cerebellar vermis enhances fronto-cerebellar resting state functional connectivity in schizophrenia with predominant negative symptoms: a randomized controlled trial. Schizophr Res [Internet]. 2021 [cited 2024 Jun 26];238:108–20. Available from: https://www.ncbi.nlm.nih.gov/pmc/articles/PMC8662658/
116. Bayrakçı A, Zorlu N, Karakılıç M, Gülyüksel F, Yalınçetin B, Oral E, et al. Negative symptoms are associated with modularity and thalamic connectivity in schizophrenia. Eur Arch Psychiatry Clin Neurosci. 2023;273(3):565–74.
117. Ohtani T, Sylvain B, Hosokawa T, Saito Y, Eckbo R, Ballinger T, Rausch AC, Melonakos ED, Kubicki M. Abnormalities in white matter connections between orbitofrontal cortex and

anterior cingulate cortex and their associations with negative symptoms in schizophrenia: a DTI study. Schizophr Res. 2014;157(1):190–7.
118. Kirschner M, Schmidt A, Hodzic-Santor B, Burrer A, Manoliu A, Zeighami Y, et al. Orbitofrontal-striatal structural alterations linked to negative symptoms at different stages of the schizophrenia spectrum. Schizophr Bull. 2021;47(3):849–63.
119. Walton E, Derrek HP, van Erp TGM, Potkin SG, Roiz Santiañez R, Crespo Facorro B, Suárez Pinilla P, van Haren NEM, de Zwarte SMC, Kahn RS, Cahn W, Doan NT, Jørgensen KN, Gurholt TP, Agartz I, Andreassen OA, Westlye LT, Melle I, Berg AO, Mørch Johnsen L, Færden A, Flyckt L, Fatouros Bergman H, Jönsson EG, Hashimoto R, Yamamori H, Fukunaga M, Jahanshad N, De Rossi P, Piras F, Banaj N, Spalletta G, Gur RE, Gur RC, Wolf DH, Satterthwaite TD, Beard LM, Sommer IEC, Koops S, Gruber O, Richter A, Krämer B, Kelly S, Donohoe G, McDonald C, Cannon DM, Corvin A, Gill M, Di Giorgio A, Bertolino A, Lawrie SM, Nickson T, Whalley HC, Neilson E, Calhoun VD, Thompson PM, Turner JA, Ehrlich S. Prefrontal cortical thinning links to negative symptoms in schizophrenia via the ENIGMA consortium. Psychol Med. 2017;48(1):82–94.
120. Roth RM, Matthew GA, Flashman LA, Koven NS, Pendergrass JC, Ford J, McAllister TW, Saykin AJ. Apathy is associated with ventral striatum volume in schizophrenia spectrum disorder. J Neuropsychiatry Clin Neurosci. 2016;28(3):191–4.
121. Caravaggio F, Gagan F, Iwata Y, Plitman E, Chung JK, Nakajima S, Mar W, Gerretsen P, Kim J, Chakravarty MM, Mulsant BH, Pollock BG, Mamo DC, Remington G, Graff Guerrero A. Amotivation is associated with smaller ventral striatum volumes in older patients with schizophrenia. Int J Geriatr Psychiatry. 2017;33(3):523–30.
122. Li Z, Yan C, Lv Q Yu, Yi Z Hui, Zhang J Ye, Wang J Hong, et al. Striatal dysfunction in patients with schizophrenia and their unaffected first-degree relatives. Schizophr Res [Internet]. 2018 [cited 2024 Jun 26];195:215–21. Available from: https://www.sciencedirect.com/science/article/pii/S0920996417305194
123. Bodnar M, Hovington CL, Buchy L, Malla AK, Joober R, Lepage M. Cortical thinning in temporo-parietal junction (TPJ) in non-affective first-episode of psychosis patients with persistent negative symptoms. PLoS One. 2014;9(6):e101372.
124. Mørch-Johnsen L, Ragnar N, Færden A, Haukvik UK, Jørgensen KN, Lange EH, Andreassen OA, Melle I, Agartz I. Brain structure abnormalities in first-episode psychosis patients with persistent apathy. Schizophr Res. 2015;164(1):59–64.
125. Kohler CG, Jeffrey WB, Martin EA, Healey KM, Moberg PJ. Facial emotion perception in schizophrenia: a meta-analytic review. Schizophr Bull. 2009;36(5):1009–19.
126. Kring AM, Ori E. Emotion deficits in people with schizophrenia. Annu Rev Clin Psychol. 2012;9(1):409–33.
127. Trémeau F. A review of emotion deficits in schizophrenia. Dialogues Clin Neurosci. 2006;8(1):59–70.
128. Tron T, Abraham P, Grinsphoon A, Weinshall D. BHI - Facial expressions and flat affect in schizophrenia, automatic analysis from depth camera data. In: 2016 IEEE-EMBS Int Conf Biomed Health Inform BHI; 2016. p. 220–3.
129. Kerr SL, John N, M. Emotion perception in schizophrenia: specific deficit or further evidence of generalized poor performance? J Abnorm Psychol. 1993;102(2):312–8.
130. Cohen AS, Sean MC, Brown LA, Minor KS. Towards a cognitive resource limitations model of diminished expression in schizotypy. J Abnorm Psychol. 2011;121(1):109–18.
131. Trémeau F, Dolores M, Duval F, Correa H, Hager Budny M, Coin Bariou L, Macher JP, Gorman JM. Facial expressiveness in patients with schizophrenia compared to depressed patients and nonpatient comparison subjects. Am J Psychiatry. 2005;162(1):92–101.
132. Varcin KJ, Phoebe BE, Henry JD. Empathic deficits in schizophrenia: the potential role of rapid facial mimicry. J Int Neuropsychol Soc JINS. 2010;16(4):621–9.
133. Lee JS, Won CJ, Yoon SY, Park HJ, Kim JJ. Involvement of the mirror neuron system in blunted affect in schizophrenia. Schizophr Res. 2013;152(1):268–74.

134. Hager OM, Matthias K, Bischof M, Riemer H, Matthias N, Kluge A, Seifritz E, Tobler P, N; Kaiser, Stefan. Reward-dependent modulation of working memory is associated with negative symptoms in schizophrenia. Schizophr Res. 2015;168(1):238–44.
135. Martínez A, Tobe RH, Gaspar PA, Malinsky D, Dias EC, Sehatpour P, et al. Disease-specific contribution of pulvinar dysfunction to impaired emotion recognition in schizophrenia. Front Behav Neurosci. 2021;15:787383.
136. Shaffer JJ, Michael PJ, McMahon MA, Bizzell J, Calhoun VD, van Erp TGM, Ford JM, Lauriello J, Lim KO, Manoach DS, McEwen S, Mathalon DH, O'Leary DS, Potkin SG, Preda A, Turner JA, Voyvodic JT, Wible CG, Belger A. Neural correlates of schizophrenia negative symptoms: distinct subtypes impact dissociable brain circuits. Mol Neuropsychiatry. 2015;1(4):191–200.
137. Gur RE, Christian KG, Ragland JD, Siegel SJ, Lesko K, Bilker WB, Gur RC. Flat affect in schizophrenia: relation to emotion processing and neurocognitive measures. Schizophr Bull. 2006;32(2):279–87.
138. Lepage M, Karine S, Benoit A, Czechowska Y, Dickie EW, Armony JL. Emotional face processing and flat affect in schizophrenia: functional and structural neural correlates. Psychol Med. 2011;41(9):1833–44.
139. Gur RE, James L, Kohler CG, Elliott MA, Lesko K, Ruparel K, Wolf DH, Bilker WB, Gur RC. Limbic activation associated with misidentification of fearful faces and flat affect in schizophrenia. Arch Gen Psychiatry. 2007;64(12):1356–66.
140. Rahm C, Benny L, Reckless GE, Ousdal OT, Melle I, Andreassen OA, Agartz I. Negative symptoms in schizophrenia show association with amygdala volumes and neural activation during affective processing. Acta Neuropsychiatr. 2015;27(4):213–20.
141. Mazza M, Alessia C, Pino MC, Giusti L, Nigri A, Pollice R, Roncone R, Casacchia M, Gallucci M. Dysfunctional neural networks associated with impaired social interactions in early psychosis: an ICA analysis. Brain Imaging Behav. 2013;7(3):248–59.
142. Orliac F, Mickael N, Joliot M, Delcroix N, Razafimandimby A, Brazo P, Dollfus S, Delamillieure P. Links among resting-state default-mode network, salience network, and symptomatology in schizophrenia. Schizophr Res. 2013;148(1):74–80.
143. Cohen AS, Jessica ME, Dinzeo TJ, Covington MA. Speech deficits in serious mental illness: a cognitive resource issue? Schizophr Res. 2014;160(1):173–9.
144. Docherty AR, Howard B, Kerns JG. Alogia and formal thought disorder: differential patterns of verbal fluency task performance. J Psychiatr Res. 2011;45(10):1352–7.
145. Joyce EM, Simon CL, Crichton P. Verbal fluency in schizophrenia: relationship with executive function, semantic memory and clinical alogia. Psychol Med. 1996;26(1):39–49.
146. Sumiyoshi C, Tomiki S, Nohara S, Yamashita I, Matsui M, Kurachi M, Niwa SI. Disorganization of semantic memory underlies alogia in schizophrenia: an analysis of verbal fluency performance in Japanese subjects. Schizophr Res. 2005;74(1):91–100.
147. Sevy S, Pierre LJ, Khan A, Ljuri I, Kulsa MKC, Jones O. Differential improvement of negative-symptom subfactors after cognitive remediation in low-functioning individuals with schizophrenia. Schizophr Res Cogn. 2019;19:100145.
148. Chan SKW, Liao Y, Hui CLM, Wong TY, Suen Y, Chang WC, et al. Longitudinal changes of cognitive function and its relationship with subdomains of negative symptoms in patients with adult-onset first-episode schizophrenia: a 4-year follow up study. Schizophr Res. 2023;252:181–8.
149. García-Mieres H, Lundin NB, Minor KS, Dimaggio G, Popolo R, Cheli S, et al. A cognitive model of diminished expression in schizophrenia: the interface of metacognition, cognitive symptoms and language disturbances. J Psychiatr Res [Internet]. 2020 [cited 2024 Apr 29];131:169–76. Available from: https://www.ncbi.nlm.nih.gov/pmc/articles/PMC8100971/
150. Bluhm R, Jodi M, Lanius RA, Osuch EA, Boksman K, Neufeld RWJ, Théberge J, Schaefer B, Williamson PC. Spontaneous low-frequency fluctuations in the BOLD signal in schizophrenic patients: anomalies in the default network. Schizophr Bull. 2007;33(4):1004–12.
151. Mingoia G, Gerd W, Langbein K, Maitra R, Smesny S, Dietzek M, Burmeister HP, Reichenbach JR, Schlösser RGM, Gaser C, Sauer H, Nenadic I. Default mode network

activity in schizophrenia studied at resting state using probabilistic ICA. Schizophr Res. 2012;138(2):143–9.
152. Geffen T, Hardikar S, Smallwood J, Kaliuzhna M, Carruzzo F, Böge K, et al. Striatal functional hypoconnectivity in patients with schizophrenia suffering from negative symptoms, longitudinal findings. Schizophr Bull. 2024;50:1337.
153. Gangl N, Conring F, Federspiel A, Wiest R, Walther S, Stegmayer K. Resting-state perfusion in motor and fronto-limbic areas is linked to diminished expression of emotion and speech in schizophrenia. Schizophr Heidelb Ger. 2023;9(1):51.
154. Kim GW, Hyeon KY, Jeong GW. Whole brain volume changes and its correlation with clinical symptom severity in patients with schizophrenia: a DARTEL-based VBM study. PLoS One. 2017;12(5):e0177251.
155. Lyne J, Brian O, Roche E, Renwick L, Cannon M, Clarke M. Negative symptoms of psychosis: a life course approach and implications for prevention and treatment. Early Interv Psychiatry. 2017;12(4):561–71.
156. Lyngstad SH, Strand GE, Engen MJ, Haatveit B, Ihler HM, Wedervang Resell K, Simonsen C, Melle I, Færden A. Trajectory and early predictors of apathy development in first-episode psychosis and healthy controls: a 10-year follow-up study. Eur Arch Psychiatry Clin Neurosci. 2020;270(6):709–22.
157. Dowd EC, Barch DM. Anhedonia and emotional experience in schizophrenia: neural and behavioral indicators. Biol Psychiatry. 2009;67(10):902–11.
158. Millan MJ, Fone K, Steckler T, Horan WP. Negative symptoms of schizophrenia: clinical characteristics, pathophysiological substrates, experimental models and prospects for improved treatment. Eur Neuropsychopharmacol [Internet]. 2014 [cited 2024 Apr 29];24(5):645–92. Available from: https://www.sciencedirect.com/science/article/pii/S0924977X14000935
159. Howes OD, Onwordi EC. The synaptic hypothesis of schizophrenia version III: a master mechanism. Mol Psychiatry [Internet]. 2023 [cited 2024 Jun 26];28(5):1843–56. Available from: https://www.nature.com/articles/s41380-023-02043-w
160. Goldsmith DR, Rapaport MH. Inflammation and negative symptoms of schizophrenia: implications for reward processing and motivational deficits. Front Psychiatry [Internet]. 2020 [cited 2024 Jun 26];11. Available from: https://www.frontiersin.org/journals/psychiatry/articles/10.3389/fpsyt.2020.00046/full
161. Goldsmith DR, Bekhbat M, Mehta ND, Felger JC. Inflammation-related functional and structural dysconnectivity as a pathway to psychopathology. Biol Psychiatry [Internet]. 2023 [cited 2024 Jun 26];93(5):405–18. Available from: https://www.sciencedirect.com/science/article/pii/S0006322322017152
162. Roalf DR, Figee M, Oathes DJ. Elevating the field for applying neuroimaging to individual patients in psychiatry. Transl Psychiatry [Internet]. 2024 [cited 2024 Jun 26];14(1):1–12. Available from: https://www.nature.com/articles/s41398-024-02781-7
163. Subbiah V. The next generation of evidence-based medicine. Nat Med [Internet]. 2023 [cited 2024 Jun 26];29(1):49–58. Available from: https://www.nature.com/articles/s41591-022-02160-z
164. Young JW. Development of cross-species translational paradigms for psychiatric research in the Research Domain Criteria era. Neurosci Biobehav Rev [Internet]. 2023 [cited 2024 Jun 26];148:105119. Available from: https://www.sciencedirect.com/science/article/pii/S014976342300088X
165. Meyer-Lindenberg A. The non-ergodic nature of mental health and psychiatric disorders: implications for biomarker and diagnostic research. World Psychiatry [Internet]. 2023 [cited 2024 Jun 26];22(2):272–4. Available from: https://onlinelibrary.wiley.com/doi/abs/10.1002/wps.21086

Negative Symptoms in Schizophrenia: The Role of Pharmacotherapy

6

Istvan Bitter and Pal Czobor

Abstract

This chapter provides psychiatrists with a synthesis of the latest evidence and clinical guidelines/guidance papers for the pharmacological treatment of negative symptoms of schizophrenia. Key definitions (e.g., primary vs. secondary negative symptoms) are reviewed in reference to companion chapters. A previous meta-analysis revealed a moderate placebo effect size (Cohen's d = 0.64) in trials examining negative symptoms. This is greater than the effect size observed in most antipsychotic trials for treating acute schizophrenia. This finding underscores the difficulty of identifying treatment effects for negative symptoms. There are limited pharmacological treatment options for primary negative symptoms. The most compelling evidence has been provided by cariprazine and amisulpride. Management of secondary negative symptoms is discussed with respect to depression, extrapyramidal side effects, and comorbidities. A frequent but poorly understood association exists between negative symptoms and diabetes mellitus. Treatment-resistant schizophrenia frequently includes persistent negative symptoms; strategies such as clozapine augmentation with dopamine partial agonists are reviewed. The chapter underscores the importance of low-dose antipsychotic regimens, individualization of care, and addressing neglected factors (e.g., hyperprolactinemia, metabolic diseases) to improve outcomes in negative symptoms of schizophrenia.

Keywords

Schizophrenia · Negative symptoms · Pharmacological treatment · Placebo response · Diabetes mellitus

I. Bitter (✉) · P. Czobor
Department of Psychiatry and Psychotherapy, Semmelweis University, Budapest, Hungary
e-mail: bitter.istvan@semmelweis.hu; czobor.pal@semmelweis.hu

A. Vita et al. (eds.), *Negative and Cognitive Symptoms in Schizophrenia*, Comprehensive Approach to Psychiatry 3,
https://doi.org/10.1007/978-3-032-06209-3_6

The primary objective of this chapter is to provide support for the work of practicing psychiatrists. The inclusion of methodological issues is intended to facilitate a better understanding of the relevant research data and to assist their translation into clinical practice. Researchers may also find this chapter useful in analyzing and planning negative symptom trials in schizophrenia and other psychotic disorders, or antipsychotic trials in any condition.

Other chapters in this book cover important aspects of negative symptoms. The first chapter of this book on "Negative symptoms in schizophrenia: definition, classification and course" by Silvana Galderisi [1] describes key terms used in this chapter, such as primary and secondary negative symptoms or persistent, predominant, or prominent symptoms. Secondary negative symptoms may be caused by antipsychotic treatment [2, 3], e.g., by producing anhedonia or other deficit symptoms, sedation, extrapyramidal symptoms, and depression. Higher striatal dopamine D2 receptor occupancy during antipsychotic treatment with second-generation antipsychotics was found to be positively associated with disturbances of subjective experiences [4] measured by the Subjective Well-Being Under Neuroleptic Treatment Scale [5] and the Subjective Deficit Syndrome Scale [6].

To prevent secondary negative symptoms and other adverse reactions of antipsychotic treatment, "…clinicians should strive to use the lowest possible dosage to control symptoms." [7]. Kapur et al. [8] confirmed earlier clinical recommendations [9] for low-dose antipsychotic treatment in schizophrenia based on their double-blind positron emission tomography scan study by concluding that a haloperidol equivalent dose in the range of 2–3 mg/day should be the optimal starting dose for first-episode patients. Low-dose antipsychotic treatment recommendations were also strongly supported by the 7-year follow-up data of a 2-year randomized clinical trial of first episode psychosis [10]. Wunderink et al. reported in this long-term follow-up study that the outcome in an early dose reduction group showed superior long-term recovery rates compared with the rates achieved with standard of care maintenance antipsychotic treatment. The haloperidol equivalent antipsychotic dose in the maintenance treatment group was between 3–4 mg/day, while in the early dose reduction group it was lower, between 2.0 and 2.5 mg/day [10].

We refer again to the description and definition of secondary negative symptoms in the chapter written by Galderisi in this book [1] and try to avoid unnecessary repetitions.

A large number of clinical trials with promising investigational drugs with good or even excellent preclinical and clinical data for the treatment of negative symptoms of schizophrenia failed or their results were negative (e.g., [11–13].

Assessing the potential causes of why so many clinical trials for the treatment of negative symptoms of schizophrenia have failed or been negative is beyond the scope of this chapter. Some methodological issues have been addressed in earlier publications, e.g., [14, 15]. Although it has been widely debated in the field, to our knowledge, no comprehensive review or analysis of the potential methodological causes of negative and failed negative symptom trials in schizophrenia has been published at the time of writing this chapter (April 2024).

In this chapter, we cover the following topics in five sections:

1. Placebo response in clinical trials of negative symptoms in schizophrenia
2. Pharmacological treatment of primary negative symptoms
3. Pharmacological treatment and prevention of secondary negative symptoms
   3.1. Treatment of negative symptoms secondary to positive symptoms
   3.2. Treatment and prevention of negative symptoms secondary to depressive symptoms
   3.3. Treatment and prevention of negative symptoms secondary to extrapyramidal symptoms (Parkinsonism)
   3.4. Treatment of negative symptoms secondary to /associated with diabetes mellitus
4. Pharmacological treatment of negative symptoms of patients with treatment-resistant schizophrenia
5. Summary of this Chapter

6.1 Placebo Response in Clinical Trials of Negative Symptoms in Schizophrenia

Significant placebo response has been repeatedly reported in negative symptom studies of schizophrenia, which seems to contradict to the generally accepted description of primary negative symptoms as persistent and poorly responding to treatment [16]. While our current reassessment of the evidence on placebo response in clinical trials on negative symptoms in schizophrenia resulted in major downward correction of an earlier effect size estimate of the placebo response in the negative symptoms of schizophrenia the effect size of placebo was still surprisingly high compared to the effect sizes of antipsychotics in the treatment of schizophrenia [17]. Our meta-analysis included 25 studies (21 add-on and 4 monotherapy studies). The pooled overall estimate for the placebo effect had a medium effect size, with a Cohen's d value of 0.64 (95% CIs: 0.46–0.83). The estimates were similar in the add-on and monotherapy studies [17]. To compare the effect size of 0.64 for the practice we cite here the magnitude of the superiority of antipsychotics over placebo in the acute phase of schizophrenia from a recent analysis, where antipsychotics were reported to have smaller effect size (0.47 (95% CIs: 0.42–0.51)) than placebo had in negative symptom trials [18]. In a large meta-analysis Fusar-Poli et al. found larger effect sizes for antipsychotic treatment of negative symptoms in schizophrenia than placebo-controlled antipsychotic trials for acute schizophrenia yielded for total symptom severity (which includes positive symptoms, negative symptoms and general psychopathology) [19]. The findings of their meta-analysis (effect sizes in placebo controlled trials in terms of a decrease of symptom severity at treatment end point compared to baseline were: "second-generation antipsychotics: −0.579 (−0.755 to −0.404); antidepressants: −0.349 (−0.551 to −0.146); combinations of pharmacological agents: −0.518 (−0.757 to −0.279)") and the opinion of the authors ("Conclusions and Relevance: Although some statistically significant effects on negative symptoms were evident, none reached the threshold for clinically significant improvement.") contributed to the already existing contradiction between the reported effect sizes in negative symptom trials and the widely held opinion of clinicians about the poor response of negative symptoms. This meta-analysis has been reviewed in detail in a European guidance paper on the treatment of negative symptoms of schizophrenia [16].

Large placebo response, as assessed by a psychometric evaluation of symptom change in negative symptoms, was the reason of the negative outcome of a current Phase III trial of pimavanserin versus placebo ("add-on" to existing antipsychotic treatment) in the treatment of negative symptoms in schizophrenia: "Pimavanserin did not demonstrate a statistically significant improvement over placebo on the study's primary endpoint, the change from baseline to week 26 on the Negative Symptom Assessment-16 (NSA-16) total score (−11.8 vs. −11.1; $p = 0.4825$; effect size = 0.07)." [13].

It has been demonstrated that both clinically and statistically significant placebo response rates can be observed in the treatment of negative symptoms of schizophrenia. Conversely, the most compelling currently available evidence indicates that "While *secondary negative symptoms* can improve as a consequence of treatment to improve symptoms in other domains (ie, positive symptoms, depressive symptoms, or extrapyramidal symptoms), *primary negative symptoms* generally do not respond well to currently available antipsychotic treatment with dopamine D2 antagonists or partial D2 agonists." [15]. (italic added by the authors of this chapter). However, a high placebo response observed in clinical trials sends an optimistic message that even a placebo might be beneficial in the treatment of negative symptoms for a significant proportion of patients. Moreover, various pharmacological and psychosocial interventions have demonstrated efficacy in the treatment of negative symptoms [15, 20, 21, 22].

6.2 Pharmacological Treatment of Primary Negative Symptoms

Unfortunately, there has been little progress in the pharmacological treatment of primary negative symptoms in the last few years. The current situation is not very encouraging either: in the first half of 2024 (when this chapter was written) bad news became available about two investigational drugs for negative symptoms. The development of pimavanserin for the treatment of negative symptoms in schizophrenia was terminated due to unfavorable study outcomes [13], and the US Food and Drug Administration did not approve the New Drug Application of roluperidone for negative symptoms of schizophrenia [23].

This section presents a brief overview of the key recommendations for *antipsychotic treatment* of primary negative symptoms. *Antidepressant treatment* of both primary and secondary negative symptoms will be described in the Sect. 6.3.

This section includes references to review and guidance papers, and to reports of clinical trials, that were specifically designed to investigate negative symptoms. Due to heterogeneity in the definition of negative symptoms, different studies may have included patients with differently defined (e.g., persistent, predominant, prominent) primary negative symptoms or with persistent secondary symptoms, as well as with a combination of primary and secondary negative symptoms. There are significant differences between the definitions of primary negative symptoms, which affect treatment outcomes; again, we refer to the chapter in this book, "Negative symptoms in schizophrenia: definition, classification, and course" written by Silvana Galderisi [1]. For consistency with the referenced publications on the efficacy of various drugs in the treatment of negative symptoms, we use the same terms for negative symptoms as they appear in the cited publications.

Based on multiple reviews and meta-analyses of the literature (e.g., [20, 24]), we briefly summarize the findings that are of interest for *predominant* negative symptoms:

1. Placebo-controlled studies: Amisulpride is the only antipsychotic that proved to be superior to placebo in the treatment of predominant negative symptoms, although the results were confounded by depression —depression scores also improved [25].
2. Active controlled ("face to face") studies:

   2.1. Olanzapine was superior to haloperidol in a very small study ($n = 35$) [26].
   2.2. Cariprazine was superior to risperidone in a study well controlled for secondary negative symptoms [27].

The results of these few positive studies support the proposal of Cerveri et al. for the treatment of negative symptoms in schizophrenia [28]. In case of first episode with predominant negative symptoms they recommend the following approach:

First line: cariprazine
Second line: amisulpride
Third line: quetiapine or olanzapine
Fourth line: consider antidepressants as add on

In case of chronic/relapsing schizophrenia with predominant negative symptoms, they propose: if the patient had received a first-generation antipsychotic, switch to a second generation antipsychotic—in case of no response, consider the use of the algorithm as described above for first episode patients (cariprazine—amisulpride—queatiapine or olanzapine—consider antidepressants as add on) [28].

We will address the pharmacological options for treatment-resistant patients with negative symptoms in Sect. 6.4 of this chapter.

6.3 Pharmacological Treatment and Prevention of Secondary Negative Symptoms

The first three subsections below summarize the treatment and possible prevention of "classic" secondary negative symptoms due to positive symptoms, due to depression, and due to extrapyramidal side effects of antipsychotics (Parkinsonism). The fourth (Sect. 6.3.4) is an emerging research topic of high importance: due to increased prevalence rates of diabetes mellitus in patients with schizophrenia and antipsychotic treatment, we would like to draw attention to this co-morbidity. It may be associated with both cognitive and negative symptoms and with an increase in the severity of cognitive symptoms in patients who had them before the onset of diabetes mellitus.

Many patients have a combination of primary and secondary negative symptoms [29] and a patient may have a cross-sectional presentation of a combination of secondary symptoms, e.g., associated with both depression and Parkinsonism. To the best of our knowledge, there is no published research on the treatment of such complex negative symptoms.

6.3.1 Treatment of Negative Symptoms Secondary to Positive Symptoms

Negative symptoms secondary to positive symptoms are synchronized with the positive symptoms (Fig. 6.1). In the case of synchronized negative symptoms, the goal is to provide effective treatment for the positive symptoms, since the negative symptoms would also remit when the positive symptoms completely remit. Desynchronized negative symptoms can have a variable long-term course, one type of which is shown in Fig. 6.2. Desynchronized negative symptoms tend to be persistent and often progressive, and are therefore generally classified as primary negative symptoms.

Careful analysis of the long-term course of positive and negative symptoms helps understand their relationship (e.g., synchronized vs. desynchronized), which may require different treatment strategies [30]. Cross-sectional analysis (or analysis of a relatively short period of time) of the relationship between positive and negative symptoms is of limited value in making treatment decisions, since a cross-sectional analysis cannot provide data for the differentiation between primary, persistent

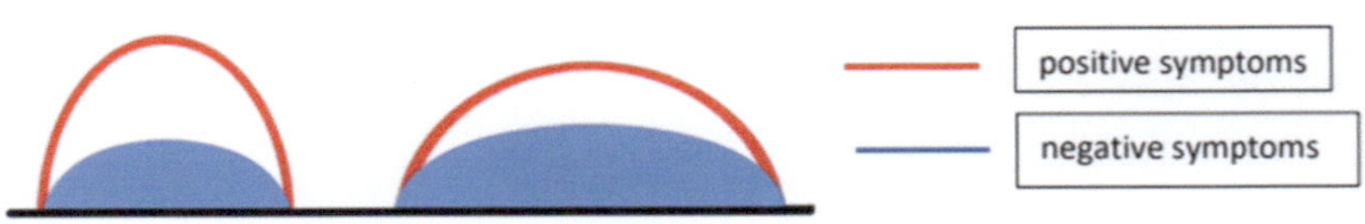

Fig. 6.1 Synchronized positive and negative symptoms during two episodes of schizophrenia with complete remission of the symptoms between the two episodes

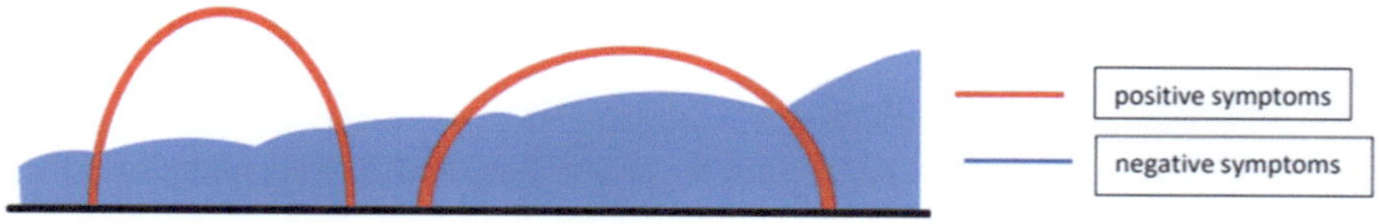

Fig. 6.2 Desynchronized positive and negative symptoms: There is a complete remission of the positive symptoms between the two episodes of positive symptoms. The negative symptoms started earlier than the positive symptoms and their course is independent of the course of the positive symptoms

negative symptoms in the presence of positive symptoms and negative symptoms secondary to positive symptoms.

Further details about the course of negative symptoms are described in Chap. 1 of this book [1].

6.3.2 Treatment and Prevention of Negative Symptoms Secondary to Depressive Symptoms

Major depressive episodes and subthreshold depressive symptoms that do not meet the full diagnostic criteria for a major depressive episode are common during the course of schizophrenia, affecting about two-thirds of patients [1, 29]. Depressive symptoms are generally associated with secondary negative symptoms. There are few studies on the pharmacological treatment of depression in schizophrenia, even fewer on the treatment of negative symptoms secondary to depression in schizophrenia. The results of treatment of secondary negative symptoms with antidepressants are controversial, and the effect sizes of such treatments are small [16, 31–33].

Before combining an antipsychotic with an antidepressant for the treatment of negative symptoms thought to be related to depression, it is recommended to check for drug–drug interactions and consider switching to another antipsychotic without liability for causing depression (i.e., no high-potency D2 dopamine receptor blockers are recommended).

Further studies are needed to confirm the usefulness of antidepressants in the treatment of negative symptoms of schizophrenia

Hyperprolactinemia has been associated with a variety of psychiatric and somatic symptoms, including anxiety, hostility, *depression,* and heart rate changes [34]. Maguire [35] reviewed papers describing a positive relationship between elevated prolactin levels and depression as well as between elevated prolactin levels and decreased libido and sexual dysfunctions, which further complicate the evaluation and treatment of negative symptoms in schizophrenia. Hyperprolactinaemia is a neglected side effect of antipsychotic treatment. The measurement of prolactin levels before the initiation of antipsychotic treatment and during antipsychotic treatment is recommended [35]. The association of prolactin with depression is further complicated by the prolactin-elevating side effects of antidepressants and by the findings that response to antidepressant treatment may be positively associated with hyperprolactinaemia [36].

Further research is needed to better understand the potential contribution of antipsychotic induced hyperprolactinemia to depression and to secondary negative symptoms due to depression during antipsychotic treatment. It is the author's recommendation that residency programs and continuing medical education programs in psychiatry incorporate available knowledge regarding the endocrinological aspects of pharmacologic treatment of psychiatric disorders into their curricula.

6.3.3 Treatment and Prevention of Negative Symptoms Secondary to Extrapyramidal Symptoms (Parkinsonism)

Motor symptoms (e.g., dyskinesia, parkinsonism, psychomotor poverty) may be part of the symptomatology of schizophrenia [37], and antipsychotic treatment may cause movement disorders (e.g., tardive dyskinesia) in addition to the parkinsonism. They are reported to be associated with negative symptoms.

In this section, we will limit our focus to Parkinsonism, given that it represents the most frequent motor side effect of antipsychotic treatment and is widely recognized as associated with modifiable secondary negative symptoms.

Mild (or "fine") motor symptoms may appear well before a patient develops "coarse" motor symptoms of Parkinsonism during antipsychotic drug treatment. Mild motor symptoms such as micrographia, difficulty buttoning a dress, changes in facial expression (hypomimia), difficulty playing sports (e.g., tennis or table tennis), or slower speech can be easily diagnosed as negative symptoms. Very low and low doses of high-potency antipsychotics (strong dopamine D2 receptor blockers), e.g., 2–3 mg/day of haloperidol, can cause motor symptoms of variable severity [8, 9].

The best strategy is to avoid extrapyramidal side effects by choosing antipsychotics with very low or low frequency of extrapyramidal side effects and using the lowest effective dose. Regular checking of extrapyramidal side effects, including mild hypomimia of the face, and manual dexterity (e.g., handwriting, buttoning clothes, tying shoes) could help in the early detection of "fine" motor side effects.

The use of anticholinergic drugs for the prevention and treatment of extrapyramidal side effects of antipsychotics is a controversial practice. Many patients do not develop extrapyramidal symptoms during antipsychotic treatment; thus, a preventive combination of anticholinergics cannot be recommended. Anticholinergic drugs are associated with a large number of central nervous system (including cognitive symptoms) and other (e.g., ocular, cardiac, gastrointestinal, urogenital) side effects. For our discussion, their frequently described cognitive side effects are of importance, since they overlap with negative symptoms [38]. A recent systematic review reported that 13 out of 17 studies "…found a statistically significant association between high anticholinergic burden and cognitive impairment (neurocognitive composite scores and individual cognitive domains such as learning and memory, executive function, processing speed), apart from a study, related to the specific characteristics of clozapine" [39].

Because of their side effect profile, anticholinergic medications should not be used for the prevention of extrapyramidal side effects of antipsychotic medications, and their use for the treatment of these side effects should be very limited and based on a careful risk-benefit analysis.

6.3.4 Treatment and Prevention of Negative Symptoms Secondary to/Associated with Diabetes Mellitus in Schizophrenia

Metabolic side effects—including diabetes mellitus—are frequent during antipsychotic treatments. The complex associations among schizophrenia, antipsychotic treatment, and diabetes mellitus are well documented in the literature, and the screening for diabetes mellitus is part of schizophrenia guidance and position papers, e.g., [40, 41]. Nevertheless, there has been a paucity of research investigating the relationship between negative symptoms of schizophrenia and diabetes mellitus. The following summary in this section is an attempt to illustrate the complex relationship between negative symptoms and type 2 diabetes mellitus and to emphasize that pharmacological treatment without psychosocial interventions may be insufficient for the treatment of negative symptoms in schizophrenia. In other words, only pharmacological treatment in combination with psychosocial interventions may be of maximum benefit to patients with negative symptoms in schizophrenia. The exclusion of psychotherapies, rehabilitation, and social interventions when testing drugs for the treatment of negative symptoms in schizophrenia is analogous to the prohibition of any exercise after surgery (including the assistance provided to patients to get out of the bed the first time after the surgery) when comparing the outcomes of a novel surgical method for hip replacement to those of an established technique. Based on the conclusions of failed and negative trials in the negative symptoms of schizophrenia over the past quarter of a century, new trial design(s) would be needed to adequately address at least the already identified problems, e.g., large placebo response of negative symptom trials. Proposals for new trial designs should be based on (re-)analyses of previous research data.

Diabetes mellitus is known to be associated with anxiety, depression, and suicide [42] and both hypoglycemia and hyperglycemia are associated with acute and chronic psychiatric symptoms, e.g., difficulties concentrating, altered consciousness, confusion, delirium, changes in psychomotor speed and thinking and cognitive impairments [43, 44].

A positive relationship was observed between inaccurate self-efficacy for managing one's diabetes *and* the presence of both diabetes mellitus and high levels of negative symptom severity and low cognitive functions [45]. In this study, self-efficacy was assessed based on hemoglobin A1C values. Wykes et al. recommended for these patients "…care management and monitoring, cognitive remediation, and skill training, to identify and correct inaccurate diabetes self-efficacy." [46].

Somatic (e.g., tiredness) and psychiatric symptoms (especially depression) of poor glycemic control are positively associated and/or are similar to negative and cognitive symptoms. Negative symptoms are known to be positively associated with inactivity, loneliness, smoking, and poor diet, which in turn are positively associated with obesity and poor glycemic control. Thus, we see a vicious circle based on the interactions between negative symptoms of schizophrenia and symptoms of diabetes mellitus. In addition to the administration of effective pharmacological agents, psychosocial interventions—including the encouragement of lifestyle modifications—are required to halt the progression of both illnesses and to facilitate improvement [46, 47].

6.4 Treatment of Negative Symptoms in Patients with Treatment Resistant Schizophrenia (TRS)

TRS has a variety of definitions, and it affects about one-third of people diagnosed with schizophrenia. A current expert consensus recommendation supported the following definition of TRS: failure of at least two treatment episodes with different antipsychotic drugs, each antipsychotic treatment episode should have lasted at least 6 weeks, at an "adequate" therapeutic dosage [48]. According to the authors of this consensus paper, "A patient's illness may meet criteria based on overall symptoms or based on specific subdomains of positive, negative, or cognitive symptoms."

Negative symptoms, especially persistent negative symptoms, are associated with poor symptomatological and functional outcomes in schizophrenia [49–51].

Clozapine remains the standard of care for TRS, even though a significant proportion (about two-thirds) of patients with TRS do not respond (or do not respond well) to clozapine and many of them have persistent negative symptoms [52]. The structure of negative symptoms remained stable during clozapine treatment of patients with TRS [53]. For clozapine non-responders, the combination of clozapine with dopamine partial agonists has been recommended [54] based on both pharmacological considerations [55] and on the excellent effectiveness of such a combination (clozapine plus aripiprazole) in preventing rehospitalization as evidenced by real-world data [56]. Aripiprazole is a frequently used dopamine D2 receptor partial agonist [57].

The successful augmentation of clozapine with cariprazine for the treatment of negative symptoms in patients with TRS has been documented in various publications, e.g., [58, 59]. Cariprazine is a partial agonist at serotonin 5-HT1A receptors and D2 and D3 receptors, with high selectivity for the D3 receptor [60]. Cariprazine was superior to risperidone in the treatment of predominant negative symptoms of schizophrenia in a randomized, double blind controlled trial [27].

The combination of cariprazine and clozapine represents a logical approach for the treatment of TRS with negative symptoms, which constitutes a challenging group of patients to treat.

6.5 Summary of This Chapter

The treatment of *persistent negative symptoms* remains a significant challenge in psychiatric care [16]. Nevertheless, the prevailing skepticism regarding the efficacy of the treatment of *all types of negative symptoms* associated with schizophrenia is inconsistent with the findings of recent research:

1. First: significant and clinically relevant *placebo responses* were detected in a large number of clinical trials of negative symptoms in schizophrenia [17].
2. Second: considerable effect sizes were found to be associated with various *pharmacological treatments* of negative symptoms [19, 20].
   and
3. Third: non-pharmacological interventions were found to be superior or significantly superior to treatment as usual [21].

1. Pharmacological treatment of primary negative symptoms

This section provides a concise overview of the evidence supporting treatment of primary (persistent) negative symptoms, emphasizing that first-generation antipsychotics should not be used for such treatment; cariprazine and, with some cautions, clozapine and amisulpride, can be considered as first-line pharmacological agents for this indication [28].

2. Pharmacological treatment and prevention of secondary negative symptoms

The guidance paper of the European Psychiatric Association on the treatment of negative symptoms in schizophrenia emphasized "it is important to note that treatment in the specific clinical situation of a patient presenting with negative symptoms, which are subsequently identified as being secondary, has not been evaluated in clinical trials. Therefore, treatment of depression, positive symptoms, and side effects should follow general principles for their management" [16].

A comprehensive overview of the potential causes of secondary negative symptoms in schizophrenia was published by Correll and Schooler in 2020 [15].

The considerable diversity of causes necessitates a comprehensive approach to intervention, encompassing medical, psychological, and social strategies. This is best achieved through a collaborative, multidisciplinary team comprising experts in various domains.

The pharmacological treatment of negative symptoms secondary to positive symptoms is recommended to follow the principle of using the lowest effective dose to prevent extrapyramidal side effects and depression due to pharmacological treatment [16]. In case of depression (fulfilling the criteria of a major depressive episode), the use of antidepressants should be considered. In order to avoid the development of Parkinsonism, it is essential that a second-generation antipsychotic drug with an appropriate dose be selected. It should be noted that the use of antiparkinsonian agents may result in a greater number of adverse effects.

This chapter introduces two new topics that are not typically included in publications on negative symptoms associated with schizophrenia: one on the relationship between diabetes mellitus and negative symptoms, the other on negative symptoms in treatment-resistant schizophrenia.

References

1. Galderisi S. Negative symptoms in schizophrenia: definition, classification and course. In: Vita A, Carpiniello B, Galderisi S, editors. Negative and cognitive symptoms in Schizophrenia. Springer, this volume.
2. Möller HJ. Neuroleptic treatment of negative symptoms in schizophrenic patients. Efficacy problems and methodological difficulties. Eur Neuropsychopharmacol. 1993;3(1):1.
3. Hartmann W. Neuroleptikabedingte pharmakogene Depressionen. In: Neuroleptika Rückschau 1952–1986, Künftige Entwicklungen: Möglichkeiten und Probleme der Neuroleptikatherapie 1987. Springer Berlin Heidelberg. pp 131–138.
4. de Haan L, Lavalaye J, Linszen D, Dingemans PM, Booij J. Subjective experience and striatal dopamine D2 receptor occupancy in patients with schizophrenia stabilized by olanzapine or risperidone. Am J Psychiatry. 2000;157(6):1019–20.
5. Naber D, Karow A, Lambert M. Subjective well-being under neuroleptic treatment and its relevance for compliance. Acta Psychiatr Scand. 2005;111:29–34.
6. Jaeger J, Bitter I, Czobor P, Volavka J. The measurement of subjective experience in schizophrenia: the Subjective Deficit Syndrome Scale. Compr Psychiatry. 1990;31(3):216–26.
7. Andreasen NC, Liu D, Ziebell S, Vora A, Ho BC. Relapse duration, treatment intensity, and brain tissue loss in schizophrenia: a prospective longitudinal MRI study. Am J Psychiatry. 2013;170(6):609–15.
8. Kapur S, Zipursky R, Jones C, Remington G, Houle S. Relationship between dopamine D2 occupancy, clinical response, and side effects: a double-blind PET study of first-episode schizophrenia. Am J Psychiatry. 2000;157(4):514–20.
9. Bitter I, Volavka J, Scheurer J. The concept of the neuroleptic threshold: an update. J Clin Psychopharmacol. 1991;11(1):28–33.
10. Wunderink L, Nieboer RM, Wiersma D, Sytema S, Nienhuis FJ. Recovery in remitted first-episode psychosis at 7 years of follow-up of an early dose reduction/discontinuation or maintenance treatment strategy: long-term follow-up of a 2-year randomized clinical trial. JAMA Psychiatry. 2013;70(9):913–20.
11. Bugarski-Kirola D, Blaettler T, Arango C, Fleischhacker WW, Garibaldi G, Wang A, Dixon M, Bressan RA, Nasrallah H, Lawrie S, Napieralski J. Bitopertin in negative symptoms of schizophrenia—results from the phase III FlashLyte and DayLyte studies. Biol Psychiatry. 2017;82(1):8–16.
12. Weiser M, Levi L, Burshtein S, Chiriţă R, Cirjaliu D, Gonen I, Yolken R, Davidson M, Zamora D, Davis JM. The effect of minocycline on symptoms in schizophrenia: results from a randomized controlled trial. Schizophr Res. 2019;(206):325–32.
13. Acadia Pharmaceuticals Announces Top-Line Results from Phase 3 ADVANCE-2 Trial of Pimavanserin in Negative Symptoms of Schizophrenia- Study Did Not Meet Primary Endpoint. https://acadia.com/media/news-releases/acadia-pharmaceuticals-announces-top-line-results-from-phase-3-advance-2-trial-of-pimavanserin-in-negative-symptoms-of-schizophrenia/. Last accessed 21 June 2024.
14. Aleman A, Lincoln TM, Bruggeman R, Melle I, Arends J, Arango C, Knegtering H. Treatment of negative symptoms: where do we stand, and where do we go? Schizophr Res. 2017;(186):55–62.

15. Correll CU, Schooler NR. Negative symptoms in schizophrenia: a review and clinical guide for recognition, assessment, and treatment. Neuropsychiatr Dis Treat. 2020;21:519–34.
16. Galderisi S, Kaiser S, Bitter I, Nordentoft M, Mucci A, Sabé M, Giordano GM, Nielsen MØ, Glenthøj LB, Pezzella P, Falkai P. EPA guidance on treatment of negative symptoms in schizophrenia. Eur Psychiatry. 2021;64(1):e21.
17. Czobor P, Kakuszi B, Bitter I. Placebo response in trials of negative symptoms in schizophrenia: a critical reassessment of the evidence. Schizophr Bull. 2022;48(6):1228–40.
18. Leucht S, Siafis S, Engel RR, Schneider-Thoma J, Bighelli I, Cipriani A, Furukawa TA, Davis JM. How efficacious are antipsychotic drugs for schizophrenia? An interpretation based on 13 effect size indices. Schizophr Bull. 2022;48(1):27–36.
19. Fusar-Poli P, Papanastasiou E, Stahl D, Rocchetti M, Carpenter W, Shergill S, McGuire P. Treatments of negative symptoms in schizophrenia: meta-analysis of 168 randomized placebo-controlled trials. Schizophr Bull. 2015;41(4):892–9.
20. Czobor P, Bitter I. Pharmacologic treatment of negative symptoms: focus on efficacy. In: Bitter I, editor. Managing negative symptoms of schizophrenia. New York: Oxford University Press; 2020. p. 67–86.
21. Savill M. Psychosocial/non-pharmacologic treatment of negative symptoms: focus on efficacy. In: Bitter I, editor. Managing negative symptoms of schizophrenia. New York: Oxford University Press; 2020. p. 87–106.
22. Pillny M, Lincoln T. Negative symptoms in schizophrenia: the role of psychosocial rehabilitation interventions. In: Vita A, Carpiniello B, Galderisi S, editors. Negative and cognitive symptoms in Schizophrenia. Springer, this volume.
23. Minerva neurosciences receives complete response letter from FDA for New drug application for Roluperidone for the treatment of negative symptoms in patients with Schizophrenia. https://ir.minervaneurosciences.com/news-releases/news-release-details/minerva-neurosciences-receives-complete-response-letter-fda-new. Last accessed 21 June 2024.
24. Krause M, Zhu Y, Huhn M, Schneider-Thoma J, Bighelli I, Nikolakopoulou A, Leucht S. Antipsychotic drugs for patients with schizophrenia and predominant or prominent negative symptoms: a systematic review and meta-analysis. Eur Arch Psychiatry Clin Neurosci. 2018;268:625–39.
25. Danion JM, Rein W, Fleurot O. Improvement of schizophrenic patients with primary negative symptoms treated with Amisulpride. Amisulpride Study Group. Am J Psychiatry. 1999;156(4):610–6.
26. Lindenmayer JP, Khan A, Iskander A, Abad MT, Parker B. A randomized controlled trial of olanzapine versus haloperidol in the treatment of primary negative symptoms and neurocognitive deficits in schizophrenia. J Clin Psychiatry. 2007;68(3):368–79.
27. Németh G, Laszlovszky I, Czobor P, Szalai E, Szatmári B, Harsányi J, Barabássy Á, Debelle M, Durgam S, Bitter I, Marder S. Cariprazine versus risperidone monotherapy for treatment of predominant negative symptoms in patients with schizophrenia: a randomised, double-blind, controlled trial. Lancet. 2017;389(10074):1103–13.
28. Cerveri G, Gesi C, Mencacci C. Pharmacological treatment of negative symptoms in schizophrenia: update and proposal of a clinical algorithm. Neuropsychiatr Dis Treat. 2019;5:1525–35.
29. Bitter I. Definitions and measurement of negative symptoms in schizophrenia. In: Bitter I, editor. Managing negative symptoms of schizophrenia. New York: Oxford University Press; 2020. p. 1–18.
30. Smulevich A, Romanov D. Long term course of negative symptoms. In: Bitter I, editor. Managing negative symptoms of schizophrenia. New York: Oxford University Press; 2020. p. 39–50.
31. Gregory A, Mallikarjun P, Upthegrove R. Treatment of depression in schizophrenia: systematic review and meta-analysis. Br J Psychiatry. 2017;211(4):198–204.
32. Helfer B, Samara MT, Huhn M, et al. Efficacy and safety of antidepressants added to antipsychotics for schizophrenia: a systematic review and meta-analysis. Am J Psychiatry. 2016;173(9):876–86.

33. Novitsky MA, De Sousa A, Asadullin AR, Gavrilyuk OA, Petrov AV, Nasyrova RF. Possibilities and limitations of antidepressant use to correct depressive and negative symptoms in schizophrenia. Pers Psychiatry Neurol. 2021;1(2):21–45.
34. Elgellaie A, Larkin T, Kaelle J, Mills J, Thomas S. Plasma prolactin is higher in major depressive disorder and females, and associated with anxiety, hostility, somatization, psychotic symptoms and heart rate. Compr Psychoneuroendocrinol. 2021;6:100049.
35. Maguire GA. Prolactin elevation with antipsychotic medications: mechanisms of action and clinical consequences. J Clin Psychiatry. 2002;63(Suppl 4):S56–62.
36. Faron-Górecka A, Kuśmider M, Solich J, Kolasa M, Szafran K, Żurawek D, Pabian P, Dziedzicka-Wasylewska M. Involvement of prolactin and somatostatin in depression and the mechanism of action of antidepressant drugs. Pharmacol Rep. 2013;65(6):1640–6.
37. Walther S, Strik W. Motor symptoms and schizophrenia. Neuropsychobiology. 2012;66(2):77–92.
38. Marzoughi S, Banerjee A, Jutzeler CR, Prado MA, Rosner J, Cragg JJ, Cashman N. Tardive neurotoxicity of anticholinergic drugs: a review. J Neurochem. 2021;158(6):1334–44.
39. Georgiou R, Lamnisos D, Giannakou K. Anticholinergic burden and cognitive performance in patients with schizophrenia: a systematic literature review. Front Psych. 2021;12:779607.
40. De Hert M, Dekker JM, Wood D, Kahl KG, Holt RI, Möller HJ. Cardiovascular disease and diabetes in people with severe mental illness position statement from the European Psychiatric Association (EPA), supported by the European Association for the Study of Diabetes (EASD) and the European Society of Cardiology (ESC). Eur Psychiatry. 2009;24(6):412–24.
41. Cooper SJ, Reynolds GP, Barnes TRE, England E, Haddad PM, Heald A, Holt RIG, Lingford-Hughes A, Osborn D, McGowan O. BAP guidelines on the management of weight gain, metabolic disturbances and cardiovascular risk associated with psychosis and antipsychotic drug treatment. J Psychopharmacol. 2016;30(8):717–48.
42. Robinson DJ, Coons M, Haensel H, Vallis M, Yale JF. DDiabetes Canada Clinical Practice Guidelines Expert Committee. Diabetes and mental health. Can J Diabetes. 2018;42:S130–41.
43. Amiel SA. The consequences of hypoglycaemia. Diabetologia. 2021;64:963–70.
44. De Groot M, Golden SH, Wagner J. Psychological conditions in adults with diabetes. Am Psychol. 2016;71(7):552.
45. Wykes TL, Lee AA, McKibbin CL, Laurent SM. Self-efficacy and hemoglobin A1C among adults with serious mental illness and type 2 diabetes: the roles of cognitive functioning and psychiatric symptom severity. Psychosom Med. 2016;78(3):263–70.
46. Vancampfort D, Firth J, Correll CU, Solmi M, Siskind D, De Hert M, Carney R, Koyanagi A, Carvalho AF, Gaughran F, Stubbs B. The impact of pharmacological and non-pharmacological interventions to improve physical health outcomes in people with schizophrenia: a meta-review of meta-analyses of randomized controlled trials. Focus. 2021;19(1):116–28.
47. Girdler SJ, Confino JE, Woesner ME. Exercise as a treatment for schizophrenia: a review. Psychopharmacol Bull. 2019;49(1):56.
48. Howes OD, McCutcheon R, Agid O, De Bartolomeis A, Van Beveren NJ, Birnbaum ML, Bloomfield MA, Bressan RA, Buchanan RW, Carpenter WT, Castle DJ. Treatment-resistant schizophrenia: treatment response and resistance in psychosis (TRRIP) working group consensus guidelines on diagnosis and terminology. Am J Psychiatry. 2017;174(3):216–29.
49. Foussias G, Remington G. Negative symptoms in schizophrenia: avolition and Occam's razor. Schizophr Bull. 2010;36(2):359–69.
50. Fervaha G, Foussias G, Agid O, Remington G. Motivational and neurocognitive deficits are central to the prediction of longitudinal functional outcome in schizophrenia. Acta Psychiatr Scand. 2014;130(4):290–9.
51. Rabinowitz J, Levine SZ, Garibaldi G, Bugarski-Kirola D, Berardo CG, Kapur S. Negative symptoms have greater impact on functioning than positive symptoms in schizophrenia: analysis of CATIE data. Schizophr Res. 2012;137(1–3):147–50.
52. Siskind D, Siskind V, Kisely S. Clozapine response rates among people with treatment-resistant schizophrenia: data from a systematic review and meta-analysis. Can J Psychiatry. 2017;62(11):772–7.

53. Wolpe N, Vituri A, Jones PB, Shahar M, Fernandez-Egea E. The longitudinal structure of negative symptoms in treatment resistant schizophrenia. Compr Psychiatry. 2024;128:152440–54.
54. Lähteenvuo M, Tiihonen J. Antipsychotic polypharmacy for the management of schizophrenia: evidence and recommendations. Drugs. 2021;81(11):1273–84.
55. Hjorth S. The more, the merrier…? Antipsychotic polypharmacy treatment strategies in schizophrenia from a pharmacology perspective. Front Psych. 2021;12:760181.
56. Tiihonen J, et al. Association of antipsychotic polypharmacy vs monotherapy with psychiatric rehospitalization among adults with schizophrenia. JAMA Psychiatry. 2019;76(5):499–507.
57. Kikuchi T, Maeda K, Suzuki M, Hirose T, Futamura T, McQuade RD. Discovery research and development history of the dopamine D2 receptor partial agonists, aripiprazole and brexpiprazole. Neuropsychopharmacol Reports. 2021;41(2):134–43.
58. Oloyede E, Clark I, Mace S, Whiskey E, Taylor D. Clozapine augmentation with cariprazine for negative symptoms: a case series and literature review. Ther Adv Psychopharmacol. 2022;12:20451253211066642.
59. Pappa S, Kalniunas A, Sharma H, Raza-Syed A, Kamal M, Larkin F. Efficacy and safety of cariprazine augmentation in patients treated with clozapine: a pilot study. Ther Adv Psychopharmacol. 2022;12:20451253221132087.
60. Kiss B, Horváth A, Némethy Z, Schmidt É, Laszlovszky I, Bugovics G, Fazekas K, Hornok K, Orosz S, Gyertyán I, Agai-Csongor E. Cariprazine (RGH-188), a dopamine D3 receptor-preferring, D3/D2 dopamine receptor antagonist–partial agonist antipsychotic candidate: in vitro and neurochemical profile. J Pharmacol Exp Ther. 2010;333(1):328–40.

Psychological Interventions for Negative Symptoms of Schizophrenia

7

Matthias Pillny, Kerem Böge, and Clementine J. Edwards

Abstract

Negative symptoms are among the most disabling aspects of schizophreniaoften persisting despite pharmacological treatment and contributing significantly to poor functional outcomes. In recent years, psychological interventions have emerged as essential tools for targeting these symptoms. Cognitive behavioural therapy (CBT) is the most widely studied approach, showing modest but consistent benefits in improving motivation and social functioning. More targeted interventions, such as the cognitive behavioural social skills training (CBSST) and the motivation and enhancement training (MOVE), have demonstrated promise in addressing low functioning . Cognitive Remediation Therapy (CRT)

M. Pillny (✉)
Clinical Psychology and Psychotherapy, Faculty for Psychology and Human Movement Science, Universität Hamburg, Hamburg, Germany
e-mail: matthias.pillny@uni-hamburg.de

K. Böge
Department of Psychiatry and Psychotherapy, Campus Benjamin Franklin, Charité – Universitätsmedizin Berlin, Berlin, Germany

German Center of Mental Health (DZPG), Berlin, Germany

Medical University Brandenburg, Brandenburg, Germany

C. J. Edwards
Department of Psychology, Institute of Psychiatry, Psychology & Neuroscience, Kings College London, London, UK

South London & Maudsley NHS Foundation Trust, London, UK

A. Vita et al. (eds.), *Negative and Cognitive Symptoms in Schizophrenia*, Comprehensive Approach to Psychiatry 3,
https://doi.org/10.1007/978-3-032-06209-3_7

has been effective in improving cognitive domains associated with negative symptoms, particularly when combined with functional skills training. Emerging strategies, including behavioural activation, mindfulness-based approaches, and motivational interviewing, are gaining traction as adjunctive therapies. Despite encouraging evidence, the heterogeneity of negative symptoms and overlap with cognitive and depressive features pose challenges in treatment design and outcome measurement. Personalized, multi-component interventions that integrate psychological therapies with functional and vocational support show the greatest potential for sustained improvement. Continued research is needed to refine these approaches and identify predictors of treatment response to optimize care for individuals with schizophrenia.

Keywords

Negative symptoms · Amotivation · Anhedonia · Psychological therapy · Evidence-based treatment

7.1 Introduction

As laid out in the previous chapters of this book, negative symptoms represent a group of symptoms in schizophrenia that are characterized by amotivational psychopathology, and lead to marked reductions in goal-directed and social behaviour [90]. These include avolition, anhedonia, and reduced affective expressions in speech, mimic, and gestures. Negative symptoms are highly prevalent, with 45–80% of people diagnosed with schizophrenia experiencing them [52] and are strongly associated with reduced engagement in daily-life activity (e.g. [74]), a decline in psychosocial functioning [30, 71], and low quality of life [91]. Nevertheless, to date, the development of psychological interventions has primarily focused on positive symptoms. Although these interventions have been proven effective in reducing positive symptoms [88, 93], it is apparent that remission of positive symptoms often does not correspond with a reduction of negative symptoms or improved psychosocial functioning [1, 15], indicating that negative symptoms constitute a separate target within the treatment of people with schizophrenia.

The development of psychosocial interventions for addressing the negative symptoms of schizophrenia has, therefore, increased significantly in recent years. Although many interventions were initially not designed explicitly to target negative symptoms, these have been shown to effectively alleviate negative symptoms, albeit yielding only small effect sizes [18, 78]. These approaches leverage interventions from cognitive behavioural therapy targeting the vicious cycle between the lack of initiation or persistence in goal-directed behaviour, negative affect, and demotivating beliefs linked to amotivational psychopathology (e.g. 'I don't have what it takes to do this'; [70]), and social skills training targeting diminished affective expression, negative interaction outcomes, and social isolation (e.g. [79]). Moreover, cognitive remediation programmes (e.g. [11, 69]), have been tested for their effectiveness on negative symptoms and are now specifically recommended for managing negative symptoms [99]. In contrast, more recent developments have shifted toward an increasing number of newly derived interventions focusing directly on negative symptoms and their underlying (neuro)psychological pathomechanisms as the primary intervention targets. Particularly, these interventions aim to target deficits in functions of the reward system, such as diminished pleasure/reward anticipation [47], mental representations of positive experiences (e.g. [68]), impairments in autobiographical memory for positive events (e.g. [27]), effort-based decision making [5, 73], and diminished positive reinforcement learning (e.g. [56]).

While these recent developments are promising in their translational and targeted approach to addressing negative symptoms in psychological therapy, it should be noted that the older yet rather unspecific interventions are based on a broad body of evidence compared to the more recent developments and, therefore, represent a solid starting point in the treatment of patients with negative symptoms of schizophrenia. This chapter aims to provide an overview of tried and tested approaches as well as to review the latest developments in psychological interventions for patients with negative symptoms. In each case, we will identify the (neuro)psychological target mechanisms, outline the basic therapeutic procedure and review the available evidence for the respective approach. Lastly, we will discuss current and future challenges in the treatment of patients with negative symptoms.

7.2 Cognitive Behavioural Therapy (CBT)

Cognitive-behavioural therapy (CBT) stands out as the most frequently employed psychological approach for addressing the negative symptoms of schizophrenia. CBT approaches integrate cognitive and behavioural techniques within a comprehensive framework that underscores the interrelated nature of thought, action, and emotion. This framework gives rise to two main targets for CBT interventions: low engagement in goal-directed behaviour and demotivating beliefs related to negative symptoms.

Reduced goal-directed behaviour may be directly addressed by behavioural techniques aiming to leverage the link between behaviour (i.e. positive/rewarding activity) and positive emotion. The most prominent technique in this domain is behavioural activation, grounded in the idea that reduced activity contributes to a lack of positive reinforcements, resulting in a dysfunctional downward spiral of negative mood, states of low motivation, and low activity engagement. Behavioural activation aims to tackle this negative cycle by actively selecting, scheduling, and monitoring daily activity with a focus on pleasurable and rewarding activities. Traditional behavioural activation involves choosing from a standardized list of pleasurable activities. Still, some newer iterations of behavioural activation advocate for a more personalized approach, encouraging the selection of activities that align with the patient's personal values and meaningful goals in life. Others propose focusing on prosocial activities or incorporating translational aspects from neurocognitive science, such as savouring and anticipating pleasure experiences (for more details, see [83]).

Demotivating beliefs can be addressed with cognitive techniques targeting these beliefs and their interrelation with action and emotion. These interventions include interview techniques such as Socratic Dialogue or a systematic consideration of pros and cons. The ultimate goal of these cognitive interventions is to empower patients to develop alternative, more adaptive interpretations or beliefs (e.g. shifting from 'I won't enjoy this anyway, so why try it?' to 'I can't know whether or not I will enjoy this, so I should try it'), that might be more helpful in facilitating positive mood and motivation to engage in goal-directed behaviour.

Behavioural activation and cognitive therapy are often integrated into behavioural experiments in CBT. For instance, patients are asked to participate in positive activities and to compare the previously anticipated pleasure against the actual level of pleasure experienced. Therapists can pinpoint mismatches, drawing attention to the cognitive triad of thought, action, and emotion. This process aims to prompt the patient to re-evaluate their automatic and negative thoughts in a given situation and to challenge their demotivating beliefs in the long run.

7.2.1 Evidence for CBT for Reducing Negative Symptoms

A meta-review comprising ten meta-analyses examining the effect of CBT approaches on negative symptoms found small to moderate effect sizes of CBT in reducing negative symptoms when compared with treatment-as-usual (TAU), but no differences between CBT and other psychological treatments [18]. Notably, just one meta-analysis included only studies in which patients were selected who exceeded a minimum criterion of elevated negative symptoms [78]. Despite a few included studies, this meta-analysis did not find a significant effect of CBT on reducing overall negative symptoms compared to TAU. Post-hoc analyses, however, revealed a significant reduction of the negative symptoms of the Experiential dimension and no effect on expressive negative symptoms associated with CBT. This suggests that the benefits that can be expected from CBT on negative symptoms might be specific to the experiential dimension of negative symptoms.

The evidence for the effect of behavioural activation as a stand-alone-treatment is vague. A recent systematic review identified only two feasibility trials examining the effect of behavioural activation on reducing negative symptoms [64]. Nevertheless, both studies found behavioural activation to be safe and well-accepted by patients with schizophrenia and indicated promising effects on the reduction of overall negative symptoms.

To date, three randomized controlled trials have tested the effect of cognitive therapies against treatment as usual on negative symptoms. All three trials revealed small to moderate effects of cognitive therapy, when compared to TAU, on negative symptoms of the experiential dimension, but not on negative symptoms of the expressive dimension [44, 45, 89].

7.3 Social Skills Training

Social Skills Training (SST) aims to enhance a person's interpersonal and communication abilities by developing the necessary skills and behaviours required for assertive social interactions in various contexts. The training typically involves interventions that address specific social challenges, such as assertiveness, active listening, non-verbal communication, and conflict solving. Therapists are encouraged to use a combination of didactic instructions, role-playing exercises, and real-world practice scenarios, in which patients can learn and refine their social skills. Generally, social skill training aims to improve social functioning, relationships, and overall quality of life.

The existing literature discusses various potential mechanisms of change in SST in alleviating negative symptoms. One assumption is that people with negative symptoms exhibit deficits in social skills that hamper and impede proactive and assertive social behaviour. Supporting evidence for this notion is derived from studies suggesting that people with relevant negative symptoms of schizophrenia demonstrate diminished social affiliation skills [4], potentially resulting in social rejection, which further increases anhedonia and amotivation in social contexts [53, 72]. As social skill deficits might not be evident in all patients with negative symptoms of schizophrenia, an alternative route may be through enhanced social engagement associated with positive affiliative experiences and improved social functioning [2]. Lastly, the extensively used role plays in SST provide a unique opportunity to test out beliefs that can interfere with successful skills practice in the community. As such, role plays can be used as behavioural experiments during sessions to elicit and test beliefs such as 'I won't be able to do it well' or 'It is not worth trying'. Consequently, SST could contribute to actual skill acquisition or activate existing but underutilized social and assertive behaviours.

7.3.1 Evidence for SST for Reducing Negative Symptoms

A fair number of meta-analyses examined the effect of SST approaches on negative symptoms, reporting moderate effects of SST on the reduction of negative symptoms when compared to TAU or active control groups [18]. Notably, one of these meta-analyses suggests that SST may even outperform other active psychological treatments in reducing negative symptoms [93]. Furthermore, SST has been found to yield similar positive effects on measures of social skills performance as well as social and daily-living skills, suggesting these as potential mechanisms of change underlying the effect of SST on negative symptoms [59].

7.4 Cognitive Remediation

Cognitive remediation (CR), also known as cognitive rehabilitation, aims to improve neuropsychological functions related to everyday functioning, such as attention, executive functions, and memory [12]. Cognitive remediation is thought to reduce negative symptoms by targeting their potential cognitive underpinnings, including reward processing abnormalities, working memory, problem-solving, and planning [16]. This is achieved by exposing patients to tasks that directly target and train the specific neuropsychological deficits related to negative symptoms [69, 96]. For example, a particular task within cognitive remediation might involve a memory exercise where patients are asked to remember and recall a series of numbers, words, or images. The difficulty level can be adjusted based on the person's current cognitive abilities, gradually increasing as they demonstrate improvement. The difficulty of these tasks should be tailored to each person to promote errorless learning through repeated practice, scaffolding, and personalized feedback. CR can be administered through traditional methods, such as paper and pencil, or via computerized platforms, including mobile and virtual reality applications [54]. It is guided by the principle of massed practice, which recommends an intensive frequency of several sessions per week. The guidance of a CR therapist is crucial, as they select tasks of appropriate difficulty, offer positive feedback, and motivate patients to participate in the treatment actively.

7.4.1 Evidence for Cognitive Remediation

A solid body of evidence attests to the efficacy of CR. The most recent and comprehensive meta-analysis included 130 studies comprising 8851 patients and reported moderate effect sizes on cognitive and functional outcomes [97], which also extends to negative symptoms [17]. Moreover, the findings of the meta-analysis by Vita et al. revealed several aspects that might improve the effect of CR on cognitive and everyday functioning. These include the presence of an active and trained therapist, the implementation of CR in structured psychiatric rehabilitation, and comprehensive training programmes addressing all core elements of neurocognitive functioning. On the patient level, Vita et al., found that fewer years of education, lower premorbid IQ, and higher baseline symptom severity were associated with larger effect sizes, suggesting that these patient subgroups might benefit most from CR and could showcase a specific treatment indication.

7.5 Mindfulness- and Acceptance-Based Interventions

Mindfulness- and acceptance-based interventions draw upon contemplative traditions, particularly mindfulness meditation, and have been adapted into secular, therapeutic approaches. Today, they are considered as an integral part of the 'third wave' of CBT. The first developments originated in the 1980s and included a large body of heterogeneous therapy modalities such as mindfulness-based stress reduction (MBSR), Mindfulness-based cognitive therapy (MBCT), acceptance- and commitment therapy (ACT), or dialectical behavioral therapy (DBT), among many others [55, 80]. The common aim of these approaches is to cultivate a non-judgmental awareness of the present moment while promoting focused attention to one's thoughts, feelings, bodily sensations, and the surrounding environment [21]. This involves an active, deliberate, and conscious perception and attitude towards internal and external, whether pleasant or unpleasant experiences, without trying to avoid or change them [50].

Mindfulness- and acceptance-based interventions can be distinguished by their respective methods of intervention, such as content focus, mechanisms of action, relevant processes, therapeutic implementation, and setting or treatment dosage [81]. MBSR- and MBCT-like programmes focus more on systematic and sustained training in formal and informal mindfulness meditation practices. These are often delivered in group formats, in conjunction with psychoeducation on attention allocation, emotion, and stress regulation, self-compassion, and metacognition. In contrast, interventions in ACT are primarily designed for individual settings. These have an increased focus on a value-driven approach that may lead to a more fulfilling life. ACT interventions target experiential avoidance, psychological inflexibility, and maladaptive beliefs and aim to foster goal-directed behaviour that aligns with the patient's personal values and goals in life [49].

7.5.1 Evidence for Mindfulness- and Acceptance-Based Interventions

Mindfulness- and acceptance-based interventions may target some of the putative psychological mechanisms related to negative symptoms. These include attenuated mindfulness and acceptance and psychological inflexibility [8]. Indeed, these interventions show a solid evidence base for a range of mental health conditions [35]. Still, although disorder-specific adaptations of mindfulness-based interventions are increasingly available, long-standing concerns persist regarding their potential to exacerbate psychotic symptoms in people with schizophrenia-spectrum disorders [10]. These apprehensions reflect deep-seated reservations of clinicians about the safety and appropriateness of such interventions for this population. In contrast, a growing body of evidence demonstrates the feasibility, acceptability, safety, and efficacy of mindfulness-based interventions for people with schizophrenia spectrum disorders [6, 66, 81]. Overall, outcomes of previous meta-analyses suggested a number of potential benefits in people with schizophrenia-spectrum disorders associated with mindfulness- and acceptance-based interventions [20, 61, 76, 82, 92]. These include reductions in negative, positive, and depressive symptoms, as well as increased well-being and social functioning, and decreased hospitalization rates. A subgroup analysis of another meta-analysis suggested that interventions delivered in groups seem superior to individual settings in reducing negative symptoms and improving social functioning (Jansen et al. 2019). In addition, sustained effects of mindfulness-based interventions have been shown for alleviating residual negative symptoms and cognitive impairment in people with schizophrenia [87], when conducted in inpatient settings [7, 9] or when combining mindfulness-based group therapy with intranasal oxytocin as a form of psychotherapeutic-pharmacological augmentation treatment [101].

However, despite these promising findings, it must be acknowledged that most studies have small sample sizes, show a high risk of bias, and investigated quite heterogeneous interventions, which limits the generalizability of the results [92, 100]. Nonetheless, the authors of a recent review concluded that the latest trials present fewer risks of bias and more homogeneous programmes suggesting an evolving field of evidence-based practice [81].

7.6 Translational and Integrated Treatment Programmes

7.6.1 Cognitive Behavioural Social Skills Training

Cognitive Behavioural Social Skills Training (CBSST) combines classical CBT techniques with social skills training. It comprises three six-session modules, namely the cognitive skills module, social skills module, and problem-solving skills module [40]. The cognitive skills module focuses on training thought-challenging skills used throughout subsequent modules. Participants are introduced to the general concepts of CBT, including the relationship between thoughts, actions, and feelings (generic cognitive model), automatic thoughts, thought challenges through behavioural experiments, and examining evidence for beliefs and mistakes in thinking. The primary thought-challenging skill trained is the 3C's: Catch It, Check It, Change It (where 'It' is an unhelpful thought). The social skills module aims to improve communication skills through behavioural role plays, including active listening, expressing positive and negative feelings, and making positive requests. Role plays include assertive interactions with co-workers, friends, and family, making new friends, and effectively interacting with case managers, doctors, and other support persons. The problem-solving skills module aims to improve basic problem-solving skills using the acronym *SCALE: S*pecify the problem, *C*onsider all possible solutions, *A*ssess the best solution, *L*ay out a plan, and *E*xecute and *E*valuate the outcome. The focus is on developing specific and feasible plans to solve real-world problems, including scheduling pleasant activities, improving living situations, handling finances, using public transportation, finding a volunteer or paid job, and enrolling in classes.

Prior research has demonstrated that, compared to treatment as usual or goal-focused supportive contacts, CBSST effectively improves functioning in people with schizophrenia at post-treatment and 1-year follow-ups [37, 38, 41–43]. In addition, the study by Granholm et al. [38] found significantly greater improvement in negative symptoms related to the experiential dimension with CBSST than with goal-directed supportive contact. Studies investigating the putative mechanisms of change in CBSST found significant reductions in defeatist performance beliefs to predict improvement in negative symptoms and functioning [37, 39]. Moreover, a recent effectiveness trial using ecological momentary assessment of outcomes found that CBSST delivered in a routine care setting by assertive community treatment teams leads to greater improvements in daily-life functioning and negative symptoms than TAU [75].

7.6.2 Motivational Interviewing Enhanced Cognitive Behavioural Therapy

Motivational interviewing aims to consolidate motivation for change by employing specifically tailored non-directive dialogue techniques [63]. This approach was originally developed to facilitate motivation for abstinence in people with substance use disorders and has a strong emphasis on fostering activation and empowering patients to engage in their behavioural change actively. Therefore, it might be particularly beneficial for alleviating negative symptoms of the experiential dimension. It has recently been integrated into a motivational interviewing augmented CBT programme focusing on building motivation for personal recovery goals, establishing a commitment to change, and resolving ambivalence with CBT-based behavioural activation, problem-solving, and cognitive restructuring exercises [62, 77]. Each session of this programme follows the same structure, including a weekly check-in to review treatment progress and homework, skills training by role-playing and problem-solving exercises, and, if applicable, cognitive restructuring exercises to identify and modify demotivating and goal setting of an individualized homework assignment of behavioural steps for the upcoming week. In a recent RCT including $N = 79$ patients with schizophrenia and moderate to severe levels of experiential negative symptoms, it was shown that this programme was superior to an adapted mindfulness-based stress reduction intervention in alleviating experiential negative symptoms and improving psychosocial functioning at the end of treatment [77]. The beneficial effects on negative symptoms, but not on functioning, were maintained at follow-up, albeit attenuated between-group effects.

7.6.3 The Positive Emotions Programme for Schizophrenia

The Positive Emotions Program for Schizophrenia (PEPS; [65]) is a group-based treatment programme aiming to improve the anticipation of positive emotions and the motivation for goal-directed activities. PEPS consists of 8 one-hour group sessions delivered using visual and audio materials presented as PowerPoint slides and is based on a psychoeducational concept built according to the experiential learning model [57]. The PEPS programme employs a collaborative and egalitarian approach. Group facilitators participate in sessions in a manner similar to that of the participants, undertaking the exercises, sharing their experiences, and carrying out all tasks. Each PEPS intervention consists of four steps (1. Experiencing, 2. Distancing and reflection, 3. Absorption, 4. Experimental phase) and aims to train patients' specific skills in experiencing pleasure. The PEPS modules comprise restructuring demotivating beliefs, maintaining positive experiences, non-verbal expression of positive emotions, talking about pleasant experiences and anticipating positive emotions in relation to upcoming activities. The skills taught include: 1. Embracing the enjoyment of present or past pleasant experiences. 2. Enhancing emotional expression through behavioural cues. 3. Making the most out of positive moments. 4. Anticipating forthcoming enjoyable experiences. Savouring a pleasurable experience entails acknowledging the pleasure or positive emotions experienced at a particular moment. For instance, participants are invited to look at a picture of the pleasant countryside or listen to soothing music, become aware of the pleasurable experience of doing this, and appreciate it. Increasing behavioural expression of emotions involves utilizing facial expressions or gestures to convey positive feelings. Participants are encouraged to mimic expressions of joy portrayed by actors, becoming attuned to the resulting sensations. Maximizing the benefits of positive experiences involves effective communication and shared celebration with others. Participants may be prompted to engage in role-playing exercises in which they describe positive events to one another, fostering affiliation and appreciation. The anticipation of pleasant moments involves the imagination of the sensations related to a positive future event. This strategy is designed to guide the participants through a series of positive feelings and emotions. It can engage the participants' different senses, for example, by having them imagining they are eating some fruit or by anticipating the emotions and the physical sensations experienced upon completing a pleasurable physical or social activity. A simple homework task is assigned to be done between each session. For instance, this could entail selecting an image or object that evokes a positive emotion or feeling in the participant, who is then required to present it to the group.

A first RCT ($N = 80$) in participants with schizophrenia and negative symptoms showed that PEPS compared to TAU yielded moderate to large reductions in negative symptoms at post-treatment and at the 6-month follow-up [28].

7.6.4 The Motivation and Enhancement Training for Schizophrenia

The Motivation and Enhancement Training for Schizophrenia (MOVE; [94]) builds on research into emotional processing, as well as techniques from a variety of behavioural and skill-building approaches. It follows an assertive therapy concept in which patients are visited at home and supported in their everyday lives. MOVE targets explicitly a broad range of negative symptoms and their functional consequences in an integrated treatment programme. It is believed that the most effective approach to MOVE is to deliver the treatment in the ecological context of the person. The therapy employs a range of techniques, including antecedent control, the identification and addressing of deficits in anticipatory pleasure, in vivo emotional processing, which focuses on the identification of emotions in others and the display of one's own emotions, CBT techniques designed to address self-defeating cognitions, and in vivo skill building to address social and other skills deficits. The antecedent control approach is realised by providing external cues and environmental supports such as signs, alarms, and the appropriate placement of supplies to initiate specific, agreed-upon behaviours within the home environment. Once the behaviour is initiated, the programme ensures clear behavioural goals by outlining the process through a step-by-step checklist or an audio tape. It reduces the number of steps required to complete each task through task analysis and the organization of belongings. The anticipation of pleasure is trained through verbal discussion with the therapist. The therapist initially assesses the anticipated pleasure of an activity through verbal discussion. Subsequently, they evaluate the actual pleasure experienced. This comparison enables the identification of discrepancies and facilitates discussion about how these differences might affect the planning of further activities. Ratings are stored for future reference, which aids the patient in reviewing past experiences when planning future activities. Patients' statements about experienced pleasure are also recorded for replay as reminders.

Furthermore, images are employed to facilitate the creation of memory aids, thereby fostering anticipatory pleasure and encouraging the repetition of similar future activities. Deficits in the processing and display of emotions, such as identifying and expressing their own emotion, identifying behavioural expressions of emotion in others, and spontaneously taking the emotional perspective of other people, are addressed using computerized emotion perception exercises. Additionally, as discussed, cognitive interventions such as behavioural experiments and skill building techniques can be implemented to challenge self-defeatist thinking and improve social and independent living skills.

A first pilot study testing the efficacy of MOVE showed a significantly stronger effect of in reducing negative symptoms compared to TAU [95]. However, a definite test of the potential benefits that can be expected from MOVE on negative symptoms is still pending.

7.6.5 The Goals in Focus Programme

The Goals in Focus programme is a translational and comprehensive treatment programme consisting of 24 individual sessions (GIF; [86]). The GIF interventions follow a customized and readily conveyable cognitive-behavioural manualized approach derived from basic research and existing therapy approaches on experiential negative symptoms. The focus is on alleviating negative symptoms related to the Experiential dimension by improving goal pursuit based on individual case formulations. 'Goals in Focus' comprises four modules: (1) Developing an individual case formulation with a focus on the Experiential negative symptoms and to deduce an individualized treatment collaboratively, (2) setting personally relevant goals and identifying first steps towards goal achievement, (3) supporting goal pursuit by increasing the salience of rewards, training anticipatory pleasure, challenging demotivating beliefs, and facilitating social and problem-solving skills, (4) facilitating autonomous goal pursuit by reflecting on effective strategies and to prepare for challenges in goal pursuit in the future, identifying helpful strategies for goal pursuit and developing new goals and plans for the future. A first pilot study evaluating the feasibility and efficacy of GIF in patients with schizophrenia is currently ongoing.

7.6.6 Memory and Prospection Trainings Addressing Mental Imagery

The memory and prospection trainings are based on findings from basic research suggesting that deficits in episodic memory [69] lead to overgeneralized autobiographical memory retrieval [3] and less detailed prospection of hypothetical future positive events [46]. These deficits are reflected in less vivid mental representations of positive experiences, which hinders the anticipation of reward and pleasure [68]. Thus, the rationale for these trainings is that patients with negative symptoms experience deficits in episodic memory that make it difficult for them to anticipate pleasure concerning upcoming activities. In memory or prospection training, patients are encouraged to recollect activities that have given them pleasure in the past and that they would like to do again. The therapist guides the patient in imagining exactly how they would perform these activities and recalling the positive feelings from the past. The common aim of these trainings is to support patients in retrieving vivid, specific and detailed mental representations of past or future experiences. One example of these training programmes that have already been tested in a pilot study with patients with schizophrenia spectrum disorders is the MemFlex autobiographical memory intervention [51]. The intervention consists of a psychoeducational video showing three vignettes demonstrating the influence of memory on motivation for certain activities. The researcher and the participant discuss the video, and the participant identifies two activities they want to engage in based on positive memories. The participant then recalls a specific memory associated with each activity, with support from the researcher, using prompts derived from previous autobiographical memory studies. These prompts include details, sensory experiences, personal identity, positive themes, generalization of positive themes to other activities, and future planning. The aim is to facilitate the recall of positive memories and use them to motivate future engagement in enjoyable activities. The results of the pilot study, testing the feasibility and proof of concept of the MemFlex intervention in $N = 31$ patients with schizophrenia spectrum disorders and experimental negative symptoms, showed that compared to a control condition in which participants were asked to recall two activities without being shown the psychoeducation video and without being prompted to recall specific details, MemFlex showed promising effects on motivation, self-efficacy and negative mood, warranting further investigation of Memflex in a larger trial [27].

7.7 Emerging and Innovative Approaches

7.7.1 Virtual Reality Augmented Programmes

The latest innovative approaches to treating negative symptoms use virtual reality (VR) environments to deliver psychosocial and cognitive interventions. A systematic review summarizing the evidence from seven studies on VR-delivered interventions found that all studies report significant effects on their primary outcome and that VR was a promising therapeutic tool for patients with negative symptoms of schizophrenia [29]. The latest and most comprehensive VR-delivered intervention is the V-NeST programme [19]. V-NeST is a 12-session therapy combining CBT principles (e.g. the link between actions and emotion) and cognitive remediation techniques (e.g. teaching metacognition and strategy used to help regulate behaviour). Treatment modules include psychoeducation, behavioural activation, and developing insight into pleasure experiences and emotions. Most importantly, V-NeST emphasizes developing awareness and recalibrating how patients use feedback to improve goal-directed behaviour. This is achieved by psychoeducation about activities (e.g. how feedback influences behaviour), the routine use of activity appraisal (e.g. enjoyment rating) and re-appraisal (e.g. outcome re-evaluation), promoting an errorless learning environment (e.g. excessive use of positive feedback) and promoting successful and enjoyable everyday life activities. The unique feature of V-NeST is that all exercises are done in purposefully designed and built VR environments. These environments present unique and tailored motivational challenges, which participants are encouraged to approach while reflecting on (and rating) their experiences. Environments include low-motivation scenarios, such as domestic rooms (e.g. lounge) where participants can engage in passive activities such as listening to music or watching TV, as well as high-motivation environments such as factory-like environments where participants are asked to perform tasks. A further scenario includes a social environment where participants are asked to converse with other avatars or a game room with multiple activities available. The first pilot trial ($N = 30$) testing V-NeST against treatment as usual in patients with schizophrenia and experiencing debilitating negative symptoms suggested possible benefits on goal-attainment and negative symptoms.

7.7.2 Ecological Momentary Interventions

Ecological Momentary Interventions (EMIs) in mental health are a form of intervention that leverages mobile technology to deliver timely and contextually relevant support to individuals in their everyday environments [36, 48]. EMIs provide support, guidance, or interventions to individuals in real-time, typically through mobile apps or wearable devices and aim to promote positive behavioural changes or coping strategies by providing timely interventions. For instance, the Motivation and Skills Support (MASS) smartphone app integrates evidence-based psychosocial treatment approaches with insights from basic affective, motivational, and cognitive science to target key contributors to impaired social functioning in schizophrenia [33]. The MASS app is personalized based on participants' identified social goals, and push notifications are used to deliver content twice daily. Participants can choose steps related to their goals and receive feedback based on their motivation and progress. The key elements include SST to address social skill challenges related to goal progress in daily life and structured social goal support and feedback to address social motivation impairments that interfere with forming and maintaining social bonds. Elements of cognitive behavioral therapy, such as including self-monitoring of progress toward social goals and social appraisals (i.e. perceptions of social competence in recent interactions), and feedback based on participants' own ratings from previous EMA instances, were incorporated to address these social motivation impairments. The app also provides validating and encouraging statements to increase anticipatory pleasure for future social experiences, consistent with the temporal experience of pleasure model of impairment in schizophrenia [58].

The Personalized Real-Time Intervention for Motivational Enhancement (PRIME) app has been developed to improve motivational impairments, particularly in the early course of schizophrenia [84]. PRIME includes a peer community, goal and achievement tracking, and cognitive behavioural therapy (CBT) based coaching. The primary focus of PRIME is to foster daily-life motivation to pursue personally meaningful goals by using social reinforcement to engage and sustain goal-directed behaviour. With PRIME, users are provided with a supportive online environment where they select small, self-determined goals related to health/wellness, social relationships, creativity, and productivity. In addition, PRIME provides users counselling with motivation coaches who are master's-level clinicians using CBT-informed interventions such as behavioural activation, mindfulness, and psychoeducation to support users in overcoming daily obstacles that hinder their goal progress.

Furthermore, the PRIME community provides a platform for users to interact with one another. Users may send messages directly to each other and can also capture and share positive, spontaneous moments in their daily lives with their peers. A first small RCT comparing PRIME to a waitlist control condition in patients with recent onset schizophrenia showed promising effects of PRIME on anticipatory pleasure, the likelihood that one experiences positive social interactions, defeatist beliefs, and depressive symptoms at both the end of treatment and at the 3-month follow-up [85].

7.8 Novel Approaches from the Field of Depression

In the field of depression treatment, a number of translational interventions and programmes have been developed in recent years that place their primary focus on promoting positive affect and positive emotional experience [83]. These interventions have been shown to yield encouraging effects on anhedonia, positive affect, and motivational outcomes in patients with major depression. They may, therefore, provide promising blueprints for transdiagnostic adaptations that can be delivered to patients with negative symptoms of schizophrenia.

For instance, the Positive Affect Treatment (PAT; [22]) aims to directly improve reward sensitivity and positive emotional experience. PAT follows a behavioural and cognitive treatment framework and includes, among other elements, behavioural activation augmented by specific 'in-the-moment' recounting of pleasurable experiences, imagining future pleasant events, reframing of neutral or negative events, and compassion-focused training such as exercises on experiencing pleasure, loving-kindness, or gratitude. In a first RCT ($N = 96$), PAT was compared to a treatment that solely focused on reducing negative affect in a sample of patients with depression or anxiety disorders [23]. At 6-month follow-up, PAT was superior in increasing positive affect and in decreasing negative affect, depression, anxiety, stress, and suicidal ideation with moderate effect sizes. Although the results of this RCT do not yet demonstrate the efficacy of PAT in patients with negative symptoms, they suggest that an explicit focus on improving the experience of positive experiences may be beneficial in improving functional and adaptive behaviours.

Similarly, Augmented Depression Therapy (ADepT) aims to improve long-term well-being and anhedonia by addressing the positive valence system [25]. ADepT consists of up to 15 acute sessions and up to five booster sessions one year after the end of treatment. The interventions of ADepT focus on supporting clients to clarify their personal values, to work towards personally relevant goals that align with their values in vocational, recreational, relational, and self-care domains, and to learn to cope with maladaptive mechanisms that inhibit the capacity to thrive during opportunities. Therapists adopt a positive, future-focused, solution-focused, and reinforcing attitude to support patients in exploring their strengths and making positive and meaningful experiences. At the end of ADepT, patients and therapists develop a 'wellbeing plan' to continue positive progress in the future. A first proof of concept feasibility trial testing ADepT against CBT in 82 people with at least a moderate current major depressive episode and exhibiting anhedonic features showed that ADepT was superior to CBT in alleviating depressive symptoms and improving well-being at the end of treatment and at 18 months follow-up [26].

7.9 Final Remarks and Future Directions

Developing effective psychological interventions for the negative symptoms of schizophrenia presents a multifaceted challenge. The most significant obstacle lies in engaging patients with these interventions, as their motivational problems often extend into treatment. Thus, designing appealing and relevant treatments for these patients is crucial, yet complex. Moreover, this population's inherently low treatment adherence poses a considerable barrier to evaluating and implementing these interventions. [98] Indeed, this makes it challenging to involve patients in treatment programmes and complicates clinical trials necessary for empirical evaluation of these interventions. The interventions presented in this chapter already represent very encouraging and promising approaches focusing on different aspects or mechanisms of negative symptoms. Future research should aim to improve those interventions that account for the unique cognitive and emotional characteristics of people with schizophrenia while simultaneously devising strategies to enhance treatment adherence and engagement.

The lack of studies evaluating interventions qualitatively or providing first-person accounts of experiences is a significant gap in the literature. This contrasts starkly with trials evaluating interventions targeting other symptom clusters (e.g. [14, 32]) and limits the development of effective interventions. Although work has been done to understand the experience of negative symptoms, highlighting issues such as isolation, low motivation, and the absence of feelings, these studies emphasize the need for interventions to address these specific challenges (e.g. [13, 34]). Research examining service user perspectives on cognitive remediation therapy (CRT) underscores the importance of the therapist's role in digital interventions, as well as the barriers and facilitators associated with technology use [31, 67]. These studies also highlight the positive impacts of interventions, including improvements in metacognition and goal-setting and align with broader reviews of service user perspectives on digital interventions for psychosis, which emphasize the importance of personalizing and individualizing interventions [24]. The V-NEST study incorporated qualitative interviews as part of its evaluation, further emphasizing the critical role of goal-setting and the experience of growing motivation through engagement in the intervention [19]. In sum, when the voice of lived experience is included, there is a consensus that goal-setting, working with a therapist, and personalization are particularly effective elements of the interventions offered. However, despite these first attempts to acknowledge the service user perspective in clinical research on negative symptoms, this perspective is largely absent in the literature and might account for low engagement with treatment programmes and high attrition in clinical trials.

Looking at recent developments in interventions for people with negative symptoms, there has been a clear shift from broad and relatively non-specific approaches, such as CBT, to specific and translational developments that attempt to address the underlying mechanisms directly. These range from comprehensive treatment programmes (e.g. The Goals in Focus Intervention; GIF) to specific and shorter interventions that address only a specific part of the mechanisms (e.g. MemFlex). Three main trends can be identified in the new developments. First, the strong focus on individualized case formulations [60]; second, the emphasis and orientation of interventions towards patients' personal values and personally relevant goals, particularly in conjunction with cognitive rehabilitation interventions; and third, the shift in focus from reducing dysfunctional thought and behaviour patterns to promoting functional and adaptive behaviours and experiences, as in the context of memory and prospection trainings. Particularly employing innovative methodologies, such as technology-assisted interventions and community-based approaches, have been shown to offer promising avenues for engaging patients with the treatment and advancing the field of psychological interventions for negative symptoms of schizophrenia.

In conclusion, psychological treatments that specifically target the relevant mechanisms of anhedonia and amotivation hold promise for treating patients with negative symptoms for whom traditional treatment approaches have been only partially effective. However, much remains to be learned about the optimal delivery of these treatments and how to engage patients appropriately. This requires a thorough assessment of the perspective of people with lived experience in future research. Finally, there is a need to better understand the mechanisms responsible for negative symptoms and their changes over time, derive clinical implications from these findings, and translate them into the treatment of patients with negative symptoms.

References

1. AlAqeel B, Margolese HC. Remission in schizophrenia: critical and systematic review. Harv Rev Psychiatry. 2012;20(6):281–97. https://doi.org/10.3109/10673229.2012.747804.
2. Bandura A. Principles of behavior modification. Holt, Rinehart, & Winston; 1969.
3. Barry TJ, Hallford DJ, Takano K. Autobiographical memory impairments as a transdiagnostic feature of mental illness: a meta-analytic review of investigations into autobiographical memory specificity and overgenerality among people with psychiatric diagnoses. Psychol Bull. 2021;147(10):1054–74. https://doi.org/10.1037/bul0000345.
4. Blanchard JJ, Park SG, Catalano LT, Bennett ME. Social affiliation and negative symptoms in schizophrenia: examining the role of behavioral skills and subjective responding. Schizophr Res. 2015;168(1):491–7. https://doi.org/10.1016/j.schres.2015.07.019.
5. Blouzard E, Pouchon A, Polosan M, Bastin J, Donde C. Effort-cost decision-making among individuals with schizophrenia: a systematic review and meta-analysis. JAMA Psychiatry. 2023;80(6):548–57. https://doi.org/10.1001/jamapsychiatry.2023.0553.
6. Böge K, Catena D, Hahn E. Mindfulness-based interventions in people with psychotic disorders: an overview of the state of research concerning efficacy and implications for clinical practice. Verhaltenstherapie. 2022;32(Suppl. 1):24–34. https://doi.org/10.1159/000524534.
7. Böge K, Hahne I, Bergmann N, Wingenfeld K, Zierhut M, Thomas N, Ta TMT, Bajbouj M, Hahn E. Mindfulness-based group therapy for in-patients with schizophrenia spectrum disorders – feasibility, acceptability, and preliminary outcomes of a rater-blinded randomized controlled trial. Schizophr Res. 2021a;228:134–44. https://doi.org/10.1016/j.schres.2020.12.008.
8. Böge K, Hallford DJ, Pillny M. Mindfulness, psychological flexibility and their relationship with psychopathology in persons with schizophrenia-spectrum-disorders and healthy controls – a multicenter cross-sectional study. Psychiatry Res. 2023;330:115591. https://doi.org/10.1016/j.psychres.2023.115591.
9. Böge K, Karadza A, Fuchs LM, Ehlen F, Ta TMT, Thomas N, Bajbouj M, Hahn E. Mindfulness-based interventions for in-patients with schizophrenia spectrum disorders-a qualitative approach. Front Psych. 2020;11:600. https://doi.org/10.3389/fpsyt.2020.00600.
10. Böge K, Thomas N, Jacobsen P. Is mindfulness for psychosis harmful? Deconstructing a myth. Br J Psychiatry. 2021b;218(2):71–2. https://doi.org/10.1192/bjp.2020.165.
11. Bora E, Binnur Akdede B, Alptekin K. Neurocognitive impairment in deficit and non-deficit schizophrenia: a meta-analysis. Psychol Med. 2017;47(14):2401–13. https://doi.org/10.1017/S0033291717000952.
12. Bowie CR, Bell MD, Fiszdon JM, Johannesen JK, Lindenmayer J-P, McGurk SR, Medalia AA, Penadés R, Saperstein AM, Twamley EW, Ueland T, Wykes T. Cognitive remediation for schizophrenia: an expert working group white paper on core techniques. Schizophr Res. 2020;215:49–53. https://doi.org/10.1016/j.schres.2019.10.047.
13. Butcher I, Berry K, Haddock G. Understanding individuals' subjective experiences of negative symptoms of schizophrenia: a qualitative study. Br J Clin Psychol. 2020;59:319–34.

14. Byrne R. Understanding psychological treatment for psychosis from the perspective of those with lived experience: 'what's important to us?'. The University of Manchester; 2014.
15. Cassidy CM, Norman R, Manchanda R, Schmitz N, Malla A. Testing definitions of symptom remission in first-episode psychosis for prediction of functional outcome at 2 years. Schizophr Bull. 2010;36(5):1001–8. https://doi.org/10.1093/schbul/sbp007.
16. Cella M, Bishara AJ, Medin E, Swan S, Reeder C, Wykes T. Identifying cognitive remediation change through computational modelling--effects on reinforcement learning in schizophrenia. Schizophr Bull. 2014;40(6):1422–32. https://doi.org/10.1093/schbul/sbt152.
17. Cella M, Preti A, Edwards C, Dow T, Wykes T. Cognitive remediation for negative symptoms of schizophrenia: a network meta-analysis. Clin Psychol Rev. 2017;52:43–51. https://doi.org/10.1016/j.cpr.2016.11.009.
18. Cella M, Roberts S, Pillny M, Riehle M, O'Donoghue B, Lyne J, Tomlin P, Valmaggia L, Preti A. Psychosocial and behavioural interventions for the negative symptoms of schizophrenia: a systematic review of efficacy meta-analyses. Br J Psychiatry. 2023;223(1):321–31. https://doi.org/10.1192/bjp.2023.21.
19. Cella M, Tomlin P, Robotham D, Green P, Griffiths H, Stahl D, Valmaggia L. Virtual reality therapy for the negative symptoms of schizophrenia (v-nest): a pilot randomised feasibility trial. Schizophr Res. 2022;248:50–7. https://doi.org/10.1016/j.schres.2022.07.013.
20. Cramer H, Lauche R, Haller H, Langhorst J, Dobos G. Mindfulness- and acceptance-based interventions for psychosis: a systematic review and meta-analysis. Glob Adv Health Med. 2016;5(1):30–43. https://doi.org/10.7453/gahmj.2015.083.
21. Crane RS, Brewer J, Feldman C, Kabat-Zinn J, Santorelli S, Williams JM, Kuyken W. What defines mindfulness-based programs? The warp and the weft. Psychol Med. 2017;47(6):990–9. https://doi.org/10.1017/s0033291716003317.
22. Craske MG, Meuret AE, Ritz T, Treanor M, Dour HJ. Treatment for anhedonia: a neuroscience driven approach. Depress Anxiety. 2016;33(10):927–38. https://doi.org/10.1002/da.22490.
23. Craske MG, Treanor M, Dour H, Meuret A, Ritz T. Positive affect treatment for depression and anxiety: a randomized clinical trial for a core feature of anhedonia. J Consult Clin Psychol. 2019;87:457–71. https://doi.org/10.1037/ccp0000396.
24. Dennard S, Patel R, Garety P, Edwards C, Gumley A. A systematic review of users experiences of using digital interventions within psychosis: a thematic synthesis of qualitative research. Soc Psychiatry Psychiatr Epidemiol. 2024; https://doi.org/10.1007/s00127-024-02692-4.
25. Dunn BD, Widnall E, Reed N, Owens C, Campbell J, Kuyken W. Bringing light into darkness: a multiple baseline mixed methods case series evaluation of augmented depression therapy (adept). Behav Res Ther. 2019;120:103418. https://doi.org/10.1016/j.brat.2019.103418.
26. Dunn BD, Widnall E, Warbrick L, Warner F, Reed N, Price A, Kock M, Courboin C, Stevens R, Wright K, Moberly NJ, Geschwind N, Owens C, Spencer A, Campbell J, Kuyken W. Preliminary clinical and cost effectiveness of augmented depression therapy versus cognitive behavioural therapy for the treatment of anhedonic depression (adept): a single-centre, open-label, parallel-group, pilot, randomised, controlled trial. eClinicalMedicine. 2023;61:102084. https://doi.org/10.1016/j.eclinm.2023.102084.
27. Edwards CJ, Garety PA, Hardy A. Remembering the past to live better in the future: a feasibility randomised controlled trial of memory specificity training for motivation in psychosis. J Behav Ther Exp Psychiatry. 2020;68:101564. https://doi.org/10.1016/j.jbtep.2020.101564.
28. Favrod J, Nguyen A, Chaix J, Pellet J, Frobert L, Fankhauser C, Ismailaj A, Brana A, Tamic G, Suter C, Rexhaj S, Golay P, Bonsack C. Improving pleasure and motivation in schizophrenia: a randomized controlled clinical trial. Psychother Psychosom. 2019;88:84–95. https://doi.org/10.1159/000496479.
29. Fernández-Sotos P, Fernández-Caballero A, Rodriguez-Jimenez R. Virtual reality for psychosocial remediation in schizophrenia: a systematic review. Eur J Psychiatry. 2020;34(1):1–10. https://doi.org/10.1016/j.ejpsy.2019.12.003.
30. Fervaha G, Foussias G, Agid O, Remington G. Amotivation and functional outcomes in early schizophrenia. Psychiatry Res. 2013;210:665–8.

31. Frawley E, Heary C, Berry C, Cella M, Fowler D, Wykes T, Donohoe G. Participant perspectives on cognitive remediation and social recovery in early psychosis (crest-r): an acceptability study. Early Interv Psychiatry. 2024;18(1):34–41. https://doi.org/10.1111/eip.13424.
32. Freeman D, Rosebrock L, Waite F, Loe BS, Kabir T, Petit A, Dudley R, Chapman K, Morrison A, O'Regan E, Aynsworth C, Jones J, Murphy E, Powling R, Peel H, Walker H, Byrne R, Freeman J, Rovira A, et al. Virtual reality (vr) therapy for patients with psychosis: satisfaction and side effects. Psychol Med. 2023;53(10):4373–84. https://doi.org/10.1017/S0033291722001167.
33. Fulford D, Gard DE, Mueser KT, Mote J, Gill K, Leung L, Mow J. Preliminary outcomes of an ecological momentary intervention for social functioning in schizophrenia: pre-post study of the motivation and skills support app. JMIR Ment Health. 2021;8(6):e27475. https://doi.org/10.2196/27475.
34. Gee B, Hodgekins J, Lavis A, Notley C, Birchwood M, Everard L, Freemantle N, Jones PB, Singh SP, Amos T, Marshall M, Sharma V, Smith J, Fowler D. Lived experiences of negative symptoms in first-episode psychosis: a qualitative secondary analysis. Early Interv Psychiatry. 2019;13(4):773–9. https://doi.org/10.1111/eip.12558.
35. Goldberg SB, Tucker RP, Greene PA, Davidson RJ, Wampold BE, Kearney DJ, Simpson TL. Mindfulness-based interventions for psychiatric disorders: a systematic review and meta-analysis. Clin Psychol Rev. 2018;59:52–60. https://doi.org/10.1016/j.cpr.2017.10.011.
36. Granholm E, Holden J, Dwyer K, Mikhael T, Link P, Depp C. Mobile-assisted cognitive behavioral therapy for negative symptoms: open single-arm trial with schizophrenia patients. JMIR Mental Health. 2020;7:e24406. https://doi.org/10.2196/24406.
37. Granholm E, Holden J, Link PC, McQuaid JR, Jeste DV. Randomized controlled trial of cognitive behavioral social skills training for older consumers with schizophrenia: defeatist performance attitudes and functional outcome. Am J Geriatr Psychiatr. 2013;21:251–62. https://doi.org/10.1016/j.jagp.2012.10.014.
38. Granholm E, Holden J, Link PC, McQuaidm JR. Randomized clinical trial of cognitive behavioral social skills training for schizophrenia: improvement in functioning and experiential negative symptoms. J Consult Clin Psychol. 2014;82:1173–85. https://doi.org/10.1037/a0037098.
39. Granholm E, Holden J, Worley M. Improvement in negative symptoms and functioning in cognitive-behavioral social skills training for schizophrenia: mediation by defeatist performance attitudes and asocial beliefs. Schizophr Bull. 2018;44:653–61. https://doi.org/10.1093/schbul/sbx099.
40. Granholm E, McQuaid J, Holden J. Cognitive-behavioral social skills training for schizophrenia: a practical treatment guide. Guilford Press; 2016.
41. Granholm E, McQuaid JR, McClure FS, Auslander LA, Perivoliotis D, Pedrelli P, Patterson T, Jeste DV. A randomized, controlled trial of cognitive behavioral social skills training for middle-aged and older outpatients with chronic schizophrenia. Am J Psychiatry. 2005;162(3):520–9. https://doi.org/10.1176/appi.ajp.162.3.520.
42. Granholm E, McQuaid JR, McClure FS, Link PC, Perivoliotis D, Gottlieb JD, Patterson TL, Jeste DV. Randomized controlled trial of cognitive behavioral social skills training for older people with schizophrenia: 12-month follow-up. J Clin Psychiatry. 2007;68:730–7. https://doi.org/10.4088/JCP.v68n0510.
43. Granholm E, Twamley EW, Mahmood Z, Keller AV, Lykins HC, Parrish EM, Thomas ML, Perivoliotis D, Holden JL. Integrated cognitive-behavioral social skills training and compensatory cognitive training for negative symptoms of psychosis: effects in a pilot randomized controlled trial. Schizophr Bull. 2021;48(2):359–70. https://doi.org/10.1093/schbul/sbab126.
44. Grant PM, Bredemeier K, Beck AT. Six-month follow-up of recovery-oriented cognitive therapy for low-functioning individuals with schizophrenia. Psychiatr Serv. 2017;68:appi.ps.2016004. https://doi.org/10.1176/appi.ps.201600413.
45. Grant PM, Huh GA, Perivoliotis D, Stolar NM, Beck AT. Randomized trial to evaluate the efficacy of cognitive therapy for low-functioning patients with schizophrenia. Arch Gen Psychiatry. 2012;69:121–7. https://doi.org/10.1001/archgenpsychiatry.2011.129.

46. Hallford DJ, Austin DW, Takano K, Raes F. Psychopathology and episodic future thinking: a systematic review and meta-analysis of specificity and episodic detail. Behav Res Ther. 2018;102:42–51. https://doi.org/10.1016/j.brat.2018.01.003.
47. Hallford DJ, Sharma MK. Anticipatory pleasure for future experiences in schizophrenia spectrum disorders and major depression: a systematic review and meta-analysis. Br J Clin Psychol. 2019;58(4):357–83. https://doi.org/10.1111/bjc.12218.
48. Hanssen E, Balvert S, Oorschot M, Borkelmans K, van Os J, Delespaul P, Fett A-K. An ecological momentary intervention incorporating personalised feedback to improve symptoms and social functioning in schizophrenia spectrum disorders. Psychiatry Res. 2020;284:112695. https://doi.org/10.1016/j.psychres.2019.112695.
49. Hayes SC. Acceptance and commitment therapy: towards a unified model of behavior change. World Psychiatry. 2019;18(2):226–7. https://doi.org/10.1002/wps.20626.
50. Hayes SC, Strosahl KD, Wilson KG. Acceptance and commitment therapy: the process and practice of mindful change. 2nd ed. Guilford Press; 2012.
51. Hitchcock C, Gormley S, Rees C, Rodrigues E, Gillard J, Panesar I, Wright IM, Hammond E, Watson P, Werner-Seidler A, Dalgleish T. A randomised controlled trial of memory flexibility training (memflex) to enhance memory flexibility and reduce depressive symptomatology in individuals with major depressive disorder. Behav Res Ther. 2018;110:22–30. https://doi.org/10.1016/j.brat.2018.08.008.
52. Husain M, Roiser JP. Neuroscience of apathy and anhedonia: a transdiagnostic approach. Nat Rev Neurosci. 2018;19(8):470–84. https://doi.org/10.1038/s41583-018-0029-9.
53. Jaya ES, Pillny M, Lincoln TM, Riehle M. Does social defeat cause negative symptoms? A prospective study in a multi-national community sample. Compr Psychiatry. 2022;113:152289.
54. Jespersen AE, Lumbye A, Vinberg M, Glenthøj L, Nordentoft M, Wæhrens EE, Knudsen GM, Makransky G, Miskowiak KW. Effect of immersive virtual reality-based cognitive remediation in patients with mood or psychosis spectrum disorders: study protocol for a randomized, controlled, double-blinded trial. Trials. 2024;25(1):82. https://doi.org/10.1186/s13063-024-07910-7.
55. Kabat-Zinn J. An outpatient program in behavioral medicine for chronic pain patients based on the practice of mindfulness meditation: theoretical considerations and preliminary results. Gen Hosp Psychiatry. 1982;4(1):33–47. https://doi.org/10.1016/0163-8343(82)90026-3.
56. Kirschner H, Nassar MR, Fischer AG, Frodl T, Meyer-Lotz G, Froböse S, Seidenbecher S, Klein TA, Ullsperger M. Transdiagnostic inflexible learning dynamics explain deficits in depression and schizophrenia. Brain. 2024;147(1):201–14. https://doi.org/10.1093/brain/awad362.
57. Kolb AY, Kolb DA. The learning way: meta-cognitive aspects of experiential learning. Simul Gaming. 2009;40(3):297–327. https://doi.org/10.1177/1046878108325713.
58. Kring AM, Caponigro JM. Emotion in schizophrenia: where feeling meets thinking. Curr Dir Psychol Sci. 2010;19:255–9. https://doi.org/10.1177/0963721410377599.
59. Kurtz MM, Mueser KT. A meta-analysis of controlled research on social skills training for schizophrenia. J Consult Clin Psychol. 2008;76:491–504. https://doi.org/10.1037/0022-006X.76.3.491.
60. Lincoln TM, Riehle M, Pillny M, Helbig-Lang S, Fladung AK, Hartmann-Riemer M, Kaiser S. Using functional analysis as a framework to guide individualized treatment for negative symptoms. Front Psychol. 2017;8
61. Louise S, Fitzpatrick M, Strauss C, Rossell SL, Thomas N. Mindfulness- and acceptance-based interventions for psychosis: our current understanding and a meta-analysis. Schizophr Res. 2018;192:57–63. https://doi.org/10.1016/j.schres.2017.05.023.
62. Martino S, Carroll K, Kostas D, Perkins J, Rounsaville B. Dual diagnosis motivational interviewing: a modification of motivational interviewing for substance-abusing patients with psychotic disorders. J Subst Abus Treat. 2002;23(4):297–308. https://doi.org/10.1016/s0740-5472(02)00295-7.
63. Miller WR, Rollnick S. Motivational interviewing: helping people change. Guilford Press; 2012.

64. Muyambi K, Walsh S, Dan B, Gray R, Dennis S, Brown E, Grimshaw M, Drummond J, Jones M. Efficacy of behavioural activation in the treatment of negative symptoms in people with schizophrenia spectrum disorders: a systematic review. Int J Nurs Stud Adv. 2023;5 https://doi.org/10.1016/j.ijnsa.2023.100132.
65. Nguyen A, Frobert L, McCluskey I, Golay P, Bonsack C, Favrod J. Development of the positive emotions program for schizophrenia: an intervention to improve pleasure and motivation in schizophrenia [methods]. Front Psych. 2016;7 https://doi.org/10.3389/fpsyt.2016.00013.
66. O'Brien-Venus B, Ellett L, Burgess-Barr S, Chadwick P. Systematic review of the safety of mindfulness-based interventions for psychosis. Clin Psychol Rev. 2024;112:102445. https://doi.org/10.1016/j.cpr.2024.102445.
67. Parri LA, Barret K, Hill R, Hoque A, Isok I, Kenny A, Markham S, Oyeleye N, Quinn R, Sweeney A, Wykes T, Cella M. Evaluating the acceptability of remote cognitive remediation from the perspective of psychosis service users. Behav Cogn Psychother. 2024;1-13 https://doi.org/10.1017/S1352465824000109.
68. Pillny M, Hallford DJ, Böge K. The nature of mental imagery and its relationship with amotivational psychopathology in people with schizophrenia-spectrum disorders. Behav Ther. 2024a; https://doi.org/10.1016/j.beth.2024.01.009.
69. Pillny M, Krkovic K, Buck L, Lincoln TM. From memories of past experiences to present motivation? A meta-analysis on the association between episodic memory and negative symptoms in people with psychosis. Schizophr Bull. 2022;48:307–24. https://doi.org/10.1093/schbul/sbab120.
70. Pillny M, Krkovic K, Lincoln TM. Development of the demotivating beliefs inventory and test of the cognitive triad of amotivation. Cogn Ther Res. 2018;42:867–77. https://doi.org/10.1007/s10608-018-9940-7.
71. Pillny M, Lincoln TM. Predictors of improved functioning in patients with psychosis: the role of amotivation and defeatist performance beliefs. Psychiatry Res. 2016;244:117–22. https://doi.org/10.1016/j.psychres.2016.07.018.
72. Pillny M, Lincoln TM. The demotivating effect of social exclusion: an experimental test of a psychosocial model on the development of negative symptoms in psychosis. Schizophr Res. 2020;215:330–6. https://doi.org/10.1016/j.schres.2019.10.005.
73. Pillny M, Renz KE, Lincoln TM, Lütje F, Hay A, Fulford D, Barch DM, Gold JM, Kaiser S. Effort-based decision-making in psychopathology: a transdiagnostic multilevel meta-analysis. 2024b; https://doi.org/10.31219/osf.io/j6asc.
74. Pillny M, Schlier B, Lincoln TM. "I just don't look forward to anything". How anticipatory pleasure and negative beliefs contribute to goal-directed activity in patients with negative symptoms of psychosis. Schizophr Res. 2020;222:429–36. https://doi.org/10.1016/j.schres.2020.03.059.
75. Pillny M, Mueser K, Holden J, Aarons G, Sommerfeld D, Perivoliotis D, Granholm E. Enhancing assertive community treatment with cognitive behavioral social skills training for schizophrenia: Ii. Ecological momentary assessment outcomes in a pragmatic randomized-controlled trial. Schizophrenia Bulletin. 2025. https://doi.org/10.1093/schbul/sbaf103.
76. Qin K, Yu Y, Cai H, Li J, Zeng J, Liang H. Effectiveness of mindfulness-based intervention in schizophrenia: a meta-analysis of randomized controlled trials. Psychiatry Res. 2024;334:115808. https://doi.org/10.1016/j.psychres.2024.115808.
77. Reddy LF, Glynn SM, McGovern JE, Sugar CA, Reavis EA, Green MF. A novel psychosocial intervention for motivational negative symptoms in schizophrenia: combined motivational interviewing and cbt. Am J Psychiatry. 2023;180(5):367–76. https://doi.org/10.1176/appi.ajp.20220243.
78. Riehle M, Böhl MC, Pillny M, Lincoln TM. Efficacy of psychological treatments for patients with schizophrenia and relevant negative symptoms: a meta-analysis. Clin Psychol Eur. 2020;2:1–23. https://doi.org/10.32872/cpe.v2i3.2899.

79. Riehle M, Mehl S, Lincoln TM. The specific social costs of expressive negative symptoms in schizophrenia: reduced smiling predicts interactional outcome. Acta Psychiatr Scand. 2018;138(2):133–44. https://doi.org/10.1111/acps.12892.
80. Sabé M, et al. Half a century of research for mindfulness-based body-mind interventions: a scientometric analysis. Mindfulness. 2025;16:1765–79.
81. Sabé M, Kohler R, Perez N, Sauvain-Sabé M, Sentissi O, Jermann F, Prada P, Perroud N, Böge K. Mindfulness-based interventions for patients with schizophrenia spectrum disorders: a systematic review of the literature. Schizophr Res. 2024;264:191–203. https://doi.org/10.1016/j.schres.2023.12.011.
82. Sabe M, Sentissi O, Kaiser S. Meditation-based mind-body therapies for negative symptoms of schizophrenia: systematic review of randomized controlled trials and meta-analysis. Schizophr Res. 2019;212:15–25. https://doi.org/10.1016/j.schres.2019.07.030.
83. Sandman CF, Craske MG. Psychological treatments for anhedonia. In: Pizzagalli D, editor. Anhedonia: preclinical, translational, and clinical integration. Berlin/Heidelberg: Springer; 2021. p. 491–513. https://doi.org/10.1007/7854_2021_291.
84. Schlosser D, Campellone T, Kim D, Truong B, Vergani S, Ward C, Vinogradov S. Feasibility of prime: a cognitive neuroscience-informed mobile app intervention to enhance motivated behavior and improve quality of life in recent onset schizophrenia. JMIR Res Protoc. 2016;5(2):e77. https://doi.org/10.2196/resprot.5450.
85. Schlosser DA, Campellone TR, Truong B, Etter K, Vergani S, Komaiko K, Vinogradov S. Efficacy of prime, a mobile app intervention designed to improve motivation in young people with schizophrenia. Schizophr Bull. 2018;44(5):1010–20. https://doi.org/10.1093/schbul/sby078.
86. Schormann AL, Pillny M, Haß K, Lincoln TM. "Goals in focus"—a targeted cbt approach for motivational negative symptoms of psychosis: study protocol for a randomized-controlled feasibility trial. Pilot Feasibility Stud. 2023;9(1):1–14. https://doi.org/10.1186/s40814-023-01284-4.
87. Shen H, Zhang L, Li Y, Zheng D, Du L, Xu F, Xu C, Liu Y, Shen J, Li Z, Cui D. Mindfulness-based intervention improves residual negative symptoms and cognitive impairment in schizophrenia: a randomized controlled follow-up study. Psychol Med. 2023;53(4):1390–9. https://doi.org/10.1017/s0033291721002944.
88. Sitko K, Bewick BM, Owens D, Masterson C. Meta-analysis and meta-regression of cognitive behavioral therapy for psychosis (cbtp) across time: the effectiveness of cbtp has improved for delusions. Schizophr Bull Open. 2020;1(1) https://doi.org/10.1093/schizbullopen/sgaa023.
89. Staring ABP, ter Huurne M-AB, van der Gaag M. Cognitive behavioral therapy for negative symptoms (cbt-n) in psychotic disorders: a pilot study. J Behav Ther Exp Psychiatry. 2013;44:300–6. https://doi.org/10.1016/j.jbtep.2013.01.004.
90. Strauss GP, Esfahlani FZ, Galderisi S, Mucci A, Rossi A, Bucci P, Rocca P, Maj M, Kirkpatrick B, Ruiz I, Sayama H. Network analysis reveals the latent structure of negative symptoms in schizophrenia. Schizophr Bull. 2019;45:1033–41.
91. Strauss GP, Sandt AR, Catalano LT, Allen DN. Negative symptoms and depression predict lower psychological well-being in individuals with schizophrenia. Compr Psychiatry. 2012;53(8):1137–44. https://doi.org/10.1016/j.comppsych.2012.05.009.
92. Su Y, Pan X, Li H, Zhang G. Effects of mind-body therapies on schizophrenia: a systematic review and network meta-analysis. Schizophr Res. 2024;264:236–47. https://doi.org/10.1016/j.schres.2023.12.030.
93. Turner DT, van der Gaag M, Karyotaki E, Cuijpers P. Psychological interventions for psychosis: a meta-analysis of comparative outcome studies. Am J Psychiatry. 2014;171(5):523–38. https://doi.org/10.1176/appi.ajp.2013.13081159.
94. Velligan D, Maples N, Roberts DL, Medellin EM. Integrated psychosocial treatment for negative symptoms. Am J Psychiatr Rehabil. 2014;17:1–19. https://doi.org/10.1080/15487768.2013.873370.

95. Velligan DI, Roberts D, Mintz J, Maples N, Li X, Medellin E, Brown M. A randomized pilot study of motivation and enhancement (move) training for negative symptoms in schizophrenia. Schizophr Res. 2015;165:175–80. https://doi.org/10.1016/j.schres.2015.04.008.
96. Ventura J, Hellemann GS, Thames AD, Koellner V, Nuechterlein KH. Symptoms as mediators of the relationship between neurocognition and functional outcome in schizophrenia: a meta-analysis. Schizophr Res. 2009;113:189–99. https://doi.org/10.1016/j.schres.2009.03.035.
97. Vita A, Barlati S, Ceraso A, Nibbio G, Ariu C, Deste G, Wykes T. Effectiveness, core elements, and moderators of response of cognitive remediation for schizophrenia: a systematic review and meta-analysis of randomized clinical trials. JAMA Psychiatry. 2021;78(8):848–58. https://doi.org/10.1001/jamapsychiatry.2021.0620.
98. Vita A, Barlati S, Ceraso A, Deste G, Nibbio G, Wykes T. Acceptability of cognitive remediation for schizophrenia: a systematic review and meta-analysis of randomized controlled trials. Psychol Med. 2023;53(8):3661–71. https://doi.org/10.1017/S0033291722000319.
99. Vita A, Gaebel W, Mucci A, Sachs G, Barlati S, Giordano GM, Nibbio G, Nordentoft M, Wykes T, Galderisi S. European psychiatric association guidance on treatment of cognitive impairment in schizophrenia. Eur Psychiatry. 2022;65(1):e57. https://doi.org/10.1192/j.eurpsy.2022.2315.
100. Wakefield S, Roebuck S, Boyden P. The evidence base of acceptance and commitment therapy (act) in psychosis: a systematic review. J Contextual Behav Sci. 2018;10:1–13. https://doi.org/10.1016/j.jcbs.2018.07.001.
101. Zierhut M, Bergmann N, Hahne I, Wohlthan J, Kraft J, Braun A, Tam Ta TM, Hellmann-Regen J, Ripke S, Bajbouj M, Hahn E, Böge K. The combination of oxytocin and mindfulness-based group therapy for empathy and negative symptoms in schizophrenia spectrum disorders – a double-blinded, randomized, placebo-controlled pilot study. J Psychiatr Res. 2024;171:222–9. https://doi.org/10.1016/j.jpsychires.2024.01.014.

Part II

Cognitive Impairment in Schizophrenia: Clinical and Management Implications

Cognitive Impairment in Schizophrenia: Definition, Classification, and Course

8

Rafael Penadés, Gisela Mezquida, Florencia Forte, and Rosa Ayesa

Abstract

Schizophrenia is considered a severe mental illness characterized by the presence of positive, negative, and cognitive symptoms. Thus, it affects various cognitive domains, including basic cognition (attention, processing speed, memory, working memory, reasoning, problem-solving), social cognition (understanding social cues, emotions, and intentions of others), and metacognition (self-awareness, insight into abilities, and the capacity to monitor and adjust cognitive processes).

The course of cognitive decline in schizophrenia remains under investigation. Different trajectories, including stable impairment, progressive decline, improvement, and fluctuating courses, have been proposed. Early intervention programs are crucial to optimize recovery and limit disability, as cognitive deficits are present during the prodromal and early stages of the illness. In chronic stages, while some studies suggest possible improvement in specific cognitive domains, others indicate stability or even decline, particularly with persistent and pharmacologically resistant symptoms.

Cognitive impairment is considered a core symptom as it is present from the illness onset and it significantly impacts daily life, hindering academic and occupational performance, social functioning, and treatment adherence. Cognitive reserve is a potential factor that may explain clinical, cognitive, and functional differences among patients. Evidence suggests that patients in both early and late stages with higher cognitive reserve show a later onset of the illness and better outcomes in cognition and functionality. Finally, psychical exercise and

R. Penadés (✉) · G. Mezquida · F. Forte · R. Ayesa
Hospital Clínic of Barcelona, University of Barcelona, IDIBAPS, CIBERSAM, Barcelona, Spain
e-mail: RPENADES@clinic.cat

A. Vita et al. (eds.), *Negative and Cognitive Symptoms in Schizophrenia*, Comprehensive Approach to Psychiatry 3,
https://doi.org/10.1007/978-3-032-06209-3_8

cognitive remediation programs are gaining traction as they address these core deficits and improve patients' ability to live independently and participate in society.

Keywords

Cognition · Cognitive impairment · Cognitive reserve · Cognitive remediation

8.1 Cognition in Schizophrenia

It is common knowledge that schizophrenia is a severe and debilitating mental condition; hence, some authors consider that it is "one of the most mysterious and costliest mental disorders in terms of human suffering and societal expenditure" [1]. Professionals working with people presenting the condition are indeed aware of the suffering that the illness represents for the patients and their families. Schizophrenia represents an extreme transformation in people's thinking, behavior, and emotional regulation, marked by delusions, hallucinations, and disorganized speech. Fortunately, those named positive symptoms respond to antipsychotic medication. In contrast, reduced emotional expression, social withdrawal, and cognitive deficits, the silent world of negative symptoms and cognitive dysfunction, do not respond to pharmacological treatments in the same good way. Particularly, cognitive impairment plays a crucial role at the onset and throughout psychotic illness, being one of the main obstacles to clinical and functional recovery [1].

Cognitive impairment restrains the functioning of people affected by psychotic conditions, determining academic and occupational performance, daily living skills, and social functioning, including treatment adherence and general well-being. Particularly, deficiencies in verbal memory and attention/vigilance prevent patients from attaining optimal adaptation and act as what Green defined as cognitive rate-limiting factors [2]. Both negative symptoms and cognitive impairment, although correlated, are distinct and have differential pathways of change with treatment [3].

Cognition in schizophrenia has not always been a priority for clinicians and researchers, although its importance was noted from the very beginning of the history of the condition. Emil Kraepelin described a decline in cognitive functions, especially in attention and associative abilities, considering both central characteristics of the syndrome, and therefore proposed the term dementia praecox. Like dementia, schizophrenia was first defined as a primarily cognitive disorder, with onset in adolescence and with progressive cognitive and functional deterioration in many cases. Kraepelin's students were the first to carry out studies of neurocognition in schizophrenia; the study of cognitive impairment in schizophrenia has existed, practically, since the moment this condition was defined [4].

Eugen Bleuler changed the name of dementia praecox to schizophrenia, basically due to two disagreements with Kraepelin [5]. The first is the need for an early age of onset, and the second is the course of progressive deterioration of the illness. Both points of disagreement do not seem doubtful at present, given that cases of schizophrenia have been frequently recorded at late ages and the course of the disease has been demonstrated not to be necessarily progressive. Bleuler made a distinction between the "core" symptoms of schizophrenia and those that were secondary or accessory. The diagnosis had to be based on the presence of the so-called "core" symptoms. Bleuler's concept of loosening associations refers broadly to a deficit that underlies the other symptoms of schizophrenia. However, the Bleulerian conception of schizophrenia ended up giving primacy to the cross-sectional analysis of other symptoms over the longitudinal analysis of cognition, forgetting to carry out studies on the course and prognosis of the disease. We understand that this lack of longitudinal studies could have influenced clinicians to end up attributing little relevance to neurocognitive alterations in schizophrenia, given that they did not help in making the diagnosis in a cross-sectional manner. Cognitive symptoms that had been considered essential ended up being accessory for the diagnosis until nowadays.

Later, with the contribution of Kurt Schneider [6], the practice of diagnosis based on the presence of the so-called "first rank" symptoms became a priority, completely forgetting neurocognitive analysis. Schneider's influence has meant that none of the subsequent diagnostic systems (RCD, DSM, or ICD) have considered the presence of neurocognitive aspects in the diagnosis of schizophrenia. Although after Schneider's contribution, there has been no shortage of examples of rigorous studies on neurocognition in schizophrenia, such as the excellent work of [6], these were very sporadic and unrelated to clinical practice. It has not been until the last three decades that interest in neurocognition in schizophrenia has been revived.

It is worth highlighting the technological improvement that different branches of neuroscience have experienced, especially cognitive psychology and neuropsychology, which have allowed a conceptual and technical development capable of attempting the study of neurocognition in an objective, quantitative, and reliable way, beyond simple clinical observation used previously [7]. Finally, the recovery of interest in longitudinal studies and the possible confirmation that it is precisely the cognitive alterations, and not the presence of other symptoms of schizophrenia, that induce a greater degree of disability in patients has given an important boost to this area, helping to relocate cognition in schizophrenia. Thus, the findings produced by cognitive studies have helped establish the prognosis of the illness in a quantifiable way. On the other hand, neuroimaging techniques have shown results that are consistent with cognitive ones, allowing a better understanding of some of the aspects of brain function in schizophrenia. All of this has allowed the reconsideration of cognitive alterations as "core" symptoms of schizophrenia [8, 9].

8.2 Basic Cognition, Social Cognition, and Metacognition

Cognition constitutes a complex and interconnected set of mental processes that underpin our perception and interaction with the world [2]. The involved domains in these mental processes encompass a wide range of functions, including neurocognitive domains, such as attention, speed of processing, memory, working memory, reasoning, and problem-solving, as well as social cognition [9, 10].

Cognitive deficits extend across multiple domains essential for daily functioning: *basic cognition*, which encompasses foundational mental processes (such as attention, memory, or problem-solving); *social cognition*, which involves understanding social cues, emotions, and the intentions of others [11] and, *metacognition*, which includes self-awareness, insight into one's abilities, and the capacity to monitor and adjust cognitive processes [12]. These deficits can collectively impact various aspects of life, from interpersonal interactions and vocational pursuits to adherence to treatment regimens and quality of life.

The initiative MATRICS (Measurement and Treatment Research to Improve Cognition in Schizophrenia) Consensus Cognitive Battery (MCCB) identified those most impaired cognitive domains in schizophrenia and established the standardization of the measurement protocols for these cognitive impairments [13, 14]. The MATRICS recognized six basic cognitive domains, comprising processing speed, attention, working memory, verbal learning and memory, visual learning and memory, and reasoning and problem solving [11], which may be intercorrelated, and boundaries between them may sometimes be blurred or overlapped.

These domains are briefly explained below. In addition, Table 8.1 delineates the principal and key basic cognitive domains impacted by schizophrenia, along with the battery of tests employed in the MCCB for their evaluation:

Table 8.1 Cognitive domains and tests used in the MATRICS Consensuated Cognition Battery

Cognitive domain	Subtest
Speed of processing	*Brief Assessment of Cognition in Schizophrenia* (BACS): Symbol-Coding *Category Fluency*: Animal Naming Trail Making Test: Part A
Attention/vigilance	*Continuous Performance Test*—Identical Pairs (CPT-IP)
Verbal working memory	*Letter–Number Span*
Visual working memory	*Wechsler Memory Scale*®—3rd Ed. (WMS®-III): Spatial Span
Verbal learning and memory	*Hopkins Verbal Learning Test—Revised*™ (HVLT-R™)
Visual learning and memory	*Brief Visuospatial Memory Test—Revised* (BVMT-R™)
Reasoning and problem solving	*Neuropsychological Assessment Battery*® (NAB®): Mazes
Emotional intelligence	*Mayer-Salovey-Caruso Emotional Intelligence Test* (MSCEIT™): Managing Emotions

Processing speed. It refers to the rate at which an individual completes perceptual-cognitive or motor tasks. An impairment in processing speed might indicate a tendency toward sequential processing rather than pre-task planning. This domain also encompasses several cognitive processes involved in perceiving, analyzing, and responding to stimuli promptly. For example, during simple reaction time tasks where participants are required to respond quickly to a stimulus (e.g., pressing a button in response to a visual cue), patients with schizophrenia often tend to demonstrate prolonged reaction times compared to individuals without the disorder. An example of extrapolating these impairments to everyday situations is that, during conversations or when following instructions, patients with schizophrenia may exhibit delays in processing and responding to verbal information.

Attention/vigilance. This domain refers to the ability to maintain focus and concentration over prolonged periods and efficiently monitor the environment for relevant information. Tasks in daily life requiring continuous attention, such as attending lectures, engaging and maintaining conversations, or reading, may be particularly challenging, and may struggle to maintain focus for lengthy periods for individuals with schizophrenia.

Working memory. It refers to the cognitive system responsible for temporarily holding and manipulating information needed for tasks such as reasoning, comprehension, and learning. The presence of deficits in this domain may interfere with the execution of complex cognitive tasks. For example, during tasks that involve manipulating verbal information, such as verbal fluency activities, individuals with schizophrenia may present a reduced capacity to hold and manipulate information or details. In real-world functioning, working memory deficits can manifest as difficulties in following conversations, remembering instructions, organizing daily activities, or completing tasks that involve multiple steps.

Verbal learning and memory. It refers to the ability to recall or recognize verbally presented material, such as words or sentences. An impairment in this domain refers to the presence of alterations in the ability to acquire, retain, and recall verbal information effectively. Real-world examples of verbal learning and memory deficits can manifest across various everyday situations, impacting the ability to function effectively in academic, occupational, and social settings; for example, forgetting instructions, having difficulties following a conversation, having challenges in absorbing and recalling lecture material, etc.

Visuospatial learning and memory. It refers to the ability tomemorize and recall visually presented information. Visuospatial deficits can impact real-world activities such as assembling furniture or reading maps. People with schizophrenia may have trouble understanding spatial instructions or visualizing how parts fit together.

Reasoning and problem-solving/executive functioning. This domain refers to the ability to analyze information, make decisions, apply alternative strategies, and apply logical thinking to solve problems or to cope with new or unexpected problems. Impairment of these abilities is part of the broader cognitive impairment observed in individuals with schizophrenia and can impact various aspects of daily functioning (executive dysfunction, logical reasoning, problem-solving skills, or cognitive inflexibility).

As previously pointed out, schizophrenia is strongly linked to significant impairments in social skills and functioning. In this context, social cognition may serve as a crucial mediator between basic cognition and real-world functioning, influencing how cognitive abilities translate into everyday outcomes [15–17]. It is defined as the mental operations underlying social interactions, including perceiving, interpreting, and generating responses to the intentions, dispositions, and behaviors of others [17]. This cognitive domain poses a significant challenge for individuals with schizophrenia, encompassing difficulties in understanding both their own and others' mental states, interpreting verbal communication, engaging in social learning, shared intentionality, and collective cognition [18].

Social cognition, akin to other aspects of cognition, is a complex and multifaceted construct comprising different sub-domains. The National Institute of Mental Health (NIMH) consensus statement on social cognition in schizophrenia identified five key domains [17]:

Theory of mind (ToM). It refers to the ability to understand and attribute mental states, such as beliefs, intentions, desires, and emotions, to oneself and others. It involves recognizing that others have thoughts and feelings different from one's own and using this understanding to predict and interpret their behavior. A real-world example of a person with schizophrenia and alterations in this ability might be struggling to understand that a friend's sarcastic comment is meant humorously, leading to confusion or feelings of hurt.

Social perception. It refers to the capacity to accurately perceive and interpret social cues, contexts, and roles. This involves understanding non-verbal signals (facial expressions, body language, tone of voice), as well as recognizing social norms and expectations in different situations. Impairments at this point can often lead to misunderstandings and inappropriate social responses. For example, a person with schizophrenia may have difficulty interpreting a friend's body language, such as crossed arms and a stern expression, as signs of frustration or disapproval, leading to confusion or inappropriate responses during a conversation.

Attributional bias/style. This sub-domain refers to the tendency and the systematic errors individuals make when explaining the causes of their own and others' behaviors, intentions or motivations, and social events. As an example, a potential illustration of hostile attributional bias could be that during a team meeting at work, a person with schizophrenia might interpret a colleague's constructive criticism as a personal attack, resulting in feelings of hostility and a defensive reaction.

Social knowledge. It refers to the understanding of social norms, roles, and expectations within different social contexts. This includes knowing how to behave appropriately in different situations, recognizing social hierarchies, and understanding cultural norms. In a real-world scenario, a person with schizophrenia and deficits in social knowledge might struggle in a formal setting like a job interview. He or she may not adhere to typical interview etiquette, such as waiting for their turn to speak or dressing appropriately. These difficulties can result in negative evaluations from interviewers.

Emotion perception and processing. This domain refers to the capacity to precisely identify, interpret, and react to facial expressions, vocal tones, or other non-verbal signals that convey emotions. A real-world example could be the scene in which an individual with schizophrenia is having a conversation with his/her mum and may misinterpret her emotional tone of voice, perceiving a neutral statement as aggressive or hostile. This misunderstanding can escalate tensions and affect communication between them.

Regarding functionality, besides the impact of the previously mentioned neurocognitive domains on daily real-world performance, social cognition seems to be more strongly related to community functioning than basic neurocognition, with the strongest association being between ToM and functional outcomes [15]. Thus, social cognition has been shown to predict social outcomes but not non-social outcomes, while neurocognition provides minimal prediction of social outcomes [19].

Another main cognitive area that is pivotal and essential in schizophrenia is metacognition, as it has been a key target of novel therapies for improving aberrant thought processes, resultant symptoms, and functional outcomes in schizophrenia spectrum disorders [20]. This concept could be defined as "thinking about thinking" or 'knowledge and cognition about cognitive phenomena [21], and it refers to a spectrum of activities. At the lower end of the spectrum, discrete metacognitive capacity involves accurately inferring another's perspective, relying on basic neurocognitive skills like recognizing information and interpreting facial expressions (skills that at the same time present certain overlap with social cognition). At the higher end, metacognitive capacity focuses less on accurate social inferences and more on creating an integrated understanding of how individuals, including oneself, interpret and act within the social world [22]. Thus, this construct involves the ability to monitor and regulate one's cognitive processes, including thinking, memory, and problem-solving. This includes aspects such as insight into illness, self-reflection, and awareness of cognitive strengths and weaknesses [12].

Therefore, metacognition is a higher-order psychological process that allows us to evaluate our way of thinking and regulate our thoughts and, consequently, our behaviors, and finally contributes to impairments in daily functioning, social interactions, and treatment adherence [23]. Within this construct, it has been highlighted two basic components [24]:

Monitoring. It refers to the ability to observe and assess one's cognitive functioning. A person with schizophrenia may struggle with accurately monitoring their cognitive processes, leading to difficulties in recognizing errors or inefficiencies in thinking.

Control. It is the ability to regulate and modify these cognitive processes based on monitoring feedback. The presence of deficits in this component can manifest as challenges in planning, decision-making, and adjusting strategies in response to changing demands.

It has been noted that individuals demonstrating poorer metacognitive performance and higher scores on measures of self-certainty (e.g., endorsing statements like 'my interpretations of my experiences are right') tend to exhibit more pronounced cognitive difficulties [25]. This observation is critical, as it informs and is linked with the delivery of cognitive treatments; individuals with subjective impairments are often more inclined to participate in therapy, whereas those with more severe objective impairments may perceive less need for intervention [8, 25].

8.3 Cognitive Impairment in Schizophrenia

Approximately 85% of people diagnosed with schizophrenia exhibit significant cognitive dysfunctions, characterized by difficulties in concentration, slowed information processing, attentional challenges, memory problems, and executive functioning deficits [26]. Similar findings have been reported among patients not undergoing antipsychotic medication. Cognitive impairment is one of the core symptoms of schizophrenia, even present at the initial phase of the illness and through all the stages, from the earliest clinical manifestations [27–29] to chronic phases [30].

Impairments can range from moderate to severe magnitude in terms of severity. When compared with other psychiatric disorders, such as bipolar disorder and other affective disorders, schizophrenia presents more severe impairments. Even when they are compared with other neurological disorders, such as Parkinson's Disease or Alzheimer's Disease, similar levels of severity have been found in different cognitive domains [31]. It is not possible to find a specific profile of cognitive impairment in patients with schizophrenia. Only with slight variations in the degree of impairment, nearly all patients present significant alterations in all the different domains. The most frequently reported are impairments in memory, executive functions, attention, and processing speed [32, 33].

It is very important to highlight the pernicious effects that cognitive deficits in the daily lives of people with schizophrenia. Cognitive impairment causes concentration problems, difficulty staying alert, slowness in processing information, learning difficulties, memory disorders, and, above all, problems in executive functions such as difficulties in planning, organizing, and carrying out daily tasks [34]. These alterations have a devastating impact on daily functioning, even interfering with the recovery process. Nonetheless, these deficits significantly affect prognosis and social functioning, with neurocognitive variables explaining up to 60% of the variance in social functioning [33]. In a recent umbrella review [32] cognitive impairment in schizophrenia is strongly related to several factors, including substance use, lack of insight, history of childhood trauma and aggressive behavior, older age, and poor outcome in symptom dimensions and functionality, substance use, insight, symptom dimensions, history of childhood trauma, older age, and aggressive behavior.

8.4 Cognitive Trajectories: Early and Chronic Stages

The course of cognition across the life span in persons with schizophrenia and other psychotic disorders is still a veritable open question. Different trajectories in cognitive functioning have been proposed: stable impairment, progressive decline, improvement, or fluctuating course. Understanding cognitive trajectories and their relationship with negative symptoms, from the early phases of psychosis to the chronic stages of schizophrenia, could lead to more effective, personalized, and proactive care strategies, ultimately enhancing the quality of life for individuals affected by these complex disorders.

Cognitive impairment is thought to be already established before the prodromal stages of the illness, and no evidence of cognitive decline is found over time [35]. Different longitudinal studies have also found stability in cognitive impairment in schizophrenia, at least over the first 5 years of illness [36]. Those data are in line with the neurodevelopmental model. However, that would not be the case for all patients. At least, those patients with persistent psychotic symptoms and resistance to pharmacotherapy seem to experience a significant cognitive decline [37]. Moreover, schizophrenia and other different psychotic disorders could follow different longitudinal patterns [38].

Impairment can be observed at illness onset, and longitudinal studies showed scarce evidence of cognitive decline, suggesting that cognitive impairment remained stable through the two first decades of the illness. However, in samples with a diagnosis of schizophrenia, a different scenario is found as it has been described that cognitive impairments remain stable during the first decades but eventually worsen in later stages. Thus, different studies found consistent cognitive decline after 10-year and even 20-year periods of follow-up [39, 40]. A recent study found a loss of 16 IQ points throughout observation, suggesting a neurodegenerative pattern [38]. Although evidence found in the studies is mixed, when longer follow-up has been carried out, then slow but significant cognitive declines are detected, particularly in schizophrenia diagnosis.

All in all, cognitive trajectories in psychotic disorders are consistent with the neurodevelopmental model, but when a schizophrenia diagnosis is established, especially with persistent and pharmacologically resistant symptoms, neurodegenerative evolution can also be detected at later stages. In general, cognitive decline began before psychosis onset, suggesting a window for primary prevention. Finally, another window for secondary prevention occurs in the third decade of illness, when a progression in cognitive impairment intensity is accelerated.

8.4.1 Early Stages

The early stages of schizophrenia encompass the period from the onset of initial symptoms to the first few years following the first episode of psychosis (FEP). Typically occurs earlier in men, with the peak age of onset in the late teens to early twenties. The peak for women is in the early to mid-twenties and may extend into the thirties, with a second peak in the perimenopause period [41]. In the words of [42], this phase is critical for diagnosis and intervention due to its impact on outcomes.

The concept of a critical period is essential to understanding that patients with FEP need specialized intervention programs to optimize their recovery, limit the risk of relapses, and prevent disability. The critical period hypothesis proposes that deterioration occurs aggressively during the early years of psychosis, the first 2–3 years, with relative stability subsequently [43]. The early phase of psychosis presents important opportunities for secondary prevention.

Cognitive impairment across different domains in terms of memory, attention, executive function, and processing speed problems is observed during the prodromal and early stages of schizophrenia. In FEP patients, the largest effects have been found in verbal memory, executive function, and IQ [42]. Several factors can predict cognitive dysfunction in the early stages, such as genetic predispositions, neurodevelopmental abnormalities, and early-life traumatic events. The correlations between cognitive deficits and negative symptoms are evident for executive functioning and theory of mind deficits in FEP [44].

8.4.2 Chronic Stages

Chronic stages of schizophrenia disorders refer to the long-term phase of the illness, which typically begins a few years after the FEP and continues throughout the individual's life. For many years, it has been maintained that a high percentage of patients with schizophrenia have very severe deficits, characterized by persistent symptoms and ongoing functional impairments. Fortunately, using Zipursky's words, 30 years of studies on individuals' FEP have changed our view of schizophrenia from a progressive and deteriorating disease to a manageable condition in the vast majority of cases [45]. This statement requires longitudinal research beyond the critical period, spanning years or even decades after a FEP. Such studies are quite challenging due to the multidimensional nature of the data, compounded by methodological questions regarding the best way to compare results from studies with different inclusion criteria and varying durations (between 5 and 30 years), questioning the validity of comparisons between studies with different follow-up periods [46].

Even when comparing prospective cohorts from early intervention programs with similar long-term follow-ups (between 8 and 10 years), several questions related to the outcomes remain unanswered (Table 8.2). An attempt to provide clarity in this aspect was the work carried out with a long-term follow-up study in a

Table 8.2 Articles of Early Intervention Programs (EIP) for psychosis with 7–12 years follow-up

Name	Start date	Article	Follow-up (years)	N baseline	N follow-up	Main results on negative symptoms and cognitive deficits
EEPIC	1997	Mihalopoulos et al.	7–8	723	65	Not significant change in negative symptoms; not reported information on cognitive deficits.
EPPIC	1997	Henry et al.	7–8	723	484	Symptomatic remission in 37–59%; not reported information on cognitive deficits.
TIPS	1997	Hegelstad et al.	10	281	174	More patients with higher negative and cognitive symptoms levels dropped out.
AESOP	1997	Zanelli et al.	10	532	106	Not reported information on negative symptoms; cognitive decline.
OPUS	1998	Austin et al.	10	496	357	Negative symptoms do not change over time; not reported information on cognitive deficits.
EASY	2001	Chan et al.	10	148	107	Negative symptoms do not change over time; not reported information on cognitive deficits.
PAFIP	2001	Rodriguez-Sanchez et al.	8–12	307	209	Stability on negative symptoms and cognitive deficits.

Spanish FEP cohort by [47]. FEP patients who had been involved in an early intervention program for 3 years were re-contacted 10 years later. The main results showed that a high percentage of cases (42%) had reached functional milestones in educational, vocational, and social areas [47]. These recovery rates confirmed the success of early intervention programs and challenged the notion that recovery is only achieved in 15% of patients with schizophrenia [48]. Only a subgroup of patients (9%) presented severe functional disability, displaying severe negative symptoms and cognitive impairment as well. Interestingly, when exploring cognitive trajectories, similar to meta-analytics findings, small improvements in certain cognitive domains were found. However, those improvements were comparable with healthy controls, suggesting ineluctable practice effects [36]. In all, the majority of patients (all but a subgroup who showed decline) do not show a significant change in cognitive performance during the first 10 years after a FEP [49].

In older adults with chronic schizophrenia, a meta-analysis by [50] investigating cognitive function revealed both generalized and specific impairments in language, immediate memory, and executive functioning. Interestingly, longitudinal studies by the same authors did not demonstrate significant cognitive decline over time [50]. These findings suggest that, while cognitive changes in chronic schizophrenia resemble typical age-related decline, they are likely more pronounced. Further research suggests these deficits might be compounded by age-related cognitive and functional decline [51, 52]. However, other studies [53] propose that these cognitive impairments begin earlier in life, potentially reflecting a form of "cognitive premature aging" in individuals with schizophrenia.

The success of early intervention programs can be compromised when a one-size-fits-all approach is accepted for all FEP, as it possibly provides adequate treatment only to a particular group of patients, while being insufficient for others. Studies using longitudinal data can contribute to a better understanding of the course of the illness and help clinicians to identify critical periods that merit specific interventions and adjust the timing, length, and intensity of the interventions to optimize primary outcomes (symptoms and functioning) for people who have experienced an FEP. Understanding the characteristics of patients and comprehending the diversity of outcomes can help design more specific and individualized interventions from which patients and their families, as well as clinicians and society in general, can benefit more.

In general, impairments across cognitive domains such as attention, working memory, language skills, and executive functioning tend to be moderate, with the most pronounced deficits repeatedly found in verbal learning and memory. All these neurocognitive domains are related to adaptive and social skills, with executive functions and verbal learning and memory showing more variance across more domains than other cognitive variables [3]. As previously mentioned, and as depicted in Fig. 8.1, negative symptoms in schizophrenia constitute another core domain alongside cognitive deficits. This symptomatology often persists through the course of the illness and tends to be less responsive to treatment compared to positive symptoms, and probably there is a complex interplay between negative symptoms and cognitive deficits. Both can mutually influence each other and contribute to functional impairment [54]. For instance, in the previously mentioned PAFIP cohort [47], when exploring the interaction of global cognitive deficit and negative symptoms, four clusters of outcomes at a 10-year follow-up were observed (see Fig. 8.2). In most of the FEP patients, cognitive functions and negative symptoms remain stable from baseline to 10-year follow-up. However, there is a subgroup (25%) that showed improvement in negative symptoms.

In summary, studying cognitive trajectories in schizophrenia, its heterogeneous presentation, and interaction with negative symptoms, not only enhances our understanding of these complex disorders but also informs better clinical practices and interventions, leading to improved outcomes for individuals affected by these conditions. This comprehensive understanding could lead to more effective, personalized, and proactive care strategies, ultimately enhancing the quality of life for individuals affected by these complex disorders.

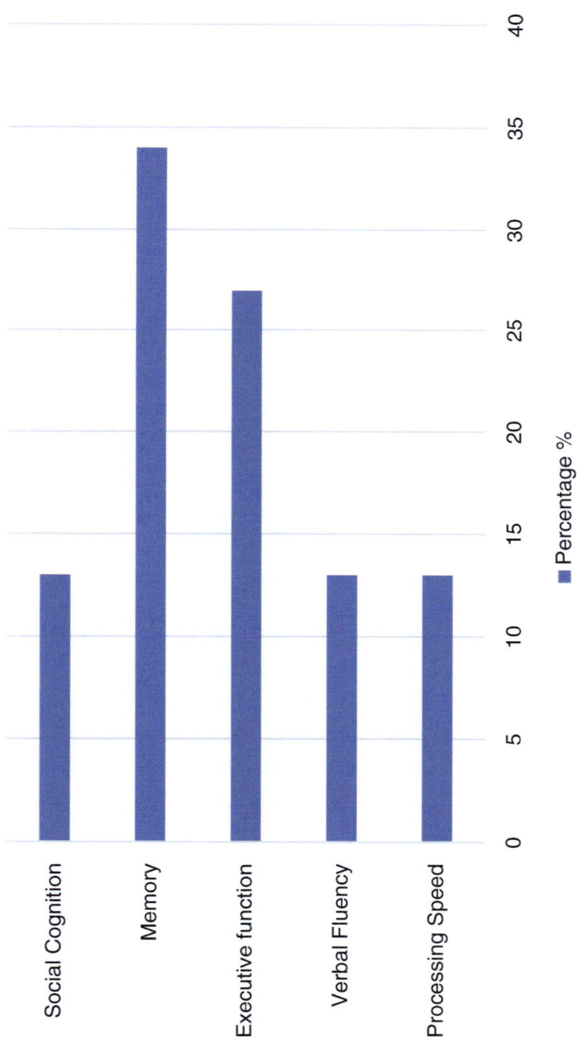

Fig. 8.1 Domain of cognition impaired in people with schizophrenia as extracted from systematic reviews of single domain reviews. (Adapted from Gebreegziabhere et al. [32])

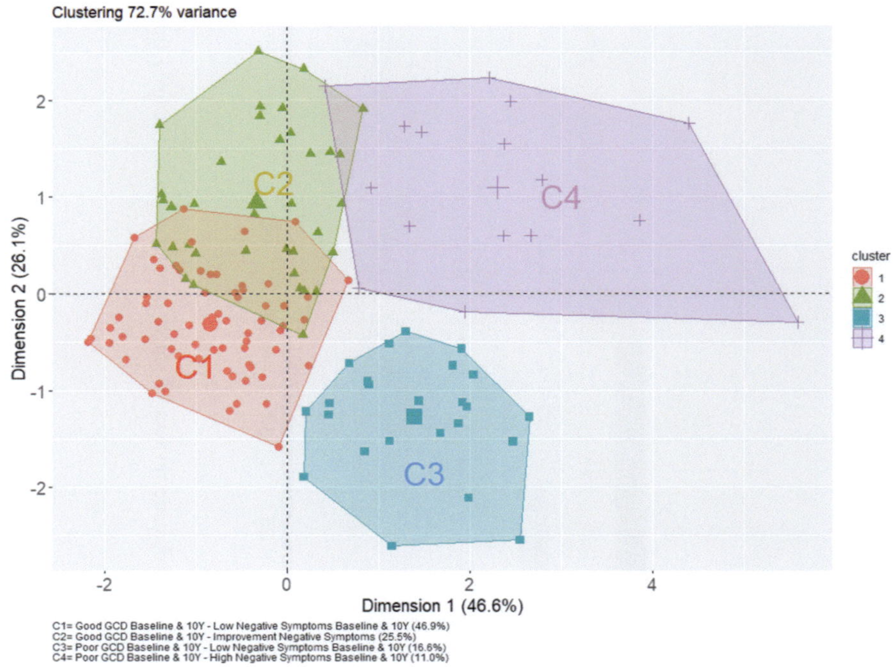

Fig. 8.2 Clustering cognitive functioning and negative symptoms trajectories in FEP patients from baseline to 10-year follow-up

8.5 Cognitive Reserve

Cognitive reserve (CR) is understood as the brain's capacity to cope with damage, delaying the onset and minimizing the severity of clinical, cognitive, and functional manifestations of the pathology. In this sense, when habitual cognitive strategies fail, CR facilitates the use of alternative approaches to achieve desired outcomes [55]. The concept of CR emerged from intriguing findings in Alzheimer's disease research, rooted in a paradigmatic study in the 1980s [56]. In this study, scientists observed a discrepancy between the severity of neuropathological changes in the brains of deceased nuns and their cognitive function during life. Some nuns with significant brain damage exhibited minimal cognitive decline. This phenomenon was attributed to the nuns' intellectually stimulating lifestyles, including extensive education, religious activities, and multilingualism. One of the major findings from the nun study was that participants' lifestyle and education might deter Alzheimer's symptoms; participants with a bachelor's degree or higher were less likely to develop the disease later in life.

The lack of a direct link between the degree of brain pathology and clinical manifestations gave rise to the term "reserve" as a moderator of this relationship [57]. In this context, there are two types of models. One is a passive model, where one of the pioneers in this area was Roth in 1955. He observed that the severity of dementia did not always correlate with the amount of brain damage measured by the number of senile plaques. To explain this discrepancy, the threshold theory was initially proposed, and the Brain Reserve (BR) capacity theory was developed based on the threshold theory [58]. This theory suggests that there is a threshold of brain damage that, when exceeded, leads to clinical symptoms and cognitive or functional deficits. BR refers to an individual's structural and morphological characteristics, such as brain size, synaptic density, and the number of neurons and synapses. Higher levels of BR are considered protective against symptoms and dysfunction, while lower levels are seen as a vulnerability factor. Later, a more complex model was proposed, suggesting that each person might have a different threshold of tolerance and that clinical manifestations depend on how the brain utilizes the remaining neural substrate [59]. According to this model, individuals with higher CR can use their neural networks more efficiently, employing alternative neural networks or cognitive strategies to compensate for the effects of brain pathology [59]. Thus, higher CR is viewed as a protective factor against the symptoms and dysfunctions of brain disease, whereas lower CR is a vulnerability factor. This model helps explain why individuals with similar brain pathologies can have different clinical presentations [57].

In recent decades, it has been suggested that the concept of CR can also be applied to mental disorders, including schizophrenia [60]. However, it has distinct characteristics. Firstly, unlike Alzheimer's disease, the psychopathology characteristic of schizophrenia itself might affect the amount of accumulated CR. Another assumption is that CR and neuropsychiatric pathology cannot be considered entirely independent variables [61].

Recent research underscores the critical role of lifestyle choices and experiences in predicting how people respond to, recover from, and manage mental disorders, including remission and relapse [62]. This highlights the importance of maintaining strong social connections, engaging in intellectually stimulating activities, and achieving higher levels of education. Interestingly, all these factors are intricately linked to CR. CR itself is not set in stone. While influenced by genetic factors and in-utero exposures, it can also be positively impacted by lifelong choices. These include education, occupation, physical exercise, leisure activities, and social engagement [55, 60, 63, 64].

Since CR is a theoretical construct, direct measurement remains challenging. However, researchers have developed a multifaceted approach to estimate CR, employing two main methods:

- Socio-behavioral proxies: These measures include intelligence quotient (IQ), education level, occupational complexity, frequency of social interaction, and participation in cultural and leisure activities. While valuable insights into potential CR contributors, they are not direct measures of CR itself, but rather indicators of the experiences that may have shaped it. The first tool specifically designed for adults with severe mental illness, the Cognitive Reserve in Health Assessment Scale (CRASH) [63], was validated in 2019. This valid, efficient, short, and ecologically sensitive scale allows for a one-time assessment of CR across various

psychiatric disorders. CRASH estimates CR by gathering information about education (including schooling, parents' education, childhood and adolescent school performance, and languages spoken), occupation (work experience and training), and intellectual and leisure activities (physical, intellectual, artistic, and cultural activities, as well as social engagement). This final domain is assessed by considering different stages of life (childhood/adolescence, adulthood, and current situation).
- Functional neuroimaging: It offers a "more direct" approach to understanding CR, to identify functional brain activation patterns, both during resting-state and task-related conditions, that may underlie CR. These patterns should exhibit two key characteristics: the expression of these functional activation patterns should correlate with established socio-behavioral indicators of CR, and these patterns should modulate the effect of brain changes on cognitive function [55].

The literature suggests that a higher CR level is associated with a later age of onset, higher insight, and lower disease severity, defined as lower symptomatology, better cognitive performance, and better psychosocial functioning [60, 61, 65]. Additionally, CR has been proposed as a protective factor that is related to the later onset of psychosis, better prognosis, greater adhesion, and fewer psychotic symptoms [60]. Amongst several possible explanations, in schizophrenia, CR might act as a protective factor against developing schizophrenia and reducing the severity of its clinical manifestations. Additionally, higher CR is associated with a more favorable course of cognitive and behavioral symptoms in schizophrenia [66].

From a clinical perspective, higher levels of CR have been associated with fewer negative symptoms after a first episode of schizophrenia during childhood or adolescence [67] and adulthood [47, 65]. Higher CR is also associated with better psychosocial functioning in both the early [65, 68] and chronic stages of non-affective psychosis [66]. Finally, regarding cognition, higher CR is associated with better neuropsychological performance in FEP [65, 68–70] and schizophrenia [66].

The fundamental role of CR in the course of the illness has also been demonstrated. Patients who relapsed but had higher CR showed less deterioration in attention, while those with higher CR who did not relapse performed better in processing speed and visual memory. Overall, CR seems to mitigate the negative effects of relapse, providing evidence for CR's protective effect throughout the illness [71].

Finally, recent studies [72] have highlighted a potential link between childhood maltreatment and cognitive and psychosocial functioning after a FEP. This suggests that CR may play a moderating role in this relationship, particularly regarding subjective cognitive complaints. Moreover, CR appears to act as a buffer, mitigating the negative effects of certain types of childhood maltreatment on specific aspects of memory, social activities, and relationships.

8.6 Cognition as a Target in the Treatment of Schizophrenia

One of the most important recent research findings in the field of schizophrenia is the discovery that cognition is strongly related to psychosocial functioning in daily life. Indeed, among all the clinical dimensions of schizophrenia, cognition has been pointed to be the key factor causing psychosocial disability [73]. Cognitive impairment affects the ability to be a student, friend, spouse, or family member. It interferes with the ability to care for themselves, to get a job, or to be able to participate in leisure and recreational activities. Additionally, it is one of the most important barriers to psychosocial rehabilitation programs. Those are important reasons to consider treating cognition as a priority objective of treatment.

Cognitive deficits have been documented at illness onset and persist for most of the patient's life without periods of spontaneous remission [74]. Symptom remission is more frequent, although many patients experience regular relapse even with adherence to oral or long-acting injectable medications. Besides, it is important to remark that even after a clinical remission of symptoms, cognitive impairment remains still a persistent clinical problem. Symptomatic remissions do not predict functional recovery, and functional deficits are associated with poor cognition, leading to very poor occupational and residential outcomes [75]. For these reasons, treating cognition is crucial to reduce the associated psychosocial disability and to facilitate recovery in schizophrenia [9].

International guidelines recommend directly treating cognition in patients with schizophrenia [9]. For those purposes, it is broadly recommended the use second-generation antipsychotics as they interfere less with cognition than first-generation antipsychotics. It has been largely demonstrated that atypical antipsychotics are superior to typical at improving overall cognitive function [76]. It is important to prevent the anticholinergic and benzodiazepine burdens, and they should be kept to a minimum. Moreover, cognitive remediation and physical exercise are recommended for the treatment of cognitive impairment in schizophrenia.

Cognitive remediation is a learning-based intervention that targets cognitive difficulties (attention, memory, executive function), to improve functional outcomes [77]. Meta-analyses have demonstrated that it is an evidence-based intervention, with improvements ranging from small to medium effects on cognition and functioning [52, 78]. It has been considered with the highest degree of recommendation according to the guidance provided by international experts [9]. However, to be successful, certain essential elements are crucial: an active and skilled therapist, the structured development of cognitive strategies, and integration with psychosocial rehabilitation programs [78].

The precise neurobiological processes underlying cognitive improvement in the context of cognitive remediation in schizophrenia have rarely been investigated. Some have assumed that those effects are related to neuroplasticity [79]. Neuroimaging studies suggest that cognitive remediation could facilitate neuroplasticity processes, as some changes in brain functioning have been described after the cognitive therapy. Improvements in patterns of activation [80], changes in neuro connectivity [81, 82], and even some structural changes in white matter [83, 84] and grey matter [85, 86] have been described. Cognitive improvement appears to be mediated by the release of neurotrophic factors such as BDNF. Some improvements in BDNF serum levels have been described [87]. Epigenetic changes in the promoter zones of the BDNF gen because of cognitive remediation outcomes have also been found [88]. Thus, cognitive remediation is a possible option to ameliorate the progressive brain changes in schizophrenia to some extent, and it could help to prevent functional impairment.

References

1. van Os J, Kapur S. Schizophrenia. Lancet. 2009;374(9690):635–45.
2. Green MF. What are the functional consequences of neurocognitive deficits in schizophrenia? Am J Psychiatry. 1996;153(3):321–30.
3. Bowie CR, Harvey PD. Cognition in schizophrenia: impairments, determinants, and functional importance. Psychiatr Clin North Am. 2005;28(3):613–33, 26.
4. Berrios GE, Luque R, Villagrán JM. Schizophrenia: a conceptual history. Int J Psychol Psychol Therapy. 2003;3(2):111–40.
5. Bleuler E. Dementia praecox or the Group of Schizophrenias. International Universities Press; 1950.
6. Schneider K. Clinical psychopathology. Grune & Stratton; 1959.
7. Lezak MD, Howieson DB, Bigler ED, Tranel D. Neuropsychological assessment, vol. xxv. 5th ed. New York: Oxford University Press; 2012. p. 1161.
8. McCutcheon RA, Keefe RSE, McGuire PK. Cognitive impairment in schizophrenia: aetiology, pathophysiology, and treatment. Mol Psychiatry. 2023;28(5):1902–18.
9. Vita A, Gaebel W, Mucci A, Sachs G, Barlati S, Giordano GM, et al. European psychiatric association guidance on treatment of cognitive impairment in schizophrenia. Eur Psychiatry. 2022;65(1):e57.
10. Ventura J, Welikson T, Ered A, Subotnik KL, Keefe RSE, Hellemann GS, et al. Virtual reality assessment of functional capacity in the early course of schizophrenia: associations with cognitive performance and daily functioning. Early Interv Psychiatry. 2020;14(1):106–14.
11. Green MF, Nuechterlein KH, Gold JM, Barch DM, Cohen J, Essock S, et al. Approaching a consensus cognitive battery for clinical trials in schizophrenia: the NIMH-MATRICS conference to select cognitive domains and test criteria. Biol Psychiatry. 2004;56(5):301–7.
12. Lysaker PH, Hamm JA, Hasson-Ohayon I, Pattison ML, Leonhardt BL. Promoting recovery from severe mental illness: implications from research on metacognition and metacognitive reflection and insight therapy. World J Psychiatry. 2018;8(1):1–11.
13. Nuechterlein KH, Green MF, Kern RS, Baade LE, Barch DM, Cohen JD, et al. The MATRICS consensus cognitive battery, part 1: test selection, reliability, and validity. Am J Psychiatry. 2008;165(2):203–13.
14. Tandon R, Nasrallah HA, Keshavan MS. Schizophrenia, "just the facts" 4. Clinical features and conceptualization. Schizophr Res. 2009;110(1–3):1–23.
15. Fett AK, Viechtbauer W, Dominguez MD, Penn DL, van Os J, Krabbendam L. The relationship between neurocognition and social cognition with functional outcomes in schizophrenia: a meta-analysis. Neurosci Biobehav Rev. 2011;35(3):573–88.
16. Frau F, Cerami C, Dodich A, Bosia M, Bambini V. Weighing the role of social cognition and executive functioning in pragmatics in the schizophrenia spectrum: a systematic review and meta-analysis. Brain Lang. 2024;252:105403.
17. Green MF, Penn DL, Bentall R, Carpenter WT, Gaebel W, Gur RC, et al. Social cognition in schizophrenia: an NIMH workshop on definitions, assessment, and research opportunities. Schizophr Bull. 2008;34(6):1211–20.

18. Ventura J, Wood RC, Hellemann GS. Symptom domains and neurocognitive functioning can help differentiate social cognitive processes in schizophrenia: a meta-analysis. Schizophr Bull. 2013;39(1):102–11.
19. Harvey PD, Isner EC. Cognition, social cognition, and functional capacity in early-onset schizophrenia. Child Adolesc Psychiatr Clin N Am. 2020;29(1):171–82.
20. Melville G, Hoffman M, Pollock A, Kurtz MM. Do metacognitive therapies for schizophrenia-spectrum disorders work? A meta-analytic investigation. Psychol Med. 2024;54(8):1510–8.
21. Flavell JH. Metacognition and cognitive monitoring: a new area of cognitive–developmental inquiry. Am Psychol. 1979;34(10):906–11.
22. McGuire N, Gumley A, Hasson-Ohayon I, Allan S, Aunjitsakul W, Aydin O, et al. Investigating the relationship between specific negative symptoms and metacognitive functioning in psychosis: a systematic review. Psychol Psychother. 2024;97(2):191–214.
23. Hasson-Ohayon I, Goldzweig G, Lavi-Rotenberg A, Luther L, Lysaker PH. The centrality of cognitive symptoms and metacognition within the interacting network of symptoms, neurocognition, social cognition and metacognition in schizophrenia. Schizophr Res. 2018;202:260–6.
24. Diaz-Cutraro L, Garcia-Mieres H, Dimaggio G, Lysaker P, Moritz S, Ochoa S. Metacognition in psychosis: what and how do we assess it? Span J Psychiatry Ment Health. 2023;16(3):206–7.
25. Moritz S, Balzan RP, Bohn F, Veckenstedt R, Kolbeck K, Bierbrodt J, et al. Subjective versus objective cognition: evidence for poor metacognitive monitoring in schizophrenia. Schizophr Res. 2016;178(1–3):74–9.
26. Kharawala S, Hastedt C, Podhorna J, Shukla H, Kappelhoff B, Harvey PD. The relationship between cognition and functioning in schizophrenia: a semi-systematic review. Schizophr Res Cogn. 2022;27:100217.
27. Fusar-Poli P, Deste G, Smieskova R, Barlati S, Yung AR, Howes O, et al. Cognitive functioning in prodromal psychosis: a meta-analysis. Arch Gen Psychiatry. 2012;69(6):562–71.
28. Bora E, Lin A, Wood SJ, Yung AR, McGorry PD, Pantelis C. Cognitive deficits in youth with familial and clinical high risk to psychosis: a systematic review and meta-analysis. Acta Psychiatr Scand. 2014;130(1):1–15.
29. Catalan A, McCutcheon RA, Aymerich C, Pedruzo B, Radua J, Rodriguez V, et al. The magnitude and variability of neurocognitive performance in first-episode psychosis: a systematic review and meta-analysis of longitudinal studies. Transl Psychiatry. 2024;14(1):15.
30. Stone WS, Phillips MR, Yang LH, Kegeles LS, Susser ES, Lieberman JA. Neurodegenerative model of schizophrenia: growing evidence to support a revisit. Schizophr Res. 2022;243:154–62.
31. Millan MJ, Agid Y, Brune M, Bullmore ET, Carter CS, Clayton NS, et al. Cognitive dysfunction in psychiatric disorders: characteristics, causes and the quest for improved therapy. Nat Rev Drug Discov. 2012;11(2):141–68.
32. Gebreegziabhere Y, Habatmu K, Mihretu A, Cella M, Alem A. Cognitive impairment in people with schizophrenia: an umbrella review. Eur Arch Psychiatry Clin Neurosci. 2022;272(7):1139–55.
33. Green MF, Kern RS, Braff DL, Mintz J. Neurocognitive deficits and functional outcome in schizophrenia: are we measuring the "right stuff"? Schizophr Bull. 2000;26(1):119–36.
34. Tschentscher N, Woll CFJ, Tafelmaier JC, Kriesche D, Bucher JC, Engel RR, et al. Neurocognitive deficits in first-episode and chronic psychotic disorders: a systematic review from 2009 to 2022. Brain Sci. 2023;13(2):299.
35. Bora E, Murray RM. Meta-analysis of cognitive deficits in ultra-high risk to psychosis and first-episode psychosis: do the cognitive deficits progress over, or after, the onset of psychosis? Schizophr Bull. 2014;40(4):744–55.
36. Watson AJ, Harrison L, Preti A, Wykes T, Cella M. Cognitive trajectories following onset of psychosis: a meta-analysis. Br J Psychiatry. 2022;221(6):714–21.
37. Spangaro M, Martini F, Bechi M, Buonocore M, Agostoni G, Cocchi F, et al. Longitudinal course of cognition in schizophrenia: does treatment resistance play a role? J Psychiatr Res. 2021;141:346–52.

38. Jonas K, Lian W, Callahan J, Ruggero CJ, Clouston S, Reichenberg A, et al. The course of general cognitive ability in individuals with psychotic disorders. JAMA Psychiatry. 2022;79(7):659–66.
39. Zanelli J, Mollon J, Sandin S, Morgan C, Dazzan P, Pilecka I, et al. Cognitive change in schizophrenia and other psychoses in the decade following the first episode. Am J Psychiatry. 2019;176(10):811–9.
40. Zanelli J, Reichenberg A, Sandin S, Morgan C, Dazzan P, Pilecka I, et al. Dynamic and static cognitive deficits in schizophrenia and bipolar disorder after the first episode. Schizophr Bull. 2022;48(3):590–8.
41. Diaz-Pons A, Gonzalez-Rodriguez A, Ortiz-Garcia de la Foz V, Seeman MV, Crespo-Facorro B, Ayesa-Arriola R. Disentangling early and late onset of psychosis in women: identifying new targets for treatment. Arch Womens Ment Health. 2022;25(2):335–44.
42. McGorry PD, Killackey E, Yung A. Early intervention in psychosis: concepts, evidence and future directions. World Psychiatry. 2008;7(3):148–56.
43. Birchwood M, Todd P, Jackson C. Early intervention in psychosis. The critical period hypothesis. Br J Psychiatry Suppl. 1998;172(33):53–9.
44. Melillo A, Caporusso E, Giordano GM, Giuliani L, Pezzella P, Perrottelli A, et al. Correlations between negative symptoms and cognitive deficits in individuals at first psychotic episode or at high risk of psychosis: a systematic review. J Clin Med. 2023;12(22):7095.
45. Zipursky RB, Agid O, Remington G. Improving outcomes in schizophrenia by preventing early relapses. Lancet Psychiatry. 2018;5(5):384–6.
46. McGorry PD, Mei C. Ultra-high-risk paradigm: lessons learnt and new directions. Evid Based Ment Health. 2018;21(4):131–3.
47. Ayesa-Arriola R, Ortiz-Garcia de la Foz V, Martinez-Garcia O, Setien-Suero E, Ramirez ML, Suarez-Pinilla P, et al. Dissecting the functional outcomes of first episode schizophrenia spectrum disorders: a 10-year follow-up study in the PAFIP cohort. Psychol Med. 2021;51(2):264–77.
48. Clemmensen L, Vernal DL, Steinhausen HC. A systematic review of the long-term outcome of early onset schizophrenia. BMC Psychiatry. 2012;12:150.
49. Rodriguez-Sanchez JM, Setien-Suero E, Suarez-Pinilla P, Mayoral Van Son J, Vazquez-Bourgon J, Gil Lopez P, et al. Ten-year course of cognition in first-episode non-affective psychosis patients: PAFIP cohort. Psychol Med. 2022;52(4):770–9.
50. Irani F, Kalkstein S, Moberg EA, Moberg PJ. Neuropsychological performance in older patients with schizophrenia: a meta-analysis of cross-sectional and longitudinal studies. Schizophr Bull. 2011;37(6):1318–26.
51. Rajji TK. Schizophrenia and cognition in late life. In: Chiu H, Shulman K, editors. Mental health and illness of the elderly. Singapore: Springer Singapore; 2017. p. 1–13.
52. Wykes T, Reeder C, Landau S, Matthiasson P, Haworth E, Hutchinson C. Does age matter? Effects of cognitive rehabilitation across the age span. Schizophr Res. 2009;113(2–3):252–8.
53. Rajji TK, Voineskos AN, Butters MA, Miranda D, Arenovich T, Menon M, et al. Cognitive performance of individuals with schizophrenia across seven decades: a study using the MATRICS consensus cognitive battery. Am J Geriatr Psychiatry. 2013;21(2):108–18.
54. Correll CU, Schooler NR. Negative symptoms in schizophrenia: a review and clinical guide for recognition, assessment, and treatment. Neuropsychiatr Dis Treat. 2020;16:519–34.
55. Stern Y, Arenaza-Urquijo EM, Bartres-Faz D, Belleville S, Cantilon M, Chetelat G, et al. Whitepaper: defining and investigating cognitive reserve, brain reserve, and brain maintenance. Alzheimers Dement. 2020;16(9):1305–11.
56. Katzman R, Terry R, DeTeresa R, Brown T, Davies P, Fuld P, et al. Clinical, pathological, and neurochemical changes in dementia: a subgroup with preserved mental status and numerous neocortical plaques. Ann Neurol. 1988;23(2):138–44.
57. Stern Y. Cognitive reserve in ageing and Alzheimer's disease. Lancet Neurol. 2012;11(11):1006–12.
58. Satz P. Brain reserve capacity on symptom onset after brain injury: a formulation and review of evidence for threshold theory. Neuropsychology. 1993;7(3):273–95.

59. Stern Y. What is cognitive reserve? Theory and research application of the reserve concept. J Int Neuropsychol Soc. 2002;8(3):448–60.
60. Barnett JH, Salmond CH, Jones PB, Sahakian BJ. Cognitive reserve in neuropsychiatry. Psychol Med. 2006;36(8):1053–64.
61. Amoretti S, Ramos-Quiroga JA. Cognitive reserve in mental disorders. Eur Neuropsychopharmacol. 2021;49:113–5.
62. Solmi M, Croatto G, Piva G, Rosson S, Fusar-Poli P, Rubio JM, et al. Efficacy and acceptability of psychosocial interventions in schizophrenia: systematic overview and quality appraisal of the meta-analytic evidence. Mol Psychiatry. 2023;28(1):354–68.
63. Amoretti S, Cabrera B, Torrent C, Bonnin CDM, Mezquida G, Garriga M, et al. Cognitive reserve assessment scale in health (CRASH): its validity and reliability. J Clin Med. 2019;8(5):586.
64. Walhovd KB, Nyberg L, Lindenberger U, Amlien IK, Sorensen O, Wang Y, et al. Brain aging differs with cognitive ability regardless of education. Sci Rep. 2022;12(1):13886.
65. Amoretti S, Bernardo M, Bonnin CM, Bioque M, Cabrera B, Mezquida G, et al. The impact of cognitive reserve in the outcome of first-episode psychoses: 2-year follow-up study. Eur Neuropsychopharmacol. 2016;26(10):1638–48.
66. Herrero P, Contador I, Stern Y, Fernandez-Calvo B, Sanchez A, Ramos F. Influence of cognitive reserve in schizophrenia: a systematic review. Neurosci Biobehav Rev. 2020;108:149–59.
67. Camprodon-Boadas P, de la Serna E, Baeza I, Puig O, Ilzarbe D, Sugranyes G, et al. Cognitive reserve in patients with first-episode psychosis as outcome predictor at 5-year follow-up. Eur Child Adolesc Psychiatry. 2021;30(12):1959–67.
68. Camprodon-Boadas P, Rosa-Justicia M, Sugranyes G, Moreno D, Baeza I, Ilzarbe D, et al. Cognitive reserve and its correlates in child and adolescent offspring of patients diagnosed with schizophrenia or bipolar disorder. Eur Child Adolesc Psychiatry. 2023;32(8):1463–73.
69. Amoretti S, Cabrera B, Torrent C, Mezquida G, Lobo A, Gonzalez-Pinto A, et al. Cognitive reserve as an outcome predictor: first-episode affective versus non-affective psychosis. Acta Psychiatr Scand. 2018;138(5):441–55.
70. de la Serna E, Andres-Perpina S, Puig O, Baeza I, Bombin I, Bartres-Faz D, et al. Cognitive reserve as a predictor of two year neuropsychological performance in early onset first-episode schizophrenia. Schizophr Res. 2013;143(1):125–31.
71. Sanchez-Torres AM, Amoretti S, Enguita-German M, Mezquida G, Moreno-Izco L, Panadero-Gomez R, et al. Relapse, cognitive reserve, and their relationship with cognition in first episode schizophrenia: a 3-year follow-up study. Eur Neuropsychopharmacol. 2023;67:53–65.
72. Fares-Otero NE, Borras R, Sole B, Torrent C, Garriga M, Serra-Navarro M, et al. Cognitive reserve moderates the relationship between childhood maltreatment, objective and subjective cognition, and psychosocial functioning in individuals with first-episode psychosis. Psychol Trauma. 2025;17(3):520–36.
73. Green MF. Impact of cognitive and social cognitive impairment on functional outcomes in patients with schizophrenia. J Clin Psychiatry. 2016;77(Suppl 2):8–11.
74. Holmen A, Juuhl-Langseth M, Thormodsen R, Melle I, Rund BR. Neuropsychological profile in early-onset schizophrenia-spectrum disorders: measured with the MATRICS battery. Schizophr Bull. 2010;36(4):852–9.
75. Harvey PD, Heaton RK, Carpenter WT Jr, Green MF, Gold JM, Schoenbaum M. Functional impairment in people with schizophrenia: focus on employability and eligibility for disability compensation. Schizophr Res. 2012;140(1–3):1–8.
76. Woodward ND, Purdon SE, Meltzer HY, Zald DH. A meta-analysis of neuropsychological change to clozapine, olanzapine, quetiapine, and risperidone in schizophrenia. Int J Neuropsychopharmacol. 2005;8(3):457–72.
77. Bowie CR, Bell MD, Fiszdon JM, Johannesen JK, Lindenmayer JP, McGurk SR, et al. Cognitive remediation for schizophrenia: an expert working group white paper on core techniques. Schizophr Res. 2020;215:49–53.

78. Vita A, Barlati S, Ceraso A, Nibbio G, Ariu C, Deste G, et al. Effectiveness, core elements, and moderators of response of cognitive remediation for schizophrenia: a systematic review and meta-analysis of randomized clinical trials. JAMA Psychiatry. 2021;78(8):848–58.
79. Matsuda Y, Makinodan M, Morimoto T, Kishimoto T. Neural changes following cognitive remediation therapy for schizophrenia. Psychiatry Clin Neurosci. 2019;73(11):676–84.
80. Mothersill D, Donohoe G. Neural effects of cognitive training in schizophrenia: a systematic review and activation likelihood estimation meta-analysis. Biol Psychiatry Cogn Neurosci Neuroimaging. 2019;4(8):688–96.
81. Penades R, Segura B, Inguanzo A, Garcia-Rizo C, Catalan R, Masana G, et al. Cognitive remediation and brain connectivity: a resting-state fMRI study in patients with schizophrenia. Psychiatry Res Neuroimaging. 2020;303:111140.
82. Ramsay IS, Nienow TM, Marggraf MP, MacDonald AW. Neuroplastic changes in patients with schizophrenia undergoing cognitive remediation: triple-blind trial. Br J Psychiatry. 2017;210(3):216–22.
83. Matsuoka K, Morimoto T, Matsuda Y, Yasuno F, Taoka T, Miyasaka T, et al. Computer-assisted cognitive remediation therapy for patients with schizophrenia induces microstructural changes in cerebellar regions involved in cognitive functions. Psychiatry Res Neuroimaging. 2019;292:41–6.
84. Penades R, Pujol N, Catalan R, Massana G, Rametti G, Garcia-Rizo C, et al. Brain effects of cognitive remediation therapy in schizophrenia: a structural and functional neuroimaging study. Biol Psychiatry. 2013;73(10):1015–23.
85. Eack SM, Hogarty GE, Cho RY, Prasad KM, Greenwald DP, Hogarty SS, et al. Neuroprotective effects of cognitive enhancement therapy against gray matter loss in early schizophrenia: results from a 2-year randomized controlled trial. Arch Gen Psychiatry. 2010;67(7):674–82.
86. Morimoto T, Matsuda Y, Matsuoka K, Yasuno F, Ikebuchi E, Kameda H, et al. Computer-assisted cognitive remediation therapy increases hippocampal volume in patients with schizophrenia: a randomized controlled trial. BMC Psychiatry. 2018;18(1):83.
87. Vinogradov S, Fisher M, Holland C, Shelly W, Wolkowitz O, Mellon SH. Is serum brain-derived neurotrophic factor a biomarker for cognitive enhancement in schizophrenia? Biol Psychiatry. 2009;66(6):549–53.
88. Penades R, Almodovar-Paya C, Garcia-Rizo C, Ruiz V, Catalan R, Valero S, et al. Changes in BDNF methylation patterns after cognitive remediation therapy in schizophrenia: a randomized and controlled trial. J Psychiatr Res. 2024;173:166–74.

Cognitive Impairment and Psychosocial Functioning in Schizophrenia

9

Noemi Sansone, Anna Boccia, Silvana Galderisi, and Armida Mucci

Abstract

The present chapter explores the relationship between cognitive impairment and psychosocial functioning in schizophrenia. Despite advances in treatments, achieving real-life functional goals remains challenging for persons with schizophrenia, while current treatment goals emphasize recovery, including clinical, social, and personal aspects. There is consolidated evidence that impairments in neurocognition and social cognition impact real-life functioning more than psychotic symptoms; however, these domains are not routinely assessed and treated in clinical practice.

Cognitive dysfunctions appear early in the course of illness, often predate the onset of psychosis, are associated with poor premorbid adjustment, and predict worse outcomes. Most cognitive domains, such as processing speed, attention, memory, and executive functions, are associated with poor functioning. Social cognition, in addition to its direct effect on real-life functioning, also mediates the relationships between neurocognitive subdomains and functioning, highlighting the need for holistic treatment. Negative symptoms and factors like social support also seem to have an impact on cognitive dysfunctions.

Many tools for the assessment of cognition are available, though there is no consensus on a specific one, especially in the context of clinical practice.

In conclusion, the present chapter highlights the importance of neurocognition and social cognition assessment and treatment in the context of integrated pathways of care to improve quality of life and the rates of recovery in people with schizophrenia.

N. Sansone · A. Boccia · S. Galderisi · A. Mucci (✉)
University of Campania, Luigi Vanvitelli, Naples, Italy

A. Vita et al. (eds.), *Negative and Cognitive Symptoms in Schizophrenia*, Comprehensive Approach to Psychiatry 3,
https://doi.org/10.1007/978-3-032-06209-3_9

Keywords

Schizophrenia · Cognitive impairment · Psychosocial functioning · Real-life functioning · Functional capacity · Recovery · Neurocognition · Social cognition · Negative symptoms · Treatment strategies

9.1 Introduction

9.1.1 Schizophrenia and Psychosocial Functioning

People suffering from schizophrenia frequently experience an impairment in psychosocial functioning, with difficulties in everyday life activities, social interactions, interpersonal relationships, and education/occupational performance. The impairment is observed in all stages of the disease [1], and the complex interplay among cognitive, affective, and behavioral dysfunctions exacerbates these difficulties that not only affect the individuals themselves but also place significant distress on their families and the healthcare systems that support them [2].

Despite the introduction of innovative medications and therapies to promote symptomatic remission [3–7], the challenge of helping individuals with schizophrenia in fulfilling their goals in real-life persists. A comprehensive, personalized, and integrated treatment addressing the different dysfunctional domains is the most appropriate strategy to improve functional outcomes and community integration [8, 9].

Functioning in schizophrenia is generally described in terms of two distinct constructs: *real-life functioning* (RLF)/community functioning (what a person does in real life, sometimes also referred to as *real-world functioning*) and *functional capacity* (FC) (what a person can do under optimal conditions) [10]. RLF includes *social functioning*, i.e., aspects like engaging in social activities as well as maintaining relationships with family and friends; *occupational or vocational functioning*, which evaluates performance at work or school, including work skills, performance and adaptability in a work environment; *independent living*, that assesses an individual's ability to manage day-to-day activities, such as housekeeping, managing finances, and transportation. Functional capacity (FC) refers to the capability to carry out tasks pertinent to daily life within a structured setting, typically under the guidance of an evaluator [11].

9.1.2 Recovery

In recent decades, the goal of schizophrenia treatment has become more ambitious and aligned with the expectations of key stakeholders [12]. Hence, in addition to achieving symptomatic remission and preventing acute relapses, the prevention and treatment of comorbidities, along with enhancing functional outcomes and subjective well-being, are now recognized as crucial objectives in the care of individuals with schizophrenia. Furthermore, many guidelines and policies cite recovery as the primary goal of care pathways, although the concept is still evolving, with various definitions proposed over the years [13–15].

Recovery comprises numerous dimensions and is a broad umbrella construct [16]. Various conceptualizations have been proposed, such as internal versus external recovery, clinical and social recovery, and subjective or personal recovery versus objective or clinical/functional recovery [13, 15, 17]. Internal factors primarily pertain to aspects like hope and health, whereas external factors relate to human and patient rights, as well as opportunities for vocational and social integration [17]. Clinical recovery corresponds to symptomatic remission and satisfactory real-life functioning [18], while social recovery corresponds to economic and social independence [17, 19]. Subjective recovery, also named personal recovery, refers to the subjective experience of recovery, defined by a satisfactory quality of life, hope, reliance on others, and not feeling overwhelmed by symptoms [20, 21]. Leamy et al. [22], through a systematic review, identified four fundamental components of personal recovery: hope, reestablishment of identity, finding meaning in life, connectedness, and empowerment.

In the past few decades, considerable progress has been achieved in studying the variables affecting real-life functioning among individuals with schizophrenia. It is now widely acknowledged that psychotic symptoms contribute minimally to the variability in psychosocial functioning [23–25], and negative symptoms, neurocognitive deficits, and impairments in social cognition have a significant impact on it [23–25], and therefore represent important targets of treatment programs. Moreover, there has been a growing emphasis on understanding the influence of resilience, internalized stigma, and contextual factors on the real-life functioning of people with schizophrenia, aiming to consider all possible factors and variables to achieve a meaningful life and increase the rates of recovery [24, 26–28].

9.2 Psychosocial Functioning Assessment

Several instruments are available for the evaluation of social functioning [29]; however, there is no unanimous consensus on the choice of the most appropriate tools. The Global Assessment of Functioning (GAF) scale, known for its efficiency and applicability in both research and clinical settings, stands out as the most utilized measure for assessing social functioning [9, 30]. Nevertheless, its reliance on a single score incorporating symptoms can introduce a bias, making it a less "pure" measure of functioning. The Social and Occupational Functioning Assessment Scale (SOFAS) [31] was developed to overcome GAF limitations [9]. In fact, the SOFAS evaluates social and occupational functioning exclusively, without taking into account psychopathological features [32, 33]. The introduction of the Personal and Social Performance scale (PSP) represented an important further development in the assessment of psychosocial functioning [34]. It provides criteria for assessing specific domains of functioning and includes specific criteria for rating the severity of dysfunction, while also considering psychopathological aspects of behavior [34]. Real-life functioning can be evaluated using the Specific Level of Functioning Scale

(SLOF), a comprehensive instrument, chosen by the participants in the Validation of Everyday Real-World Outcomes (VALERO) study as the best tool for this aim. The SLOF examines various dimensions of functioning and relies on the primary caregiver's assessment of the patient's behavior and functioning [13]. Recently, the Schizophrenia Outcomes Functioning Interview (SOFI) was designed with the purpose of assessing changes in functional outcomes due to patient psychopathology and cognitive impairment [35].

Performance-based assessments of social and daily living skills, such as the University of California, San Diego, Performance-Based Skills Assessment (UPSA), are increasingly utilized to evaluate functional capacity [36]. They also represent outcome measures in pharmacological and cognitive training studies in schizophrenia. While these assessments show potential for predicting independent living and vocational outcomes, they demand considerable time and specialized resources. To overcome this limitation, a brief version of the UPSA, the UPSA-B, focusing on financial skills and communication skills subscales (counting change, telephone calls, and paying bills) was developed. This abbreviated version takes about 15 minutes to complete and accurately predicts a patient's ability to live independently, similar to the full version of the UPSA [9, 37].

9.3 Relationship Between Cognition and Psychosocial Functioning

The association between neurocognition and functional outcomes in schizophrenia has been documented in several prior studies [38, 39], which provided consistent evidence that cognitive impairment (CI) persists across the acute, chronic, and residual phases of the disease and represents both a correlate and a predictor of functional outcome or degree of disability [27, 40, 41].

9.3.1 Cognitive Impairment and Real-Life Functioning in the Early Stages of Schizophrenia

CI as compared with psychotic symptoms, exhibits greater predictive value of the functional impairment in schizophrenia. There is evidence that detectable levels of neurocognitive and social cognitive impairment are present both in the premorbid phase in persons who later develop schizophrenia and in non-affected first-degree relatives of persons with schizophrenia. These observations indicate that cognitive abnormalities reflect an impaired neurodevelopment and a possible genetic predisposition and/or early developmental challenges, or their interplay [42, 43]. Poorer premorbid adjustment (PA) (which is generally distinguished in two different subdomains, academic and social) correlates with the severity of neurocognitive impairment, negative symptoms, and worse functional outcomes [44–49]. Neurocognitive impairment has often been associated with the academic domain of premorbid adjustment [46, 50–54], however, recent studies have highlighted a direct link between some cognitive domains, such as attention and executive functioning, and the social subdomain of PA [55–57].

Seidman and co-authors [58] observed that "there is ample evidence of significant but milder impairments during the premorbid phase, greater deficits during the prodromal or clinical high-risk (CHR) period [of psychosis], culminating in relatively severe deficits in the first episode and chronic phases." According to some studies, attention might be one of the least affected domains, even though further analysis revealed that this domain consistently declined following resolution of the first-episode psychosis, particularly when compared to neurotypical children [59].

In the early-onset schizophrenia (EOS) population, memory consistently exhibits impairment both prior to the onset of psychosis and following clinical stabilization. Several studies have identified working memory, verbal memory, and learning as significantly impaired domains. While working memory tends to remain stable over time, learning demonstrates a decline when compared with healthy children. In patients with early-onset schizophrenia (EOS), executive functioning impairment appears less pronounced during prodromal phases but becomes evident at the onset of the first episode and persists thereafter (Table 9.1).

Regarding social cognition, we might hypothesize that the decline in social cognitive abilities during crucial prodromal periods may also contribute to the deterioration in social functioning preceding the onset of psychosis.

A cognitive profile characterized by the presence of both neurocognitive and social cognition deficits might predict the transition from prodromal states to early-onset psychosis. However, this remains to be demonstrated; more studies incorporating social cognition measures are needed before drawing conclusions, as the available ones present important methodological limitations such as small sample sizes and/or a lack of longitudinal social cognition data [43, 60].

Table 9.1 Cognitive profile before and after the onset of psychosis in EOS

Cognitive profile before psychosis onset	
Large effect size	General intellectual ability
	Processing speed
	Working memory
	Verbal memory and learning
Medium effect size	Rule discovery and perseveration
	Planning and problem solving
Minimal effect size	Sustained attention
	Selective attention
Cognitive profile after psychosis onset	
Evidence of increasing deviance	Verbal memory and learning
	Attention
	Perseverative errors
Evidence of stability	General intellectual ability
	Working memory
	Rule discovery
Limited or conflicting evidence	Processing speed
	Planning and problem solving
	Response inhibition

9.3.2 Cognitive Impairment and Real-Life Functioning in Chronic Schizophrenia

The progression of CI following the onset of schizophrenia remains a debated topic. Some studies support a neurodevelopmental theory, claiming that CI is due to altered brain development and therefore reflects a compromised acquisition of function, rather than a decline [61]. In contrast, other studies suggest a neurodegenerative component, alongside the developmental CI, possibly associated with inflammatory processes or the chronicity of psychotic symptoms, which contribute to further cognitive deterioration following the onset of schizophrenia [62]. Unfortunately, few studies have provided longitudinal observations in patients with schizophrenia, with an appropriate duration of follow-up and taking into account comparison groups and the role of confounding factors. For instance, Bora and Murray conducted a meta-analysis of longitudinal studies examining cognition in individuals affected by schizophrenia and found an overall improvement in cognitive functions following schizophrenia onset, with the exception of working memory and verbal fluency [63]. Moreover, the Italian Network for Research on Psychoses (NIRP) performed a 4-year follow-up evaluation in a large sample of patients affected by chronic schizophrenia and found that the majority of social cognition variables showed improvement, whereas neurocognition variables remained relatively stable, with notable changes observed only in verbal learning (slightly improved) and working memory (slightly worsened) [64].

9.4 Studies Investigating Associations Between Cognitive Domains and Psychosocial Functioning

As previously reported, a large body of literature suggests significant relationships between specific cognitive and functional domains, while some studies suggest that global cognitive functioning predicts global functional outcome [8, 65, 66].

A systematic review and meta-analysis identified small to large correlations of both social cognition and neurocognition with functional outcome [67]. This meta-analysis, in particular, revealed that the strength of these associations tended to be moderate when examining individual domains of neurocognition, with more robust connections observed when utilizing global measures of neurocognition.

The strength of the correlations between neurocognitive impairment and real-life functioning depends on several factors, including the analyzed cognitive domains, the utilization of composite scores (which, as previously explained, tend to show higher correlations compared to individual cognitive domains), and the source of the functional assessment (with clinician ratings typically showing stronger correlations than patient's self-reports) [68, 69].

The examination of individual neurocognitive domains revealed that verbal memory could be a significant predictor of psychosocial functioning, especially associated with skills acquisition, and shows no dependence on gender [9, 70, 71]. Immediate verbal memory has been linked to the acquisition of psychosocial skills, such as conversational and leisure skills. Executive functioning has been associated with the execution of daily activities, while sustained attention has been connected to social problem-solving. It is worth noting that processing speed and sustained attention can interfere with the development of other cognitive domains. These basic cognitive processes underlie more complex domains such as memory, working memory, and executive functioning. Consequently, impairments in processing speed or sustained attention can lead to greater challenges in daily activities requiring complex cognitive processes, such as using transportation (e.g., understanding public transportation schedules, using maps), managing finances (e.g., using ATMs or budgeting), and occupational functioning (e.g., employment, work skills, and work behavior) [72–75]. Some studies reported associations between attention/working memory and work skills or executive functions and interpersonal behavior [8, 18, 66].

However, neurocognitive impairment alone does not account for the decline in social functioning. In fact, its impact on functioning can be mediated or worsened by other factors, such as social cognition or intrinsic motivation [41, 76–83].

Several studies show that neurocognition and social cognition are directly and indirectly linked to functional outcomes. For example, social cognition mediates the relationship between certain neurocognitive abilities (e.g., processing speed [72], sustained attention [72] and verbal memory [84]), and functional outcomes. Additionally, social perception connects early visual processing to functional outcomes [85]. Theory of mind (ToM) links neurocognition to social competence [86] and emotion recognition ties it to social functioning in individuals with schizophrenia and first-episode psychoses [87]. On the other hand, neurocognition also affects social cognition. Therefore, interventions focusing on just one cognitive domain may have little or no effect on functional outcomes in schizophrenia [82, 88].

Recent research has indicated that the effect of neurocognitive impairment on real-life functioning may be mediated by functional capacity, and some studies have found that the impact of cognitive impairment on real-life functioning becomes insignificant when functional capacity is taken into account [66, 77]. However, there are also conflicting findings, such as no correlation between neurocognitive measures or functional capacity and self-reported real-life functioning [89], or a significant impact of neurocognitive dysfunction on everyday functioning despite no influence of functional capacity [23, 90]. Through a network analysis, the NIRP provided clear evidence of the critical role played by functional capacity and everyday life skills, which exhibited the highest centrality within the network. Functional capacity acts as a pivotal node connecting neurocognition and social cognition with real-life functioning nodes, particularly those related to everyday life skills such as household activities, financial management, and use of transportation [40]. Within neurocognition, the working memory node emerges as central, strongly linked to all other neurocognition nodes, and serving as a bridge to social cognition and functional capacity. Additionally, verbal learning, visuospatial learning, and attention exhibit direct connections with functional capacity. The association between emotion recognition and functional capacity underscores the importance of social and emotional processing in overall psychosocial functioning.

Furthermore, the network analysis reveals that social competence, but not social cognition, is directly linked to interpersonal functioning [40].

Non-cognitive mediators have also been found to influence the impact of CI on psychosocial functioning and will be described in the next paragraph.

9.4.1 Negative Symptoms

In addition to their relationships with real-life functioning independent of CI [86, 91], negative symptoms have been identified as mediators of the effect of neurocognition on functioning [86, 91–95]. A large study conducted in adolescents with an early-onset psychosis showed that verbal learning was positively associated with global functioning and that this association was mediated by negative and disorganized symptoms [94]. Furthermore, impairments in executive functioning, for instance, may disrupt goal-directed behavior required to achieve rewards, which is conceptualized as a pathophysiological mechanism underlying the negative symptoms motivational deficit domain (including avolition, anhedonia, and asociality) [96]. Similarly, deficits in social cognition may either result from or contribute to asociality [96, 97]. Additionally, the expressive deficit domain of negative symptoms (blunted affect and alogia) might be rooted in deficiencies in social cognition and neurocognition. Blunted affect may arise from difficulties in identifying and discriminating emotions, as well as abnormalities in perceiving nonverbal social cues, leading to an inability to infer meaning from social situations and behaviors and respond appropriately [98, 99]. Alogia may be associated to poor verbal fluency, indicating a deficit in the ability to retrieve information from memory [98, 99].

9.5 Conclusions

Cognitive deficits significantly influence various aspects of psychosocial functioning, including social interaction, occupational performance, and independent living. Moreover, the presence of cognitive impairment often exacerbates challenges related to maintaining interpersonal relationships and a job, and achieving personal goals. Conversely, psychosocial factors, such as social support, environmental adaptations, and rehabilitation interventions, play a pivotal role in mitigating the impact of cognitive deficits and enhancing overall functioning. Effective treatment approaches should thus adopt a holistic perspective that addresses both cognitive and psychosocial domains, aiming not only to alleviate symptoms but also to promote meaningful engagement in daily activities and improve quality of life. Further research is necessary to identify components of personalized and integrated interventions that optimize outcomes of persons with schizophrenia. By integrating advances in cognitive rehabilitation, psychosocial interventions, and personalized medicine, we can implement more comprehensive and effective approaches to support the diverse needs of each patient.

References

1. Apiquian R, et al. Validity of the Spanish version of the Personal and Social Performance scale in schizophrenia. Schizophr Res. 2009;112(1–3):181–6.
2. Kossmann C, et al. Assessment of psychosocial functioning in a large cohort of patients with schizophrenia. Psychiatry Q. 2021;92(1):177–91.
3. Bond GR, Drake RE, Becker DR. An update on individual placement and support. World Psychiatry. 2020;19(3):390–1.
4. Huhn M, et al. Comparative efficacy and tolerability of 32 oral antipsychotics for the acute treatment of adults with multi-episode schizophrenia: a systematic review and network meta-analysis. Lancet. 2019;394(10202):939–51.
5. Kishimoto T, et al. Long-term effectiveness of oral second-generation antipsychotics in patients with schizophrenia and related disorders: a systematic review and meta-analysis of direct head-to-head comparisons. World Psychiatry. 2019;18(2):208–24.
6. McKenna P, et al. The controversy about cognitive behavioural therapy for schizophrenia. World Psychiatry. 2019;18(2):235–6.
7. Gaebel W, Falkai P, Hasan A. The revised German evidence- and consensus-based schizophrenia guideline. World Psychiatry. 2020;19(1):117–9.
8. Green MF, et al. Neurocognitive deficits and functional outcome in schizophrenia: are we measuring the "right stuff"? Schizophr Bull. 2000;26(1):119–36.
9. Brissos S, et al. The importance of measuring psychosocial functioning in schizophrenia. Ann General Psychiatry. 2011;10:18.
10. Gupta M, et al. Functional outcomes in schizophrenia: understanding the competence-performance discrepancy. J Psychiatr Res. 2012;46(2):205–11.
11. Kharawala S, et al. The relationship between cognition and functioning in schizophrenia: a semi-systematic review. Schizophr Res Cogn. 2022;27:100217.
12. Harvey PD, Bellack AS. Toward a terminology for functional recovery in schizophrenia: is functional remission a viable concept? Schizophr Bull. 2009;35(2):300–6.
13. Vita A, Barlati S. Recovery from schizophrenia: is it possible? Curr Opin Psychiatry. 2018;31(3):246–55.
14. Frese FJ III, Knight EL, Saks E. Recovery from schizophrenia: with views of psychiatrists, psychologists, and others diagnosed with this disorder. Schizophr Bull. 2009;35(2):370–80.
15. Van Eck RM, et al. The relationship between clinical and personal recovery in patients with schizophrenia spectrum disorders: a systematic review and meta-analysis. Schizophr Bull. 2017;44(3):631–42.
16. Slade M, et al. Uses and abuses of recovery: implementing recovery-oriented practices in mental health systems. World Psychiatry. 2014;13(1):12–20.
17. Amering M, Schmolke M. Recovery in mental health: reshaping scientific and clinical responsibilities. 2009. pp1–267.
18. Giordano GM, et al. Determinants of clinical recovery in schizophrenia. In: Carpiniello B, Vita A, Mencacci C, editors. Recovery and major mental disorders. Cham: Springer International Publishing; 2022. p. 23–43.
19. Shrivastava A, et al. Redefining outcome measures in schizophrenia: integrating social and clinical parameters. Curr Opin Psychiatry. 2010;23(2):120–6.
20. Silverstein SM, Bellack AS. A scientific agenda for the concept of recovery as it applies to schizophrenia. Clin Psychol Rev. 2008;28(7):1108–24.
21. Roe D, Mashiach-Eizenberg M, Lysaker PH. The relation between objective and subjective domains of recovery among persons with schizophrenia-related disorders. Schizophr Res. 2011;131(1):133–8.
22. Leamy M, et al. Conceptual framework for personal recovery in mental health: systematic review and narrative synthesis. Br J Psychiatry. 2011;199(6):445–52.

23. Bowie CR, et al. Determinants of real-world functional performance in schizophrenia subjects: correlations with cognition, functional capacity, and symptoms. Am J Psychiatry. 2006;163(3):418–25.
24. Leifker FR, Bowie CR, Harvey PD. Determinants of everyday outcomes in schizophrenia: the influences of cognitive impairment, functional capacity, and symptoms. Schizophr Res. 2009;115(1):82–7.
25. Galderisi S, et al. Categorical and dimensional approaches to negative symptoms of schizophrenia: focus on long-term stability and functional outcome. Schizophr Res. 2013;147(1):157–62.
26. Harvey PD, Strassnig M. Predicting the severity of everyday functional disability in people with schizophrenia: cognitive deficits, functional capacity, symptoms, and health status. World Psychiatry. 2012;11(2):73–9.
27. Galderisi S, et al. The influence of illness-related variables, personal resources and context-related factors on real-life functioning of people with schizophrenia. World Psychiatry. 2014;13(3):275–87.
28. Galderisi S, et al. Pathways to functional outcome in subjects with schizophrenia living in the community and their unaffected first-degree relatives. Schizophr Res. 2016;175(1):154–60.
29. Figueira ML, Brissos S. Measuring psychosocial outcomes in schizophrenia patients. Curr Opin Psychiatry. 2011;24(2):91–9.
30. Endicott J, et al. The global assessment scale. A procedure for measuring overall severity of psychiatric disturbance. Arch Gen Psychiatry. 1976;33(6):766–71.
31. Diagnostic and statistical manual of mental disorders : DSM-IV. 1994, Fourth edition. Washington, DC : American Psychiatric Association; 1994 ©1994.
32. Goldman HH, Skodol AE, Lave TR. Revising axis V for DSM-IV: a review of measures of social functioning. Am J Psychiatry. 1992;149(9):1148–56.
33. Morosini PL, et al. Development, reliability and acceptability of a new version of the DSM-IV Social and Occupational Functioning Assessment Scale (SOFAS) to assess routine social functioning. Acta Psychiatr Scand. 2000;101(4):323–9.
34. Nasrallah H, Morosini P, Gagnon DD. Reliability, validity and ability to detect change of the Personal and Social Performance scale in patients with stable schizophrenia. Psychiatry Res. 2008;161(2):213–24.
35. Kleinman L, et al. Development and psychometric performance of the schizophrenia objective functioning instrument: an interviewer administered measure of function. Schizophr Res. 2009;107(2–3):275–85.
36. Mausbach BT, et al. A review of instruments for measuring functional recovery in those diagnosed with psychosis. Schizophr Bull. 2009;35(2):307–18.
37. Mausbach BT, et al. Development of a brief scale of everyday functioning in persons with serious mental illness. Schizophr Bull. 2007;33(6):1364–72.
38. Green MF. Impact of cognitive and social cognitive impairment on functional outcomes in patients with schizophrenia. J Clin Psychiatry. 2016;77(Suppl 2):8–11.
39. Javed A, Charles A. The importance of social cognition in improving functional outcomes in schizophrenia. Front Psych. 2018;9:157.
40. Galderisi S, et al. Interplay among psychopathologic variables, personal resources, context-related factors, and real-life functioning in individuals with schizophrenia: a network analysis. JAMA Psychiatry. 2018;75(4):396–404.
41. Mucci A, et al. Factors associated with real-life functioning in persons with schizophrenia in a 4-year follow-up study of the Italian network for research on psychoses. JAMA Psychiatry. 2021;78(5):550–9.
42. Rapoport JL, Giedd JN, Gogtay N. Neurodevelopmental model of schizophrenia: update 2012. Mol Psychiatry. 2012;17(12):1228–38.
43. Harvey PD, et al. Cognitive dysfunction in schizophrenia: an expert group paper on the current state of the art. Schizophr Res. 2022;29:100249.
44. Bucci P, et al. Characterization of premorbid functioning during childhood in patients with deficit vs. non-deficit schizophrenia and in their healthy siblings. Schizophr Res. 2016;174(1–3):172–6.

45. Galderisi S, et al. Historical, psychopathological, neurological, and neuropsychological aspects of deficit schizophrenia: a multicenter study. Am J Psychiatry. 2002;159(6):983–90.
46. Larsen TK, et al. Premorbid adjustment in first-episode non-affective psychosis: distinct patterns of pre-onset course. Br J Psychiatry. 2004;185(2):108–15.
47. McGlashan TH. Premorbid adjustment, onset types, and prognostic scaling: still informative? Schizophr Bull. 2008;34(5):801–5.
48. Cuesta MJ, et al. Premorbid adjustment and clinical correlates of cognitive impairment in first-episode psychosis. PEPsCog Study Schizophr Res. 2015;164(1–3):65–73.
49. Stentebjerg-Olesen M, et al. Clinical characteristics and predictors of outcome of schizophrenia-spectrum psychosis in children and adolescents: a systematic review. J Child Adolesc Psychopharmacol. 2016;26(5):410–27.
50. Chang WC, et al. The relationship of early premorbid adjustment with negative symptoms and cognitive functions in first-episode schizophrenia: a prospective three-year follow-up study. Psychiatry Res. 2013;209(3):353–60.
51. Barajas A, et al. Three-factor model of premorbid adjustment in a sample with chronic schizophrenia and first-episode psychosis. Schizophr Res. 2013;151(1–3):252–8.
52. Norman RM, et al. Premorbid adjustment in first episode schizophrenia and schizoaffective disorders: a comparison of social and academic domains. Acta Psychiatr Scand. 2005;112(1):30–9.
53. Rund BR, et al. The course of neurocognitive functioning in first-episode psychosis and its relation to premorbid adjustment, duration of untreated psychosis, and relapse. Schizophr Res. 2007;91(1–3):132–40.
54. Rund BR, et al. Neurocognitive dysfunction in first-episode psychosis: correlates with symptoms, premorbid adjustment, and duration of untreated psychosis. Am J Psychiatry. 2004;161(3):466–72.
55. Silverstein ML, Mavrolefteros G, Close D. Premorbid adjustment and neuropsychological performance in schizophrenia. Schizophr Bull. 2002;28(1):157–65.
56. González-Blanch C, et al. Pretreatment predictors of cognitive deficits in early psychosis. Psychol Med. 2008;38(5):737–46.
57. Bucci P, et al. Premorbid academic and social functioning in patients with schizophrenia and its associations with negative symptoms and cognition. Acta Psychiatr Scand. 2018;138(3):253–66.
58. Seidman LJ, et al. Association of Neurocognition with transition to psychosis: baseline functioning in the second phase of the North American Prodrome Longitudinal Study. JAMA Psychiatry. 2016;73(12):1239–48.
59. Frangou S. Neurocognition in early-onset schizophrenia. Child Adolesc Psychiatr Clin N Am. 2013;22(4):715–26.
60. Harvey PD, Isner EC. Cognition, social cognition, and functional capacity in early-onset schizophrenia. Child Adolesc Psychiatr Clin N Am. 2020;29(1):171–82.
61. Zipursky RB, Reilly TJ, Murray RM. The myth of schizophrenia as a progressive brain disease. Schizophr Bull. 2012;39(6):1363–72.
62. Monji A, et al. Neuroinflammation in schizophrenia especially focused on the role of microglia. Prog Neuro-Psychopharmacol Biol Psychiatry. 2013;42:115–21.
63. Bora E, Murray RM. Meta-analysis of cognitive deficits in ultra-high risk to psychosis and first-episode psychosis: do the cognitive deficits progress over, or after, the onset of psychosis? Schizophr Bull. 2013;40(4):744–55.
64. Galderisi S, et al. The interplay among psychopathology, personal resources, context-related factors and real-life functioning in schizophrenia: stability in relationships after 4 years and differences in network structure between recovered and non-recovered patients. World Psychiatry. 2020;19(1):81–91.
65. Laes JR, Sponheim SR. Does cognition predict community function only in schizophrenia?: a study of schizophrenia patients, bipolar affective disorder patients, and community control subjects. Schizophr Res. 2006;84(1):121–31.

66. Bowie CR, et al. Predicting schizophrenia patients' real-world behavior with specific neuropsychological and functional capacity measures. Biol Psychiatry. 2008;63(5):505–11.
67. Fett A-KJ, et al. The relationship between neurocognition and social cognition with functional outcomes in schizophrenia: a meta-analysis. Neurosci Biobehav Rev. 2011;35(3):573–88.
68. McClure MM, et al. Correlations of functional capacity and neuropsychological performance in older patients with schizophrenia: evidence for specificity of relationships? Schizophr Res. 2007;89(1–3):330–8.
69. Leifker FR, et al. Validating measures of real-world outcome: the results of the VALERO expert survey and RAND panel. Schizophr Bull. 2011;37(2):334–43.
70. Puig O, et al. Verbal memory, negative symptomatology and prediction of psychosocial functioning in schizophrenia. Psychiatry Res. 2008;158(1):11–7.
71. Green MF. What are the functional consequences of neurocognitive deficits in schizophrenia? Am J Psychiatry. 1996;153(3):321–30.
72. Ruiz-Toca A, et al. Social cognition mediates the impact of processing speed and sustained attention on global functioning in schizophrenia. Psicothema. 2023;35(1):87–97.
73. Agnew-Blais J, Seidman LJ. Neurocognition in youth and young adults under age 30 at familial risk for schizophrenia: a quantitative and qualitative review. Cogn Neuropsychiatry. 2013;18(1–2):44–82.
74. Rodríguez-Sánchez JM, et al. Cognitive dysfunction in first-episode psychosis: the processing speed hypothesis. Br J Psychiatry. 2007;191(S51):s107–10.
75. Shi C, et al. An ecologically valid performance-based social functioning assessment battery for schizophrenia. Psychiatry Res. 2013;210(3):787–93.
76. Uchino T, et al. Mediation effects of social cognition on the relationship between neurocognition and social functioning in major depressive disorder and schizophrenia Spectrum disorders. J Personal Med. 2023;13(4):683.
77. Bowie CR, et al. Prediction of real-world functional disability in chronic mental disorders: a comparison of schizophrenia and bipolar disorder. Am J Psychiatry. 2010;167(9):1116–24.
78. Pinkham AE, Harvey PD, Penn DL. Social cognition psychometric evaluation: results of the final validation study. Schizophr Bull. 2018;44(4):737–48.
79. Okano H, et al. Evaluation of social cognition measures for Japanese patients with schizophrenia using an expert panel and modified Delphi method. J Pers Med. 2021;11(4):275.
80. Lim K, et al. Evaluation of social cognitive measures in an Asian schizophrenia sample. Schizophr Res Cogn. 2020;20:100169.
81. Kubota R, et al. Study protocol: the evaluation study for social cognition measures in Japan (ESCoM). J Pers Med. 2021;11(7):667.
82. Halverson TF, et al. Pathways to functional outcomes in schizophrenia spectrum disorders: meta-analysis of social cognitive and neurocognitive predictors. Neurosci Biobehav Rev. 2019;105:212–9.
83. Nemoto T, et al. Impact of changes in social anxiety on social functioning and quality of life in outpatients with schizophrenia: a naturalistic longitudinal study. J Psychiatr Res. 2020;131:15–21.
84. González-Ortega I, et al. Influence of social cognition as a mediator between cognitive reserve and psychosocial functioning in patients with first episode psychosis. Psychol Med. 2020;50(16):2702–10.
85. Sergi MJ, et al. Social perception as a mediator of the influence of early visual processing on functional status in schizophrenia. Am J Psychiatry. 2006;163(3):448–54.
86. Couture SM, Penn DL, Roberts DL. The functional significance of social cognition in schizophrenia: a review. Schizophr Bull. 2006;32(suppl_1):S44–63.
87. Addington J, Saeedi H, Addington D. Facial affect recognition: a mediator between cognitive and social functioning in psychosis? Schizophr Res. 2006;85(1):142–50.
88. Bell M, et al. Neurocognition, social cognition, perceived social discomfort, and vocational outcomes in schizophrenia. Schizophr Bull. 2009;35(4):738–47.
89. McKibbin CL, et al. Direct assessment of functional abilities: relevance to persons with schizophrenia. Schizophr Res. 2004;72(1):53–67.

90. Heinrichs RW, et al. Cognitive performance and functional competence as predictors of community independence in schizophrenia. Schizophr Bull. 2010;36(2):381–7.
91. Ventura J, et al. Symptoms as mediators of the relationship between neurocognition and functional outcome in schizophrenia: a meta-analysis. Schizophr Res. 2009;113(2–3):189–99.
92. Giordano GM, et al. Negative symptoms and social cognition as mediators of the relationship between neurocognition and functional outcome in schizophrenia. Front Psychiatry. 2024;15:1333711.
93. Luther L, et al. Clarifying the direction of impact of negative symptoms and neurocognition on prospective work functioning in psychosis: a 20-year longitudinal study. Schizophr Res. 2020;220:232–9.
94. Smelror RE, et al. Negative and disorganized symptoms mediate the relationship between verbal learning and global functioning in adolescents with early-onset psychosis. Eur Child Adolesc Psychiatry. 2020;29(12):1693–703.
95. Starzer M, et al. 20-year trajectories of positive and negative symptoms after the first psychotic episode in patients with schizophrenia spectrum disorder: results from the OPUS study. World Psychiatry. 2023;22(3):424–32.
96. Marder SR, Galderisi S. The current conceptualization of negative symptoms in schizophrenia. World Psychiatry. 2017;16(1):14–24.
97. Rocca P, et al. Does social cognition change? Evidence after 4 years from the Italian Network for Research on Psychoses. Eur Psychiatry. 2023;66(1):e10.
98. Galderisi S, et al. Negative symptoms of schizophrenia: new developments and unanswered research questions. Lancet Psychiatry. 2018;5(8):664–77.
99. Kaiser S, et al. Individual negative symptoms and domains—relevance for assessment, pathomechanisms and treatment. Schizophr Res. 2017;186:39–45.

Feasible and Acceptable Approaches to Assessment of Cognitive Health Need

10

Alice M. Saperstein, Lauren Gonzales, and Alice Medalia

Abstract

Cognitive impairment associated with schizophrenia is prevalent, persistent, and has strong implications for real-world functioning. Assessment of cognitive health needs is critical for creating a shared understanding of clients' experiences and behaviors and sets the stage for collaborative, personalized treatment planning. Identifying areas of cognitive strength and relative need informs appropriate intervention and supports. Mental health services research has highlighted the complex challenges of obtaining assessments of cognitive health needs that are feasible, acceptable, and clinically informative in routine practice settings. It is useful to consider three types of assessments—screening, brief targeted assessment, and comprehensive evaluation—each with a unique purpose and associated challenges impacting implementation. A number of practical solutions are recommended to address these challenges, improve uptake of assessment practices, and better meet client needs.

Keywords

Cognitive Health · Cognitive Screening · Toolkit · Cognitive Health Stigma · Shared Decision-Making

A. M. Saperstein (✉)
New York State Psychiatric Institute, Department of Psychiatry, Columbia University Vagelos College of Physicians & Surgeons, and New York-Presbyterian, New York, NY, USA

Columbia University Irving Medical Center, New York, NY, USA
e-mail: Alice.Saperstein@nyspi.columbia.edu

L. Gonzales · A. Medalia
New York State Psychiatric Institute, Department of Psychiatry, Columbia University Vagelos College of Physicians & Surgeons, and New York-Presbyterian, New York, NY, USA

A. Vita et al. (eds.), *Negative and Cognitive Symptoms in Schizophrenia*, Comprehensive Approach to Psychiatry 3,
https://doi.org/10.1007/978-3-032-06209-3_10

10.1 Introduction

Schizophrenia, from prodrome to end of life, is associated with cognitive impairments that are persistent, unresponsive to pharmacologic treatment, and constrain psychosocial achievement at tremendous financial, social, familial, and personal cost [1–4]. Difficulties with skills such as memory, attention, and problem-solving add to illness burden by imposing limitations on the ability to adhere to pharmacotherapy, benefit from psychosocial programs, and function in the social, vocational, and educational venues integral to community integration [5–7]. Multinational treatment guidelines for schizophrenia recognize the importance of addressing cognitive health [8–10], and cognition has been identified as a valued target of treatment by service participants [11–13] and providers [14]. Timely assessment is critical for the identification of cognitive health needs and informs service planning, with the ultimate goal of optimizing mental health and functional outcomes.

Mental health services research has highlighted the complex challenges of obtaining cognitive health assessments that are acceptable and clinically informative. This chapter will review relevant assessment issues facing clinicians and clients, and present feasible and acceptable solutions to assessing cognition in people with schizophrenia.

10.2 The Utility of Cognitive Assessment

In mental health settings, cognitive assessment is highly valued by clinicians and clients [15, 16]. Cognitive assessment is sought to better understand how cognition impacts role functioning and the likely etiology of cognitive deficits, thereby contributing to a diagnostic formulation. Assessments vary in comprehensiveness depending on the goals, but typically consist of standardized neuropsychological tests, which can be considered in conjunction with behavioral observation and clinical interviews. Test results are interpreted considering an individual's developmental, medical, and learning history, current clinical presentation, and functioning. A holistic understanding of a person's cognitive health, with respect to areas of personal strength as well as specific areas of cognitive difficulty, aims to support case formulation and inform the provision of appropriate treatments or supports [15, 17, 18]. The following example illustrates how a person's current behavioral challenges may be best understood when multiple aspects of their cognitive health are considered together:

> Mario G was admitted to an outpatient community clinic for clinical management and independent living support. Mario's psychiatric and education histories taken at intake documented a self-reported history of inattention and learning difficulty throughout childhood and adolescence, though records of prior formal evaluation were unavailable. He achieved a 12th-grade education and pursued vocational training thereafter. Mario recalled increasing difficulty staying focused and managing coursework with marked distress, followed by a first episode of psychosis and hospitalization at age 19. Now, age 32, Mario lives in supported housing in his own apartment. Mario's attendance at the outpatient clinic is sporadic.

Scheduled appointments are often missed, or Mario arrives for services on the wrong dates or times. When prompted, Mario endorsed multiple cognitive complaints—forgetting, losing track of time, misplacing items of importance—which make it difficult to stay organized, self-manage his medications, and keep to a schedule. Mario indicated a future goal to return to vocational training though he declined supported employment services, citing substantial self-doubt regarding his ability to manage. He stated, "I've never been good at learning, so I just don't see the point." Mario was referred for cognitive assessment to gain clarity on barriers to learning and areas of current cognitive health need.

Appreciation of a person's cognitive health history, current cognitive challenges, and their collective behavioral and emotional impact, has critical implications for case formulation and treatment planning. Recommendations from a cognitive assessment might include adjustments to a service plan, such as pharmacological optimization, work/school/living accommodations, cognitive skills training, and Cognitive Behavioral Therapy for psychosis. Understanding the prospect of improved cognition with appropriate treatment can bring optimism and hope. The ultimate goal of pairing assessment and cognitive health services is to help the person better negotiate daily tasks and progress toward meeting recovery goals.

10.3 Challenges to Cognitive Assessment in Clinical Practice

Despite its clinical utility, the commonly faced challenges in clinical practice settings can limit the accessibility of cognitive health assessment. Insufficient availability of professionals trained to conduct assessments, coupled with a lack of understanding about who and when to refer for evaluation, limits appropriate referrals and service access for those who might benefit [17–20]. Further, the cost of a cognitive evaluation and the time required for completion can be deterrents, both at the agency and consumer levels. Acceptability is an additional challenge to obtaining a cognitive assessment. Clinical care providers have noted concern about the risk of stigma, labelling, and their potential impact on individuals' self-perception and well-being [21]. However, the potential risk that assessment may heighten stigma due to its inherent process of identifying and labeling mental and cognitive difficulties is offset by research suggesting that it can also be a source of relief, validation, legitimization, and hope for clients, thereby facilitating communication and increased self-understanding [22].

Further, clinicians may hold stigmatizing attitudes toward cognition in people with schizophrenia, which may play a role in the accessibility of cognitive assessment. Schizophrenia has historically been considered a chronic and progressive brain disorder, despite current knowledge that cognitive impairment in schizophrenia is dynamic and can improve with treatment [23–25]. If providers do not believe that clients' cognitive impairments can improve, they may perceive cognitive assessments to have low utility and may not make referrals for evaluation or treatment services as a result. Also, when clinical services and accommodation workflows to address cognitive health are unavailable, clinicians may question the utility of an assessment. More challenges are posed when the assessment does not provide recommendations, leaving it unclear how to interpret and use the data. Although clinical follow-up after a cognitive assessment is warranted, neither feedback sessions nor clinical recommendations are routinely provided [17]. Addressing these service limitations has the potential to improve the feasibility, acceptability, and clinical utility of cognitive health assessment.

10.4 Practical Solutions to Support Feasible and Acceptable Cognitive Assessment

Increasingly, cognitive health services are becoming available in large systems of care serving people with serious mental illnesses [26]. Such services bridge assessment and treatment through specific workflows that are designed to facilitate uptake and sustainability of practices that holistically address cognitive health. Here, we name several practical solutions to the challenges of routine cognitive assessment in clinical settings.

10.4.1 Solution 1: Foster a Culture of Cognitive Health

When practice settings foster a culture that emphasizes the value of cognitive health, it brings awareness to cognitive impairment as a barrier to recovery and as a key target of intervention. Implementation work highlights the utility of educational campaigns that build awareness and understanding of what is meant by cognitive symptoms and their impact in daily life [27]. When a clinician uses a "cognitive lens" through which to understand the challenges underlying a person's experiences and behaviors, areas of cognitive health need can then be identified and addressed. Greater awareness about cognitive health among clinicians may facilitate the dissemination of information to clients and their families. Educational materials can play an important role in aligning cognitive health promotion with clients' and family members' concerns and goals. A culture that places cognitive health in the context of recovery may help reduce potential stigma associated with assessing and addressing cognitive health needs [20, 21, 28].

10.4.2 Solution 2: Adopt a Cognitive Health Toolkit

Concurrent with a culture that values good cognitive health for overall wellness and recovery, clinical practice settings benefit from workflows that systematically address cognitive health. A collection of resources in the form of a Cognitive Health Toolkit can help address clinical knowledge-practice gaps, such as how to talk to individuals about their cognitive health, when to screen for cognitive health needs, who might be appropriate for cognitive health services, and under what circumstances a more comprehensive assessment might be clinically useful. Workflows can be adapted for local service settings to delineate pathways to cognitive assessment. Different types of assessment tools can be included in a workflow, depending on the setting and purpose the assessment serves.

A toolkit also provides a standardized mechanism for ensuring relevant and timely follow-up, such as a review of assessment findings and referral for specific services. Discussion guides can facilitate collaborative decision-making about cognitive health services to support personalized treatment planning. Aligned with contemporary recovery-focused models of mental healthcare, identifying cognitive strengths alongside challenges may foster self-efficacy, reduce self-stigma, and promote engagement in the therapeutic milieu.

10.4.3 Solution 3: Choose Appropriate Assessment Tools

Assessment tools vary depending upon their purpose, which can range from screening to targeted to comprehensive assessment (Table 10.1).

Feasible and acceptable screening tools are a first step in assisting clinicians to identify for whom and in what ways cognitive health services may be useful in mental health practice settings. Screening tools might provide an indication of areas of need, guide clinical management, or prompt referral for a more comprehensive

Table 10.1 Types of cognitive health assessments

	Data collected	Clinical utility
Learning history screening	Academic achievement Diagnostic history of learning difficulties Perceived learning challenges and strengths Formal accommodations granted at school or work	Understand potential facilitators and barriers to learning Identify useful learning supports or formal accommodations to facilitate goal attainment
Single word reading test	IQ estimate Reading grade level	Rule out intellectual disability Flag need to evaluate for learning disability (e.g., dyslexia) Define baseline cognitive ability against which current cognitive skill performance can be compared
Cognitive health needs screening	Client and collateral perceived utility of improving cognitive skills	Identify need for evaluation Inform shared decision making for clinical management, including cognitive health intervention
Brief targeted cognitive assessment	Standardized scores on a discrete set of performance-based cognitive tests Behavioral observation	Identify areas of cognitive strength and relative weakness Identify focus areas for skills training
Comprehensive neuropsychological evaluation	Standardized neuropsychological test scores Symptom ratings Behavioral observation Adaptive functioning	Identify historical and current areas of cognitive strength and relative weakness Identify or rule out etiologies of cognitive impairment Identify treatment, support, and accommodation priorities to facilitate improved functioning and recovery goal attainment

assessment. Published guidelines for cognitive screening highlight the need for brevity (less than 10 minutes), sensitivity to the range of cognitive difficulties experienced by people with serious mental health conditions, and the feasibility of administration by a range of healthcare practitioners [29, 30]. Existing measures widely accepted and used for detecting cognitive impairment in clinical practice settings [31] include the Mini-Mental State Examination (MMSE; [32]) and Montreal Cognitive Assessment (MoCA; [33]). However, these are typically used in people with known or suspected neuropathology and may not be sensitive to the nature of cognitive impairment in people with serious mental health conditions such as schizophrenia. Additionally, these instruments are typically used in older adults (i.e., age 55+) and thus may not be appropriate for use in younger populations. A comprehensive review of the literature to date yielded no consensus on which published brief performance-based screening measures are most appropriate for youth and young adults (age 25 and under). Cognitive health screening is a critical area of clinical need in youth-oriented service settings such as early intervention services, which have the opportunity to address cognitive health at a time of significant vulnerability [34].

In light of the need for screening tools with broader clinical utility [35], qualitative studies surveying mental health clinicians have sought to provide guidance for the development and implementation of brief cognition measures. From an acceptability standpoint, an assessment tool should have ecological validity, and the goal of assessing cognitive health needs should be explicitly and clearly communicated to the client [15]. From a feasibility standpoint, implementation of an assessment tool needs to be sensitive to the time and resource constraints of practitioners and clients. Despite the published guidelines for screening, there is some opinion that even the timeframe of 10 minutes is too long to be feasible and acceptable. Brief self-report questionnaires may have an advantage over performance-based measures as screening tools; questions that can be woven into routine intake or other information-gathering practices may encourage better uptake.

A Self-Assessment of Cognitive Functioning [36] is one such screening tool that has demonstrated utility in identifying areas of cognitive health need in routine practice settings (Table 10.2). This brief self-report questionnaire asks whether a client perceives improvement in memory, attention, and/or critical thinking skills as useful, e.g., "Do you think you could accomplish more (at valued activities) if you could stay more organized, plan ahead, or manage your time better?" Memory,

Table 10.2 Self-assessment of cognitive functioning

Cognitive skill domain	Query
Memory	Would it be helpful to you if you remembered better what you've read, heard, learned? Or about appointments, events, things to do?
Attention	Would it be helpful to you if you could concentrate better or stay focused over a longer period of time?
Critical Thinking	Do you think you could accomplish more if you could stay more organized, plan ahead, or manage your time better?

attention, and critical thinking skills are the designated focus of the screening tool, given that objective testing commonly finds deficits in these areas [37, 38]. They have been linked to functional outcomes [6, 7] and, importantly, they have previously been self-identified as areas of concern among people with a schizophrenia diagnosis [12]. Follow-up questions query whether a change in cognitive ability from a previous time is perceived, and when and why that change may have occurred. When changes in cognitive health are perceived in relation to the onset of significant stress, medication adjustments, sleep difficulty, or other mental health symptoms, appropriate next steps, including further cognitive or clinical assessment, consultation with a psychiatric care provider (MD/RN) can be determined.

Perceptions of cognitive health from collateral sources (e.g., clinicians, family) are complementary to self-report. A family member, for example, might share information regarding when cognitive challenges began to emerge and their impact as observed at home or when with friends. A clinician might report on a client's appointment attendance, skills deployment, and goal progress, which can indicate how one is able to manage time, remember and learn new information, plan and execute goal-oriented tasks. Multiple, unique perspectives facilitate a detailed understanding of how cognitive challenges manifest, and which areas of cognitive skill, if improved, would help a person better negotiate daily demands.

The Self-Assessment of Cognitive Functioning screening tool was implemented in a large network of early intervention service programs for psychosis in the United States. To support feasibility, its use was integrated within a standard set of data collection tools for the purposes of clinical management and tracking, with guidance to administer the questionnaire at least once within the first 3 months following admission. To support acceptability, questions are framed in the positive, thereby avoiding a deficit focus. Reference to the use of cognition in valued activities (e.g., school, work, socializing) highlights the practical implications of having better cognitive health. Its brevity and simplicity support repeat administration, which facilitates understanding of cognitive health needs as they may evolve or be prioritized differently over time.

An analysis of program evaluation data indicated that across four consecutive years, the majority of program participants who were assessed wanted better memory (78%), attention (85%), and critical thinking skills (85%). These rates of self-identified need approximate rates of cognitive impairment in schizophrenia reported in the literature based on performance-based assessment batteries [39]. Additionally, as a result of screening, 68.7% of those program participants indicated they wished to learn more about ways to improve their cognitive health [40]. A workflow in which the Self-Assessment of Cognitive Functioning is embedded guides decision-making as to when a more comprehensive assessment is needed. For clinic-based management of cognitive health challenges, the Self-Assessment may be sufficient, particularly when considered alongside clinicians' observation and information garnered from past records or family members' reports. When prior testing is unavailable or a more nuanced understanding of cognitive health needs is sought, a performance-based cognitive assessment may be clinically advantageous.

When screening from patient or collateral sources indicates cognitive health need, a referral for targeted or comprehensive assessment may also be indicated. These assessments can take 30 minutes to several hours, and use standardized performance-based tests to provide objective data about cognitive strengths and weaknesses. Choice of assessment tools depends on the questions that need to be addressed, the setting, and available resources. Shorter but statistically robust test batteries are useful for evaluating the magnitude of clinically meaningful cognitive impairment and parsing areas of cognitive strength and relative weakness, which informs personalized treatment planning. Longer assessments, which may include neuropsychological, IQ, and academic testing, are often required when there are medical comorbidities, such as head trauma or brain disease, or when accommodations for school or work are being sought.

In mental health clinics, demand is greatest for the brief targeted assessments that serve to diagnose cognitive impairment associated with schizophrenia. Targeted assessment batteries typically evaluate domains of cognition considered key in the presentation and prediction of outcomes of people with schizophrenia, including processing speed, verbal learning and memory, working memory, attention/vigilance, reasoning, and problem solving. Global and domain-specific standardized scores, expressed as either T-Scores or Z-Scores, can be interpreted in two ways. First, a standardized score may be interpreted with respect to the normative comparison population, i.e., those without a schizophrenia diagnosis of the same age. A second method is to compare a standardized score to an estimate of the individual's own intellectual ability, which may be obtained as a part of the assessment via a brief single word reading test such as the Wide Range Achievement Test—Reading Subtest (WRAT; [41]) or Test of Premorbid Functioning (TOPF; [42]). This comparison, between current cognitive performance and an estimate of cognitive capacity, identifies areas of preserved ability and areas of decline while the magnitude of the discrepancy indicates the severity of the impairment [43]. This method provides a more personalized means of interpreting a global or domain-specific test score.

Numerous tools exist for the measurement of current cognitive ability for the specific use with people diagnosed with schizophrenia. The development and evaluation of such batteries as the MATRICS Consensus Cognitive Battery (MCCB; [44]) and Brief Assessment of Cognition in Schizophrenia (BACS; [45]), as well as discussion of their relative merits (e.g., psychometric properties, sensitivity to change, language availability) have been described in several publications [46–49]. The benefits of a brief targeted assessment are numerous. A cognitive performance profile can be used to delineate areas of strength and relative need, which become focus areas for interventions such as cognitive skills training. Repeat assessment can be used to evaluate progress in response to cognitive treatments. Such performance-based assessments also provide a unique, focused opportunity for behavioral observation. A person's approach to cognitive tasks during an evaluation likely reflects how cognitively demanding tasks and novel situations are approached in other environments as well. Degree of cognitive effort, frustration tolerance, task persistence, and strategy use are all informative for understanding areas of strength and need for support and skill building.

To integrate targeted assessment batteries into routine practice, testers must be trained and licensed appropriately. Batteries can be costly to purchase, and the time to administer, score, and interpret the results may be prohibitive; there is some opinion that even a 30-minute test battery is too burdensome in busy clinic settings [50]. For these reasons, systematic evaluation of cognitive health using even a brief performance-based battery is rarely routine in clinical practice [51, 52]. However, when several facilitating conditions are present, assessment can be implemented as a standard of care. An example of this is the cognitive health service established by the New York State Office of Mental Health in the United States.

Cognitive Remediation to Promote Recovery (CR2PR) is the first known statewide level implementation of cognitive health services for adults with serious mental illness in the United States [27, 53]. CR2PR program initiation was preceded by a systemwide educational campaign that embedded the value of good cognitive health in the context of recovery. CR2PR implementation, from 2017 to the present day, is facilitated by systematic and cross-disciplinary staff training on who may benefit from the service and how to refer for a cognitive evaluation. A 25-minute battery comprised of subtests from the BACS and MCCB evaluates key cognitive domains: verbal memory, working memory, processing speed, problem solving, and attention/vigilance. The battery is administered alongside the WRAT Reading subtest, which provides an estimated IQ score for comparison. One licensed psychologist at each facility is trained on test administration, scoring, and how to use results for cognitive remediation treatment planning. A scoring template yields a visual depiction of results, which facilitates interpretation, such as identification of areas of cognitive strength and relative need, and provision of feedback (Fig. 10.1). Collaborative decision making is used to identify treatment focus areas, based on a combination of cognitive test results and client-identified goals. A treatment plan template facilitates the translation of assessment results to concrete action steps for skill building. Repeat assessment post-treatment is offered but not required; client satisfaction with the service, perceived impact on cognition, and on daily functioning are valued and cost-effective treatment outcomes [54] (Fig. 10.1).

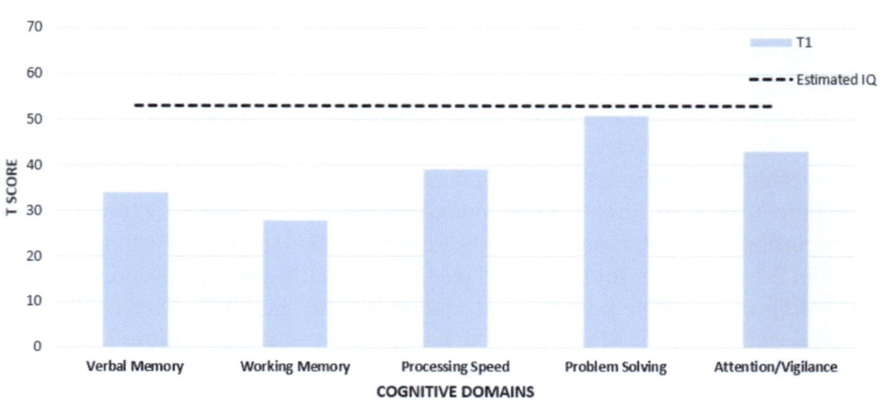

Fig. 10.1 Cognitive Remediation to Promote Recovry (CR2PR): brief cognitive battery

Factors facilitating the feasible integration of performance-based assessments in he CR2PR initiative include staff who value cognitive assessment, availability of a brief test battery, trained and licensed staff to administer it, as well as administrative support for billing. Acceptability of assessment procedures is supported by explicit communication with clients about the purpose as well as feedback on testing results. Collaboration with clients around assessment balances perceived benefits with clients' concerns about time and potential stigma. The CR2PR battery and its implementation methods may be considered scalable for routine use in the context of a large-scale cognitive health service.

10.4.4 Solution 4: Routinely Deliver Feedback

Acceptability of cognitive assessment is enhanced when personalized feedback is offered [55, 56]. When communicating test results, it is important to reference the original purpose of the testing, using the data to address the concerns as conveyed by the client or referring clinician. Written and visual representation of results, summarizing evidence of both cognitive strengths and relative weaknesses, will facilitate understanding as well as satisfaction with the service. Further, perceived utility is enhanced when results and recommendations are framed in a contextualized and personally meaningful way. Recommendations might include the use of cognitive strategies such as those to aid memory for verbal information (e.g., chunking, verbal rehearsal) and techniques to compensate for areas of difficulty (e.g., keeping to-do lists, programming alarms) with the ultimate goal of improving functioning in a personally relevant context (e.g., school, work). Recommendations for clinical service provision may include modifications to therapies that capitalize on an individual's strengths (e.g., visual memory) and accommodate areas of weakness (e.g., provide handouts, encourage note-taking) to optimize treatment participation and opportunity for clinical gain. Service plan changes may be recommended, such as the addition of cognitive remediation to address areas of cognitive health need in conjunction with additional services such as educational or vocational support.

Cognitive assessments completed in routine mental health service settings may provide an indication that a specialist referral is warranted [57, 58]. A comprehensive neuropsychological assessment can address a need for diagnostic clarity or the identification of the likely causes of cognitive impairment, e.g., acquired brain injury versus psychiatric diagnosis. A referral may also be warranted if deterioration in cognitive health is observed or reported. Individuals with schizophrenia are at disproportionately high risk of developing dementia, with published estimates ranging from 2.1-fold to 5.8-fold increased risk across the lifespan [59]. Routine clinical monitoring of cognitive health needs among middle-aged and older adults may facilitate early detection of decline, which can inform recommendations to mitigate modifiable risk factors, alleviate cognitive burden, and preserve independent functioning.

The feedback process for cognitive assessments should also consider the potential influence of stigma in its approach. There is evidence that self-perceived cognitive deficits are associated with both internalized mental illness stigma and quality of life among individuals with schizophrenia spectrum disorders [60]. Initial research also suggests that regardless of scores on performance-based measures of cognition, when individuals with schizophrenia spectrum disorders perceive that they have cognitive deficits, they also report experiencing stigma from others related to these deficits. Stigma may influence their participation in goal-oriented activities and behavior [61]. The goal of assessment feedback is to inform individuals of the findings and provide recommendations; however, because this feedback may increase an individual's awareness of their cognitive symptoms, this also presents the potential for increasing both internalized mental illness stigma and anticipated cognitive health stigma from others.

To minimize its potential contribution to stigma, assessment feedback should (1) incorporate a verbal feedback session in addition to a written summary with visual aids to understand the data; (2) check for client understanding of findings in the verbal feedback session, and clarify any misunderstandings, including those derived from common and inaccurate negative stereotypes of schizophrenia; (3) clarify how and whether cognitive test findings fit with the client's real-world functional experience; and (4) provide psychoeducation regarding the nature of cognitive deficits in schizophrenia and their ability to improve with treatment.

10.5 Ongoing Work and Future Directions

Shared decision-making (SDM) is a formal process in which clinicians and clients work together to select tests, treatments, management, or support packages, weighing clinical evidence and the client's values and informed preferences. SDM in a mental health context has garnered growing interest with increasing efforts to expand implementation in routine mental health care settings where individuals with serious mental illness participate in services [62, 63]. A variety of SDM interventions have been tested among people with serious mental illnesses [64, 65], but none explicitly features decisions on addressing cognitive health. A SDM framework is, however, warranted and aligns with the person-centered orientation and collaborative model of cognitive health services as described in this chapter. As the scope of SDM expands beyond psychopharmacological decision-making (a vestige of SDM's origin in physical healthcare; [64, 66]), SDM interventions that target mental health goal setting and treatment planning might integrate discussion around cognitive assessment. Screening tools like the Self-Assessment of Cognitive Functioning may be clinically useful in this regard as the tool itself queries what the individual might find helpful so that they can do the activities and pursue the goals they care most about. Assessment of client needs and preferences is intended to guide decision-making about cognitive health supports and pursuit of further evaluation, whether brief or comprehensive. Further, at continuous care decision-making points, cognitive assessment results can guide choice of intervention, using the

SDM framework and decision aids to consider when, why, and how addressing cognitive health will be personally meaningful. Formalizing the integration of cognitive assessment within SDM interventions and tools may facilitate the feasible and acceptable implementation of cognitive health assessments in clinical practice.

Another area of ongoing work is the development of an anti-stigma stance in cognitive health services, inclusive of assessment practices. Cognitive assessments may be the first time a client learns about cognitive health, and/or their own cognitive strengths and challenges, and how these translate into real-world functioning. This presents a critical opportunity for psychoeducation that takes an anti-stigma approach and actively challenges inaccurate negative stereotypes regarding cognitive health and functioning in schizophrenia. One avenue of exploration is the development of a strengths-based cognitive assessment that features a more explicit focus on identifying a client's intact repertoires, environmental resources, and interpersonal strengths to inform treatment planning and delivery. Building on existing strengths and applying them in practical ways aims to facilitate a person's functional recovery, promote autonomous participation in cognitive health services, enhance perceived self-efficacy, and reduce the potential for cognitive health stigma [28]. Foundational work still needs to be conducted to conceptualize cognitive strengths and appropriate methods of measurement, accounting for multiple stakeholder perspectives in the process.

Work is also needed regarding how the concept of cognitive health may be understood within various cultural conceptualizations of health, and how culture may be considered in the cognitive health assessment process for people with schizophrenia. Research in aging and cognition, for example, tells us that culture is an important part of how we conceptualize, think about, and prioritize cognitive health that should be considered in cognitive health services research. There is evidence for racial, ethnic, and gender differences in appraisals of cognitive health and related concerns among older adults [67–69]. In addition to cultural differences in how cognitive health is conceptualized or addressed, different cultural groups can have substantially different experiences interfacing with mental health services more generally, which should also be considered for the delivery of cognitive health services.

Finally, as cognitive health services are becoming increasingly available in large systems of care serving people with serious mental illness, methods of cognitive assessment continue to evolve in order to better fit service models and meet clients' needs. Increasingly, web-based cognitive assessments are being developed to circumvent implementation challenges such as clinician training, time, and space. Web-based testing options are currently available for brief cognitive batteries validated for use in some neuropsychiatric populations, including dementia, attention deficit hyperactivity disorder, unipolar depression, and bipolar disorder [70]. Batteries can be completed independently on a web-based platform, enabling cost-effective administration [47, 71]. An automatic data capture system removes threats to test validity, such as scoring and data entry errors. Further, electronic results may be easily shared across clinical care providers, thereby facilitating communication of an individual's cognitive health needs [71]. Validation studies comparing independently completed web-based assessments to those that are supervised in a laboratory have yielded promising results with high correlations and equivalent internal reliability [72–74]. Emerging data suggest web-based cognitive testing may be feasible, acceptable, and valid when conducted with individuals with schizophrenia, though clinical utility remains to be demonstrated [75].

10.6 Summary

Cognitive assessment plays an important role in personalized and recovery-oriented services. This chapter has identified numerous challenges to assessment feasibility and acceptability and has put forth four recommendations to facilitate the uptake of assessment practices in mental health settings where people with schizophrenia participate in services. These recommendations stem from the extant literature and ongoing services research.

1. *Foster a Culture of Cognitive Health*: Educational campaigns oriented toward organizations and providers, and educational materials for clients and their support network, can facilitate the use of a cognitive lens through which to understand a client's experiences and behaviors. A culture that encourages the use of assessments to create shared understanding of an individual's cognitive health can help validate and de-stigmatize cognitive health concerns. This is critical for improving assessment acceptability.
2. *Adopt a Cognitive Health Toolkit*: Workflows that address cognitive health serve to guide who, when, and how to refer for cognitive assessment. Decision-making tools offering a standardized mechanism for ensuring relevant and timely follow-up will improve assessment utility and make treatments more accessible to those who might benefit. Future work may seek to formally integrate Shared-Decision Making tools and processes to reinforce the alignment of cognitive health assessment and treatment services with a person-centered orientation.
3. *Choose Appropriate Assessment Tools*: Brief screening tools can be feasibly implemented in large systems of care and create a first point of cognitive health need identification. When indicated, a referral for targeted or comprehensive assessment may be pursued. Brief standardized test batteries scalable for routine use can facilitate personalized treatment planning. Practitioners should remain abreast of technological advances in assessment practices that may improve the feasibility of routine evaluation. Ongoing work developing a strengths-based approach to cognitive assessment will further support the anti-stigma stance of recovery-oriented cognitive health services.
4. *Routinely Deliver Feedback*: Timely feedback on assessment results provides further opportunity to create a shared understanding of an individual's cognitive strengths and areas of need. Feedback that links results to a client's real-world functional experience, is culturally sensitive, clarifies misconceptions about the nature of cognitive deficits in schizophrenia, and emphasizes their ability to

improve with treatment is de-stigmatizing and promotes collaboration in treatment planning.

Feasible, acceptable, and timely cognitive assessment is critical for addressing the cognitive health needs of people with schizophrenia, with the ultimate goal of optimizing mental health and personally meaningful functional outcomes.

References

1. Fett AJ, Velthorst E, Reichenberg A, Ruggero CJ, Callahan JL, Fochtmann LJ, Carlson GA, Perlman G, Bromet EJ, Kotov R. Long-term changes in cognitive functioning in individuals with psychotic disorders: findings from the Suffolk County Mental Health Project. JAMA Psychiatry. 2020;77(4):387–96. https://doi.org/10.1001/jamapsychiatry.2019.3993.
2. Fusar-Poli P, Deste G, Smieskova R, Barlati S, Yung AR, Howes O, Stieglitz RD, Vita A, McGuire P, Borgwardt S. Cognitive functioning in prodromal psychosis: a meta-analysis. Archives of General Psychiatry. 2012;69(6):562–71. https://doi.org/10.1001/archgenpsychiatry.2011.1592.
3. Green MF, Llerena K, Kern RS. The "Right Stuff" revisited: what have we learned about the determinants of daily functioning in schizophrenia? Schizophr Bull. 2015;41(4):781–5. https://doi.org/10.1093/schbul/sbv018.
4. Santesteban-Echarri O, Paino M, Rice S, González-Blanch C, McGorry P, Gleeson J, Alvarez-Jimenez M. Predictors of functional recovery in first-episode psychosis: a systematic review and meta-analysis of longitudinal studies. Clin Psychol Rev. 2017;58:59–75. https://doi.org/10.1016/j.cpr.2017.09.007.
5. Bowie CR, Harvey PD. Cognitive deficits and functional outcome in schizophrenia. Neuropsychiatr Dis Treat. 2006;2(4):531–6. https://doi.org/10.2147/nedt.2006.2.4.531.
6. Cowman M, Holleran L, Lonergan E, O'Connor K, Birchwood M, Donohoe G. Cognitive predictors of social and occupational functioning in early psychosis: a systematic review and meta-analysis of cross-sectional and longitudinal data. Schizophr Bull. 2021;47(5):1243–53. https://doi.org/10.1093/schbul/sbab033.
7. Nuechterlein KH, Subotnik KL, Green MF, Ventura J, Asarnow RF, Gitlin MJ, Yee CM, Gretchen-Doorly D, Mintz J. Neurocognitive predictors of work outcome in recent-onset schizophrenia. Schizophr Bull. 2011;37 Suppl 2(Suppl 2):S33–40. https://doi.org/10.1093/schbul/sbr084.
8. Galletly C, Castle D, Dark F, Humberstone V, Jablensky A, Killackey E, Kulkarni J, McGorry P, Nielssen O, Tran N. Royal Australian and New Zealand College of Psychiatrists clinical practice guidelines for the management of schizophrenia and related disorders. Aust N Z J Psychiatry. 2016;50(5):410–72. https://doi.org/10.1177/0004867416641195.
9. Norman R, Lecomte T, Addington D, Anderson E. Canadian treatment guidelines on psychosocial treatment of schizophrenia in adults. Can J Psychiatr. 2017;62(9):617–23. 10.1177%2F0706743717719894
10. Vita A, Gaebel W, Mucci A, Sachs G, Barlati S, Giordano GM, Nibbio G, Nordentoft M, Wykes T, Galderisi S. European psychiatric association guidance on treatment of cognitive impairment in schizophrenia. Eur Psychiatry. 2022;65(1):e57. https://doi.org/10.1192/j.eurpsy.2022.2315.
11. Bryce S, Cheng N, Dalton A, Ojinnaka A, Stainton A, Zbukvic I, Ratheesh A, O'Halloran C, Uren J, Gates J, Daglas-Georgiou R, Wood SJ, Allott K. Cognitive health treatment priorities and preferences among young people with mental illness: the your mind, your choice survey. Early Interv Psychiatry. 2024;18(2):94–101. https://doi.org/10.1111/eip.13436.

12. Moritz S, Berna F, Jaeger S, Westermann S, Nagel M. The customer is always right? Subjective target symptoms and treatment preferences in patients with psychosis. Eur Arch Psychiatry Clin Neurosci. 2017;267(4):335–9. https://doi.org/10.1007/s00406-016-0694-5.
13. Rose D, Wykes T, Farrier D, Doran AM, Sporle T, Bogner D. What do clients think of cognitive remediation therapy?: a consumer-led investigation of satisfaction and side effects. Am J Psychiatr Rehabil. 2008;11(2):181–204. https://doi.org/10.1080/15487760801963694.
14. Saperstein AM, Medalia A, Bello I, Dixon LB. Addressing cognitive health in coordinated specialty care for early psychosis: real-world perspectives. Early Interv Psychiatry. 2021a;15(2):374–9. https://doi.org/10.1111/eip.12966.
15. Allott K, Brewer W, McGorry PD, Proffitt TM. Referrers' perceived utility and outcomes of clinical neuropsychological assessment in an adolescent and young adult public mental health service. Aust Psychol. 2011;46(1):15–24. https://doi.org/10.1111/j.1742-9544.2010.00002.x.
16. Westervelt HJ, Brown LB, Tremont G, Javorsky DJ, Stern RA. Patient and family perceptions of the neuropsychological evaluation: how are we doing? Clin Neuropsychol. 2007;21(2):263–73. https://doi.org/10.1080/13854040500519745.
17. Delagneau G, Bowden SC, Bryce S, van-der-El K, Hamilton M, Adams S, Burgat L, Killackey E, Rickwood D, Allott K. Thematic analysis of youth mental health providers' perceptions of neuropsychological assessment services. Early Interv Psychiatry. 2020;14(2):220–7.
18. Fisher CA, Hetrick SE, Merrett Z, Parrish EM, Allott K. Neuropsychology and youth mental health in Victoria: the results of a clinical service audit. Aust Psychol. 2017;52(6):453–60. https://doi.org/10.1111/ap.12259.
19. Allott K, Van-der-el K, Bryce S, Hamilton M, Adams S, Burgat L, et al. Need for clinical neuropsychological assessment in headspace youth mental health services: a national survey of providers. Aust J Psychol. 2019;71(2):108–16. https://doi.org/10.1111/ajpy.12225.
20. Temple RO, Carvalho J, Tremont G. A national survey of physicians' use of and satisfaction with neuropsychological services. Arch Clin Neuropsychol. 2006;21(5):371–82. https://doi.org/10.1016/j.acn.2006.05.002.
21. Corrigan P. How stigma interferes with mental health care. Am Psychol. 2004;59(7):614–25. https://doi.org/10.1037/0003-066X.59.7.614.
22. Sims R, Michaleff ZA, Glasziou P, Thomas R. Consequences of a diagnostic label: a systematic scoping review and thematic framework. Front Public Health. 2021;9:*725877*. https://doi.org/10.3389/fpubh.2021.725877.
23. Lejeune JA, Northrop A, Kurtz MM. A meta-analysis of cognitive remediation for schizophrenia: efficacy and the role of participant and treatment factors. Schizophr Bull. 2021;47(4):997–1006. https://doi.org/10.1093/schbul/sbab022.
24. Vita A, Barlati S, Ceraso A, Nibbio G, Ariu C, Deste G, Wykes T. Effectiveness, core elements, and moderators of response of cognitive remediation for schizophrenia: a systematic review and meta-analysis of randomized clinical trials. JAMA Psychiatry. 2021;78(8):848–58. https://doi.org/10.1001/jamapsychiatry.2021.0620.
25. Wykes T, Huddy V, Cellard C, McGurk SR, Czobor P. A meta-analysis of cognitive remediation for schizophrenia: methodology and effect sizes. Am J Psychiatry. 2011;168(5):472–85. https://doi.org/10.1176/appi.ajp.2010.10060855.
26. Dark FL, Amado I, Erlich MD, Ikezawa S. International experience of implementing CRT for people with psychotic disorders. Manuscript submitted for publication. 2024.
27. Medalia A, Saperstein AM, Erlich MD, Sederer LI. Cognitive remediation in large systems of psychiatric care. CNS Spectr. 2019a;24(1):163–73. https://doi.org/10.1017/s1092852918000822.
28. Allott K, Steele P, Boyer F, de Winter A, Bryce S, Alvarez-Jimenez M, Phillips L. Cognitive strengths-based assessment and intervention in first-episode psychosis: a complementary approach to addressing functional recovery? Clin Psychol Rev. 2020;79:101871. https://doi.org/10.1016/j.cpr.2020.101871.
29. American Psychological Association Practice Organization. Distinguishing between screening and assessment for mental and behavioral health problems: a statement from an American Psychological Association Practice Organization Workgroup on Screening and psychological

Assessment. 2014, December. Retrieved April 16, 2024, from https://www.apaservices.org/practice/reimbursement/billing/assessment-screening.
30. Roebuck-Spencer TM, Glen T, Puente AE, Denney RL, Ruff RM, Hostetter G, Bianchini KJ. Cognitive screening tests versus comprehensive neuropsychological test batteries: a National Academy of neuropsychology education paper. Arch Clin Neuropsychol. 2017;32(4):491–8. https://doi.org/10.1093/arclin/acx021.
31. McIntyre RS, Anderson N, Baune BT, Brietzke E, Burdick K, Fossati P, Gorwood P, Harmer C, Harrison J, Harvey P, Mansur RB, Medalia A, Miskowiak K, Ramey T, Rong C, Rosenblat JD, Young A, Stahl SM. Expert consensus on screening and assessment of cognition in psychiatry. CNS Spectr. 2019;24(1):154–62. https://doi.org/10.1017/S1092852918001189.
32. Folstein MF, Folstein SE, McHugh PR. "Mini-mental state". A practical method for grading the cognitive state of patients for the clinician. J Psychiatr Res. 1975;12(3):189–98. https://doi.org/10.1016/0022-3956(75)90026-6.
33. Nasreddine ZS, Phillips NA, Bédirian V, Charbonneau S, Whitehead V, Collin I, Cummings JL, Chertkow H. The Montreal Cognitive Assessment, MoCA: a brief screening tool for mild cognitive impairment. J Am Geriatr Soc. 2005;53(4):695–9. https://doi.org/10.1111/j.1532-5415.2005.53221.x.
34. Bryce SD, Bowden SC, Wood SJ, Allott K. Brief, performance-based cognitive screening in youth aged 12-25: a systematic review. J Int Neuropsychol Soc. 2021;27(8):835–54. https://doi.org/10.1017/S1355617720001241.
35. Bryce S, Allott K. Cognitive screening: a significant unmet need in youth mental health. Aust N Z J Psychiatry. 2019;53(8):813. https://doi.org/10.1177/0004867419834356.
36. Saperstein AM, Medalia A, Malinovsky I, Bello I, Dixon LB. Toolkit for assessing and addressing cognitive health in early psychosis: evaluation of feasibility and utility in a coordinated specialty care setting. Early Interv Psychiatry. 2021b;15(5):1376–81. https://doi.org/10.1111/eip.13070.
37. McCleery A, Ventura J, Kern RS, Subotnik KL, Gretchen-Doorly D, Green MF, Hellemann GS, Nuechterlein KH. Cognitive functioning in first-episode schizophrenia: MATRICS Consensus Cognitive Battery (MCCB) profile of impairment. Schizophr Res. 2014;157(1–3):33–9. https://doi.org/10.1016/j.schres.2014.04.039.
38. Mesholam-Gately RI, Giuliano AJ, Goff KP, Faraone SV, Seidman LJ. Neurocognition in first-episode schizophrenia: a meta-analytic review. Neuropsychology. 2009;23(3):315–36. https://doi.org/10.1037/a0014708.
39. McCleery A, Nuechterlein KH. Cognitive impairment in psychotic illness: prevalence, profile of impairment, developmental course, and treatment considerations. Dialogues Clin Neurosci. 2019;21(3):239–48. https://doi.org/10.31887/dcns.2019.21.3/amccleery.
40. Saperstein AM, Bello I, Nossel I, Dixon LB, Medalia A. Implementation of cognitive health services in large systems of care: highlights from coordinated specialty care for first episode psychosis. Schizophr Bull. 2024; sbae030. Advance online publication.
41. Wilkinson GS, Robertson GJ. Wide Range Achievement Test, Fifth Edition. Psychological Assessment Resources; 2017.
42. Wechsler D. Test of premorbid functioning: advanced clinical solutions for use with WAIS-IV and WMS-IV. San Antonio: The Psychological Corporation; 2011.
43. Lezak MD, Howieson DB, Loring DW. Neuropsychological assessment. 4th ed. New York: Oxford University Press; 2004.
44. Nuechterlein KH, Green MF, Kern RS, Baade LE, Barch DM, Cohen JD, Essock S, Fenton WS, Frese FJ 3rd, Gold JM, Goldberg T, Heaton RK, Keefe RS, Kraemer H, Mesholam-Gately R, Seidman LJ, Stover E, Weinberger DR, Young AS, Zalcman S, et al. The MATRICS consensus cognitive battery, part 1: test selection, reliability, and validity. Am J Psychiatry. 2008;165(2):203–13. https://doi.org/10.1176/appi.ajp.2007.07010042.
45. Keefe RS, Goldberg TE, Harvey PD, Gold JM, Poe MP, Coughenour L. The brief assessment of cognition in schizophrenia: reliability, sensitivity, and comparison with a standard neurocognitive battery. Schizophr Res. 2004;68(2–3):283–97. https://doi.org/10.1016/j.schres.2003.09.011.

46. Bowie CR, Grossman M, Gupta M, Oyewumi LK, Harvey PD. Cognitive remediation in schizophrenia: efficacy and effectiveness in patients with early versus long-term course of illness. Early Interv Psychiatry. 2014;8(1):32–8. https://doi.org/10.1111/eip.12029.
47. Keefe RS, Haig GM, Marder SR, Harvey PD, Dunayevich E, Medalia A, Davidson M, Lombardo I, Bowie CR, Buchanan RW, Bugarski-Kirola D, Carpenter WT, Csernansky JT, Dago PL, Durand DM, Frese FJ, Goff DC, Gold JM, Hooker CI, Kopelowicz A, et al. Report on ISCTM consensus meeting on clinical assessment of response to treatment of cognitive impairment in schizophrenia. Schizophr Bull. 2016;42(1):19–33. https://doi.org/10.1093/schbul/sbv111.
48. Keefe RS, Poe M, Walker TM, Harvey PD. The relationship of the brief assessment of cognition in schizophrenia (BACS) to functional capacity and real-world functional outcome. J Clin Exp Neuropsychol. 2006;28(2):260–9. https://doi.org/10.1080/13803390500360539.
49. Kern RS, Nuechterlein KH, Green MF, Baade LE, Fenton WS, Gold JM, Keefe RS, Mesholam-Gately R, Mintz J, Seidman LJ, Stover E, Marder SR. The MATRICS consensus cognitive battery, part 2: co-norming and standardization. Am J Psychiatry. 2008;165(2):214–20. https://doi.org/10.1176/appi.ajp.2007.07010043.
50. Rossell SL. Response to: Bryce and Allott—Cognitive screening: a significant unmet need in youth mental health. 30 minutes is too long! Aust N Z J Psychiatry. 2019;53(8):814. https://doi.org/10.1177/0004867419832652.
51. Belgaied W, Samp J, Vimont A, Rémuzat C, Aballéa S, El Hammi E, Kooli A, Toumi M, Akhras K. Routine clinical assessment of cognitive functioning in schizophrenia, major depressive disorder, and bipolar disorder. Eur Neuropsychopharmacol. 2014;24(1):133–41. https://doi.org/10.1016/j.euroneuro.2013.11.001.
52. Green MF, Barnes TR, Danion JM, Gallhofer B, Meltzer HY, Pantelis C. The FOCIS international survey on psychiatrists' opinions on cognition in schizophrenia. Schizophr Res. 2005;74(2–3):253–61. https://doi.org/10.1016/j.schres.2004.05.023.
53. Medalia A, Erlich M. Why cognitive health matters. Am J Public Health. 2017;107(1):45–7. https://doi.org/10.2105/ajph.2016.303544.
54. Medalia A, Erlich MD, Soumet-Leman C, Saperstein AM. Translating cognitive behavioral interventions from bench to bedside: the feasibility and acceptability of cognitive remediation in research as compared to clinical settings. Schizophr Res. 2019b;203:49–54. https://doi.org/10.1016/j.schres.2017.07.044.
55. Evans CL, Pulsifer MB, Grieco JA. Communication of neuropsychological results to caregivers of Latino immigrant children. Transl Issues Psychol Sci. 2019;5(1):42–50. https://psycnet.apa.org/doi/10.1037/tps0000178.
56. Gruters AA, Ramakers IH, Verhey FR, Kessels RP, de Vugt ME. A scoping review of communicating neuropsychological test results to patients and family members. Neuropsychol Rev. 2022;32(2):294–315. https://doi.org/10.1007/s11065-021-09507-2.
57. Braun M, Tupper D, Kaufmann P, McCrea M, Postal K, Westerveld M, Wills K, Deer T. Neuropsychological assessment: a valuable tool in the diagnosis and management of neurological, neurodevelopmental, medical, and psychiatric disorders. Cogn Behav Neurol. 2011;24(3):107–14. https://doi.org/10.1097/wnn.0b013e3182351289.
58. Proffitt TM, Brewer WJ, Parrish EM, McGorry PD, Allott KA. Reasons for referral and findings of clinical neuropsychological assessment in youth with mental illness: a clinical file audit. Appl Neuropsychol Child. 2018;7(2):164–74. https://doi.org/10.1080/21622965.2017.1284663.
59. Stroup TS, Olfson M, Huang C, Wall MM, Goldberg T, Devanand DP, Gerhard T. Age-specific prevalence and incidence of dementia diagnoses among older US adults with schizophrenia. JAMA Psychiatry. 2021;78(6):632–41. https://doi.org/10.1001/jamapsychiatry.2021.0042.
60. Shin YJ, Joo YH, Kim JH. Self-perceived cognitive deficits and their relationship with internalized stigma and quality of life in patients with schizophrenia. Neuropsychiatr Dis Treat. 2016;12:1411–7. https://doi.org/10.2147/NDT.S108537.
61. Gonzales L, Saperstein AM, Jones N, Erlich MD, Medalia A. Perceived stigma toward cognitive impairment in people with schizophrenia. Psychiatr Serv. 2025;76(1):86–89.

62. Chmielowska M, Zisman-Ilani Y, Saunders R, Pilling S. Trends, challenges, and priorities for shared decision making in mental health: the first umbrella review. Int J Soc Psychiatry. 2023;69(4):823–40. https://doi.org/10.1177/00207640221140291.
63. Salyers MP, Zisman-Ilani Y. Shared decision-making and self-directed care. In: Goldman HH, Frank RG, Morrissey JP, editors. The Palgrave handbook of American mental health policy. Springer International Publishing; 2020. p. 197–228.
64. Thomas EC, Ben-David S, Treichler E, Roth S, Dixon LB, Salzer M, Zisman-Ilani Y. A systematic review of shared decision-making interventions for service users with serious mental illnesses: state of the science and future directions. Psychiatr Serv. 2021;72(11):1288–300. https://doi.org/10.1176/appi.ps.202000429.
65. Zisman-Ilani Y, Barnett E, Harik J, Pavlo A, O'Connell M. Expanding the concept of shared decision making for mental health: systematic search and scoping review of interventions. Ment Health Rev J. 2017;22(3):191–213. https://doi.org/10.1108/MHRJ-01-2017-0002.
66. Deegan PE, Rapp C, Holter M, Riefer M. Best practices: a program to support shared decision making in an outpatient psychiatric medication clinic. Psychiatr Serv. 2008;59(6):603–5. https://doi.org/10.1176/ps.2008.59.6.603.
67. Jang Y, Choi EY, Franco Y, Park NS, Chiriboga DA, Kim MT. Racial and ethnic differences in cognitive health appraisals: a comparison of non-Hispanic white, non-Hispanic Black, and Hispanic older adults. Cognition Aging. 2020;26(3):519–25. https://doi.org/10.1093/geroni/igab046.2652.
68. Latidka JN, Laditka SB, Liu R, Price AE, Wu B, Friedman DB, Corwin SJ, Sharkey JR, Tseng W, Hunter R, Logsdon RG. Older adults' concerns about cognitive health: commonalities and differences among six United States ethnic groups. Aging Soc. 2011;31(7):1202–28. https://doi.org/10.1017/S0144686X10001273.
69. Wu B, Goins RT, Laditka JN, Ignatenko V, Goedereis E. Gender differences in views about cognitive health and healthy lifestyle behaviors among rural older adults. The Gerontologist. 2009;49(S1):S72–8. https://doi.org/10.1093/geront/gnp077.
70. Miskowiak KW, Jespersen AE, Obenhausen K, Hafiz P, Hestbæk E, Gulyas L, Kessing LV, Bardram JE. Internet-based cognitive assessment tool: sensitivity and validity of a new online cognition screening tool for patients with bipolar disorder. J Affect Disord. 2021;289:125–34. https://doi.org/10.1016/j.jad.2021.04.039.
71. Feenstra HE, Vermeulen IE, Murre JM, Schagen SB. Online cognition: factors facilitating reliable online neuropsychological test results. Clin Neuropsychol. 2017;31(1):59–84. https://doi.org/10.1080/13854046.2016.1190405.
72. Backx R, Skirrow C, Dente P, Barnett JH, Cormack FK. Comparing web-based and lab-based cognitive assessment using the Cambridge neuropsychological test automated battery: a within-subjects counterbalanced study. J Med Internet Res. 2020;22(8):e16792. https://www.jmir.org/2020/8/e16792.
73. Feenstra HEM, Murre JMJ, Vermeulen IE, Kieffer JM, Schagen SB. Reliability and validity of a self-administered tool for online neuropsychological testing: the Amsterdam Cognition Scan. J Clin Exp Neuropsychol. 2018;40(3):253–73. https://doi.org/10.1080/13803395.2017.1339017.
74. Germine L, Nakayama K, Duchaine BC, Chabris CF, Chatterjee G, Wilmer JB. Is the web as good as the lab? Comparable performance from web and lab in cognitive/perceptual experiments. Psychon Bull Rev. 2012;19(5):847–57. https://doi.org/10.3758/s13423-012-0296-9.
75. Lynham AJ, Jones IR, Walters JTR. Web-based cognitive testing in psychiatric research: validation and usability study. J Med Internet Res. 2022;24(2):e28233. https://doi.org/10.2196/28233.

Cognitive Impairment in Schizophrenia: Neurobiological Bases 11

Francesca Martini, Marco Spangaro, Jacopo Sapienza, Marta Bosia, and Roberto Cavallaro

Abstract

Cognitive impairment represents a hallmark feature of schizophrenia, affecting multiple domains with varying degrees of severity and significantly hampering functional recovery. Unraveling the neurobiological underpinnings of cognitive deficits is crucial for advancing our understanding and treatment of this debilitating disorder.

Multiple neurobiological mechanisms have been associated with cognitive impairment in schizophrenia, including genetic factors, disrupted neurotransmission, regional and whole-brain structural and functional alterations, as well as, more recently, neuroinflammation and metabolic dysfunction. The past decades witnessed significant knowledge advancements, and the focus shifted to disentangling their intricate interplay, which collectively contribute to cognitive impairment.

This chapter provides an overview of the main findings on neurobiological bases of cognitive impairment in schizophrenia, filling in the dots to highlight their interactions, pointing out open questions, and discussing clinical implications.

F. Martini · M. Spangaro
Department of Clinical Neurosciences, IRCCS San Raffaele Scientific Institute, Milan, Italy

J. Sapienza (✉)
Department of Clinical Neurosciences, IRCCS San Raffaele Scientific Institute, Milan, Italy

Department of Humanities and Life Sciences, University School for Advanced Studies IUSS, Pavia, Italy
e-mail: sapienza.jacopo@hsr.it

M. Bosia · R. Cavallaro
Department of Clinical Neurosciences, IRCCS San Raffaele Scientific Institute, Milan, Italy

School of Medicine, Vita -Salute San Raffaele University, Milan, Italy

A. Vita et al. (eds.), *Negative and Cognitive Symptoms in Schizophrenia*, Comprehensive Approach to Psychiatry 3,
https://doi.org/10.1007/978-3-032-06209-3_11

Keywords

Cognition · Psychosis · Genetics · Neurotransmission · Neuroinflammation · Neuroimaging · EEG

11.1 Introduction

Cognitive impairment is the silent enemy still to be defeated in schizophrenia. It subtly emerges prior to the onset of the disease, persists, and even worsens during the course and severely undermines daily functioning and quality of life. Despite being widely acknowledged, cognitive deficits are still silent in many ways: they may be less openly manifest compared to other symptoms and need to be accurately assessed with neuropsychological batteries, or in most severe cases, they may literally take a person's voice away, significantly hampering communication. Further, cognitive deficits are still absent from diagnostic criteria and, more importantly, their biological fingerprints remain elusive. Over a century ago, Emil Kraepelin stated, "*Like a tree whose roots do not find nourishment in their soil dies, in the same way, when psychic deficits occur, we can hypothesize underlying alterations in the structure and functioning of the brain cortex*". Nowadays, far from being a hypothesis, the presence of underlying alterations in the structure and functioning of the brain in people with schizophrenia is a certainty, supported by multiple levels of evidence. However, the issue is how to translate the large amount of knowledge on the biological bases into significant clinical changes. Findings are complex and heterogenous and for many observed alterations is even still unclear if they are causal or compensatory. While it would be naive to expect a single key mechanism accounting for all the biological and behavioral alterations, tackling the neurobiological underpinning of cognitive dysfunction in schizophrenia is still worth. Indeed, a greater understanding holds the potential to drive a treatment shift, testing new targets and allowing for more personalized approaches.

With this goal in mind, here, we summarize the main neurobiological mechanisms implicated in the cognitive impairment in schizophrenia, starting from genetic factors, neurotransmission's alterations and neuroinflammation, to the structural and functional brain anomalies detectable through neuroimaging techniques.

11.2 Genetic Factors

Extensive evidence shows that both schizophrenia and cognitive impairment have a high heritability, ranging between 70% and 90% and 24% and 55% respectively [89].

Twin studies [38, 177, 223] and molecular genetic studies [74, 137] have reported a strong negative genetic correlation between liability for schizophrenia and cognitive ability, indicating that they share common genetic factors [151]. Particularly, a series of studies showed significant negative correlations between the genetic risk of schizophrenia and different cognitive traits including intelligence [212], general cognitive function [55], verbal-numerical reasoning and reaction time [96, 106, 117, 225], known to be affected in schizophrenia [1, 63, 88, 211]. Consistently, unaffected relatives of schizophrenia patients show significantly lower performance in many cognitive tasks [4, 21, 73, 107, 213].

Over the last decade, extensive Genome-Wide Association Studies (GWASs) of general cognitive function and schizophrenia have shown that a significant proportion of the heritability of cognitive function and schizophrenia can be attributed to a polygenic component consisting of numerous common genetic variants with minor individual effects [173]. In detail, large-scale studies of community-based samples confirmed that higher PRSs for schizophrenia are associated with poorer cognitive functioning in healthy populations, while low PRSs for cognition are associated with an increased risk of developing the disorder [96, 117, 137, 153]. Among individuals diagnosed with schizophrenia, cognitive performance correlates with PRSs for IQ and educational attainment, but not for schizophrenia itself [143, 194]. These findings suggest that, although cognitive impairment in schizophrenia may not be directly due to genetic liability for the disorder, cognition could serve as a mediating factor through which genetic risk exerts its effects [224]. Additionally, while higher PRSs for schizophrenia are associated with lower baseline cognitive performance in healthy subjects, they are not associated with greater cognitive decline, implying a neurodevelopmental rather than neurodegenerative nature of the process [125, 151]. However, it is crucial to recognize that PRS analyses typically explain less than 10% of risk variance and are currently only applicable to individuals of European ancestry [151].

GWASs have identified more than 20 genomic loci that jointly influence cognitive function in schizophrenia. All implicated genes are globally expressed in the developmental and adult human brain, suggesting that they may influence brain function across the life span, consistent with the neurodevelopmental hypothesis. The strongest overlapping signal detected was shared between schizophrenia, general cognitive function, and verbal-numerical reasoning. This locus is located on chromosome 22q13.2 and contains many genes, including TCF20, also implicated in neurodevelopmental disability [222], CYP2D6, and NAGA. CYP2D6 encodes cytochrome P450 enzymes metabolizing drugs, while NAGA encodes lysosomal enzymes. These genes could represent potential drug targets for the treatment of cognitive functioning in schizophrenia [211]. Other identified shared loci were a variant at CACNA1C, which encodes a subunit of a voltage-gated calcium channel strongly linked to schizophrenia, and an AKAP6 variant, which appears to influence multiple cognitive traits in addition to schizophrenia ([55, 56]). For the majority of loci (18/21) jointly influencing schizophrenia risk and cognitive function, the schizophrenia risk alleles were associated with worse cognitive performance, confirming that common gene variants may underlie both schizophrenia and cognitive impairment. Notably, three loci with the same effect directions in schizophrenia and cognitive function have also been identified, suggesting that some schizophrenia risk variants may even improve cognitive performance. This could help elucidate why some patients with schizophrenia show normal cognitive functioning [211].

Overall, most of the identified risk genes for both cognitive impairment and schizophrenia participate in neurodevelopment, neuroplasticity, and neurotransmission.

11.2.1 Neurodevelopmental and Neuroplasticity Genes

Many associations have been reported with genes coding for molecules, which are key in synaptic development, neurotransmission, and plasticity. Among these, dystrobrevin binding protein 1 (DTNBP1), disrupted in schizophrenia 1 (DISC1), and brain-derived neurotrophic factor (BDNF) genes are he most investigated, although with mixed findings [74, 249].

DTNBP1 gene encodes dysbindin, which regulates protein trafficking and cell-surface expression of neurotransmitter receptors, modulates PFC activity via glutamatergic neurotransmission [171, 219], and it is involved in hippocampal formation [147]. In subjects with schizophrenia, a substantial reduction of DTNBP1 in glutamatergic neuronal terminal fields within the hippocampus has been observed, supporting that glutamatergic dysconnectivity may contribute to the cognitive impairment related to the disorder [219]. Several studies have explored the association between DTNBP1 variants and cognitive functioning (IQ, attention/vigilance, spatial working memory) in schizophrenia, with mixed results, suggesting a complex relationship [14, 36, 66, 181].

DISC1 gene plays a crucial role in neurogenesis and neuroplasticity, and its downstream cascade has implications for learning, memory, and mood regulation. Studies have reported associations between DISC1 variants and several cognitive domains (short- and long-term memory, visual memory, attention, working memory, verbal fluency) in schizophrenia, further highlighting its potential involvement in neurocognitive deficits [35, 37, 39, 104, 167].

BDNF gene has been extensively studied in schizophrenia because of its involvement in cell differentiation, survival, synaptic plasticity, long-term potentiation, learning, and memory [51, 91, 136, 172]. The Val66Met polymorphism of BDNF has been linked both to memory impairment and schizophrenia susceptibility. However, findings regarding its association with cognitive deficits in schizophrenia have been inconsistent across studies [249].

11.2.2 Dopaminergic System Genes

A decrease in dopaminergic neurotransmission in specific brain regions, including the dorso-lateral prefrontal cortex (PFC) and the anterior cingulate cortex (ACC), has been proposed to affect cognition in schizophrenia, specifically working memory and executive function [142].

The most investigated candidate gene in neurocognition of schizophrenia is the catechol-O-methyltransferase (COMT) gene. In particular, the Val158Met polymorphism has garnered significant attention due to its influence on dopaminergic neurotransmission in the PFC [249]. However, findings regarding the impact of COMT Val158Met on cognitive performance in schizophrenia have exhibited inconsistency across studies [17, 68, 142].

Similarly, studies on other dopamine-related genes such as the dopamine transporter (DAT), dopamine receptors (DRD1, DRD2, DRD3, DRD4, DRD5), dopamine beta-hydroxylase (DBH), vesicular monoamine transporter 2 (SLC18A2), ANKK1, and PPP1R1B, yielded mixed results, with some indicating associations with cognitive measures in schizophrenia and others reporting no significant effects [249].

Overall, although dopamine-related genes may contribute to neurocognitive deficits in schizophrenia, particularly in memory, attention, and executive function, the evidence remains limited and inconclusive. Moreover, recent GWASs of cognitive performance in schizophrenia have not significantly implicated dopamine-related genes, suggesting the involvement of additional mechanisms and interactions in the genetic underpinnings of neurocognitive impairments in schizophrenia [64, 74, 98, 137].

11.2.3 Glutamatergic System Genes

The glutamatergic neurotransmitter system gained significant interest in schizophrenia research due to its neuronal excitatory properties, particularly in the cerebral cortex, and its association with both psychotic symptoms and cognitive impairment [69, 72]. Glutamate represents the major excitatory pathway in the central nervous system, and it is involved in critical processes such as neural development, synaptic plasticity, learning, and memory. Reduced glutamate levels have been observed in individuals with schizophrenia and altered cognitive control functioning, but not in healthy subjects [72].

Despite the crucial role of the glutamatergic system both in schizophrenia and cognitive functioning, only a few genes within this system have been studied in relation to neurocognitive impairments in schizophrenia, such as the glutamate receptor, ionotropic, N-methyl-D-aspartate, subunit 2B (GRIN2B), GRIN2A, metabotropic 3 (GRM3) [119, 167], and the glutamate transporter (excitatory amino acid transporter 2—EAAT2). Notably, genetic variability of EAAT2 has been associated with executive functions and working memory in both individuals with schizophrenia and healthy controls [214]. While some authors also reported a significant association between the GRIN2B rs220599T allele with poorer immediate and delayed recall on the Rey Auditory Verbal Learning Test and enhanced cognitive performance with the GRM3 rs2189814C allele but not with GRIN2A, others did not detect any positive findings [119, 167, 249].

11.2.4 Serotonergic System Genes

The serotonergic system plays a crucial role in the regulation of morphogenesis in brain development, neuronal proliferation, migration, differentiation, and cognition [249]. The serotonin 2A receptor (HTR2A) is particularly abundant in the frontal cortex and ACC. A few studies have investigated the HTR2A T102C polymorphism in relation to cognitive deficits in schizophrenia, once again with mixed results. Some authors found an association between the T allele and reduced attention, together with an association between the combination of the T allele and the Met BDNF allele and higher scores for involuntary attention and lower scores for voluntary attention [7]. Similarly, significant associations between the high-expression T allele with a reduced hit rate in the Continuous Performance Task (CPT) have been reported. Furthermore, the T/C genotype was correlated with a worse performance on the CPT and on the Wisconsin Card Sorting Test (WCST) [228, 249]. Again, a significant association between T/T homozygotes and reduced verbal fluency was described in male patients with schizophrenia [6]. However, other studies did not find significant associations with cognitive deficits [44, 167].

Similarly, investigations into serotonin 1A receptor (HTR1A) and serotonin transporter (5HTT, SLC6A4) genetic variations have yielded conflicting findings. While two studies found negative results [13, 41], Bosia and colleagues found that schizophrenia patients with the low-expression CC genotype of the HTR1A −1019C/G polymorphism performed better on Theory of Mind tasks [24, 249]. Additionally, a significant association between the HTTLPR polymorphism (high-activity long allele) and better executive function and poorer sustained attention in patients with schizophrenia was reported [23].

11.3 Dysfunction of Neurotransmitter Systems

In subjects with schizophrenia, the development of aberrant connections in the brain arises from abnormal synaptic plasticity and network dysconnectivity, which involves key neuromodulatory neurotransmitters, including dopamine, serotonin, and acetylcholine. In particular, imbalances in dopaminergic, cholinergic, glutamatergic, and gamma-aminobutyric acid (GABA) activity—each of which is deeply interconnected—are hypothesized to play a significant role in cognitive deficits in schizophrenia [151, 159].

11.3.1 Glutamate and GABA (Excitatory/Inhibitory [E/I] Balance)

As previously noted, GWAS have shown that single nucleotide polymorphisms (SNPs) associated with schizophrenia are mainly concentrated in genes expressed in E/I neurons [151] and the E/I balance disruption has been linked to cognitive deficits.

In schizophrenia patients with cognitive deficits, a reduction in the glutamatergic stimulation of NMDA receptors in pyramidal neurons has been reported. Activation of these receptors relies on co-agonists like D-serine and glycine. However, in schizophrenia, higher levels of kynurenic acid in the dorsal PFC block glycine sites on NMDA receptors. Despite reduced NMDA receptor sensitivity in pyramidal neurons, glutamate levels in the dorsolateral PFC of individuals with schizophrenia are not decreased, as glutaminase, the enzyme involved in glutamate synthesis, is overexpressed [147].

Changes in GABAergic inputs, crucial for pyramidal neuron synchronicity, have also been observed. Post-mortem studies, supporting a model of cortical disinhibition, described reductions in glutamate decarboxylase GAD65 and GAD67 enzymes, which are essential for GABA synthesis, diminished excitatory input to inhibitory neurons, and decreased numbers of interneurons [122, 139]. This correlates with reduced levels of parvalbumin (PV), a marker for GABAergic interneurons in the DLPFC that modulates pyramidal neuron activity. Additionally, a reduction in other GABAergic neuron subpopulations, such as somatostatin-expressing neurons, has been reported [147].

Overall, these findings highlight significant alterations in both glutamatergic and GABAergic signaling pathways in the DLPFC of schizophrenia patients, contributing to cognitive deficits. [147].

Several neuroimaging studies, using different techniques, reported consistent findings. Magnetic resonance spectroscopy (MRS) can estimate glutamate and GABA concentrations at rest, and functional MRS (fMRS) can investigate dynamic changes in these concentrations during cognitive challenges [151]. Studies in individuals with schizophrenia have shown increased Glx (combined glutamate and glutamine signal) [155] and reduced GABA levels [165], consistent with the cortical disinhibition hypothesis. Despite heterogeneity in results, possibly due to variations in acquisition methods and exposure to medication as well as pathophysiological heterogeneity [155, 156], increased Glx in patients unresponsive to standard dopamine antagonist treatments is a relatively robust finding [60, 70, 165]. fMRS indicates that cognitive challenges increase the glutamate: GABA concentration ratio, with preliminary evidence suggesting that this increase is diminished in schizophrenia [120, 221]. Again, PET studies suggest reductions in both GABA and NMDA receptors [18, 183]. Finally, diminished availability of metabotropic glutamate receptor type 5 has been associated with cognitive symptoms in schizophrenia [191].

11.3.2 Dopamine

Dopamine dysfunction is pivotal in schizophrenia's pathology, evident from preclinical models showing associations between increased dopamine signaling and schizophrenia-like behaviors, to post-mortem studies revealing dopamine receptor upregulation in patients, to neuroimaging studies indicating presynaptic hyperdopaminergia in the dorsal striatum of patients [149–151, 208].

A cortical hypodopaminergic state is reported in schizophrenia, potentially linked to striatal hyperdopaminergia [151]. Cortical dopamine signaling affects cognitive functions, including attention, working memory, and executive functions, so altered dopamine function in schizophrenia may affect these processes [50, 78, 132, 169]. Moreover, increased spontaneous dopamine transients in the striatum may promote abnormal learning for neutral stimuli, that would be erroneously identified as salient, contributing to the reinforcement learning paradox [150].

Experimental studies in Parkinson's disease and non-human primates with cortical dopamine alterations indicate that dopamine signaling deficits contribute to working memory and executive functioning deficits [42, 175]. On the other hand, pro-dopaminergic compounds used in ADHD treatment exert cognitive benefits through the augmentation of cortical dopamine signaling [57]. Finally, the partial dopaminergic agonist cariprazine has shown some advantages in cognitive symptoms [77].

However, pro-dopaminergic treatments have not been approved for cognitive deficits in schizophrenia, possibly due to the complexity of dopamine effects. The activity of dopamine in the cortex differs from that in the striatum, with only a subset of cortical pyramidal neurons expressing dopamine receptors, and cortical interneurons exhibiting a relatively higher expression. In both pyramidal cells and interneurons, dopamine's presynaptic effects mainly decrease neurotransmitter release. Postsynaptically, dopamine increases the excitability of deep-layer pyramidal neurons and enhances the frequency of interneuron spiking. The overall impact of these mechanisms is that cortical dopamine reduces pyramidal cell firing through its influence on interneurons. Therefore, dopamine plays a crucial role in the E/I balance and diminished cortical dopamine signaling, as that observed in schizophrenia, is linked to a decreased net inhibitory input, known as cortical disinhibition [151, 227].

11.3.3 Acetylcholine

Acetylcholine's crucial role in attention and memory is evident from the cognitive impairments caused or worsened by anticholinergic drugs and the benefits of promoting cholinergic signaling in Alzheimer's treatment [101, 148]. Central cholinergic innervation is divided into two main networks, which include the forebrain complex, projecting to the cortex, and the pedunculopontine cholinergic complex, projecting to the midbrain. Acetylcholine effects are exerted through G protein-coupled muscarinic receptors and ionotropic nicotinic receptors. The nicotinic system's link to schizophrenia is highlighted by the high prevalence of smoking among patients [79]. Nicotine administration acutely ameliorates sensory gating alterations and improves cognitive performance in schizophrenia [3, 65]. However, chronic use does not maintain this enhancement and is rather associated with poorer cognitive performance and increased risk of late-life cognitive decline [105, 235]. The immediate benefits of smoking likely result from initial agonism at cholinergic receptors, while the adverse effects in the long term may result from receptor desensitization [151].

The muscarinic system is implicated in schizophrenia as well. Muscarinic antagonists' acute administration can exacerbate both positive symptoms and cognitive deficits in individuals with schizophrenia. G protein-coupled muscarinic receptors exhibit primarily excitatory effects through the M1, M3, and M5 subtypes, while the M2 and M4 subtypes have predominantly inhibitory effects. Post-mortem studies show reductions in M1 and M4 receptors in the hippocampus and cortex of subjects with schizophrenia [40, 53]. Moreover, PET studies have shown decreased muscarinic receptor density in schizophrenia, which has been linked to the presence of cognitive symptoms [16, 187, 202]. M1 receptors are crucial in learning and memory [247], and M1 agonism may improve both positive symptoms and cognitive performance in schizophrenia [31, 205, 248].

Both nicotinic and muscarinic receptor systems influence GABAergic and glutamatergic signaling, contributing to the E/I balance [247] through complex effects on pyramidal neurons and interneurons. While the net effect of nicotinic stimulation is to increase pyramidal cell firing, the net effect of M1 receptor activation may be both excitatory or inhibitory on cortical pyramidal neurons. The overall effect of modulating M1 receptor activity likely depends on the level of endogenous tone: at typical levels, effects may be primarily inhibitory. Muscarinic agonism may therefore offer cognitive benefits in a state of cortical disinhibition, as occurs in schizophrenia [31, 151, 205, 248].

Finally, cholinergic signaling also affects the dopamine system, with nicotinic receptor stimulation inhibiting dopamine release at low concentrations and increasing dopamine neuron firing at higher concentrations [141]. M4 receptor agonism may reduce striatal dopamine release, while M1 agonism appears to increase cortical dopamine release in conditions of reduced cortical signaling, such as schizophrenia [151].

11.4 Neuroinflammation

In recent years, research has shed light on the intricate interplay between neuroinflammation, metabolic dysfunction, and aberrant neurotransmission in the pathophysiology of schizophrenia and particularly on cognitive symptoms [22, 25, 129, 178, 197, 199].

Neuroinflammation in schizophrenia is characterized by the activation of microglia and astrocytes and the release of pro-inflammatory cytokines in the central nervous system (CNS) [111] and the immune cross-talk between the CNS and the periphery is noteworthy [190]. Indeed, a systemic sub-inflammatory state has also been widely reported in schizophrenia [178], particularly within the context metabolic comorbidities, such as the metabolic syndrome (MetS), which affects around 40% of patients with schizophrenia [152]. Then, a negative synergistic impact of the two conditions on the cognitive outcome was also hypothesized [246].

Mounting evidence suggests that metabolic dysfunction contributes to the pathogenesis of schizophrenia and exacerbates cognitive impairment through various mechanisms [27, 28, 140]. Adipose tissue-derived cytokines, such as leptin, adiponectin, and resistin may modulate neuroinflammatory processes and synaptic transmission, further exacerbating cognitive deficits in schizophrenia [45, 154]. Moreover, insulin resistance and dysregulated glucose metabolism can impair neuronal energy production and compromise synaptic plasticity, leading to cognitive dysfunction [133, 241].

Both systemic and neuroinflammation are able to induce the kynurenine pathway (KP), responsible for the metabolism of tryptophan [47, 49]. Indeed, a recent meta-analysis on the biomarkers of the cerebrospinal fluid (CSF) in schizophrenia found increased levels of either kynurenines and proinflammatory cytokines [195]. The KP plays a pivotal role in the modulation of immune response, synaptic transmission, and neuronal homeostasis. In schizophrenia, an aberrant activation of the KP has been linked to the pathophysiology of the illness and cognitive impairment [196–200]. Specifically, the imbalance between the two neuroactive and neuroactive/neurotoxic metabolites of the pathway (quinolinic acid, QUIN; and kynurenic acid, KYNA) is of interest, as both modulate glutamatergic and cholinergic neurotransmission [198].

Overall, the interplay between metS, systemic sub-inflammation and neuroinflammation, and KP dysregulation can exert profound effects on neurotransmission, contributing to cognitive dysfunction in schizophrenia.

11.4.1 Neuroinflammation and Cognition in Schizophrenia

Neuroinflammation is characterized by the activation of microglia (activation of the pro-inflammatory M1 phenotype) and astrocytes and the release of pro-inflammatory cytokines in the central nervous system (CNS) [111]. Notably, proinflammatory cytokines have been shown to activate microglia and astrocytes, increasing the secretion of proinflammatory cytokines, creating a vicious circle [134, 207]. Postmortem studies [59] and PET studies targeting the translocator protein (TPSO) [144, 231] in schizophrenia have consistently reported increased microglial activation, whereas clinical studies found elevated levels of cytokines, particularly interleukin 1-β (IL-1-β), interleukin-6 (IL-6), and tumor necrosis factor-alpha (TNF-α) [178]. Perinatal infections or complications, maternal immune activation, or childhood traumas serve as established risk factors able to trigger/sensitize microglia and the immune system to future "hits", typically represented by environmental risk factors [111, 114]. Cytokine release can disrupt neurodevelopmental processes critical for synaptic homeostasis, myelination, and connectivity, leading to long-lasting alterations in brain structure and function [46, 158, 197]. These alterations may predispose individuals to the development of schizophrenia later in life and contribute to cognitive decline along the course of the illness. Inflammatory states during neurodevelopment may lead to abnormal synaptic plasticity, increased synaptic pruning, altered synaptic homeostasis, volume loss, and cognitive deficits [100,

170, 226, 236]. Furthermore, such activation of glial cells under inflammatory conditions is related to the secretion of reactive oxygen species, ultimately leading to neuronal cell death [26, 163]. Indeed, positive associations have been observed between cortical thickness in specific brain regions and cognitive performance in schizophrenia patients [8], although this alone cannot fully explain cognitive deficits. Longitudinal studies have shown associations between systemic inflammation during adulthood and cognitive decline in healthy adults [239]. Elevated peripheral inflammatory biomarkers have been linked to decreased cortical thickness and brain volumes in regions responsible for cognitive functions (e.g., attention, visual learning, verbal fluency) [75, 170]. Furthermore, significant negative correlations have been found between inflammatory biomarkers and cognitive performances in schizophrenia patients [110, 193], suggesting a critical role of inflammatory markers in cognitive deficits observed in schizophrenia. The associations between inflammatory biomarkers and cognitive domains have been investigated by multiple studies. A recent meta-analysis by Patlolla and colleagues focused on the studies reporting associations between CRP, the three most involved cytokines in the pathogenesis of schizophrenia (IL-6, IL-1β, and TNF-α), and cognitive domains found that all these inflammatory biomarkers were significantly associated with altered cognitive performances in multiple cognitive domains [178]. The five cognitive domains assessed by the meta-analysis were attention and processing speed (APS), executive function (EF), verbal learning and memory (VLM), visual learning and memory (VisLM), and working memory (WM). Summarizing, the authors reported significant negative associations between IL-6 and APS, EF, VLM, and WM. TNF-α was negatively associated with APS, EF, and VisLM, whereas CRP showed strong inverse associations with all the domains except for EF. Finally, IL-1β showed strong negative correlations with APS, VisLM, and WM [178]. In summary, all these findings suggested a decline in cognitive functioning with an increase in inflammatory biomarker levels. However, the extent of this relationship cannot be definitively determined, given the small effect sizes observed in some domains. In addition, there are several confounding factors such as age, duration of illness, BMI, and other co-morbid conditions characterized by an inflammatory pattern, and interestingly, MetS is an important factor to consider. A possible interpretation of these findings is that a sub-inflammatory state, overactive towards multiple stressors/risk factors/stimuli, characterized schizophrenia since childhood [111]. Cytokines and CRP-induced microglial activation cause increased pruning of dendrites and axons and neuronal death during the early stages of the illness prior to the onset of the illness [109, 111, 114, 126], contributing to the decreased cognitive performance that predisposes the onset. Then, along the course of the illness, relapses and the perpetuation of the inflammatory state lead to the progressive volume loss, particularly in frontal and temporal areas, maybe mainly due to the progressive loss of synapses as recently highlighted by PET studies targeting the SVA2 protein, a marker of synaptic density [112, 174]. The role of the dysregulated dopaminergic system in the prefrontal cortex, probably linked to altered synaptic homeostasis and inflammation should also be acknowledged in the pathogenesis of cognitive symptoms [113].

11.4.2 The Kynurenine Pathway in Schizophrenia: Implications for Cognition

Under physiological conditions, tryptophan (Trp) undergoes catabolism via the kynurenine pathway (KP) to produce nicotinic acid, a precursor of nicotinamide adenine dinucleotide (NAD+), a crucial coenzyme involved in regulating various metabolic pathways [49, 203]. During inflammatory states, Trp catabolism through the KP leads to the formation of several bioactive metabolites, some of which have neurotoxic, neuroprotective, and neuroactive properties and can impact cholinergic and glutamatergic neurotransmission [203]. The reason why the KP pathway might serve as a nexus between inflammation and neurotransmission lies in the fact that inflammatory cytokines can upregulate Trp catabolism via the KP, leading to an imbalance between neurotoxic and neuroprotective metabolites [49]. Specifically, pro-inflammatory cytokines like IFN-γ, IL-6, along with markers of oxidative and nitrosative stress such as ROS and RNS, as well as lipopolysaccharide (LPS), induce the enzyme Indoleamine 2,3-dioxygenase (IDO), thereby increasing Trp catabolism into KYN. Notably, elevated levels of these various IDO inducers have been consistently reported in schizophrenia [9, 76, 230]. Regarding the pharmacodynamic properties of KP metabolites, kynurenic acid (KYNA) and quinolinic acid (QUIN) modulate NMDA receptor activity by acting respectively as an antagonist at the glycine allosteric site and a potent agonist [81]. Moreover, KYNA inhibits nicotinic cholinergic receptors containing α7 subunits (α7nAChR) [15] and notably, the cholinergic system plays a crucial role in cognitive modulation. Multiple risk factors, such as perinatal complications or childhood trauma, have been associated with an overactive immune system in schizophrenia, with microglia playing a significant role through the activation of the pro-inflammatory M1 phenotype [111]. Therefore, the heightened activation of the KP pathway could partly be attributed to (1) the production of cytokines capable of activating the pathway, such as TNF-α and IFN-γ, associated with the inflammatory state; and (2) the increased demand for NAD+ to sustain the immune response. This, in turn, could partially elucidate the pathophysiology and the biological underpinnings of cognitive symptoms, establishing a connection between inflammation and symptomatology [47, 201]. Preclinical studies conducted in rodents have demonstrated that metabolites of the kynurenine (KYN) pathway affect cognition by inducing neurochemical and morphological changes similar to those often observed in schizophrenia [71, 180, 184]. Specifically, exposure to elevated levels of KYN and kynurenic acid (KYNA) during embryonic to postnatal development stages leads to alterations in dendritic spine density, glutamate levels, and expression of α7 nicotinic acetylcholine receptors (α7nAChR) in the prefrontal cortex (PFC) of adult rats, mirroring aspects of the pathophysiology of schizophrenia [180]. Moreover, elevated KYNA levels in rats have been associated with deficits in behavioral tasks involving prefrontal functions, as well as impairments in contextual memory and passive avoidance [71, 180, 184]. Concerning studies conducted on humans, clinical studies to examine the possible link between the KP and cognitive functioning of individuals with schizophrenia are limited. Kindler and colleagues found that attention and Rostral Middle Frontal Gyrus volume were negatively correlated with the KYN/TRP ratio, and index of KP activation, in patients with schizophrenia displaying high levels of cytokines [129]. Huang X and colleagues added a piece to the puzzle; indeed they found an increased activation of the KP (KYN/Trp ratio) and increased KYNA levels that were associated with deficits in attention-vigilance and social cognition tasks, suggesting a detrimental effect of KYNA on cognitive functions [115]. Moreover, a detrimental role of QUIN on cognition was found by Cathomas and colleagues, who reported an inverse correlation between QUIN levels and composite cognitive score in patients with schizophrenia [43]. Huang J et al. found an increase in 5-HIAA/KYNA and QUIN/KYNA ratios in patients with schizophrenia and, in turn, positive associations between 5-HIAA/KYNA ratio, working memory, and left lateral orbitofrontal cortex thickness in the schizophrenia subgroup. Moreover, left lateral orbitofrontal cortex thickness was positively associated with several cognitive functions [116]. Overall, several preclinical and clinical findings showed that the activation of the KP is associated with greater cognitive deficits in individuals affected by schizophrenia, with both elevations and reductions of KP metabolites being correlated with cognition, depending on the specific metabolite [196, 197]. However, the number of studies addressing this topic is very limited and shows high heterogeneity due to high methodological variability encompassing the cognitive evaluations, the KP metabolites analyzed, and the techniques used to quantify them; therefore, much evidence is needed to draw firm conclusions.

11.4.3 Metabolic Syndrome

MetS is a complex condition characterized by several metabolic abnormalities that tend to co-occur, increasing the risk of multiple comorbidities. The criteria used to define MetS encompass high blood pressure, abdominal obesity (increased waist circumference), elevated blood glucose and triglyceride levels, and low levels of high-density lipoprotein (HDL) cholesterol. The diagnosis typically is based on the presence of three or more risk factors among those abovementioned [5]. Evidence from the Clinical Antipsychotic Trials of Intervention Effectiveness (CATIE) indicates that individuals with schizophrenia are more likely to develop metabolic syndrome compared to the general population [152]. This susceptibility is attributed to factors like obesity, antipsychotic medication use, and lifestyle choices, leading to higher rates of diabetes, hypertension, and other cardiovascular risk factors. Indeed, MetS is a frequent comorbidity of schizophrenia affecting over 40% of people with schizophrenia, and it is associated with increased peripheral levels of cytokines and chemokines due to insulin resistance, which causes the secretion of resistin by the adipose tissue, and dyslipidemia, creating a vicious circle [25, 54, 90, 152, 217]. The systemic sub-inflammation through a cross-talk with the brain worsens or induces the pattern of neuroinflammation, oxidative stress, cytotoxicity, enhanced synaptic pruning due to microglia activation, and white matter (WM) disruption [112, 114, 170, 178, 197, 200, 214, 230]. All these mechanisms could affect the cognitive outcome; indeed, patients with metabolic syndrome show lower cognitive performances across multiple domains compared to those without metabolic syndrome [140]. Lindenmayer and colleagues found that patients with MetS performed significantly worse on tests measuring processing speed, attention/vigilance, working memory, and problem solving/reasoning compared to those without MetS. Specifically, wider waist circumference was associated with lower scores on attention/vigilance, HDL and triglycerides were respectively positively and negatively related to attention/vigilance scores [140]. Indeed, according to a meta-analysis by Bora and colleagues, patients with schizophrenia and MetS or DM have significantly more severe cognitive deficits compared to the other patients with schizophrenia. Interestingly, a relationship between each of the components of MetS and cognitive impairment in schizophrenia has been reported among several studies [27, 140]. It is important to note that the effect sizes were small, and most of the cognitive deficits in schizophrenia are mainly related to neurodevelopmental abnormalities [19]. However, these findings suggest that MetS and every single component of metS can induce further small cognitive decline and [27]. Concerning the brain functional substrate underlying the relationship between MetS and cognitive impairment in schizophrenia, Boyer and colleagues found that patients with MetS performed significantly worse than patients without MetS in executive functions, attention and memory, exhibiting significant hypoperfusion at the cortical level in the left orbital prefrontal cortex [28]. It should not be neglected that the opposite could also be possible, and cognitive deficits could contribute to higher prevalence of MetS rather than being a consequence of them. Indeed, cognitive deficits could underly poor decision making, lower socioeconomic status and unhealthy lifestyle. Concerning this, de Nijs and colleagues found that patients with MetS showed impaired premorbid cognitive functioning in adolescence and lower educational achievement, irrespective of the parental socioeconomic status. These findings corroborate the hypothesis that poor premorbid cognitive abilities represent a risk factor for metabolic complications later in life [58]. The role of antipsychotic treatments is an important confounding factor as Mitchell and colleagues found that the prevalence of MetS drastically increases with treatments comparing unmedicated patients, first episode psychosis and patients established on antipsychotic medication, highlighting a significantly lower cardiovascular risk in early schizophrenia than in chronic schizophrenia [161]. Also, the role of gender is very important to consider in relation to the comorbidity between MetS and schizophrenia [25]. MetS is more common among females than males according to the CATIE study, with an overall prevalence of 51.6% for females compared to 36.0% for males [152]. This represents an important confounder, given that male gender is associated with more severe cognitive deficits in schizophrenia [20, 138]. Interestingly, one of the few studies that did not find a significant effect of MetS on cognition in schizophrenia [157] had significantly more females in the schizophrenia/MetS group. Moreover, despite only considering the presence/absence of diabetes, Zhang and colleagues found that male patients with schizophrenia performed worse in terms of cognitive performance than their female counterpart, irrespective of whether they had diabetes. However, the presence of diabetes reduced gender differences in terms of cognitive outcome, especially in attention, despite they still persisted [250]. To disentangle all the variables able to modulate the impact of MetS on cognition in schizophrenia, longitudinal studies are required in order to investigate the effects of the development of MetS and its components starting from early years. Longitudinal studies could also help to distinguish the cause-and-effect relationship between MetS and cognitive impairment and identify the cognitive deficits that predate the onset of MetS from those that are a consequence of cardiovascular risk factors.

11.5 Structural and Functional Brain Alterations

Structural abnormalities in the brain have been extensively reported in the context of schizophrenia, with comprehensive meta-analyses revealing a reduced volume of the hippocampus in affected individuals relative to control subjects, accompanied by reductions in the volumes of the amygdala, thalamus, nucleus accumbens, and overall intracranial volume, alongside an increase in the volumes of the pallidum and lateral ventricles. Furthermore, individuals diagnosed with schizophrenia exhibit extensive cortical thinning and a diminished cortical surface area, with the most significant alterations observed in the frontal and temporal lobe regions [127]. Variations in cortical thickness tend to be more localized, while the effects on cortical surface area are observed to be more widespread. Notably, reductions in cortical thickness are markedly more evident in patients undergoing antipsychotic treatment, with a negative correlation identified between these reductions and both the dosage of medication and the severity and duration of symptoms. Recent progress in neuroimaging techniques has contributed to the formulation of a dysconnectivity hypothesis regarding schizophrenia, suggesting that the disorder may be characterized by aberrant or inefficient communication between functional brain regions, which is further influenced by abnormalities in the underlying white matter connections. Diffusion tensor imaging allows in vivo examination of the microstructural characteristics of white matter. Research utilizing diffusion tensor imaging in the context of schizophrenia predominantly indicates reduced fractional anisotropy (FA) in affected individuals in comparison to control subjects, particularly within the fiber pathways that interconnect the prefrontal and temporal lobes. Recent meta-analytic studies evidenced that reductions in FA are widespread in schizophrenia, affecting nearly all white matter regions, with the most pronounced alterations noted in the cortical–thalamic and interhemispheric pathways. Altered connectivity in schizophrenia is also suggested by in vivo functional MR imaging (fMRI), Single-photon emission computed tomography (SPECT), Positron emission tomography (PET), and Electroencephalography (EEG) studies, evidencing task-activated hypofrontality and abnormal resting state brain activity.

Given these premises, numerous studies over the years have explored the neuroimaging correlates of cognitive deficits in schizophrenia, collectively contributing to the development of a substantial body of literature.

11.5.1 Structural Brain-Imaging Findings

As previously mentioned, studies employing structural imaging in individuals with schizophrenia have consistently demonstrated a broad reduction in cortical gray matter volume when compared to healthy controls, with prominent deficits in frontal and medial temporal areas [206, 233]. Additionally, subcortical structures such as the hippocampus and amygdala have also shown volume reductions [232]. Investigations focusing on cortical thickness and surface area, which collectively define brain volume, have revealed significant cortical thinning and diminished surface area throughout the cortical mantle in schizophrenia [233]. Given the widespread neural and cognitive changes observed in schizophrenia, extensive research has explored the relationship between structural brain abnormalities (e.g., reduced volume, cortical thinning, decreased surface area) and cognitive impairments in this condition, leading to multiple reviews and meta-analyses [128].

In this context, deficits in processing speed have been linked to diminished temporal lobe volume and increased ventricular size [12]. Attentional deficits have been associated with structural changes, such as decreased volume and gray matter density, along with cortical thinning in the frontal and temporal lobes [124], as well as with widespread reductions in the volume of subcortical regions [12]. Reduced working memory has been correlated with decreased volumes and cortical thickness in the frontal and temporal lobes [11, 52], while deficits in verbal learning and memory have been associated with reductions in hippocampal volume [12, 124]. Impaired visual learning and memory have been related to hippocampal volume loss as well as reduced volumes in the frontal and primary visual occipital areas [12, 124]. Furthermore, deficits in reasoning and problem-solving, which encompass higher-order executive functions, have been associated with a reduction in frontal lobe volume, particularly in the dorsolateral prefrontal cortex [12, 94].

Some studies have focused on the longitudinal correlation over time between gray matter and cognitive functions, reporting that progressive alterations in brain structure throughout the course of the illness are also associated with cognitive performance. For instance, Andreasen et al. [10] examined structural brain measures over periods extending up to 15 years following the onset of the illness and found that the progressive deterioration of brain structure in patients correlated with cognitive deficits, with the most significant impacts observed in attention, as well as verbal learning, problem-solving, and working memory. Notably, losses in frontal, temporal, and parietal white matter were linked to impaired attention performance, along with reductions in total frontal and parietal volumes. Subcortical structures, particularly the hippocampus, have also been identified as relevant to working memory performance in individuals with schizophrenia. A decrease in hippocampal gray matter volume has been associated with poorer working memory task performance in patients during both early [92, 93] and late stages of the illness [244], although findings across studies have shown some inconsistencies [99]. Similarly, reductions in caudate volume have been connected to impaired working memory performance in schizophrenia [135]. Of note, patients with severe cognitive deficit have been associated with significantly greater enlargement of the lateral ventricles and reductions in total hippocampal volume [242].

White matter has also been extensively investigated in patients with schizophrenia, with findings of reduced fractional anisotropy, suggesting a decline in white matter structural integrity [108]. A comprehensive Diffusion tensor imaging (DTI) study conducted by the ENIGMA group revealed extensive white matter microstructural abnormalities in chronic schizophrenia [123]. These widespread DTI abnormalities were subsequently identified also in patients with early-stage schizophrenia. Focusing on the association between white matter alterations and cognitive impairment, fronto-temporal white matter has been linked to attention deficits in schizophrenia. Research by Singh et al. [209], utilizing DTI demonstrated significantly lower fractional anisotropy (FA) values in the uncinate fasciculus compared to control subjects. In patients with schizophrenia, a positive correlation was found between attention, spatial memory, sensorimotor dexterity, and emotion with the FA of the uncinate fasciculus. The uncinate fasciculus is particularly noteworthy as it links the amygdala, a region associated with mood, to areas responsible for executive functions, namely the medial and orbitofrontal cortices. Fronto-parietal white matter has been also associated with working memory in recent-onset schizophrenia, with the FA of the superior longitudinal fasciculus correlating with verbal working memory performance in both patients and control groups.

Kochunov et al. [130] conducted a mega-analysis using ENIGMA database in order to explore the relationship between processing speed, working memory, and white matter integrity in schizophrenia, reporting that white matter impairment in schizophrenia is regional tract-specific, particularly in tracts normally supporting processing speed performance, such as the body of the corpus callosum and anterior corona radiata.

11.5.2 Functional Brain-Imaging Findings

Task-based fMRI has proven invaluable for conducting real-time evaluations of brain function while patients engage in cognitive tasks within the scanner. Early studies involving small patient cohorts showed consistent patterns of either increased or reduced neural activation; however, more recent findings indicate considerable heterogeneity among patients concerning which circuits or networks are activated during tasks, similar to the variability observed in individuals without psychiatric disorders [124]. fMRI investigations focused on regional activity during specific cognitive tasks indicated the presence of hypofrontality and deficits in the modulation of fronto-parietal effective connectivity related to working memory in the early stages of schizophrenia, even when working memory performance appears intact [168]. Furthermore, in a cohort of first-episode patients, Tan et al. [220] found that the manipulation of information in working memory is more significantly impaired than the maintenance of information, and this impairment correlates with dysfunctional and compensatory prefrontal responses observed in the dorsolateral and ventrolateral prefrontal cortices. Several initial fMRI studies revealed aberrant activation in the DLPFC during working memory tasks in individuals with schizophrenia compared to healthy controls. Functional MRI studies have great results variability with both increases and decreases in DLPFC activation during working memory tasks, prompting meta-analyses to consolidate results and identify potential moderating factors [234]. Meta-analyses of fMRI studies have primarily focused on specific domains of neurocognition, such as working memory, episodic memory, and executive functioning. An early meta-analysis highlighted abnormal activation in the DLPFC, anterior cingulate cortex, and insula in schizophrenia patients compared to healthy controls [87]. More recent meta-analyses have confirmed dysfunction in these areas, along with the posterior parietal cortex and supplementary motor area, indicating that these regions are key nodes within the cognitive control and salience networks [245]. Additionally, meta-analyses investigating episodic memory in schizophrenia have found aberrant activation in regions such as the left inferior prefrontal cortex, hippocampus, and left cerebellum relative to healthy controls [2]. A meta-analysis encompassing 41 functional neuroimaging studies on executive functioning (often referred to as cognitive control) in schizophrenia revealed decreased activation in the DLPFC, anterior cingulate cortex, and thalamus [160]. These findings have been largely corroborated in reviews examining neural correlates across various neurocognitive domains throughout different stages of schizophrenia, suggesting that many neural abnormalities evident in chronic schizophrenia may be present to some extent prior to the onset of illness [124]. In this context, a meta-analysis of fMRI studies involving neurocognitive tasks in individuals at clinical high risk (CHR) showed reduced activation in the inferior parietal lobule and medial frontal gyrus compared to healthy controls, with reduced activation in the inferior parietal lobule noted in a subset of studies focusing on working memory tasks [67]. The brain regions implicated in these cognitive functions are widely distributed and often overlap, suggesting that these deficits may not be isolated. The DLPFC has been proposed as a potential common substrate for numerous cognitive impairments. As previously mentioned, neuroimaging research in schizophrenia indicates that cognitive performance relies on distributed brain systems or networks rather than isolated regions. A systematic review examining the relationships between resting state functional connectivity and neurocognition across different domains found that abnormal connectivity between cortical and subcortical regions (part of the cortico-cerebellar-striatal-thalamic loop) was associated with deficits in executive functioning, working memory, and processing speed. Furthermore, disrupted connectivity between the default mode network (DMN) and the frontoparietal (e.g., DLPFC) and cingulo-opercular (e.g., anterior cingulate cortex) networks was related to impairments across multiple cognitive domains [204]. Notably, unique associations between specific cognitive domains and distinct abnormalities in functional connectivity were not observed, reinforcing the notion of disrupted shared mechanisms across neurocognitive domains that contribute to the generalized cognitive deficits observed in individuals with schizophrenia. A recent meta-analysis also reviewed studies investigating the relationship between structural brain metrics and cognitive domains in schizophrenia, mapping these structural findings onto resting-state functional brain networks [128]. The frontoparietal (cognitive control) network was associated with the greatest number of cognitive domains, while the somatomotor, dorsal attention, and ventral attention networks were implicated in several cognitive domains. Generally, more complex cognitive processes, such as reasoning and executive function, as well as social cognition, were associated with a greater number of networks [238].

Finally, a recent meta-analysis conducted by Picó-Pérez et al. [182] examined structural and functional studies involving 4789 patients across various domains, including processing speed, attention/vigilance, working memory, verbal learning and memory, visual learning and memory, executive function/reasoning, problem-solving, verbal fluency, verbal comprehension, emotional perception, social perception and knowledge, theory of mind, and attributional bias. The authors observed reductions in activation in the dorsomedial prefrontal cortex, supplementary motor area, and right inferior frontal gyrus when patients engaged in cognitive tasks (data from 3446 patients). Notably, the authors identified attention/vigilance tasks as those that consistently demonstrated deficits in individuals with schizophrenia across multiple studies.

PET studies also suggested that hypofrontality in schizophrenia may account for the observed impairments in cognition and memory performance among patients (e.g., [162]). These investigations have also underscored the significance of glucose metabolism and reduced cerebral blood flow in fronto-temporal regions as they relate to verbal learning performance (e.g., [102, 189]). Additionally, PET studies have been crucial in enhancing our understanding of the dopaminergic system and its association with cognitive impairments in schizophrenia. For example, dopamine deficits within the DLPFC have been associated with abnormal activation in this area during tasks that demand working memory and cognitive control [210]. Moreover, several PET studies have explored the involvement of the serotonin system and neuroinflammation in cognitive deficits among first-episode individuals and high-risk populations. However, more research is required to draw definitive conclusions (e.g., [82, 95]).

Alterations in EEG measures, mainly using event-related potentials (ERPs) and frequency bands, have been consistently linked to cognitive dysfunction in schizophrenia [179]. ERPs ensure a temporal resolution of milliseconds, able to capture the electrophysiology underlying the cognitive processing of sounds, images, and decision-making. The most replicated findings in relation to cognitive deficits regard Mismatch negativity (MMN) and P300 [179]. MMN is an index of pre-attentive processing and sensory memory and is represented by a negative component 150–250 ms after the presentation of a deviant tone (unexpected stimulus) in a sequence of standard tones. Reduced MMN amplitude is associated with impairments in several cognitive domains [84, 164]. P300 occurs 300 ms after the presentation of a deviant stimulus (visual, visuospatial, or auditory) and it is considered an index of attention shifts in relation to environmental changes and their cognitive processing [103, 185]. Altered P300 amplitude and latency have been extensively reported in different stages of schizophrenia [29, 30, 83, 97, 188]. Reduced P300 amplitude has been linked to impaired visual attention [33, 48, 216], working memory [62, 240]), language processing [240], and executive functioning [80].

Resting-state analyses in schizophrenia revealed increased delta/theta and reduced alpha activity [83, 85, 118, 166], indicative of a pattern of "hypofrontality." Moreover, disrupted synchronization across frequency bands, particularly in beta and gamma activities, during both resting state and cognitive tasks, is also reported [215, 229].

Concerning delta activity at rest and its relation to cognitive abilities, weak evidence has emerged. In task-related studies, reduced delta activity was consistently found to be associated with impairments in working memory [186]. Moreover, evoked delta activity was positively associated to attention, speed of processing, visual learning [145], and social cognition [146].

The theta band, especially in the slow activity range, showed the strongest association with a wide range of cognitive deficits, in particular verbal memory and processing speed. Indeed, several studies found significant negative associations between increased theta activity at rest and working memory [131], visuospatial memory [243], verbal learning [131, 243], emotion recognition [86], and executive functioning [243]. Theta activity evoked during working memory or visual tasks was also related to working memory [121, 186], attention [61], visual learning [145], speed of processing [61, 145, 186], and social perception [146].

Considering the alpha band, most studies reported a decrease in the absolute power during resting-state recordings [166], but studies on the association between resting-state or evoked alpha activity and cognitive functions were inconsistent [86, 121, 146, 186].

Focusing on the two bands characterized by the highest frequency, namely the beta and gamma bands, dysfunctions in their activity/synchronization were widely reported in schizophrenia [34, 176, 192]. Considering beta activity, some evidence pointed to a positive correlation between resting-state beta power and working memory, visual and verbal memory, vigilance, and emotion recognition [86]. Similar findings emerged on task-related activity, highlighting positive associations between evoked beta power and working memory [32, 121]. However, a lack of associations between beta activity and different cognitive functions, such as working memory, verbal learning [131], and metacognitive functions [237] was also reported.

Considering gamma activity, representing a measure of E/I balance, abnormalities within this band were linked to impairments in several cognitive domains. Specifically, dysfunctional gamma-band responses (GBR), mainly characterized by reduced amplitudes or reduced phase locking at the prefrontal cortex (PFC) level, were related to impairments in working memory (WM) and several other cognitive functions. Diminished "gamma synchrony" was also described in first-episode patients [218], reinforcing the evidence on the presence of cognitive deficits starting from early years, and probably suggesting that early synchrony disruptions within neural networks can contribute to the onset of the illness.

11.6 Conclusions

Cognitive deficits in schizophrenia are highly heteterogeneous both between and within individuals, representing a perfect storm of complexity, and unraveling the neurobiological bases is key to reversing them.

Cognitive dysfunction clearly has a genetic component, and the evidence from genetic association studies, converging in neurodevelopment, neuroplasticity, and neurotransmission, allowed a better understanding of the implicated pathways, supporting previous hypotheses and highlighting potential novel targets. However, the weight of each single gene on cognition is very limited and the overall impact of multiple risk genes is difficult to estimate as it is likely to vary at the individual level, due to complex interplays.

The link between cognitive deficits and alterations in neurotransmission is widely acknowledged and in the past decades, the focus on E/I balance disruption highlighted novel treatment targets within the glutamatergic and GABAergic pathways, leading to the development of new drugs.

The role of neuroinflammation in schizophrenia and its impact on cognition has also been established, although the efficacy of the currently studied anti-inflammatory agents on cognition is still yet to be supported by results. The kynurenine pathway appears a promising target being at the cross-road between neuroinflammation, peripheral inflammation, and neurotransmission.

Finally, neuromaging findings point to diffuse alterations encompassing different brain regions and structures, and especially affecting functional connectivity. So far, imaging markers do not reach significant prognostic value in the single patient; however, their integration for personalized therapy holds promise.

While crucial advancements have been made, there are still many open questions on the nature of cognitive deficits in schizophrenia. Starting from the shared genetic liability between psychotic disorders and cognitive disturbances, is it schizophrenia that causes cognitive impairment or cognitive impairment that leads to schizophrenia? Are the biological features underlying cognitive deficits in schizophrenia distinguished from those of other neuropsychiatric disorders, or, more likely, are they shared across diagnoses? Which alteration is upstream and which is downstream?

Future research should prioritize a deeper understanding of the neurobiological changes associated with available interventions aiming at improving cognition, as well as transdiagnostic approaches.

On the clinical side, the integration, together with neuropsychological assessments of the already available biomarkers in the diagnostic and therapeutic process, will allow a more fine-grained stratification of patients and thus personalized treatment with potential high benefits in terms of cognitive outcome.

References

1. Aas M, Dazzan P, Mondelli V, Melle I, Murray RM, Pariante CM. A systematic review of cognitive function in first- episode psychosis, including a discussion on childhood trauma, stress, and inflammation. Front Psych. 2014;4:182.
2. Achim AM, Lepage M. Episodic memory-related activation in schizophrenia: meta analysis. Br J Psychiatry. 2005;187(6):500–9.
3. Adler E, Hoffer LD, Wiser A. Normalization of auditory physiology by cigarette smoking in schizophrenic patients. Am J Psychiatry. 1993;150:1856–61.
4. Agnew-Blais J, Seidman LJ. Neurocognition in youth and young adults under age 30 at familial risk for schizophrenia: a quantitative and qualitative review. Cogn Neuropsychiatry. 2013;18(1–2):44–82. https://doi.org/10.1080/13546805.2012.676309. Epub 2012 Sep 21. PMID: 22998599; PMCID: PMC3577989.
5. Alberti KG, Zimmet P, Shaw J. Metabolic syndrome—a new world-wide definition. A Consensus Statement from the International Diabetes Federation. Diabet Med. 2006;23(5):469–80. https://doi.org/10.1111/j.1464-5491.2006.01858.x.
6. Alfimova MV, Golimbet VE, Mitiushina NG. Polymorphism of the serotonin receptor (5-HTR2A) gene and verbal fluency in normalcy and schizophrenia. Mol Biol. 2003;37(1):68–73.
7. Alfimova MV, Lezheĭko TV, Golimbet VE, Korovaĭtseva GI, Lavrushkina OM, Kolesina NI, Frolova LP, Muratova AA, Abramova LI, Kaleda VG. Investigation of association of the brain-derived neurotrophic factor (BDNF) and a serotonin receptor 2A (5-HTR2A) genes with voluntary and involuntary attention in schizophrenia. Zh Nevrol Psikhiatr Im S S Korsakova. 2008;108(4):62–9. Russian.
8. Alkan E, Davies G, Evans SL. Cognitive impairment in schizophrenia: relationships with cortical thickness in fronto-temporal regions, and dissociability from symptom severity. NPJ Schizophr. 2021;7(1):20. https://doi.org/10.1038/s41537-021-00149-0. PMID: 33737508; PMCID: PMC7973472.
9. Anderson G, Maes M. Schizophrenia: linking prenatal infection to cytokines, the tryptophan catabolite (TRYCAT) pathway, NMDA receptor hypofunction, neurodevelopment and neuroprogression. Prog Neuro-Psychopharmacol Biol Psychiatry. 2013;42:5–19. https://doi.org/10.1016/j.pnpbp.2012.06.01443.
10. Andreasen NC, Nopoulos P, Magnotta V, Pierson R, Ziebell S, Ho B-C. Progressive brain change in schizophrenia: a prospective longitudinal study of first-episode schizophrenia. Biol Psychiatry. 2011;70(7):672–9.
11. Antoniades M, Schoeler T, Radua J, Valli I, Allen P, Kempton MJ, McGuire P. Verbal learning and hippocampal dysfunction in schizophrenia: a meta-analysis. Neurosci Biobehav Rev. 2018;86:166–75.
12. Antonova E. The relationship between brain structure and neurocognition in schizophrenia: a selective review. Schizophr Res. 2004;70(2–3):117–45.
13. Ayoub MA, Angelicheva D, Vile D, Chandler D, Morar B, Cavanaugh JA, et al. Deleterious GRM1 mutations in schizophrenia. PLoS One. 2012;7(3):e32849.
14. Baek JH, Kim JS, Ryu S, Oh S, Noh J, Lee WK, et al. Association of genetic variations in DTNBP1 with cognitive function in schizophrenia patients and healthy subjects. Am J Med Genet Part B Neuropsychiatr Genet. 2012;159B(7):841–9.
15. Bai MY, Lovejoy DB, Guillemin GJ, et al. Galantamine– memantine combination and kynurenine pathway enzyme inhibitors in the treatment of neuropsychiatric disorders. Complex Psychiatry. 2021;7:19–33. https://doi.org/10.1159/000515066.
16. Bakker G, Vingerhoets C, Boucherie D, Caan M, Bloemen O, Eersels J, et al. Relationship between muscarinic M1 receptor binding and cognition in medication-free subjects with psychosis. Neuroimage Clin. 2018;18:713–9.

17. Barnett JH, Jones PB, Robbins TW, Muller U. Effects of the catechol-O-methyltransferase Val158Met polymorphism on executive function: a meta-analysis of the Wisconsin Card Sort Test in schizophrenia and healthy controls. Mol Psychiatry. 2007;12(5):502–9.
18. Beck K, Arumuham A, Veronese M, Santangelo B, McGinnity CJ, Dunn J, et al. N- methyl-D-aspartate receptor availability in first-episode psychosis: a PET-MR brain imaging study. Transl Psychiatry. 2021;11:1–8.
19. Bora E. Neurodevelopmental origin of cognitive impairment in schizophrenia. Psychol Med. 2015;45(1):1–9. https://doi.org/10.1017/S0033291714001263. Epub 2014 May 30.
20. Bora E, Yucel M, Pantelis C. Cognitive functioning in schizophrenia, schizoaffective disorder and affective psychoses: meta-analytic study. Br J Psychiatry. 2009;195(6):475–82. https://doi.org/10.1192/bjp.bp.108.055731. PMID: 19949193.
21. Bora E, Lin A, Wood SJ, Yung AR, Mcgorry PD, Pantelis C. Cognitive deficits in youth with familial and clinical high risk to psychosis: a systematic review and meta-analysis. Acta Psychiatr Scand. 2014;130:1–15.
22. Bora E, Akdede BB, Alptekin K. The relationship between cognitive impairment in schizophrenia and metabolic syndrome: a systematic review and meta-analysis. Psychol Med. 2017;47(6):1030–40. https://doi.org/10.1017/S0033291716003366. Epub 2016 Dec 29. Erratum in: Psychol Med. 2018 May;48(7):1224. https://doi.org/10.1017/S0033291717003932. PMID: 28032535.
23. Bosia M, Anselmetti S, Pirovano A, Ermoli E, Marino E, Bramanti P, et al. HTTLPR functional polymorphism in schizophrenia: executive functions vs: sustained attention dissociation. Prog Neuro-Psychopharm Biol Psychiatry. 2010;34(1):81–5.
24. Bosia M, Anselmetti S, Bechi M, Lorenzi C, Pirovano A, Cocchi F, et al. Effect of 5-HT1A-receptor functional polymorphism on theory of mind performances in schizophrenia. Psychiatry Res. 2011;188(2):187–90.
25. Bosia M, Spangaro M, Sapienza J, Martini F, Civardi S, Buonocore M, Bechi M, Lorenzi C, Cocchi F, Bianchi L, Guglielmino C, Cavallaro R. Cognition in schizophrenia: modeling the interplay between interleukin-1β C-511T polymorphism, metabolic syndrome, and sex. Neuropsychobiology. 2021;80(4):321–32. https://doi.org/10.1159/000512082. Epub 2021 Jan 4.
26. Bourgognon JM, Cavanagh J. The role of cytokines in modulating learning and memory and brain plasticity. Brain Neurosci Adv. 2020;4:2398212820979802. https://doi.org/10.1177/2398212820979802. PMID: 33415308; PMCID: PMC7750764.
27. Boyer L, Richieri R, Dassa D, Boucekine M, Fernandez J, Vaillant F, Padovani R, Auquier P, Lancon C. Association of metabolic syndrome and inflammation with neurocognition in patients with schizophrenia. Psychiatry Res. 2013;210(2):381–6. https://doi.org/10.1016/j.psychres.2013.06.020. Epub 2013 Jul 10.
28. Boyer L, Testart J, Michel P, Richieri R, Faget-Agius C, Vanoye V, Auquier P, Lancon C, Guedj E. Neurophysiological correlates of metabolic syndrome and cognitive impairment in schizophrenia: a structural equation modeling approach. Psychoneuroendocrinology. 2014;50:95–105. https://doi.org/10.1016/j.psyneuen.2014.07.019. Epub 2014 Aug 10.
29. Bramon E, Rabe-Hesketh S, Sham P, Murray RM, Frangou S. Meta-analysis of the P300 and P50 waveforms in schizophrenia. Schizophr Res. 2004;70:315–29.
30. Bramon E, McDonald C, Croft RJ, Landau S, Filbey F, Gruzelier JH, Sham PC, Frangou S, Murray RM. Is the P300 wave an endophenotype for schizophrenia? A meta-analysis and a family study. NeuroImage. 2005;27:960–8.
31. Brannan SK, Sawchak S, Miller AC, Lieberman JA, Paul SM, Breier A. Muscarinic cholinergic receptor agonist and peripheral antagonist for schizophrenia. N Engl J Med. 2021;384:717–26.
32. Briley PM, Liddle EB, Simmonite M, Jansen M, White TP, Balain V, Palaniyappan L, Bowtell R, Mullinger KJ, Liddle PF. Regional brain correlates of beta bursts in health and psychosis: a concurrent electroencephalography and functional magnetic resonance imaging study. Biol Psychiatry Cogn Neurosci Neuroimaging. 2021;6:1145–56.

33. Bruder G, Kayser J, Tenke C, Rabinowicz E, Friedman M, Amador X, Sharif Z, Gorman J. The time course of visuospatial processing deficits in schizophrenia: an event-related brain potential study. J Abnorm Psychol. 1998;107:399–411.
34. Bucci P, Mucci A, Merlotti E, Volpe U, Galderisi S. Induced gamma activity and event-related coherence in schizophrenia. Clin EEG Neurosci. 2007;38:96–104.
35. Burdick KE, Hodgkinson CA, Szeszko PR, Lencz T, Ekholm JM, Kane JM, et al. DISC1 and neurocognitive function in schizophrenia. Neuroreport. 2005;16(12):1399–402.
36. Burdick KE, Goldberg TE, Funke B, Bates JA, Lencz T, Kucherlapati R, et al. DTNBP1 genotype influences cognitive decline in schizophrenia. Schizophr Res. 2007;89(1–3):169–72.
37. Callicott JH, Straub RE, Pezawas L, Egan MF, Mattay VS, Hariri AR, et al. Variation in DISC1 affects hippocampal structure and function and increases risk for schizophrenia. Proc Natl Acad Sci USA. 2005;102(24):8627–32.
38. Cannon M, Jones P, Huttunen MO, Tanskanen A, Huttunen T, Rabe-Hesketh S, Murray RM. School performance in Finnish children and later development of schizophrenia: A population-based longitudinal study. Arch Gen Psychiatry. 1999;56:457–63.
39. Cannon TD, Hennah W, van Erp TG, Thompson PM, Lonnqvist J, Huttunen M, et al. Association of DISC1/TRAX haplotypes with schizophrenia, reduced prefrontal gray matter, and impaired short- and long-term memory. Arch Gen Psychiatry. 2005;62(11):1205–13.
40. Carruthers SP, Gurvich CT, Rossell SL. The muscarinic system, cognition and schizophrenia. Neurosci Biobehav Rev. 2015;55:393–402.
41. Cassidy C, Buchy L, Bodnar M, Dell'elce J, Choudhry Z, Fathalli F, et al. Association of a risk allele of ANK3 with cognitive performance and cortical thickness in patients with first-episode psychosis. J Psychiatry Neurosci: JPN. 2014;39(1):31–9.
42. Castner SA, Goldman-Rakic PS, Williams GV. Animal models of working memory: insights for targeting cognitive dysfunction in schizophrenia. Psychopharmacology. 2004;174:111–25.
43. Cathomas F, Guetter K, Seifritz E, Klaus F, Kaiser S. Quinolinic acid is associated with cognitive deficits in schizophrenia but not major depressive disorder. Sci Rep. 2021;11(1):9992. https://doi.org/10.1038/s41598-021-89335-9. PMID: 33976271; PMCID: PMC8113521.
44. Chen RY, Sham P, Chen EY, Li T, Cheung EF, Hui TC, et al. No association between T102C polymorphism of serotonin-2A receptor gene and clinical phenotypes of Chinese schizophrenic patients. Psychiatry Res. 2001;105(3):175–85.
45. Chen VC, Chen CH, Chiu YH, Lin TY, Li FC, Lu ML. Leptin/adiponectin ratio as a potential biomarker for metabolic syndrome in patients with schizophrenia. Psychoneuroendocrinology. 2018;92:34–40. https://doi.org/10.1016/j.psyneuen.2018.03.021. Epub 2018 Mar 29.
46. Chew LJ, Fusar-Poli P, Schmitz T. Oligodendroglial alterations and the role of microglia in white matter injury: relevance to schizophrenia. Dev Neurosci. 2013;35(2–3):102–29. https://doi.org/10.1159/000346157. Epub 2013 Feb 27. PMID: 23446060; PMCID: PMC4531048.
47. Chiappelli J, Notarangelo FM, Pocivavsek A, Thomas MAR, Rowland LM, Schwarcz R, Hong LE. Influence of plasma cytokines on kynurenine and kynurenic acid in schizophrenia. Neuropsychopharmacology. 2018;43(8):1675–80. https://doi.org/10.1038/s41386-018-0038-4. Epub 2018 Feb 27. PMID: 29520060; PMCID: PMC6006321.
48. Clementz BA, Wang J, Keil A. Normal electrocortical facilitation but abnormal target identification during visual sustained attention in schizophrenia. J Neurosci. 2008;28:13411–8.
49. Comai S, Bertazzo A, Brughera M, Crotti S. Tryptophan in health and disease. Adv Clin Chem. 2020;95:165–218. https://doi.org/10.1016/bs.acc.2019.08.005. Epub 2019 Oct 15.
50. Cools R, D'Esposito M. Inverted-U-shaped dopamine actions on human working memory and cognitive control. Biol Psychiatry. 2011;69:e113–25.
51. Cowansage KK, LeDoux JE, Monfils MH. Brain-derived neurotrophic factor: a dynamic gatekeeper of neural plasticity. Curr Mol Pharmacol. 2010;3(1):12–29.
52. Crespo-Facorro B, Barbadillo L, Pelayo-Terán JM, Rodríguez-Sánchez JM. Neuropsychological functioning and brain structure in schizophrenia. Int Rev Psychiatry. 2007;19(4):325–36.

53. Crook JM, Tomaskovic-Crook E, Copolov DL, Dean B. Decreased muscarinic receptor binding in subjects with schizophrenia: a study of the human hippo- campal formation. Biol Psychiatry. 2000;48:381–8.
54. Cuoco F, Agostoni G, Lesmo S, Sapienza J, Buonocore M, Bechi M, Martini F, Ferri I, Spangaro M, Bigai G, Seghi F, Guglielmino C, Cocchi F, Cavallaro R, Bosia M. Get up! Functional mobility and metabolic syndrome in chronic schizophrenia: effects on cognition and quality of life. Schizophr Res Cogn. 2022;28:100245. https://doi.org/10.1016/j.scog.2022.1002456. PMID: 35251942; PMCID: PMC889214.
55. Davies G, Armstrong N, Bis JC, Bressler J, Chouraki V, Giddaluru S, et al. Genetic contributions to variation in general cognitive function: a meta-analysis of genome–wide association studies in the CHARGE consortium (N=53949). Mol Psychiatry. 2015;20:183–92.
56. Davies G, Lam M, Harris SE, Trampush JW, Luciano M, Hill WD, et al. Study of 300,486 individuals identifies 148 independent genetic loci influencing general cognitive function. Nat Commun. 2018;9:1–16.
57. De Crescenzo F, Cortese S, Adamo N, Janiri L. Pharmacological and non-pharmacological treatment of adults with ADHD: a meta-review. Evid Based Ment Health. 2017;20:4–11.
58. de Nijs J, Pet MA, GROUP Investigators. Metabolic syndrome in schizophrenia patients associated with poor premorbid school performance in early adolescence. Acta Psychiatr Scand. 2016;133(4):289–97. https://doi.org/10.1111/acps.12528. Epub 2015 Nov 12.
59. De Picker LJ, Victoriano GM, Richards R, Gorvett AJ, Lyons S, Buckland GR, Tofani T, Norman JL, Chatelet DS, Nicoll JAR, Boche D. Immune environment of the brain in schizophrenia and during the psychotic episode: a human post-mortem study. Brain Behav Immun. 2021;97:319–27. https://doi.org/10.1016/j.bbi.2021.07.017. Epub 2021 Jul 30. PMID: 34339805; PMCID: PMC8475749.
60. Demjaha A, Egerton A, Murray RM, Kapur S, Howes OD, Stone JM, et al. Anti-psychotic treatment resistance in schizophrenia associated with elevated glutamate levels but normal dopamine function. Biol Psychiatry. 2014;75:e11–3.
61. Dias EC, Van Voorhis AC, Braga F, Todd J, Lopez-Calderon J, Martinez A, Javitt DC. Impaired fixation-related theta modulation predicts reduced visual span and guided search deficits in schizophrenia. Cereb Cortex. 2020;30:2823–33.
62. Dichter GS, van der Stelt O, Boch JL, Belger A. Relations among intelligence, executive function, and P300 event related potentials in schizophrenia. J Nerv Ment Dis. 2006;194:179–87.
63. Dickinson D, Ragland JD, Gold JM, Gur RC. General and specific cognitive deficits in schizophrenia: goliath defeats David? Biol Psychiatry. 2008;64:823–7.
64. Dickinson D, Straub RE, Trampush JW, Gao Y, Feng N, Xie B, et al. Differential effects of common variants in SCN2A on general cognitive ability, brain physiology, and messenger RNA expression in schizophrenia cases and control individuals. JAMA Psychiatry. 2014;71(6):647–56.
65. Dondé C, Brunelin J, Mondino M, Cellard C, Rolland B, Haesebaert F. The effects of acute nicotine administration on cognitive and early sensory processes in schizophrenia: a systematic review. Neurosci Biobehav Rev. 2020;118:121–33.
66. Donohoe G, Morris DW, Clarke S, McGhee KA, Schwaiger S, Nangle JM, et al. Variance in neurocognitive performance is associated with dysbindin-1 in schizophrenia: a preliminary study. Neuropsychologia. 2007;45(2):454–8.
67. Dutt A, Tseng H-H, Fonville L, Drakesmith M, Su L, Evans J, Zammit S, et al. Exploring neural dysfunction in 'clinical high risk' for psychosis: a quantitative review of fMRI studies. J Psychiatr Res. 2015;61:122–34.
68. Egan MF, Goldberg TE, Kolachana BS, Callicott JH, Mazzanti CM, Straub RE, et al. Effect of COMT Val108/158 Met genotype on frontal lobe function and risk for schizophrenia. Proc Natl Acad Sci USA. 2001;98(12):6917–22.
69. Egerton A, Stone JM. The glutamate hypothesis of schizophrenia: neuroimaging and drug development. Curr Pharm Biotechnol. 2012;13(8):1500–12.

70. Egerton A, Broberg BV, Van Haren N, Merritt K, Barker GJ, Lythgoe DJ, et al. Response to initial antipsychotic treatment in first episode psychosis is related to anterior cingulate glutamate levels: a multicentre 1 H-MRS study (OPTiMiSE). Mol Psychiatry. 2018;23:2145–55.
71. Erhardt S, Pocivavsek A, Repici M, et al. Adaptive and behavioral changes in kynurenine 3-monooxygenase knockout mice: relevance to psychotic disorders. Biol Psychiatry. 2017;82:756–65. https://doi.org/10.1016/J.BIOPSYCH.2016.12.011.
72. Falkenberg LE, Westerhausen R, Craven AR, Johnsen E, Kroken RA, Berg EML, Specht K, Hugdahl K. Impact of glutamate levels on neuronal response and cognitive abilities in schizophrenia. NeuroImage Clin. 2014;4:576–84.
73. Faraone SV, Seidman LJ, Kremen WS, Toomey R, Pepple JR, Tsuang MT. Neuropsychologic functioning among the nonpsychotic relatives of schizophrenic patients: the effect of genetic loading. Biol Psychiatry. 2000;48(2):120–6. https://doi.org/10.1016/s0006-3223(99)00263-2.
74. Fernandes CP, Christoforou A, Giddaluru S, Ersland KM, Djurovic S, Mattheisen M, Lundervold AJ, Reinvang I, Nöthen MM, Rietschel M, Ophoff RA, Genetic Risk and Outcome of Psychosis (GROUP), Hofman A, Uitterlinden AG, Werge T, Cichon S, Espeseth T, Andreassen OA, Steen VM, Le Hellard S. A genetic deconstruction of neurocognitive traits in schizophrenia and bipolar disorder. PLoS One. 2013;8(12):e81052. https://doi.org/10.1371/journal.pone.0081052. PMID: 24349030; PMCID: PMC3861303.
75. Fillman SG, Weickert TW, Lenroot RK, Catts SV, Bruggemann JM, Catts VS, Weickert CS. Elevated peripheral cytokines characterize a subgroup of people with schizophrenia displaying poor verbal fluency and reduced Broca's area volume. Mol Psychiatry. 2016;21(8):1090–8. https://doi.org/10.1038/mp.2015.90. Epub 2015 Jul 21. PMID: 26194183; PMCID: PMC4960447.
76. Flatow J, Buckley P, Miller BJ. Meta-analysis of oxidative stress in schizophrenia. Biol Psychiatry. 2013;74:400–9. https://doi.org/10.1016/j.biopsych.2013.03.018.
77. Fleischhacker W, Galderisi S, Laszlovszky I, Szatmári B, Barabássy Á, Acsai K, et al. The efficacy of cariprazine in negative symptoms of schizophrenia: post hoc analyses of PANSS individual items and PANSS-derived factors. Eur Psychiatry. 2019;58:1–9.
78. Floresco SB, Magyar O. Mesocortical dopamine modulation of executive functions: beyond working memory. Psychopharmacology. 2006;188:567–85.
79. Fornaro M, Carvalho AF, De Prisco M, Mondin AM, Billeci M, Selby P, et al. The prevalence, odds, predictors, and management of tobacco use disorder or nicotine dependence among people with severe mental illness: systematic review and meta-analysis. Neurosci Biobehav Rev. 2022;132:289–303.
80. Francisco AA, Horsthuis DJ, Popiel M, Foxe JJ, Molholm S. Atypical response inhibition and error processing in 22q11.2 Deletion Syndrome and schizophrenia: towards neuromarkers of disease progression and risk. NeuroImage: Clin. 2020;27:102351.
81. Fujigaki H, Mouri A, Yamamoto Y, et al. Linking phencyclidine intoxication to the tryptophan–kynurenine pathway: therapeutic implications for schizophrenia. Neurochem Int. 2019;125:1–6. https://doi.org/10.1016/j.neuint.2019.02.001.
82. Fusar-Poli P, Broome MR, Matthiasson P, Woolley JB, Johns LC, Tabraham P, Bramon E, et al. Spatial working memory in individuals at high risk for psychosis: longitudinal fMRI study. Schizophr Res. 2010;123(1):45–52.
83. Galderisi S, Mucci A, Volpe U, Boutros N. Evidence-based medicine and electrophysiology in schizophrenia. Clin EEG Neurosci. 2009;40:62–77.
84. Garrido MI, Kilner JM, Stephan KE, Friston KJ. The mismatch negativity: a review of underlying mechanisms. Clin Neurophysiol. 2009;120:453–63.
85. Gattaz WF, Mayer S, Ziegler P, Platz M, Gasser T. Hypofrontality on topographic EEG in schizophrenia. Correlations with neuropsychological and psychopathological parameters. Eur Arch Psychiatry Clin Neurosci. 1992;241(6):328–32. https://doi.org/10.1007/BF02191956.
86. Gica S, Poyraz BC, Gulec H. Are emotion recognition deficits in patients with schizophrenia states or traits? A 6-month follow-up study. Indian J Psychiatry. 2019;61:45–52.

87. Glahn DC, Ragland JD, Abramoff A, Barrett J, Laird AR, Bearden CE, Velligan DI. Beyond hypofrontality: a quantitative meta-analysis of functional neuroimaging studies of working memory in schizophrenia. Human Brain Mapping. 2005;25(1):60–9.
88. Green MF, Kern RS, Braff DL, Mintz J. Neurocognitive deficits and functional outcome in schizophrenia: are we measuring the "right stuff"? Schizophr Bull. 2000;26:119–36.
89. Greenwood TA, Braff DL, Light GA, Cadenhead KS, Calkins ME, Dobie DJ, Freedman R, Green MF, Gur RE, Gur RC, Mintz J, Nuechterlein KH, Olincy A, Radant AD, Seidman LJ, Siever LJ, Silverman JM, Stone WS, Swerdlow NR, Tsuang DW, Tsuang MT, Turetsky BI, Schork NJ. Initial heritability analyses of endophenotypic measures for schizophrenia: the consortium on the genetics of schizophrenia. Arch Gen Psychiatry. 2007;64(11):1242–50. https://doi.org/10.1001/archpsyc.64.11.1242. PMID: 17984393; PMCID: PMC10588564.
90. Grover SRP, Sahoo S, Gopal S, Nehra R, Ganesh A, Raghavan V, Sankaranarayan A. Relationship of metabolic syndrome and neurocognitive deficits in patients with schizophrenia. Psychiatry Res. 2019;278:56–64. https://doi.org/10.1016/j.psychres.2019.05.023. Epub 2019 May 15.
91. Guillin O, Diaz J, Carroll P, Griffon N, Schwartz JC, Sokoloff P. BDNF controls dopamine D3 receptor expression and triggers behavioural sensitization. Nature. 2001;411(6833):86–9.
92. Guo W, Hu M, Fan X, Liu F, Wu R, Chen J, Guo X, et al. Decreased gray matter volume in the left middle temporal gyrus as a candidate biomarker for schizophrenia: a study of drug naive, first-episode schizophrenia patients and unaffected siblings. Schizophr Res. 2014a;159(1):43–50.
93. Guo X, Li J, Wang J, Fan X, Hu M, Shen Y, Chen H, et al. Hippocampal and orbital inferior frontal gray matter volume abnormalities and cognitive deficit in treatment-naive, first-episode patients with schizophrenia. Schizophr Res. 2014b;152(2–3):339–43.
94. Gur C, Ragland JD, Gur RE. Cognitive changes in schizophrenia-a critical look. Int Rev Psychiatry. 1997;9(4):449–58.
95. Hafizi S, Tseng H-H, Rao N, Selvanathan T, Kenk M, Bazinet RP, Suridjan I, et al. Imaging microglial activation in untreated first-episode psychosis: a PET study with [^{18} F]FEPPA. Am J Psychiatry. 2017;174(2):118–24.
96. Hagenaars SP, Harris SE, Davies G, Hill WD, Liewald DC, Ritchie SJ, Marioni RE, Fawns-Ritchie C, Cullen B, Malik R, METASTROKE Consortium, International Consortium for Blood Pressure GWAS; SpiroMeta Consortium; CHARGE Consortium Pulmonary Group, CHARGE Consortium Aging and Longevity Group, Worrall BB, Sudlow CL, Wardlaw JM, Gallacher J, Pell J, McIntosh AM, Smith DJ, Gale CR, Deary IJ. Shared genetic aetiology between cognitive functions and physical and mental health in UK Biobank (N=112 151) and 24 GWAS consortia. Mol Psychiatry. 2016;21(11):1624–32. https://doi.org/10.1038/mp.2015.225. Epub 2016 Jan 26. PMID: 26809841; PMCID: PMC5078856.
97. Hamilton HK, Roach BJ, Mathalon DH. Forecasting remission from the psychosis risk syndrome with mismatch negativity and P300: potentials and pitfalls. Biol Psychiatry: Cogn Neurosci Neuroimaging. 2021;6:178–87.
98. Hargreaves A, Anney R, O'Dushlaine C, Nicodemus KK, Gill M, Corvin A, et al. The one and the many: effects of the cell adhesion molecule pathway on neuropsychological function in psychosis. Psychol Med. 2013;44:1–11.
99. Harms MP, Wang L, Csernansky JG, Barch DM. Structure–function relationship of working memory activity with hippocampal and prefrontal cortex volumes. Brain Struct Funct. 2013;218(1):173–86.
100. Harvey PD, Bosia M, Cavallaro R, Howes OD, Kahn RS, Leucht S, Müller DR, Penadés R, Vita A. Cognitive dysfunction in schizophrenia: an expert group paper on the current state of the art. Schizophr Res Cogn. 2022;29:100249. https://doi.org/10.1016/j.scog.2022.100249. PMID: 35345598; PMCID: PMC8956816.
101. Hasselmo ME. The role of acetylcholine in learning and memory. Curr Opin Neurobiol. 2006;16:710–5.

102. Hazlett EA, Buchsbaum MS, Jeu LA, Nenadic I, Fleischman MB, Shihabuddin L, Haznedar MM, et al. Hypofrontality in unmedicated schizophrenia patients studied with PET during performance of a serial verbal learning task. Schizophr Res. 2000;43(1):33–46.
103. Helfrich RF, Knight RT. Cognitive neurophysiology: event-related potentials. Handb Clin Neurol. 2019;160:543–58.
104. Hennah W, Tuulio-Henriksson A, Paunio T, Ekelund J, Varilo T, Partonen T, et al. A haplotype within the DISC1 gene is associated with visual memory functions in families with a high density of schizophrenia. Mol Psychiatry. 2005;10(12):1097–103.
105. Hickling LM, Perez-Iglesias R, Ortiz-García de la Foz V, Balanzá-Martínez V, McGuire P, Crespo-Facorro B, et al. Tobacco smoking and its association with cognition in first episode psychosis patients. Schizophr Res. 2018;192:269–73.
106. Hill WD, Davies G, Group CCW, Liewald DC, McIntosh AM, Deary IJ. Age-dependent pleiotropy between general cognitive function and major psychiatric disorders. Biol Psychiatry. 2016;80:266–73.
107. Hilti CC, Hilti LM, Heinemann D, Robbins T, Seifritz E, Cattapan-Ludewig K. Impaired performance on the Rapid Visual Information Processing task (RVIP) could be an endophenotype of schizophrenia. Psychiatry Res. 2010;177(1–2):60–4. https://doi.org/10.1016/j.psychres.2009.12.012. Epub 2010 Jan 27.
108. Ho NF, Chong PLH, Lee DR, Chew QH, Chen G, Sim K. The amygdala in schizophrenia and bipolar disorder: a synthesis of structural MRI, diffusion tensor imaging, and resting-state functional connectivity findings. Harv Rev Psychiatry. 2019;27(3):150–64.
109. Hoffman RE, Dobscha SK. Cortical pruning and the development of schizophrenia: a computer model. Schizophr Bull. 1989;15(3):477–90. https://doi.org/10.1093/schbul/15.3.477.
110. Hope S, Hoseth E, Dieset I, Mørch RH, Aas M, Aukrust P, Djurovic S, Melle I, Ueland T, Agartz I, Ueland T, Westlye LT, Andreassen OA. Inflammatory markers are associated with general cognitive abilities in schizophrenia and bipolar disorder patients and healthy controls. Schizophr Res. 2015;165(2–3):188–94. https://doi.org/10.1016/j.schres.2015.04.004. Epub 2015 May 5.
111. Howes OD, McCutcheon R. Inflammation and the neural diathesis-stress hypothesis of schizophrenia: a reconceptualization. Transl Psychiatry. 2017;7(2):e1024. https://doi.org/10.1038/tp.2016.278. PMID: 28170004; PMCID: PMC5438023.
112. Howes OD, Onwordi EC. The synaptic hypothesis of schizophrenia version III: a master mechanism. Mol Psychiatry. 2023;28(5):1843–56. https://doi.org/10.1038/s41380-023-02043-w. Epub 2023 Apr 11. PMID: 37041418; PMCID: PMC10575788.
113. Howes OD, Fusar-Poli P, Bloomfield M, Selvaraj S, McGuire P. From the prodrome to chronic schizophrenia: the neurobiology underlying psychotic symptoms and cognitive impairments. Curr Pharm Des. 2012;18(4):459–65. https://doi.org/10.2174/138161212799316217. PMID: 22239576; PMCID: PMC3662992.
114. Howes OD, McCutcheon R, Owen MJ, Murray RM. The role of genes, stress, and dopamine in the development of schizophrenia. Biol Psychiatry. 2017;81(1):9–20. https://doi.org/10.1016/j.biopsych.2016.07.014. Epub 2016 Aug 6. PMID: 27720198; PMCID: PMC5675052.
115. Huang X, Ding W, Wu F, Zhou S, Deng S, Ning Y. Increased plasma kynurenic acid levels are associated with impaired attention/vigilance and social cognition in patients with schizophrenia. Neuropsychiatr Dis Treat. 2020;16:263–71. https://doi.org/10.2147/NDT.S239763. PMID: 32158211; PMCID: PMC6986175.
116. Huang J, Tong J, Zhang P, Zhou Y, Cui Y, Tan S, Wang Z, Yang F, Kochunov P, Chiappelli J, Tian B, Tian L, Tan Y, Hong LE. Effects of neuroactive metabolites of the tryptophan pathway on working memory and cortical thickness in schizophrenia. Transl Psychiatry. 2021;11(1):198. https://doi.org/10.1038/s41398-021-01311-z. PMID: 33795641; PMCID: PMC8016899.
117. Hubbard L, Tansey KE, Rai D, Jones P, Ripke S, Chambert KD, Moran JL, McCarroll SA, Linden DE, Owen MJ, O'Donovan MC, Walters JT, Zammit S. Evidence of common genetic overlap between schizophrenia and cognition. Schizophr Bull. 2016;42(3):832–42.

https://doi.org/10.1093/schbul/sbv168. Epub 2015 Dec 16. PMID: 26678674; PMCID: PMC4838093.

118. Hughes JR, John ER. Conventional and quantitative electroencephalography in psychiatry. J Neuropsychiatry Clin Neurosci. 1999;11(2):190–208. https://doi.org/10.1176/jnp.11.2.190.
119. Jablensky A, Morar B, Wiltshire S, Carter K, Dragovic M, Badcock JC, et al. Polymorphisms associated with normal memory variation also affect memory impairment in schizophrenia. Genes Brain Behav. 2011;10(4):410–7.
120. Jelen LA, King S, Horne CM, Lythgoe DJ, Young AH, Stone JM. Functional magnetic resonance spectroscopy in patients with schizophrenia and bipolar affective disorder: glutamate dynamics in the anterior cingulate cortex during a working memory task. Eur Neuropsychopharmacol. 2019;29:222–34.
121. Johannesen JK, Bi J, Jiang R, Kenney JG, Chen C-MA. Machine learning identification of EEG features predicting working memory performance in schizophrenia and healthy adults. Neuropsychiatr Electrophysiol. 2016;2:3.
122. Kaar SJ, Angelescu I, Marques TR, Howes OD. Pre-frontal parvalbumin interneurons in schizophrenia: a meta-analysis of post-mortem studies. J Neural Transm. 2019;126:1637–51.
123. Kelly S, Jahanshad N, Zalesky A, Kochunov P, Agartz I, Alloza C, Andreassen OA, et al. Widespread white matter microstructural differences in schizophrenia across 4322 individuals: results from the ENIGMA Schizophrenia DTI Working Group. Mol Psychiatry. 2018;23(5):1261–9.
124. Kelly S, Guimond S, Lyall A, Stone WS, Shenton ME, Keshavan M, Seidman LJ. Neural correlates of cognitive deficits across developmental phases of schizophrenia. Neurobiol Dis. 2019;131:104353.
125. Kępińska AP, MacCabe JH, Cadar D, Steptoe A, Murray RM, Ajnakina O. Schizophrenia polygenic risk predicts general cognitive deficit but not cognitive decline in healthy older adults. Transl Psychiatry. 2020;10(1):422. https://doi.org/10.1038/s41398-020-01114-8. PMID: 33293510; PMCID: PMC7722936.
126. Keshavan MS, Anderson S, Pettegrew JW. Is schizophrenia due to excessive synaptic pruning in the prefrontal cortex? The Feinberg hypothesis revisited. J Psychiatr Res. 1994;28(3):239–65. https://doi.org/10.1016/0022-3956(94)90009-4.
127. Keshavan MS, Collin G, Guimond S, Kelly S, Prasad KM, Lizano P. Neuroimaging in schizophrenia. Neuroimaging Clin N Am. 2020;30(1):73–83.
128. Khalil M, Hollander P, Raucher-Chéné D, Lepage M, Lavigne KM. Structural brain correlates of cognitive function in schizophrenia: a meta-analysis. Neurosci Biobehav Rev. 2022;132:37–49.
129. Kindler J, Lim CK, Weickert CS, Boerrigter D, Galletly C, Liu D, Jacobs KR, Balzan R, Bruggemann J, O'Donnell M, Lenroot R, Guillemin GJ, Weickert TW. Dysregulation of kynurenine metabolism is related to proinflammatory cytokines, attention, and prefrontal cortex volume in schizophrenia. Mol Psychiatry. 2020;25(11):2860–72. https://doi.org/10.1038/s41380-019-0401-9. Epub 2019 Apr 3. PMID: 30940904; PMCID: PMC7577855.
130. Kochunov P, Coyle TR, Rowland LM, Jahanshad N, Thompson PM, Kelly S, Du X, et al. Association of white matter with core cognitive deficits in patients with schizophrenia. JAMA Psychiatry. 2017;74(9):958–66.
131. Koshiyama D, Miyakoshi M, Tanaka-Koshiyama K, Joshi YB, Sprock J, Braff DL, Light GA. Abnormal phase discontinuity of alpha- and theta-frequency oscillations in schizophrenia. Schizophr Res. 2021;231:73–81.
132. Krystal JH, Anticevic A, Yang GJ, Dragoi G, Driesen NR, Wang XJ, et al. Impaired tuning of neural ensembles and the pathophysiology of schizophrenia: a translational and computational neuroscience perspective. Biol Psychiatry. 2017;81:874–85.
133. Kullmann S, Heni M, Hallschmid M, Fritsche A, Preissl H, Häring HU. Brain insulin resistance at the crossroads of metabolic and cognitive disorders in humans. Physiol Rev. 2016;96(4):1169–209. https://doi.org/10.1152/physrev.00032.2015. Epub 2016 Aug 3.

134. Kwon HS, Koh SH. Neuroinflammation in neurodegenerative disorders: the roles of microglia and astrocytes. Transl Neurodegener. 2020;9(1):42. https://doi.org/10.1186/s40035-020-00221-2. PMID: 33239064; PMCID: PMC7689983.
135. Lawyer G, Nyman H, Agartz I, Arnborg S, Jönsson EG, Sedvall GC, Hall H. Morphological correlates to cognitive dysfunction in schizophrenia as studied with Bayesian regression. BMC Psychiatry. 2006;6(1):31.
136. Lee JL, Everitt BJ, Thomas KL. Independent cellular processes for hippocampal memory consolidation and reconsolidation. Science. 2004;304(5672):839–43.
137. Lencz T, Knowles E, Davies G, Guha S, Liewald DC, Starr JM, et al. Molecular genetic evidence for overlap between general cognitive ability and risk for schizophrenia: a report from the Cognitive Genomics consorTium (COGENT). Mol Psychiatry. 2014;19:168–74.
138. Leung A, Chue P. Sex differences in schizophrenia, a review of the literature. Acta Psychiatr Scand Suppl. 2000;401:3–38. https://doi.org/10.1111/j.0065-1591.2000.0ap25.x. PMID: 10887978.
139. Lewis DA, Curley AA, Glausier JR, Volk DW. Cortical parvalbumin interneurons and cognitive dysfunction in schizophrenia. Trends Neurosci. 2012;35:57–67.
140. Lindenmayer JP, Khan A, Kaushik S, Thanju A, Praveen R, Hoffman L, Cherath L, Valdez G, Wance D. Relationship between metabolic syndrome and cognition in patients with schizophrenia. Schizophr Res. 2012;142(1–3):171–6. https://doi.org/10.1016/j.schres.2012.09.019. Epub 2012 Oct 26.
141. Mackowick KM, Barr MS, Wing VC, Rabin RA, Ouellet-Plamondon C, George TP. Neurocognitive endophenotypes in schizophrenia: modulation by nicotinic receptor systems. Prog Neuro-Psychopharmacol Biol Psychiatry. 2014;52:79–85.
142. Malhotra AK, Kestler LJ, Mazzanti C, Bates JA, Goldberg T, Goldman D. A functional polymorphism in the COMT gene and performance on a test of prefrontal cognition. Am J Psychiatry. 2002;159(4):652–4.
143. Mallet J, Le Strat Y, Dubertret C, Gorwood P. Polygenic risk scores shed light on the relationship between schizophrenia and cognitive functioning: review and meta-analysis. J Clin Med. 2020;9(2):341. https://doi.org/10.3390/jcm9020341. PMID: 31991840; PMCID: PMC7074036.
144. Marques TR, Ashok AH, Pillinger T, Veronese M, Turkheimer FE, Dazzan P, Sommer IEC, Howes OD. Neuroinflammation in schizophrenia: meta-analysis of in vivo microglial imaging studies. Psychol Med. 2019;49(13):2186–96. https://doi.org/10.1017/S0033291718003057. Epub 2018 Oct 25. PMID: 30355368; PMCID: PMC6366560.
145. Martínez A, Gaspar PA, Hillyard SA, Andersen SK, Lopez-Calderon J, Corcoran CM, Javitt DC. Impaired motion processing in schizophrenia and the Attenuated Psychosis syndrome: etiological and clinical implications. Am J Psychiatry. 2018;175:1243–54.
146. Martínez A, Tobe R, Dias EC, Ardekani BA, Veenstra-VanderWeele J, Patel G, Breland M, Lieval A, Silipo G, Javitt DC. Differential patterns of visual sensory alteration underlying face emotion recognition impairment and motion perception deficits in schizophrenia and Autism Spectrum Disorder. Biol Psychiatry. 2019;86:557–67.
147. Martínez AL, Brea J, Rico S, de los Frailes MT, Loza MI. Cognitive deficit in schizophrenia: from etiology to novel treatments. Int J Mol Sci. 2021;22:9905. https://doi.org/10.3390/ijms22189905.
148. Marucci G, Buccioni M, Ben DD, Lambertucci C, Volpini R, Amenta F. Efficacy of acetylcholinesterase inhibitors in Alzheimer's disease. Neuropharmacology. 2021;190:108352.
149. McCutcheon R, Beck K, Jauhar S, Howes OD. Defining the locus of dopaminergic dysfunction in schizophrenia: a meta-analysis and test of the mesolimbic hypothesis. Schizophr Bull. 2018;44(6):1301–11. https://doi.org/10.1093/schbul/sbx180. PMID: 29301039; PMCID: PMC5933516.
150. McCutcheon RA, Krystal JH, Howes OD. Dopamine and glutamate in schizophrenia: biology, symptoms and treatment. World Psychiatry. 2020;19:15–33. https://doi.org/10.1002/wps.20693.

151. McCutcheon RA, Keefe RSE, McGuire PK. Cognitive impairment in schizophrenia: aetiology, pathophysiology, and treatment. Mol Psychiatry. 2023;28(5):1902–18. https://doi.org/10.1038/s41380-023-01949-9. Epub 2023 Jan 23. Erratum in: Mol Psychiatry. 2023 May;28(5):1919. https://doi.org/10.1038/s41380-023-01984-6. PMID: 36690793; PMCID: PMC10575791.
152. McEvoy JP, Meyer JM, Goff DC, Nasrallah HA, Davis SM, Sullivan L, Meltzer HY, Hsiao J, Scott Stroup T, Lieberman JA. Prevalence of the metabolic syndrome in patients with schizophrenia: baseline results from the Clinical Antipsychotic Trials of Intervention Effectiveness (CATIE) schizophrenia trial and comparison with national estimates from NHANES III. Schizophr Res. 2005;80(1):19–32. https://doi.org/10.1016/j.schres.2005.07.014. Epub 2005 Aug 30.
153. McIntosh AM, Gow A, Luciano M, Davies G, Liewald DC, Harris SE, Corley J, Hall J, Starr JM, Porteous DJ, Tenesa A, Visscher PM, Deary IJ. Polygenic risk for schizophrenia is associated with cognitive change between childhood and old age. Biol Psychiatry. 2013;73(10):938–43. https://doi.org/10.1016/j.biopsych.2013.01.011. Epub 2013 Feb 16.
154. Mednova IA, Boiko AS, Kornetova EG, Parshukova DA, Semke AV, Bokhan NA, Loonen AJM, Ivanova SA. Adipocytokines and metabolic syndrome in patients with schizophrenia. Metabolites. 2020;10(10):410. https://doi.org/10.3390/metabo10100410. PMID: 33066473; PMCID: PMC7602179.
155. Merritt K, Egerton A, Kempton MJ, Taylor MJ, McGuire PK. Nature of glutamate alterations in schizophrenia a meta-analysis of proton magnetic resonance spectroscopy studies. JAMA Psychiatry. 2016;73:665–74. https://doi.org/10.1001/jamapsychiatry.2016.0442.
156. Merritt K, McGuire PK, Egerton A, Aleman A, Block W, Bloemen OJN, et al. Association of age, antipsychotic medication, and symptom severity in schizophrenia with proton magnetic resonance spectroscopy brain glutamate level: a mega-analysis of individual participant-level data. JAMA Psychiatry. 2021;78:667–81.
157. Meyer JM, Nasrallah HA, McEvoy JP, Goff DC, Davis SM, Chakos M, Patel JK, Keefe RS, Stroup TS, Lieberman JA. The Clinical Antipsychotic Trials of Intervention Effectiveness (CATIE) schizophrenia trial: clinical comparison of subgroups with and without the metabolic syndrome. Schizophr Res. 2005;80(1):9–18. https://doi.org/10.1016/j.schres.2005.07.015. Epub 2005 Aug 24. PMID: 16125372.
158. Michalczyk A, Tyburski E, Podwalski P, Waszczuk K, Rudkowski K, Kucharska-Mazur J, Mak M, Rek-Owodziń K, Plichta P, Bielecki M, Andrusewicz W, Cecerska-Heryć E, Samochowiec A, Misiak B, Sagan L, Samochowiec J. Serum inflammatory markers and their associations with white matter integrity of the corpus callosum in schizophrenia patients and healthy controls. Prog Neuro-Psychopharmacol Biol Psychiatry. 2022;116:110510. https://doi.org/10.1016/j.pnpbp.2022.110510. Epub 2022 Jan 19.
159. Mihaljević-Peleš A, Bajs Janović M, Šagud M, Živković M, Janović Š, Jevtović S. Cognitive deficit in schizophrenia: an overview. Psychiatr Danub. 2019;31(Suppl 2):139–42.
160. Minzenberg MJ, Laird AR, Thelen S, Carter CS, Glahn DC. Meta-analysis of 41 functional neuroimaging studies of executive function in schizophrenia. Arch Gen Psychiatry. 2009;66(8):811.
161. Mitchell AJ, Vancampfort D, De Herdt A, Yu W, De Hert M. Is the prevalence of metabolic syndrome and metabolic abnormalities increased in early schizophrenia? A comparative meta-analysis of first episode, untreated and treated patients. Schizophr Bull. 2013;39(2):295–305. https://doi.org/10.1093/schbul/sbs082. Epub 2012 Aug 27. PMID: 22927670; PMCID: PMC3576152.
162. Molina V, Solera S, Sanz J, Sarramea F, Luque R, Rodríguez R, Jiménez-Arriero MA, et al. Association between cerebral metabolic and structural abnormalities and cognitive performance in schizophrenia. Psychiatry Res: Neuroimaging. 2009;173(2):88–93.
163. Muzio G, Barrera G, Pizzimenti S. Peroxisome proliferator-activated receptors (PPARs) and oxidative stress in physiological conditions and in cancer. Antioxidants (Basel). 2021;10(11):1734. https://doi.org/10.3390/antiox10111734. PMID: 34829605; PMCID: PMC8614822.

164. Näätänen R, Paavilainen P, Rinne T, Alho K. The mismatch negativity (MMN) in basic research of central auditory processing: a review. Clin Neurophysiol. 2007;118:2544–90.
165. Nakahara T, Tsugawa S, Noda Y, Ueno F, Honda S, Kinjo M, et al. Glutamatergic and GABAergic metabolite levels in schizophrenia-spectrum disorders: a meta-analysis of 1H-magnetic resonance spectroscopy studies. Mol Psychiatry. 2022;27:744–57. https://doi.org/10.1038/s41380-021-01297-6.
166. Newson JJ, Thiagarajan TC. EEG frequency bands in psychiatric disorders: a review of resting state studies. Front Hum Neurosci. 2019;12:521.
167. Nicodemus KK, Elvevag B, Foltz PW, Rosenstein M, Diaz-Asper C, Weinberger DR. Category fluency, latent semantic analysis and schizophrenia: a candidate gene approach. Cortex: J Devoted Study Nervous Syst Behav. 2013.
168. Nielsen JD, Madsen KH, Wang Z, Liu Z, Friston KJ, Zhou Y. Working memory modulation of frontoparietal network connectivity in first-episode schizophrenia. Cerebral Cortex. 2017;27(7):3832–41.
169. Nieoullon A. Dopamine and the regulation of cognition and attention. Prog Neurobiol. 2002;67:53–83.
170. North HF, Bruggemann J, Cropley V, Swaminathan V, Sundram S, Lenroot R, Pereira AM, Zalesky A, Bousman C, Pantelis C, Weickert TW, Shannon Weickert C. Increased peripheral inflammation in schizophrenia is associated with worse cognitive performance and related cortical thickness reductions. Eur Arch Psychiatry Clin Neurosci. 2021;271(4):595–607. https://doi.org/10.1007/s00406-021-01237-z. Epub 2021 Mar 24. PMID: 33760971.
171. Numakawa T, Yagasaki Y, Ishimoto T, Okada T, Suzuki T, Iwata N, et al. Evidence of novel neuronal functions of dysbindin, a susceptibility gene for schizophrenia. Hum Mol Genet. 2004;13(21):2699–708.
172. Numakawa T, Suzuki S, Kumamaru E, Adachi N, Richards M, Kunugi H. BDNF function and intracellular signaling in neurons. Histol Histopathol. 2010;25(2):237–58.
173. Ohi K, Sumiyoshi C, Fujino H, Yasuda Y, Yamamori H, Fujimoto M, Shiino T, Sumiyoshi T, Hashimoto R. Genetic overlap between general cognitive function and schizophrenia: a review of cognitive GWASs. Int J Mol Sci. 2018;19(12):3822. https://doi.org/10.3390/ijms19123822. PMID: 30513630; PMCID: PMC6320986.
174. Onwordi EC, Whitehurst T, Shatalina E, Mansur A, Arumuham A, Osugo M, Marques TR, Jauhar S, Gupta S, Mehrotra R, Rabiner EA, Gunn RN, Natesan S, Howes OD. Synaptic terminal density early in the course of schizophrenia: an in vivo UCB-J positron emission tomographic imaging study of SV2A. Biol Psychiatry. 2024;95(7):639–46. https://doi.org/10.1016/j.biopsych.2023.05.022. Epub 2023 Jun 15. PMID: 37330164; PMCID: PMC10923626.
175. Owen AM, Iddon JL, Hodges JR, Summers BA, Robbins TW. Spatial and non-spatial working memory at different stages of Parkinson's disease. Neuropsychologia. 1997;35:519–32.
176. Parciauskaite V, Bjekic J, Griskova-Bulanova I. Gamma-range auditory steady-state responses and cognitive performance: a systematic review. Brain Sci. 2021;11:217.
177. Pardo PJ, Knesevich MA, Vogler GP, Pardo JV, Towne B, Cloninger CR, Posner MI. Genetic and state variables of neurocognitive dysfunction in schizophrenia: a twin study. Schizophr Bull. 2000;26(2):459–77. https://doi.org/10.1093/oxfordjournals.schbul.a033466.
178. Patlola SR, Donohoe G, McKernan DP. The relationship between inflammatory biomarkers and cognitive dysfunction in patients with schizophrenia: a systematic review and meta-analysis. Prog Neuro-Psychopharmacol Biol Psychiatry. 2023;121:110668. https://doi.org/10.1016/j.pnpbp.2022.110668. Epub 2022 Oct 23.
179. Perrottelli A, Giordano GM, Brando F, Giuliani L, Pezzella P, Mucci A, Galderisi S. Unveiling the associations between EEG indices and cognitive deficits in schizophrenia-Spectrum disorders: a systematic review. Diagnostics (Basel). 2022;12(9):2193. https://doi.org/10.3390/diagnostics12092193. PMID: 36140594; PMCID: PMC9498272.
180. Pershing ML, Bortz DM, Pocivavsek A, et al. Elevated levels of kynurenic acid during gestation produce neurochemical, morphological, and cognitive defcits in

adulthood: implications for schizophrenia. Neuropharmacology. 2015;90:33–41. https://doi.org/10.1016/J.NEUROPHARM.2014.10.017.
181. Peters K, Wiltshire S, Henders AK, Dragovic M, Badcock JC, Chandler D, et al. Comprehensive analysis of tagging sequence variants in DTNBP1 shows no association with schizophrenia or with its composite neurocognitive endophenotypes. Am J Med Genet Part B Neuropsychiatric Genet. 2008;147B(7):1159–66.
182. Picó-Pérez M, Vieira R, Fernández-Rodríguez M, De Barros MAP, Radua J, Morgado P. Multimodal meta-analysis of structural gray matter, neurocognitive and social cognitive fMRI findings in schizophrenia patients. Psychol Med. 2022;52(4):614–24.
183. Pilowsky LS, Bressan RA, Stone JM, Erlandsson K, Mulligan RS, Krystal JH, et al. First in vivo evidence of an NMDA receptor deficit in medication-free schizophrenic patients. Mol Psychiatry. 2006;11:118–9.
184. Pocivavsek A, Thomas MAR, Elmer GI, et al. Continuous kynurenine administration during the prenatal period, but not during adolescence, causes learning and memory deficits in adult rats. Psychopharmacology. 2014;231:2799–809. https://doi.org/10.1007/S00213-014-3452-251.
185. Polich J. Updating P300: an integrative theory of P3a and P3b. Clin Neurophysiol. 2007;118:2128–48.
186. Prieto M, Iglesias-Parro S, Soriano MF, Ibáñez-Molina A. Local oscillatory brain dynamics of mind wandering in schizophrenia. Brain Sci. 2021;11:910.
187. Raedler TJ, Knable MB, Jones DW, Urbina RA, Gorey JG, Lee KS, et al. In vivo determination of muscarinic acetylcholine receptor availability in schizophrenia. Am J Psychiatry. 2003;160:118–27.
188. Raghavan DV, Bharathi P, Shanmugiah A, Jeyaprakash R. P300 and neuropsychological measurements in patients with schizophrenia and their healthy biological siblings. Indian J Psychiatry. 2016;58:454–8.
189. Ragland JD, Gur RC, Raz J, Schroeder L, Kohler CG, Smith RJ, Alavi A, et al. Effect of schizophrenia on frontotemporal activity during word encoding and recognition: a PET cerebral blood flow study. Am J Psychiatry. 2001;158(7):1114–25.
190. Reale M, Costantini E, Greig NH. Cytokine imbalance in schizophrenia. From research to clinic: potential implications for treatment. Front Psychiatry. 2021;12:536257. https://doi.org/10.3389/fpsyt.2021.536257. PMID: 33746786; PMCID: PMC7973221.
191. Régio Brambilla C, Veselinović T, Rajkumar R, Mauler J, Orth L, Ruch A, et al. mGluR5 receptor availability is associated with lower levels of negative symptoms and better cognition in male patients with chronic schizophrenia. Hum Brain Mapp. 2020;41:2762–81.
192. Reilly TJ, Nottage JF, Studerus E, Rutigliano G, De Micheli AI, Fusar-Poli P, McGuire P. Gamma band oscillations in the early phase of psychosis: a systematic review. Neurosci Biobehav Rev. 2018;90:381–99.
193. Ribeiro-Santos R, de Campos-Carli SM, Ferretjans R, Teixeira-Carvalho A, Martins-Filho OA, Teixeira AL, Salgado JV. The association of cognitive performance and IL-6 levels in schizophrenia is influenced by age and antipsychotic treatment. Nord J Psychiatry. 2020;74(3):187–93. https://doi.org/10.1080/08039488.2019.1688389. Epub 2019 Nov 18.
194. Richards AL, Pardiñas AF, Frizzati A, Tansey KE, Lynham AJ, Holmans P, Legge SE, Savage JE, Agartz I, Andreassen OA, GAM B, Corvin A, Cosgrove D, Degenhardt F, Djurovic S, Espeseth T, Ferraro L, Gayer-Anderson C, Giegling I, van Haren NE, Hartmann AM, Hubert JJ, Jönsson EG, Konte B, Lennertz L, Olde Loohuis LM, Melle I, Morgan C, Morris DW, Murray RM, Nyman H, Ophoff RA, GROUP Investigators; van Os J; EUGEI WP2 Group; Schizophrenia Working Group of the Psychiatric Genomics Consortium, Petryshen TL, Quattrone D, Rietschel M, Rujescu D, Rutten BPF, Streit F, Strohmaier J, Sullivan PF, Sundet K, Wagner M, Escott-Price V, Owen MJ, Donohoe G, O'Donovan MC, Walters JTR. The relationship between polygenic risk scores and cognition in schizophrenia. Schizophr Bull. 2020;46(2):336–44. https://doi.org/10.1093/schbul/sbz061. PMID: 31206164; PMCID: PMC7442352.

195. Rømer TB, Jeppesen R, Christensen RHB, Benros ME. Biomarkers in the cerebrospinal fluid of patients with psychotic disorders compared to healthy controls: a systematic review and meta-analysis. Mol Psychiatry. 2023;28(6):2277–90. https://doi.org/10.1038/s41380-023-02059-2. Epub 2023 May 11.
196. Sapienza J, Spangaro M, Guillemin GJ, Comai S, Bosia M. Importance of the dysregulation of the kynurenine pathway on cognition in schizophrenia: a systematic review of clinical studies. Eur Arch Psychiatry Clin Neurosci. 2023a;273(6):1317–28. https://doi.org/10.1007/s00406-022-01519-0. Epub 2022 Dec 2.
197. Sapienza J, Bosia M, Spangaro M, Martini F, Agostoni G, Cuoco F, Cocchi F, Cavallaro R. Schizophrenia and psychedelic state: dysconnection versus hyper-connection. A perspective on two different models of psychosis stemming from dysfunctional integration processes. Mol Psychiatry. 2023b;28(1):59–67. https://doi.org/10.1038/s41380-022-01721-5. Epub 2022 Aug 5. PMID: 35931756.
198. Sapienza J, Agostoni G, Dall'Acqua S, Sut S, Nasini S, Martini F, Marchesi A, Bechi M, Buonocore M, Cocchi F, Cavallaro R, Spangaro M, Comai S, Bosia M. The kynurenine pathway in treatment-resistant schizophrenia at the crossroads between pathophysiology and pharmacotherapy. Schizophr Res. 2024a;264:71–80. https://doi.org/10.1016/j.schres.2023.12.005. Epub 2023 Dec 14.
199. Sapienza J, Agostoni G, Comai S, Nasini S, Dall'Acqua S, Sut S, Spangaro M, Martini F, Bechi M, Buonocore M, Bigai G, Repaci F, Nocera D, Ave C, Guglielmino C, Cocchi F, Cavallaro R, Deste G, Bosia M. Neuroinflammation and kynurenines in schizophrenia: impact on cognition depending on cognitive functioning and modulatory properties in relation to cognitive remediation and aerobic exercise. Schizophr Res Cogn. 2024b;38:100328. https://doi.org/10.1016/j.scog.2024.100328. PMID: 39281320; PMCID: PMC11399803.
200. Sapienza J, Pacchioni F, Spangaro M, Bosia M. Dysconnection in schizophrenia: filling the dots from old to new evidence. Clin Neurophysiol. 2024c;162:226–8. https://doi.org/10.1016/j.clinph.2024.03.013. Epub 2024 Mar 21.
201. Savitz J. The kynurenine pathway: a finger in every pie. Mol Psychiatry. 2020;25:131–47. https://doi.org/10.1038/S41380-019-0414-4.
202. Scarr E, Cowie TF, Kanellakis S, Sundram S, Pantelis C, Dean B. Decreased cortical muscarinic receptors define a subgroup of subjects with schizophrenia. Mol Psychiatry. 2009;14:1017–23.
203. Schwarcz R, Stone TW. The kynurenine pathway and the brain: challenges, controversies and promises. Neuropharmacology. 2017;112:237–47. https://doi.org/10.1016/J.NEUROPHARM.2016.08.003.
204. Sheffield JM, Barch DM. Cognition and resting-state functional connectivity in schizophrenia. Neurosci Biobehav Rev. 2016;61:108–20.
205. Shekhar A, Potter WZ, Lightfoot J, Lienemann J, Dubé S, Mallinckrodt C, et al. Selective muscarinic receptor agonist xanomeline as a novel treatment approach for schizophrenia. Am J Psychiatry. 2008;165:1033–9.
206. Shenton ME, Dickey CC, Frumin M, McCarley RW. A review of MRI findings in schizophrenia. Schizophr Res. 2001;49(1–2):1–52.
207. Simi A, Tsakiri N, Wang P, Rothwell NJ. Interleukin-1 and inflammatory neurodegeneration. Biochem Soc Trans. 2007;35(Pt 5):1122–6. https://doi.org/10.1042/BST0351122.
208. Simpson EH, Kellendonk C. Insights about striatal circuit function and schizophrenia from a mouse model of dopamine D2 receptor upregulation. Biol Psychiatry. 2017;81:21–30.
209. Singh S, Singh K, Trivedi R, Goyal S, Kaur P, Singh N, Bhatia T, et al. Microstructural abnormalities of uncinate fasciculus as a function of impaired cognition in schizophrenia: a DTI study. J Biosci. 2016;41(3):419–26.
210. Slifstein M, van de Giessen E, Van Snellenberg J, Thompson JL, Narendran R, Gil R, Hackett E, et al. Deficits in prefrontal cortical and extrastriatal dopamine release in schizophrenia. JAMA Psychiatry. 2015;72(4):316.
211. Smeland OB, Andreassen OA. How can genetics help understand the relationship between cognitive dysfunction and schizophrenia? Scand J Psychol. 2018;59:26–31.

212. Sniekers S, Stringer S, Watanabe K, Jansen PR, Coleman JRI, Krapohl E, et al. Genome-wide association meta-analysis of 78,308 individuals identifies new loci and genes influencing human intelligence. Nat Genet. 2017;49:1107–12.
213. Snitz BE, MacDonald AW, Carter CS. Cognitive deficits in unaffected first-degree relatives of schizophrenia patients: a meta-analytic review of putative endophenotypes. Schizophr Bull. 2006;32:179–94.
214. Spangaro M, Mazza E, Poletti S, Cavallaro R, Benedetti F. Obesity influences white matter integrity in schizophrenia. Psychoneuroendocrinology. 2018;97:135–42. https://doi.org/10.1016/j.psyneuen.2018.07.017. Epub 2018 Jul 17.
215. Stephan KE, Friston KJ, Frith CD. Dysconnection in schizophrenia: from abnormal synaptic plasticity to failures of self-monitoring. Schizophr Bull. 2009;35(3):509–27. https://doi.org/10.1093/schbul/sbn176. Epub 2009 Jan 20. PMID: 19155345; PMCID: PMC2669579.
216. Strandburg RJ, Marsh JT, Brown WS, Asarnow RF, Guthrie D, Higa J, Yee-Bradburay CM, Nuechterlein KH. Reduced attention-related negative potentials in schizophrenic adults. Psychophysiology. 1994;31:272–81.
217. Strassnig M, Signorile J, Gonzalez C, Harvey PD. Physical performance and disability in schizophrenia. Schizophr Res Cogn. 2014;1(2):112–21. https://doi.org/10.1016/j.scog.2014.06.002. PMID: 25254158; PMCID: PMC4171045.
218. Symond MP, Harris AW, Gordon E, Williams LM. "Gamma synchrony" in first-episode schizophrenia: a disorder of temporal connectivity? Am J Psychiatry. 2005;162(3):459–65. https://doi.org/10.1176/appi.ajp.162.3.459. Erratum in: Am J Psychiatry. 2005 May;162(5):1042. Symond, Matthew B [corrected to Symond, Matthew P]. PMID: 15741462.
219. Talbot K, Eidem WL, Tinsley CL, Benson MA, Thompson EW, Smith RJ, et al. Dysbindin-1 is reduced in intrinsic, glutamatergic terminals of the hippocampal formation in schizophrenia. J Clin Invest. 2004;113(9):1353–63.
220. Tan H-Y, Choo W-C, Fones CSL, Chee MWL. fMRI study of maintenance and manipulation processes within working memory in first-episode schizophrenia. Am J Psychiatry. 2005;162(10):1849–58.
221. Taylor R, Neufeld RWJ, Schaefer B, Densmore M, Rajakumar N, Osuch EA, et al. Functional magnetic resonance spectroscopy of glutamate in schizophrenia and major depressive disorder: anterior cingulate activity during a color-word Stroop task. NPJ Schizophr. 2015;1:15028.
222. Torti E, Keren B, Palmer EE, Zhu Z, Afenjar A, Anderson IJ, Andrews MV, Atkinson C, Au M, Berry SA, Bowling KM, Boyle J, Buratti J, Cathey SS, Charles P, Cogne B, Courtin T, Escobar LF, Finley SL, Graham JM Jr, Grange DK, Heron D, Hewson S, Hiatt SM, Hibbs KA, Jayakar P, Kalsner L, Larcher L, Lesca G, Mark PR, Miller K, Nava C, Nizon M, Pai GS, Pappas J, Parsons G, Payne K, Putoux A, Rabin R, Sabatier I, Shinawi M, Shur N, Skinner SA, Valence S, Warren H, Whalen S, Crunk A, Douglas G, Monaghan KG, Person RE, Willaert R, Solomon BD, Juusola J. Variants in TCF20 in neurodevelopmental disability: description of 27 new patients and review of literature. Genet Med. 2019;21(9):2036–42. https://doi.org/10.1038/s41436-019-0454-9. Epub 2019 Feb 11. PMID: 30739909; PMCID: PMC7171701.
223. Toulopoulou T, Picchioni M, Rijsdijk F, Hua-Hall M, Ettinger U, Sham P, Murray R. Substantial genetic overlap between neurocognition and schizophrenia: genetic modeling in twin samples. Arch Gen Psychiatry. 2007;64(12):1348–55. https://doi.org/10.1001/archpsyc.64.12.1348.
224. Toulopoulou T, Zhang X, Cherny S, Dickinson D, Berman KF, Straub RE, Sham P, Weinberger DR. Polygenic risk score increases schizophrenia liability through cognition-relevant pathways. Brain. 2019;142(2):471–85. https://doi.org/10.1093/brain/awy279. PMID: 30535067; PMCID: PMC6359897.
225. Trampush JW, Yang ML, Yu J, Knowles E, Davies G, Liewald DC, et al. GWAS meta-analysis reveals novel loci and genetic correlates for general cognitive function: a report from the COGENT consortium. Mol Psychiatry. 2017;22:336–45.

226. Tripathi A, Kar SK, Shukla R. Cognitive deficits in schizophrenia: understanding the biological correlates and remediation strategies. Clin Psychopharmacol Neurosci. 2018;16(1):7–17. https://doi.org/10.9758/cpn.2018.16.1.7. PMID: 29397662; PMCID: PMC5810454.
227. Tritsch NX, Sabatini BL. Dopaminergic modulation of synaptic transmission in cortex and striatum. Neuron. 2012;76:33–50.
228. Ucok A, Alpsan H, Cakir S, Saruhan-Direskeneli G. Association of a serotonin receptor 2A gene polymorphism with cognitive functions in patients with schizophrenia. Am J Med Genet Part B Neuropsychiatr Genet. 2007;144B(5):704–7.
229. Uhlhaas PJ, Singer W. Abnormal neural oscillations and synchrony in schizophrenia. Nat Rev Neurosci. 2010;11(2):100–13. https://doi.org/10.1038/nrn2774.
230. Upthegrove R, Khandaker GM. Cytokines, oxidative stress and cellular markers of infammation in schizophrenia. Curr Top Behav Neurosci. 2020;44:49–66. https://doi.org/10.1007/7854_2018_8844.
231. Uranova NA, Vikhreva OV, Rakhmanova VI, Orlovskaya DD. Ultrastructural pathology of oligodendrocytes adjacent to microglia in prefrontal white matter in schizophrenia. NPJ Schizophr. 2018;4(1):26. https://doi.org/10.1038/s41537-018-0068-2. PMID: 30546020; PMCID: PMC6292874.
232. van Erp TGM, Hibar DP, Rasmussen JM, Glahn DC, Pearlson GD, Andreassen OA, Agartz I, et al. Subcortical brain volume abnormalities in 2028 individuals with schizophrenia and 2540 healthy controls via the ENIGMA consortium. Mol Psychiatry. 2016;21(4):547–53.
233. van Erp TGM, Walton E, Hibar DP, Schmaal L, Jiang W, Glahn DC, Pearlson GD, et al. Cortical brain abnormalities in 4474 individuals with schizophrenia and 5098 control subjects via the enhancing neuro imaging genetics through meta analysis (ENIGMA) consortium. Biol Psychiatry. 2018;84(9):644–54.
234. Van Snellenberg JX, Torres IJ, Thornton AE. Functional neuroimaging of working memory in schizophrenia: task performance as a moderating variable. Neuropsychology. 2006;20(5):497–510.
235. Vermeulen JM, Schirmbeck F, Blankers M, Van Tricht M, Bruggeman R, Van Den Brink W, et al. Association between smoking behavior and cognitive functioning in patients with psychosis, siblings, and healthy control subjects: results from a prospective 6-year follow-up study. Am J Psychiatry. 2018;175:1121–8.
236. Vita A, De Peri L, Deste G, Sacchetti E. Progressive loss of cortical gray matter in schizophrenia: a meta-analysis and meta-regression of longitudinal MRI studies. Transl Psychiatry. 2012;2(11):e190. https://doi.org/10.1038/tp.2012.116. Erratum in: Transl Psychiatry. 2013;3:e275. PMID: 23168990; PMCID: PMC3565772.
237. Vohs JL, Leonhardt BL, Francis MM, Westfall D, Howell J, Bolbecker AR, O'Donnell BF, Hetrick WP, Lysaker PH. A preliminary study of the association among metacognition and resting state EEG in schizophrenia. J Psychophysiol. 2015;30:47–54.
238. Voineskos AN, Hawco C, Neufeld NH, Turner JA, Ameis SH, Anticevic A, Buchanan RW, et al. Functional magnetic resonance imaging in schizophrenia: current evidence, methodological advances, limitations and future directions. World Psychiatry. 2024;23(1):26–51.
239. Walker KA, Gottesman RF, Wu A, Knopman DS, Gross AL, Mosley TH Jr, Selvin E, Windham BG. Systemic inflammation during midlife and cognitive change over 20 years: the ARIC study. Neurology. 2019;92(11):e1256–67. https://doi.org/10.1212/WNL.0000000000007094. Epub 2019 Feb 13. PMID: 30760633; PMCID: PMC6511107.
240. Wang J, Liu Q, Wydell TN, Liao J, Wang F, Quan W, Tian J, Wang P, Liu J, Dong W. Electrophysiological basis of reading related phonological impairment in Chinese speakers with schizophrenia: an ERP study. Psychiatry Res Neuroimaging. 2017;261:65–71.
241. Wang J, Li L, Zhang Z, Zhang X, Zhu Y, Zhang C, Bi Y. Extracellular vesicles mediate the communication of adipose tissue with brain and promote cognitive impairment associated with insulin resistance. Cell Metab. 2022;34(9):1264–1279.e8. https://doi.org/10.1016/j.cmet.2022.08.004.

242. Weinberg D, Lenroot R, Jacomb I, Allen K, Bruggemann J, Wells R, Balzan R, et al. Cognitive subtypes of schizophrenia characterized by differential brain volumetric reductions and cognitive decline. JAMA Psychiatry. 2016;73(12):1251.
243. Wichniak A, Okruszek Ł, Linke M, Jarkiewicz M, Jedrasik-Styła M, Ciołkiewicz A, Wierzbicka A, Jernajczyk W, Jarema M. Electroencephalographic theta activity and cognition in schizophrenia: Preliminary results. World J Biol Psychiatry. 2015;16:206–10.
244. Wolf RC, Höse A, Frasch K, Walter H, Vasic N. Volumetric abnormalities associated with cognitive deficits in patients with schizophrenia. Eur Psychiatry. 2008;23(8):541–8.
245. Wu D, Jiang T. Schizophrenia-related abnormalities in the triple network: a meta-analysis of working memory studies. Brain Imaging Behav. 2020;14(4):971–80.
246. Yaffe K, Kanaya A, Lindquist K, Simonsick EM, Harris T, Shorr RI, Tylavsky FA, Newman AB. The metabolic syndrome, inflammation, and risk of cognitive decline. JAMA. 2004;292(18):2237–42. https://doi.org/10.1001/jama.292.18.2237.
247. Yi F, Ball J, Stoll KE, Satpute VC, Mitchell SM, Pauli JL, et al. Direct excitation of parvalbumin-positive interneurons by M1 muscarinic acetylcholine receptors: roles in cellular excitability, inhibitory transmission and cognition. J Physiol. 2014;592:3463–94.
248. Yohn SE, Weiden PW, Felder CC, Stahl SM. Muscarinic acetylcholine receptors for psychotic disorders: bench-side to clinic. Trends Pharmacol Sci. 2022;43:1098–112. https://doi.org/10.1016/j.tips.2022.09.006.
249. Zai G, Robbins TW, Sahakian BJ, Kennedy JL. A review of molecular genetic studies of neurocognitive deficits in schizophrenia. Neurosci Biobehav Rev. 2017;72:50–67. https://doi.org/10.1016/j.neubiorev.2016.10.024. Epub 2016 Nov 17.
250. Zhang BH, Han M, Zhang XY, Hui L, Jiang SR, Yang FD, Tan YL, Wang ZR, Li J, Huang XF. Gender differences in cognitive deficits in schizophrenia with and without diabetes. Compr Psychiatry. 2015;63:1–9. https://doi.org/10.1016/j.comppsych.2015.07.003. Epub 2015 Jul 14. PMID: 26555485.

Pharmacological Treatment of Cognitive Impairment in Schizophrenia 12

Megan A. Neff and Philip D. Harvey

Abstract

Cognitive deficits in schizophrenia are wide-ranging and associated with functional disability. Pharmacological treatment efforts specifically targeting cognition have been underway for over 20 years, with efforts shaped by the collaborative Measurement and Treatment Research to Improve Cognition in Schizophrenia effort. Despite considerable enthusiasm and dozens of attempts, there have only been two treatments that advanced to phase 3 approval studies. There are several possible reasons for these large numbers of failures and we will examine the importance of these efforts, the challenges associated with both schizophrenia itself and the recommended research strategies, and previous strategies that guided drug development. We will evaluate several "near misses" in terms of approval and the latest developments with new pharmacological mechanisms that may be promising for different reasons, when compared to previous drug development targets.

Keywords

Cognition · Schizophrenia · Functional Capacity · Pharmacological Treatment · MATRICS

12.1 The Importance of Cognition in Schizophrenia

The prevalence and prognostic burden of cognitive impairment in schizophrenia are considerable. Over 98% of patients with schizophrenia perform at some degree below expected levels of cognitive functioning [56], based on their own prior

M. A. Neff · P. D. Harvey (✉)
University of Miami Miller School of Medicine, Miami, FL, USA
e-mail: pharvey@miami.edu

A. Vita et al. (eds.), *Negative and Cognitive Symptoms in Schizophrenia*, Comprehensive Approach to Psychiatry 3,
https://doi.org/10.1007/978-3-032-06209-3_12

functioning and at least 70% manifest current performance below the standard cut-off for impairment [68]. These impairments commonly precede the onset of psychosis [77] and persist even when psychoses are temporarily controlled [2]. The course and correlates of cognitive impairment have prompted experts to consider schizophrenia a primarily cognitive disorder [50]. Long-term reduction or elimination of psychotic symptoms with antipsychotic pharmacotherapy can correlate with improvements in everyday functioning but rarely with recovery, likely because cognitive dysfunction remains in the background [43]. While antipsychotic medications have improved patient quality of life to some degree, they not produce substantial improvements in cognition [55] and the persistence of cognitive and social cognitive symptoms [25, 33] remains one of the two primary obstacles to clinical and functional recovery, along with negative symptoms [31].

There is significant evidence to suggest that cognitive deficits manifest even before the onset of illness and may serve as the first sign of pathology. Individuals in the premorbid phase of schizophrenia have consistently demonstrated deficits in IQ, as well as impairments in specific neurocognitive domains, such as attention, memory, reasoning, executive functioning, and processing speed [79]. A recent meta-analysis of studies comparing the cognitive performance of subjects meeting criteria for Clinical High Risk for Psychosis (CHR-P) to that of healthy controls, showed that CHR-P was consistently associated with worse performance on measures of executive functioning, attention, general intelligence level, processing speed, verbal fluency, and visual, verbal and working memory [17]. Cognitive functioning does not appear to worsen progressively during the prodrome [15], although large-scale studies [77] and meta-analyses [76] suggest that the earlier appearance and greater severity of cognitive impairment is a marker of the highest risk to develop psychosis.

Among the features of prodromal functioning is the progressive development of disability, including social and role functioning difficulties [16]. The linkage between cognition and disability in the prodrome is most clearly revealed longitudinally, where it has been reported that individuals whose attenuated psychotic prodromal symptoms remit also manifest remission of cognitive deficits and disability [59]. In cases that do not convert to psychosis and do not remit, cognitive impairments and disability remain persistent, despite a lack of conversion to psychosis [66]. These findings underscore the close connection between cognition and disability, even in cases that are not manifesting psychotic symptoms.

As would be expected from the consistent detection of cognitive deficits in prodromal cases before initiation of antipsychotic medications, studies of cognitive impairments in never-treated patients find consistent impairments as well. Fatouros-Bergman et al. [24] published a meta-analysis of 23 studies exploring cognitive abilities in drug-naïve patients with schizophrenia, with data from a total of 1106 patients. Those patients demonstrated greater impairments in verbal memory, speed of processing, and working memory when compared to healthy controls. Additionally, the deficits seen in drug-naïve patients were comparable to those reported in antipsychotic-treated patients, suggesting that the cognitive deficits seen in FEP are inherent to schizophrenia and not a product of antipsychotic pharmacotherapy.

Cognitive and social cognitive impairments typically are found to manifest a mediated impact on disability, through their influence on functional capacity, the ability to perform critical everyday living skills [61]. These skills, in social, vocational, and everyday functioning domains, appear proximal to everyday disability and are strongly influenced by cognitive performance. In some large-scale studies, the best fitting predictor models for everyday disability suggest that there is no direct, non-mediated impact of cognitive impairments on everyday disability [30, 82]. Impairments in functional capacity are also detectable in CHR-P cases, who have evidence of real-world functional deficits. In fact, deficits on a performance-based measure of functional capacity during the prodrome were found to be a more significant predictor of conversion to psychosis than concurrent impairments in cognitive performance [64]. Thus, treatment efforts need to consider the impact of cognitive improvements on indices of functional capacity. Complicating this issue, however, is the fact that functional skills are acquired through exposure and experience, and many people with schizophrenia report that they have never performed some of the skills measured by common functional capacity measures [46].

Although there is heterogeneity in the severity and affected domains, cognitive impairment is pervasive in schizophrenia across all stages, and its effect on disability is well understood. Clearly, treatment of cognitive impairments at the time of the first identified psychotic episode seems important and targeting prodromal cases may also be a viable strategy. Disability associated with cognitive impairments develops at or before the first psychotic episode and there is no evidence that delaying efforts to improve cognition has any value, much like the well-understood urgency to treat psychosis effectively as soon as it appears. Given that there is evidence of persistence of cognitive impairment and possibly even slight worsening over the lifespan in schizophrenia [26], a lifespan approach to the treatment of cognitive deficits seems highly appropriate.

12.2 Pharmacological Treatment Strategies: Rationale

The use of medication to enhance cognition in schizophrenia has a long history, but a call to action on the part of the director of the NIMH and director of schizophrenia research at the institute [49] solidified the idea that this was an idea that needed attention. As a result, consensus guidance was developed through a consortium, the Measurement and Treatment Research to Improve Cognition in Schizophrenia (MATRICS; [34]) initiative. The MATRICS initiative was a collaboration between the US Food and Drug Administration (FDA), the National Institute of Mental Health (NIMH), participating companies from the Pharmaceutical Industry, and a cohort of academic scientists interested in schizophrenia. Among the products of the MATRICS Initiative was an assessment strategy, the MATRICS Consensus Cognitive Battery (MCCB; [67]), and recommendations regarding research design [12].

12.3 Pharmacological Treatment Strategies: Regulatory Requirements

As with any potential drug or device product, FDA approval is required for its use. Consumer safety and efficacy of treatments are the typical standards for evaluation of potential new treatments and sustained assessment of existing treatments. In the case of cognition in schizophrenia, this situation can have complications because of several factors.

- Not all people with schizophrenia may have current cognitive impairments, so agreement on whether studies should be "enriched" for participants with impairment is required. Historically, enrollment used an all-comers approach [54], but several recent studies have found larger improvement in participants with evidence of greater baseline impairment [75].
- Consistent assessment strategies across clinical studies are required, because there are many ways to assess cognition and there is considerable variability in the psychometric quality of the tests. This has resulted in one battery, the MATRICS Consensus Cognitive Battery (MCCB; [67]) being agreed upon as uniformly acceptable, with the option in place to substitute with alternatives, particularly in early-stage studies.
- Other factors, such as mood state or psychosis, could potentially influence cognitive impairments or make their assessment difficult. Thus, high levels of current depression or psychosis are potentially exclusionary, with these exclusions generally applied across trials.
- As cognitive impairment produces functional disability and the goal of treatment is disability reduction, understanding he extent to which the treatment is successful in that regard is important. Given the fact that everyday functional skills are learned and environmental factors exert a significant influence on real-world functioning [72], immediate changes could not be expected, but evidence of changes in potential would be required.

12.3.1 The MATRICS Consensus Cognitive Battery (MCCB)

The MCCB was developed through a multi-step process and resulted in an assessment battery with 11 performance-based tests. The FDA has agreed that clinical trials that employ the MCCB will not have to further justify their outcome assessments. There are clearly other strategies for the assessment of cognition in schizophrenia, including stand-alone assessments aimed at capturing the important features of cognition in schizophrenia, which will be discussed below in the context of challenges with the current strategies.

12.3.2 MATRICS Consensus Standards for Trial Design

12.3.2.1 Initial Proposal

Prior to the MATRICS Initiative, the lack of consensus guidelines for clinical trial design was a limitation for drug developers seeking to demonstrate the therapeutic efficacy of cognitive-enhancing drugs to the FDA. The guidelines from the MATRICS Initiative were a consensus product based upon evidence-based expert suggestions deliberated amongst a panel of expert presenters and representatives from the FDA, NIMH, MATRICS scientists, and the Pharmaceutical Industry; for aspects of clinical trial design where evidence was insufficient, the guideline reflects expert panelist opinion. The initial suggestions [12] focused on three issue groups and will be discussed as follows: (1) inclusion criteria; (2) outcome measures; and (3) other design and statistical issues.

The FDA's standards for diagnostic specificity in the approval/labeling of pharmacologic agents suggest that clinical trials should initially limit study populations to include only patients with schizophrenia. The FDA perspective is that there may be features of cognitive impairment specific to schizophrenia which could be distinguishable from the patterns of cognitive impairment seen in mood disorders and Alzheimer's disease. While perspective is debatable given results since that time [42], a homogenous population does have advantages for the sensitivity of outcome measures.

Participants should be clinically stable in the residual phase of their illness and on their current medication regimen for a specified length of time and exhibit no more than a moderate level of positive and negative symptoms, with minimal depressive and extrapyramidal symptoms. These parameters were established to better isolate cognitive changes from other symptom variabilities during the clinical trials.

In studies of adjunctive/co-treatment agents, subjects should ideally be limited to those taking only one antipsychotic medication. Permissible antipsychotics should be stratified by stage of study. When establishing proof-of-concept in stage I, drug developers should be more selective to avoid pharmacodynamic or pharmacokinetic interactions between antipsychotics and the cognitive-enhancing agent being studied; this inclusion criterion may then be relaxed in stage II, where the goal is to evaluate drug efficacy under typical clinical conditions. Despite these suggestions, it is quite common for participants in studies to date to have received more than one antipsychotic treatment; some evidence suggests that this does have a detrimental effect [10].

Subjects who would not benefit from cognitive enhancement or whose data may not be valid should be considered for exclusion. As noted above, studies have proceeded to date with the assumption that most, if not all, participants with schizophrenia will manifest sufficient cognitive impairment to receive a benefit from treatment.

The MCCB was developed to serve as the standard assessment for the primary outcome measure in trials of cognitive-enhancing agents for schizophrenia: change in cognitive performance. However, the FDA also requires drug developers to demonstrate concurrent change in a co-primary measure of real-world functional outcome for drug approval [5]. This measure of functioning can be a measure of functional potential, but it is extremely hard to separate potential from current performance, and there are challenges in this area for several reasons.

There was no consensus on the preferred co-primary measure in the MATRICS initiative, as no validated assessments existed at that time. Subsequent studies evaluated potential candidates, but none were definitively recommended [35, 36]. The concluding guideline recommends against using a measure of real-world community functioning, as this is dependent on psychosocial support and community resources that are beyond the control of a clinical trial. In place of community functioning, the FDA requires a proxy measure of functional outcome, whereupon subjects are scored on their ability to perform simulated activities of real-world functioning, such as taking public transportation or preparing a meal. Constructing a uniform measure is handicapped by the very fact that people with schizophrenia, manifesting cognitive impairments and disability starting from premorbid stages, are less likely to have learned critical functional skills. As an alternative, an interview-based measure of cognition may be substituted. In this, informants would be asked to provide their assessment of the subject's abilities to perform cognitively relevant functional tasks. We will return to the issue of co-primary measures and their empirically validated importance later.

When considering comparator agents, a placebo should be used in the study of an adjunctive/co-treatment agent. This should allow the interpretation of analyses comparing the combination of antipsychotic and adjunctive/co-treatment agent to that of antipsychotic and adjunctive placebo to be relatively straightforward. On the other hand, there was a lack of consensus regarding the choice of comparator for what were referred to as "broad spectrum agents," for which both changes in cognitive impairment and other schizophrenia symptoms would be expected. The ideal comparator agent would be a cognitively-neutral antipsychotic; however, there is not adequate information on impacts on cognition, adverse or beneficial, so preliminary studies would be required. While second-generation antipsychotics may be considered a superior comparator choice to first-generation antipsychotics, they are not conclusively without adverse cognitive effects, and their choice as a comparator would need to be substantiated. The perspective is that to be approved as a cognitive enhancer, a broad-spectrum agent should improve cognition from baseline as well as other symptoms of the illness and not simply have reduced adverse cognitive effects relative to the comparator (See [60]; for a potential example of this phenomenon).

Trial duration is sufficient in length to ensure that benefits are persistent over time. At the time of the MATRICS recommendations, input from the US Food and Drug ADministration (FDA) suggested a trial duration of at least 6 months to assess lasting cognitive effects. The primary outcome measure should be repeated on multiple occasions throughout the trial, and statistical analyses should accommodate data from multiple assessment times. We will revisit the impact of trial duration and repeated cognitive assessments later.

Buchanan et al. [13] presented a 5-year update on the MATRICS initiative guidelines that reviewed the experience of trials using the suggested guidelines and proposed several changes to the inclusion criteria. These included increasing the accepted threshold for severity of hallucinations and delusions from moderate to moderately severe, but maintaining the criterion that subjects have no more than a moderate severity rating on disorganization symptoms to ensure adequate ability to respond to test stimuli. Notably, the intention of the original guideline for positive symptom severity was to ensure that participants were in the residual phase of their illness.

12.4 How Do We Approach Identification of Cognitive Enhancing Medications?

There are multiple medications that have the potential to improve cognitive functioning, based on both knowledge of the neurobiology of cognition and on treatments for other conditions such as attention deficit disorder. These include an array of medications and potential mechanisms, such as stimulants, cholinergic, monoaminergic, GABA-ergic, and glutamatergic agents, and other medications that could beneficially impact performance. One of the theoretical advantages of pharmacological strategies to treat cognition is that they do not require the same level of sustained effort on the part of the treated population as rehabilitation-focused interventions, such as computerized cognitive training. The challenges of adherence to medication self-administration may be underestimated by this perspective, because such adherence would still be required to achieve efficacy unless a long-acting cognitive enhancer became available.

Efforts associated with the MATRICS initiative identified several possible treatment targets comprising two separate but related strategies. The first is to focus on domains of neurotransmission and circuit systems that have previously been shown to be abnormal in schizophrenia. Such systems include glutamate receptors, including NMDA [20] and AMPA [21], GABA [32], nicotinic acetylcholinergic subsystems including α-7 [62] and $\alpha 4\beta 2$ (Tanzer et al. 2020), and monoamines, including dopamine receptor subtypes such as D1 [1]. As considerable evidence supports alterations in these systems, the logic is that these should be the primary strategies for enhancing cognition, which is ubiquitously impaired in schizophrenia.

The schizophrenia-related pharmacological targets attempted have been quite diverse and have reflected the incredible complexity of the neurobiology of schizophrenia. Hypoactivity of the NMDA receptor, as induced by drugs such as PCP and ketamine, leads to a psychotic syndrome in healthy people that highly resembles schizophrenia and promotes exacerbations in individuals with a history of psychosis [20]. For instance, the NMDA Glutamate receptor has an obligatory co-transmitter, glycine, and low levels of glycine can also induce hypoactivity of the NMDA receptor. Recent studies have suggested that naturally occurring hypoactivity of the NMDA system is also associated with schizophrenia, including in first episode patients [4]. As this hypoactivity was detected in the hippocampus, this suggests that cognitive functions, many of which are linked to hippocampal functioning, could be affected by excessively low levels of NMDA receptor activity. Challenges in this drug development approach are associated with newly and rapidly developing information, suggesting that cognitive functioning is highly similar in schizophrenia and other conditions, such as bipolar disorder.

The second strategy is to identify the cognitive impairments most seen in schizophrenia and to identify their potential neurobiological underpinnings in a more general sense [28]. As impairments in cognitive domains such as processing speed, working memory, episodic memory, reasoning and problem solving, and attention can also be simulated and tested in animal models [86] and in healthy controls or individuals with different cognitively relevant illnesses, an ongoing strategy for drug discovery is to develop models of the cognitive impairments themselves and identify their neurobiological underpinnings [3]. Such a strategy can be developed in a disease-agnostic manner, which has advantages given the overlap of cognitive impairments across different conditions, particularly bipolar disorder and major depression, with schizophrenia.

Comparative studies have found that despite modest differences in the level of impairment in participants with schizophrenia and bipolar disorder, there are substantial similarities in the profile of relative impairments across domains compared to healthy controls, in the temporal stability of performance across clinical state, presence of impairment in unaffected relatives, in correlations with indices of functional outcome [40], and in the correlations of performance indices with polygenic scores for cognition, intelligence, and educational attainment obtained in general population samples [6]. Differences in relative impairments may also be a function of differences in premorbid functioning detectable at or before the first episode [71, 87] Given the similarities in cognition across conditions, the question must be raised as to whether focusing on neurobiological factors unique to schizophrenia as a treatment strategy for impairments that are shared with other conditions is questionable.

12.5 Outcomes in Previous Studies

Drug development commonly proceeds in phases, with phase 1 being safety, phase 2 being preliminary efficacy, signal detection, and dose finding, followed by phase 3 efficacy studies designed to be pivotal by virtue of being adequately powered and well controlled. There have been many studies done, with Keefe et al. [57] identifying 118 such studies in clinicaltrials.gov as of 2011, and many more launched since then. Some studies had methodological challenges, including small sample sizes and designs that were not fully congruent with the MATRICS guidelines. Out of this huge mass of studies, 2 drugs have met success in suitable phase 2 trials and advanced to phase 3 studies. Results of these studies have been reviewed before, with the long list of unsuccessful mechanisms reviewed by Harvey [38], updated by Keefe et al. [57], and again by Horan et al. [47].

As of now, two phase 2 studies using MATRICS methods reported positive results for cognition in schizophrenia. One of these was encenicline, a muscarinic α-7 partial agonist, which separated from placebo for cognition and interview-based assessments of cognition [57]. The second was iclepertin, a glycine transport antagonist, targeting increases in NMDA receptor activity. As encenicline did not separate from placebo in a phase 3 study and its development was stopped, we will focus on the results for iclepertin, subject of a very large phase 3 research program.

12.5.1 Iclepertin Phase 2

Glycine transporter-1 (GlyT1) inhibitors have been a recent drug target of interest for the potential treatment of cognitive impairment and negative symptoms associated with schizophrenia. By increasing the synaptic concentration of glycine, GlyT1 inhibitors are thought to augment the neurotransmission of glutamate at NMDA receptors, thereby improving downstream neuroplasticity and ultimately, cognitive function. Previous studies investigating the GlyT1 inhibitor sarcosine produced inadequate results regarding its efficacy as a pro-cognitive agent in schizophrenia [18]. A large-scale program of studies of bitopertin for negative symptoms in schizophrenia was also unsuccessful [14]. However, a more recent Phase 2 clinical trial of novel agent iclepertin (BI 425809) demonstrated potential as a viable treatment option for cognitive impairment associated with schizophrenia [27].

The trial examined dose-dependent outcomes associated with iclepertin as an add-on agent in stable patients with schizophrenia treated with standard-of-care medication. A total of 509 subjects were assigned to one of five treatment groups: iclepertin at doses of 2 mg, 5 mg, 10 mg, or 25 mg, or placebo. Using the MCCB scores as an outcome at baseline and week 12, iclepertin was associated with a significant improvement in cognitive performance relative to placebo, with the effect size being greatest for subjects in the 10 mg and 25 mg groups and no difference between the 25 mg group versus those taking 10 mg. The cognitive domains with significant changes included processing speed, working memory, and reasoning and problem-solving. Notably, improvements in cognition did not coincide with changes in positive or negative symptoms, suggesting that the pro-cognitive effects of iclepertin are an independent finding. There is a current large-scale set of worldwide studies examining iclepertin vs. placebo. These three similar studies are scheduled to be completed in November 2024 and will be the definitive test of the GLYT-1 mechanism as a cognitive enhancer.

In an issue that may arise later, although iclepertin separated from placebo on cognition, it did not on an interview-based measure of functional capacity. It will be a defining moment for the field if iclepertin clearly enhances cognition but does not improve the co-primary measure, forcing the issue of the necessity and validity of co-primary measures in such clinical trials.

12.6 Xanomeline-Trospium: KarXT Phase 2 and 3 Studies

Another recently promising drug target is the dual agonism of M1 and M4 muscarinic acetylcholine receptors (mAChRs). These two targets are associated with cognitive performance and the potential for reduction of psychotic symptoms, respectively. Xanomeline was previously investigated as a potential pharmacologic therapy for the cognitive impairments described in patients with Alzheimer's Disease (AD). In these early investigations, xanomeline was associated with improved cognition relative to placebo [7], and with considerably greater improvements seen in patients with moderate AD compared to those with mild AD [84]. Unexpectedly, xanomeline was also found to improve psychotic-like symptoms. A small, phase 2 proof-of-concept study attempted to bridge these findings with their potential utility in the treatment of schizophrenia and notably demonstrated significant improvements in cognitive, positive, and negative symptoms in acute schizophrenia inpatients prescribed xanomeline ($n = 10$) compared to those in the placebo group ($n = 10$) [80]. Despite these encouraging benefits of xanomeline, the drug was poorly tolerated due to gastrointestinal side effects ascribed to the peripheral agonism of mAChRs. As a result, further investigation of xanomeline and its clinical utility in the treatment of schizophrenia was stalled.

Now over two decades since xanomeline was initially investigated as a potentially cognitive-enhancing drug for use in AD, novel agent KarXT (xanomeline–trospium; Karuna Therapeutics TM, Boston, MA, USA) has made recent strides toward a breakthrough in the pharmacologic treatment of schizophrenia. Kar-XT is a combination of xanomeline and the peripherally active muscarinic antagonist trospium. Because trospium is limited in crossing the blood-brain barrier and competes with xanomeline for peripheral mAChRs binding sites, KarXT is expected to offer the formerly described benefits of xanomeline, but with mitigation of the unfavorable side effects that had previously limited its clinical utility.

Xanomeline was successful in the treatment of psychotic symptoms in three successive randomized placebo-controlled trials [11, 51, 52]. Thus, substantial evidence for the efficacy of the M4 mechanism is evident. In a 5-week randomized, double-blind, placebo-controlled phase 2 trial, primary analyses revealed that Kar-XT monotherapy significantly improved positive and negative symptoms in patients experiencing an acute exacerbation of schizophrenia. Additional exploratory analyses using data from an abbreviated version of the CogState computerized cognitive battery suggested that Kar-XT also had a modest, but not significant, overall pro-cognitive effect. In a series of post-hoc analyses, investigators sought to address these limitations by defining clinically meaningful baseline cognitive impairment as one standard deviation below normative standards, which is a cut-off commonly used for patients with schizophrenia and more generally in clinical neuropsychology [75]. In the subjects with meaningful baseline cognitive impairment, Kar-XT was associated with a significant improvement in cognitive performance of medium-large effect size, both from baseline and compared to placebo. On the other hand, there was no significant difference in cognitive performance in the subjects who were only minimally impaired at baseline.

The investigators noted that the original trial design was not optimized to assess the efficacy of Kar-XT as a pro-cognitive agent. For instance, because cognitive impairment is strongly associated with poor functional capacity, the demands of a 5-week phase 2 clinical trial might have inadvertently selected for participants with better baseline cognitive performance; however, these participants may also have been less likely to show meaningful improvements in cognitive performance.

These analyses were later replicated using pooled data from the two similarly designed phase 3 trials of Kar-XT monotherapy in acute exacerbations of schizophrenia reported above. This time, subjects in both studies completed cognitive assessments using an abbreviated version of the Cambridge Neuropsychological Test Automated Battery (CANTAB). Similar to the previous post-hoc analyses, when including the whole sample, Kar-XT was not associated with a significant change in cognitive performance [48]. However, after applying the same cut-off used in phase 2, cognitively impaired subjects showed a significant improvement in CANTAB performance with Kar-XT, with both significant changes from baseline and compared to placebo. Subjects with minimal cognitive impairment at baseline again did not significantly improve with either Kar-XT or placebo.

Importantly, in both phases 2 and 3 analyses, improvements in cognitive performance were independent of improvements in positive and negative symptoms, suggesting that the pro-cognitive effects of Kar-XT are not directly mediated by changes in psychosis. Going forward, the cognitive-enhancing qualities of Kar-XT seen in trials of acutely psychotic patients should attempt to be replicated in clinically stable patients, in order to fully characterize its therapeutic potential for cognitive impairment associated with schizophrenia.

It is important not to have excessive enthusiasm or optimism regarding the potential for approval for cognitive impairment in schizophrenia for KarXT. The research design is not consistent with the MATRICS suggestions and FDA regulators may require a cumbersome or impractical (switching stable patients onto placebo) trial design to confirm these particularly straightforward results.. However, these results may be a turning point in the search for cognitive-enhancing agents in schizophrenia, because several similar medications are also in development.

12.7 Why So Many Failures to Date?

There are likely several reasons for these challenges. Some may come from treatments targeting mechanisms that are simply not related to cognitive impairment, although abnormal in schizophrenia compared to healthy controls. The complex multidimensional nature of schizophrenia could mean that neurobiological factors that correlate with some illness features are simply not relevant to cognition.

It may be that drugs themselves do not work as expected. There were early studies (e.g., [22]) of dopamine D1 partial (SKF38393) and full (dihydrexidine) agonists that were later confirmed to have minimal oral brain bioavailability [74]. Correct dosing is also an issue and is affected by many factors, but even phase 2 trials are so expensive that many drugs may be being tested at very few doses.

Keefe [53] recently outlined several reasons for the lack of success to date, with a specific focus on the clinical trials themselves.

Noise Larger sample sizes have more statistical power, but require strategies that create noise and weaken power. Single-site studies with dedicated investigators are better at eliminating noise. Later trials with tens or even hundreds of site investigators could not be expected to be as effective at noise reduction. Noise can originate from many causes, but poor assessment competence and recruitment of participants who do not have schizophrenia or who are not going to take their medication is a major issue.

Regulation As described above, the MATRICS process is highly detailed and results in large-scale studies with many outcomes collected. Having everything work at the same time is very hard and attrition may increase, because visits may be too long for outpatients to tolerate.

Selection of performance-based endpoints A treatment signal is more likely when investigators can match a mechanism of action with an appropriate endpoint (e.g., choosing a processing speed endpoint with a short time frame of follow-up for a stimulant trial), but more difficult to detect with an endpoint selected by regulators for general relevance. Although composite endpoints are most reliable and most highly correlated with disability, this does not guarantee that they are most sensitive to treatment effects. Some cognitive measures are extremely reliable, such as vocabulary, but long-term fund of knowledge is not going to change in a clinical setting with enhancement of fluid cognition.

Simple regression to the mean Early phase studies with positive results are forwarded to the next phase, but not negative studies. Because of the small sample sizes of these trials necessitated by cost concerns, statistical power is low, and some sponsors lower the threshold for statistical significance of these studies for their go/no-go decisions. Oftentimes, analyses may include multiple comparisons and post hoc analyses without correction. In the next phase of trial, with improved statistical power and the enhanced precision that it brings, there is more power to detect a positive drug effect, but also more power for negative effects as well.

There are several features of the MATRICS regulatory recommendations that may also contribute to high risk for failures.

12.7.1 Practical Issues Associated with the MCCB

The complexity of this traditional 10-subtest paper-and-pencil-based cognitive assessment battery necessitates substantial training at start-up for raters meeting educational/experience requirements (and for replacement raters during the course of a trial), as well as substantial ongoing efforts to maintain rater fidelity. The administration time is approximately 90 minutes, which contributes to long study visits. Requirements for concurrent assessments (e.g. co-primary measure, general and specific symptoms (total, negative, and depressive), extrapyramidal symptoms, and labs are considerable; a baseline assessment can span two consecutive days.

There are challenges in standardized administration and scoring of older paper-and-pencil assessments, leading to scoring errors and lags in identifying problem data, as well as challenges in handling paper materials and documents. Raters need to remember to use correct alternate forms, and there is commonly a requirement to scan or fax paper documents for centralized review and scoring confirmation. While momentary electronic data capture (EDC) is the norm for clinical assessments, the requirements for testers to scan or fax documents are aspects of outdated technology with which testers may lack familiarity. Site- and region-based differences, in clinical symptoms [19] and in performance-based assessments, are not uncommon [37]. Such differences highlight the need for continuous real-time data monitoring and quality assurance procedures, which are inefficient with paper-based data.

12.7.2 Co-primary Requirement

The co-primary requirement was carried forward by regulators from previous studies attempting to treat cognition and functioning in dementia. However, the MATRICS group did not view any existing measure as sufficiently reliable, valid, and sensitive to recommend a "gold standard" to provide evidence of clinical meaningfulness for CIAS ([35]). There are major challenges associated with the two general types of measures available.

Interview-based measures of cognition These assessments require a patient and informant to answer questions about the patient's cognitive impairments and the degree to which they impaired daily functioning during the past few weeks. For most people with schizophrenia, self-reports of level of cognitive ability and functioning are notoriously inaccurate ([8]; [23]). Although high-quality informant ratings can be highly convergent with other information [41], many patients do not have an informant who sees them with sufficient frequency to provide meaningful ratings (or is willing to complete interviews, meaning that an informant requirement can be a substantial barrier to recruitment. When informants are available, they are often limited by a highly selective observational window into the patient's behavior and can be prone to biases [73], including cultural differences. There is also the possibility of reactive effects of nominating an individual as an informant, in that their increased attention to the participant during the trial may lead to them discovering other elements of impairment that they had not observed prior to the baseline assessment. All these factors can increase variability and noise in the data.

Performance-based measures of functional capacity These tests comprise simulations of real-word activities such as shopping and making other purchases, taking public transportation, or maintaining a conversation [58, 69]. While these measures do not rely on patient or informant reports and demonstrate a closer association with cognitive test data than the interview-based measures [44], typical paper-and-pencil versions have limitations for repeated use in multi-site clinical trials. Such limitations include the requirement for a large set of props, lack of alternative forms, extensive off-line scoring and data management procedures, poor psychometrics (e.g., ceiling effects, large practice effects), absence of established norms, outdated item content, and vulnerability to cultural and practical variations in how such activities are done outside of the United States [83]. Further, many participants have never performed the component skills that are assessed by these measures, suggesting that improvement with cognitive changes should not be assumed [45].

Six-month trial duration The consensus guidelines recommend a 6-month trial duration. This makes sense for studies in dementia, where prevention or slowing of decline, which may require at least 6 months to be able to detect a difference, is a favorable outcome in addition to improvement from baseline. The applicability of a dementia model to CIAS is debatable. In addition, many cognitive enhancing agents have meaningful effects in much shorter time frames. At the time of the enactment of this requirement, a leading candidate medication was a cholinergic alpha-7 agonist, which manifested potential for loss of efficacy because of tachyphylaxis [29]. Tachyphylaxis is not an issue for many drugs; however, requiring all drugs to be tested over an extended duration because some drugs may manifest loss of efficacy is a very broad requirement.

12.7.3 Summary of Learnings from CIAS Trials

Through 20 years of industry experience, it has become clear that the measures and methods prescribed by the MATRICS guidelines are challenging to implement in real-world, multi-site, multi-national trials due to the associated practical complexities, patient/site burden, and trial duration. These factors present barriers to participant enrollment, adherence, and retention, and the current scientific basis of some of the guidelines is debatable. In combination, these factors make trials slow to recruit for and expensive to implement, frequently resulting in pressure on drug developers to "speed-up" trials by relaxing standards (e.g., by adding sites or countries with limited clinical trial experience, lowering requirements for qualified raters), which leads to increased variability and compromised data quality [10].

12.7.4 Alternative Treatment Strategies

Computerized cognitive training has been shown across multiple studies to have benefits that are large enough to be detectable and have clinical meaning, as evidenced in very large-scale meta-analyses [85]. Other strategies, including augmentation of cognitive training with pharmacological add-on therapies, have shown success as well [39]. Skills training, either general (targeting multiple social and functional domains; [9]) or more specifically targeted at work outcomes [63], has been shown to improve functional capacity and everyday functioning, particularly when combined with computerized cognitive training These findings suggest that drug development as part of combination psychosocial and pharmacological interventions (e.g., varenicline and smoking cessation therapy) could be viable.

12.8 Closing Comments

Although pharmacological treatment of cognitive impairment was associated with high levels of optimism at the time of the introduction of the MATRICS consensus guideline for trial design, outcomes have been very disappointing. We have identified several reasons why the hundreds of studies have come up empty, including challenges in drug development, regulatory hurdles, and a very difficult research design that is matched to a population on which it is hard to do research. Some studies have failed because of poor decisions on the part of the sponsor of the studies, where business concerns were superimposed on research procedures, leading to collapses late in trials that had every other indication of being about to succeed [10] or because nonadherence was substantial.

It seems time to suggest some reconsideration of the standards attached to trials for pharmacological treatment of cognitive enhancement, including the length of the trial and the assessments required, with a focus on recruitment of participants with the most to gain from participation. Regulatory standards for other treatment targets in schizophrenia, such as negative symptoms, have been revealed to exclude an extraordinary proportion of potential research candidates, as many as 92% [70]. The selection criteria for participants in MATRICS cognition trials, other than diagnosis, are also selections based entirely on exclusions: symptoms and depression can't be too severe; treatments and patients should be clinically stable; can't be unwilling to be a study for half a year with very common visits; other factors can't be the origin of cognitive impairments (e.g., severe negative symptoms). In addition to the lack of knowledge as to how common participants with these features are in the overall population of people with schizophrenia (probably more than 8%, but we can't be sure), the procedures required to confirm continued adherence to the exclusion criteria are burdensome, as described above. It seems like shorter trials, shorter assessments, and reduced focus on co-morbidities (e.g., negative symptoms), which have proven impossible to treat to date as well, could lead to a streamlined process and a higher chance of success.

References

1. Abi-Dargham A, Javitch JA, Slifstein M, et al. Dopamine D1R receptor stimulation as a mechanistic pro-cognitive target for schizophrenia. Schizophr Bull. 2022;48(1):199–210. https://doi.org/10.1093/schbul/sbab09.
2. Altshuler LL, Ventura J, van Gorp WG, et al. Neurocognitive function in clinically stable men with bipolar I disorder or schizophrenia and normal control subjects. Biol Psychiatry. 2004;56(8):560–9. https://doi.org/10.1016/j.biopsych.2004.08.002.
3. Barch DM, Mathalon DH. Using brain imaging measures in studies of procognitive pharmacologic agents in schizophrenia: psychometric and quality assurance considerations. Biol Psychiatry. 2011;70(1):13–8. https://doi.org/10.1016/j.biopsych.2011.01.004.
4. Beck K, Arumuham A, Veronese M, et al. N-methyl-D-aspartate receptor availability in first-episode psychosis: a PET-MR brain imaging study. Transl Psychiatry. 2021;11(1):425. https://doi.org/10.1038/s41398-021-01540-2.
5. Bellack AS, Green MF, Cook JA, et al. Assessment of community functioning in people with schizophrenia and other severe mental illnesses: a white paper based on an NIMH-sponsored workshop. Schizophr Bull. 2007;33(3):805–22. https://doi.org/10.1093/schbul/sbl035.
6. Bigdeli TB, Fanous AH, Li Y. Et al. genome-wide association studies of schizophrenia and bipolar disorder in a diverse cohort of US veterans. Schizophr Bull. 2021;47(2):517–29. https://doi.org/10.1093/schbul/sbaa133.
7. Bodick NC, Offen WW, Levey AL, et al. Effects of xanomeline, a selective muscarinic receptor agonist, on cognitive function and behavioral symptoms in Alzheimer disease. Arch Neurol. 1997;54:465–73.
8. Bowie CR, Twamley EW, Anderson H, et al. Self-assessment of functional status in schizophrenia. J Psychiatr Res. 2007;41(12):1012–8.
9. Bowie CR, McGurk SR, Mausbach B, et al. Combined cognitive remediation and functional skills training for schizophrenia: effects on cognition, functional competence, and real-world behavior. Am J Psychiatry. 2012;169(7):710–8. https://doi.org/10.1176/appi.ajp.2012.11091337.
10. Brannan SK. The difficulties in going from P2 to P3 in CNS trials: red flags from a recent cognitive impairment in schizophrenia program. Washington, DC: International Society of CNS Clincial Trials and Methods; 2017.
11. Brannan SK, Sawchak S, Miller AC, et al. (2021) muscarinic cholinergic receptor agonist and peripheral antagonist for schizophrenia. N Engl J Med. 2021;384(8):717–26. https://doi.org/10.1056/NEJMoa2017015.
12. Buchanan RW, Davis M, Goff D, et al. A summary of the FDA-NIMH-MATRICS workshop on clinical trial design for neurocognitive drugs for schizophrenia. Schizophr Bull. 2005;31(1):5–19. https://doi.org/10.1093/schbul/sbi020.
13. Buchanan RW, Keefe RS, Umbricht D, et al. The FDA-NIMH-MATRICS guidelines for clinical trial design of cognitive-enhancing drugs: what do we know 5 years later? Schizophr Bull. 2011;37(6):1209–17. https://doi.org/10.1093/schbul/sbq038.
14. Bugarski-Kirola D, Blaettler T, Arango C, et al. Bitopertin in negative symptoms of schizophrenia-results from the phase III FlashLyte and DayLyte studies. Biol Psychiatry. 2017;82(1):8–16. https://doi.org/10.1016/j.biopsych.2016.11.014.
15. Carrión RE, Walder DJ, Auther AM, et al. (2018). From the psychosis prodrome to the first-episode of psychosis: no evidence of a cognitive decline. J Psychiatr Res. 2018;96:231–8. https://doi.org/10.1016/j.jpsychires.2017.10.014.
16. Carrión RE, Auther AM, McLaughlin D, et al. The global functioning: social and role scales-further validation in a large sample of adolescents and Young adults at clinical high risk for psychosis. Schizophr Bull. 2019;45(4):763–72. https://doi.org/10.1093/schbul/sby126.
17. Catalan A, Salazar De Pablo G, Aymerich C, et al. Neurocognitive functioning in individuals at clinical high risk for psychosis: a systematic review and meta-analysis. JAMA Psychiatry. 2021;78:859–67.

18. Chang CH, Lin CH, Liu CY, et al. Efficacy and cognitive effect of sarcosine (N-methylglycine) in patients with schizophrenia: a systematic review and meta-analysis of double-blind randomised controlled trials. J Psychopharmacol. 2020;34(5):495–505. https://doi.org/10.1177/0269881120908016.
19. Chen YF, Wang SJ, Khin NA, et al. Trial design issues and treatment effect modeling in multiregional schizophrenia trials. Pharm Stat. 2010;9(3):217–29. https://doi.org/10.1002/pst.439.
20. Coyle JT. The glutamatergic dysfunction hypothesis for schizophrenia. Harv Rev Psychiatry. 1996;3(5):241–53. https://doi.org/10.3109/10673229609017192.
21. Danysz W. Positive modulators of AMPA receptors as a potential treatment for schizophrenia. Curr Opin Investig Drugs. 2002;3(7):1062–6.
22. Davidson M, Harvey PD, Bergman RL, et al. Effects of the D-1 agonist SKF-38393 combined with haloperidol in schizophrenic patients. Arch Gen Psychiatry. 1990;47(2):190–1. https://doi.org/10.1001/archpsyc.1990.01810140090014.
23. Durand D, Strassnig M, Sabbag S, et al. Factors influencing self-assessment of cognition and functioning in schizophrenia: implications for treatment studies. Eur Neuropsychopharmacol. 2015;25(2):185–91. https://doi.org/10.1016/j.euroneuro.2014.07.008.
24. Fatouros-Bergman H, Cervenka S, Flyckt L, et al. Meta-analysis of cognitive performance in drug-naïve patients with schizophrenia. Schizophr Res. 2014;158:156–62.
25. Fett AK, Viechtbauer W, Dominguez MD, et al. The relationship between neurocognition and social cognition with functional outcomes in schizophrenia: a meta-analysis. Neurosci Biobehav Rev. 2011;35(3):573–88. https://doi.org/10.1016/j.neubiorev.2010.07.001.
26. Fett AKJ, Velthorst E, Reichenberg A, et al. Long-term changes in cognitive functioning in individuals with psychotic disorders: findings from the Suffolk County mental health project. JAMA Psychiatry. 2020;77:387–96.
27. Fleischhacker WW, Podhorna J, Gröschl M, et al. Efficacy and safety of the novel glycine transporter inhibitor BI 425809 once daily in patients with schizophrenia: a double-blind, randomised, placebo-controlled phase 2 study. Lancet Psychiatry. 2021;8(3):191–201.
28. Floresco SB, Geyer MA, Gold LH, Grace AA. Developing predictive animal models and establishing a preclinical trials network for assessing treatment effects on cognition in schizophrenia. Schizophr Bull. 2005;31(4):888–94. https://doi.org/10.1093/schbul/sbi041.
29. Freedman R, Olincy A, Buchanan RW, et al. Initial phase 2 trial of a nicotinic agonist in schizophrenia. Am J Psychiatry. 2008;165(8):1040–7.
30. Galderisi S, Rossi A, Rocca P, et al. Italian Network For Research on Psychoses. The influence of illness-related variables, personal resources and context-related factors on real-life functioning of people with schizophrenia. World Psychiatry. 2014;13(3):275–87. https://doi.org/10.1002/wps.20167.
31. Galderisi S, Mucci A, Buchanan RW, Arango C. Negative symptoms of schizophrenia: new developments and unanswered research questions. Lancet Psychiatry. 2018;5(8):664–77. https://doi.org/10.1016/S2215-0366(18)30050-6.
32. Gonzalez-Burgos G, Fish KN, Lewis DA. (2011) GABA neuron alterations, cortical circuit dysfunction and cognitive deficits in schizophrenia. Neural Plast:723184. https://doi.org/10.1155/2011/723184
33. Green MF, Kern RS, Braff DL, Mintz J. Neurocognitive deficits and functional outcome in schizophrenia: are we measuring the "right stuff?" Schizophr. Dengue Bull. 2000;26:119–36.
34. Green MF, Nuechterlein KH, Gold JM, et al. Approaching a consensus cognitive battery for clinical trials in schizophrenia: the NIMH-MATRICS conference to select cognitive domains and test criteria. Biol Psychiatry. 2004;56(5):301–7. https://doi.org/10.1016/j.biopsych.2004.06.023.
35. Green MF, Nuechterlein KH, Kern RS, et al. Functional co-primary measures for clinical trials in schizophrenia: results from the MATRICS Psychometric and Standardization Study. Am J Psychiatry. 2008;165(2):221–8.
36. Green MF, Schooler NR, Kern RS, et al. Evaluation of functionally meaningful measures for clinical trials of cognition enhancement in schizophrenia. Am J Psychiatry. 2011;168(4):400–7. https://doi.org/10.1176/appi.ajp.2010.10030414.

37. Guzman A, Klein H, Keefe RSE, et al. Location-based differences in cognition and functional capacity: consistent levels of relative impairment in participants with schizophrenia compared to healthy controls. Psychiatry Res. 2024, In press.
38. Harvey PD. Pharmacological cognitive enhancement in schizophrenia. Neuropsychol Rev. 2009;19(3):324–35. https://doi.org/10.1007/s11065-009-9103-4.
39. Harvey PD, Sand M. Pharmacological augmentation of psychosocial and remediation training efforts in schizophrenia. Front Psych. 2017;8:177. https://doi.org/10.3389/fpsyt.2017.00177.
40. Harvey PD, Wingo AP, Burdick KE, Baldessarini RJ. Cognition and disability in bipolar disorder: lessons from schizophrenia research. Bipolar Disord. 2010;12(4):364–75. https://doi.org/10.1111/j.1399-5618.2010.00831.x.
41. Harvey PD, Raykov T, Twamley EW, et al. Validating the measurement of real-world functional outcomes: phase I results of the VALERO study. Am J Psychiatry. 2011;168(11):1195–201. https://doi.org/10.1176/appi.ajp.2011.10121723.
42. Harvey PD, Aslan M, Du M, et al. Factor structure of cognition and functional capacity in two studies of schizophrenia and bipolar disorder: implications for genomic studies. Neuropsychology. 2016;30(1):28–39. https://doi.org/10.1037/neu0000245.
43. Harvey PD, Strassnig MT, Silberstein J. Prediction of disability in schizophrenia: symptoms, cognition, and self-assessment. J Exp Psychopathol. 2019;10(3):2043808719865693.
44. Harvey PD, Horan WP, Atkins AS, et al. Factor structure of cognitive performance and functional capacity in schizophrenia: evidence for differences across functional capacity measures. Schizophr Res. 2020;223:297–304. https://doi.org/10.1016/j.schres.2020.08.010.
45. Harvey PD, Bosia M, Cavallaro R, et al. Cognitive dysfunction in schizophrenia: an expert group paper on the current state of the art. Schizophr Res. 2022;29:100249.
46. Holshausen K, Bowie CR, Mausbach BT, et al. Neurocognition, functional capacity, and functional outcomes: the cost of inexperience. Schizophr Res. 2014;152(2–3):430–4.
47. Horan WP, Catalano LT, Green MF. An update on treatment of cognitive impairment associated with schizophrenia. Curr Top Behav Neurosci. 2023;63:407–36. https://doi.org/10.1007/7854_2022_382.
48. Horan WP, Sauder, C, Harvey PD, et al. The impact of KarXT on cognitive impairment in acute schizophrenia: replication in pooled data from two phase 3 trials. Am J Psychiatry. 2024, in press.
49. Hyman SE, Fenton WS. Medicine. What are the right targets for psychopharmacology? Science. 2003;299(5605):350–1. https://doi.org/10.1126/science.1077141
50. Kahn RS, Keefe RS. Schizophrenia is a cognitive illness: time for a change in focus. JAMA Psychiatry. 2013;70(10):1107–12. https://doi.org/10.1001/jamapsychiatry.2013.155.
51. Kaul I, Sawchak S, Correll CU, et al. Efficacy and safety of the muscarinic receptor agonist KarXT (xanomeline-trospium) in schizophrenia (EMERGENT-2) in the USA: results from a randomised, double-blind, placebo-controlled, flexible-dose phase 3 trial. Lancet. 2014;403(10422):160–70. https://doi.org/10.1016/S0140-6736(23)02190-6
52. Kaul I, Sawchak S, Walling DP, et al. Efficacy and safety of xanomeline-trospium chloride in schizophrenia: a randomized clinical trial. JAMA Psychiatry. 2024. https://doi.org/10.1001/jamapsychiatry.2024.0785
53. Keefe RSE. Why are there no approved treatments for cognitive impairment in schizophrenia? World Psychiatry. 2019;18(2):167–8. https://doi.org/10.1002/wps.20648.
54. Keefe RSE. Where do we cut the bell-shaped curve of CIAS? Schizophr Res. 2021;228:633–4. https://doi.org/10.1016/j.schres.2021.01.013.
55. Keefe RS, Bilder RM, Davis SM, Harvey PD, Palmer BW, Gold JM, Meltzer HY, Green MF, Capuano G, Stroup TS, McEvoy JP, Swartz MS, Rosenheck RA, Perkins DO, Davis CE, Hsiao JK, Lieberman JA; CATIE Investigators; Neurocognitive Working Group. Neurocognitive effects of antipsychotic medications in patients with chronic schizophrenia in the CATIE Trial. Arch Gen Psychiatry. 2007;64(6):633–47. https://doi.org/10.1001/archpsyc.64.6.633. PMID: 17548746.
56. Keefe RS, Eesley CE, Poe MP. Defining a cognitive function decrement in schizophrenia. Biol Psychiatry. 2005;57:688–91.

57. Keefe RS, Davis VG, Spagnola NB, et al. Reliability, validity and treatment sensitivity of the schizophrenia cognition rating scale. Eur Neuropsychopharmacol. 2015;25(2):176–84. https://doi.org/10.1016/j.euroneuro.2014.06.009.
58. Keefe RSE, Davis VG, Atkins AS, et al. Validation of a computerized test of functional capacity. Schizophr Res. 2016;175(1–3):90–6.
59. Lam M, Lee J, Rapisarda A, et al. Longitudinal cognitive changes in Young individuals at ultrahigh risk for psychosis. JAMA Psychiatry. 2018;75(9):929–39. https://doi.org/10.1001/jamapsychiatry.2018.1668.
60. Loebel AD, Siu CO, Cucchiaro JB, et al. Daytime sleepiness associated with lurasidone and quetiapine XR: results from a randomized double-blind, placebo-controlled trial in patients with schizophrenia. CNS Spectr. 2014;19(2):197–205. https://doi.org/10.1017/S1092852913000904.
61. Mantovani LM, Teixeira AL, Salgado JV. Functional capacity: a new framework for the assessment of everyday functioning in schizophrenia. Braz J Psychiatry. 2015;37(3):249–55. https://doi.org/10.1590/1516-4446-2014-1551.
62. Martin LF, Kem WR, Freedman R. Alpha-7 nicotinic receptor agonists: potential new candidates for the treatment of schizophrenia. Psychopharmacology. 2004;174(1):54–64. https://doi.org/10.1007/s00213-003-1750-1.
63. McGurk SR, Mueser KT, Xie H, et al. Cognitive remediation for vocational rehabilitation nonresponders. Schizophr Res. 2016;175(1–3):48–56. https://doi.org/10.1016/j.schres.2016.04.045.
64. McLaughlin D, Carrión RE, Auther AM, et al. Functional capacity assessed by the map task in individuals at clinical high-risk for psychosis. Schizophr Bull. 2016;42(5):1234–42. https://doi.org/10.1093/schbul/sbw039.
65. Millgate E, Smart SE, Pardiñas AF, et al. STRATA Consortium. Cognitive performance at first episode of psychosis and the relationship with future treatment resistance: Evidence from an international prospective cohort study. Schizophr Res. 2023;255:173–81. https://doi.org/10.1016/j.schres.2023.03.020.
66. Nelson B, Yuen HP, Wood SJ, et al. Long-term follow-up of a group at ultra high risk ("prodromal") for psychosis: the PACE 400 study. JAMA Psychiatry. 2013;70(8):793–802.
67. Nuechterlein KH, Green MF, Kern RS, et al. The MATRICS Consensus Cognitive Battery, part 1: test selection, reliability, and validity. Am J Psychiatry. 2008;165(2):203–13. https://doi.org/10.1176/appi.ajp.2007.07010042.
68. Palmer BW, Heaton RK, Paulsen JS, et al. Is it possible to be schizophrenic yet neuropsychologically normal? Neuropsychology. 1997;11(3):437–46. https://doi.org/10.1037/0894-4105.11.3.437.
69. Patterson TL, Goldman S, McKibbin CL, et al. UCSD Performance-Based Skills Assessment: development of a new measure of everyday functioning for severely mentally ill adults. Schizophr Bull. 2001;27(2):235–45.
70. Rabinowitz J, Werbeloff N, Caers I, et al. Negative symptoms in schizophrenia--the remarkable impact of inclusion definitions in clinical trials and their consequences. Schizophr Res. 2013;150:334–8. https://doi.org/10.1016/j.schres.2013.06.023.
71. Reichenberg A, Harvey PD, Bowie CR, et al. Neuropsychological function and dysfunction in schizophrenia and psychotic affective disorders. Schizophr Bull. 2009;35(5):1022–9. https://doi.org/10.1093/schbul/sbn044.
72. Rosenheck R, Leslie D, Keefe R, et al. Barriers to employment for people with schizophrenia. Am J Psychiatry. 2006;163(3):411–7. https://doi.org/10.1176/appi.ajp.163.3.411.
73. Sabbag S, Twamley EM, Vella L, et al. Assessing everyday functioning in schizophrenia: not all informants seem equally informative. Schizophr Res. 2011;131(1–3):250–5. https://doi.org/10.1016/j.schres.2011.05.003.
74. Salmi P, Isaacson R, Kull B. Dihydrexidine--the first full dopamine D1 receptor agonist. CNS Drug Rev. 2004;10(3):230–42. https://doi.org/10.1111/j.1527-3458.2004.tb00024.x.
75. Sauder C, Allen LA, Baker E, et al. Effectiveness of KarXT (xanomeline-trospium) for cognitive impairment in schizophrenia: post hoc analyses from a randomised, double-blind, placebo-controlled phase 2 study. Trans Psychiatry. 2023;12(1):491.

76. Seabury RD, Cannon TD. Memory Impairments and Psychosis Prediction: A Scoping Review and Theoretical Overview. Neuropsychol Rev. 2020;30(4):521–45. https://doi.org/10.1007/s11065-020-09464-2. Epub 2020 Nov 23. PMID: 33226539.
77. Seidman LJ, Shapiro DI, Stone WS, et al. Association of neurocognition with transition to psychosis: baseline functioning in the second phase of the North American prodrome longitudinal study. JAMA Psychiatry. 2016;73(12):1239–48. https://doi.org/10.1001/jamapsychiatry.2016.2479.
78. Shah S, Benson C, et al. Varenicline for cognitive impairment in people with schizophrenia: systematic review and meta-analysis. Psychopharmacology. 2020;237(1):11–9. https://doi.org/10.1007/s00213-019-05396-9.
79. Sheffield JM, Karcher NR, Barch DM. Cognitive deficits in psychotic disorders: a lifespan perspective. Neuropsychol Rev. 2018;28:509–33.
80. Shekhar A, Potter WZ, Lightfoot J, et al. Selective muscarinic receptor agonist xanomeline as a novel treatment approach for schizophrenia. Am J Psychiatry. 2008;165:1033–9.
81. Simon AE, Cattapan-Ludewig K, Zmilacher S, et al. Cognitive functioning in the schizophrenia prodrome. Schizophr Bull. 2007;33:761–71.
82. Strassnig MT, Raykov T, O'Gorman C, et al. Determinants of different aspects of everyday outcome in schizophrenia: the roles of negative symptoms, cognition, and functional capacity. Schizophr Res. 2015;165(1):76–82. https://doi.org/10.1016/j.schres.2015.03.033.
83. Velligan DI, Rubin M, Fredrick MM, et al. The cultural adaptability of intermediate measures of functional outcome in schizophrenia. Schizophr Bull. 2012;38(3):630–41.
84. Veroff AE, Bodick NC, Offen WW, et al. Efficacy of xanomeline in Alzheimer disease: cognitive improvement measured using the Computerized Neuropsychological Test Battery (CNTB). Alzheimer Dis Assoc Disord. 1998;12(4):304–12.
85. Vita A, Barlati S, Ceraso A, et al. Effectiveness, Core elements, and moderators of response of cognitive remediation for schizophrenia: a systematic review and meta-analysis of randomized clinical trials. JAMA Psychiatry. 2021;78(8):848–58. https://doi.org/10.1001/jamapsychiatry.2021.0620.
86. Young JW, Powell SB, Risbrough V, et al. Using the MATRICS to guide development of a preclinical cognitive test battery for research in schizophrenia. Pharmacol Ther. 2009;122(2):150–202. https://doi.org/10.1016/j.pharmthera.2009.02.004.
87. Zanelli J, Mollon J, Sandin S, et al. Cognitive change in schizophrenia and other psychoses in the decade following the first episode. Am J Psychiatry. 2019;176:811–9.

Cognitive Impairment in Schizophrenia: The Role of Cognitive Rehabilitation

13

Antonio Vita, Gabriele Nibbio, Irene Calzavara-Pinton, Lorenzo Bertoni, Stefano Paolini, Daniela Zardini, Antonio Baglioni, Laura Poddighe, Viola Bulgari, Anna Ceraso, Jacopo Lisoni, Giacomo Deste, and Stefano Barlati

Abstract

Cognitive impairment represents one of the core features of schizophrenia, with an important negative impact on several real-world functional outcomes: as such, it has become a key target in the treatment and in the rehabilitation process of schizophrenia spectrum disorders. Antipsychotic medications that are consistently effective in managing positive symptoms do not provide substantial benefits as regards cognitive performance. Furthermore, new medications that specifically target cognitive impairment in schizophrenia are not yet available for use in everyday clinical settings. In this scenario, psychosocial interventions that target cognitive performance represent the most effective treatment for cognitive impairment in people with schizophrenia. Among these, cognitive remediation represents the intervention with the most consistent evidence of effectiveness. In the present chapter, the features, the evidence of effectiveness, and the core elements of cognitive remediation, as well as future research and clinical implications, will be reported and discussed. Cognitive remediation currently represents the psychosocial intervention with the highest degree of recommendation in the European Psychiatric Association guidance dedicated to the treatment of

A. Vita (✉) · S. Barlati
Department of Clinical and Experimental Sciences, University of Brescia, Brescia, Italy

Department of Mental Health and Addiction Services, ASST Spedali Civili of Brescia, Brescia, Italy
e-mail: antonio.vita@unibs.it

G. Nibbio · L. Bertoni · S. Paolini · A. Baglioni · G. Deste
Department of Clinical and Experimental Sciences, University of Brescia, Brescia, Italy

I. Calzavara-Pinton · D. Zardini · L. Poddighe · V. Bulgari · A. Ceraso · J. Lisoni
Department of Mental Health and Addiction Services, ASST Spedali Civili of Brescia, Brescia, Italy

A. Vita et al. (eds.), *Negative and Cognitive Symptoms in Schizophrenia*, Comprehensive Approach to Psychiatry 3,
https://doi.org/10.1007/978-3-032-06209-3_13

cognitive impairment in schizophrenia. These recommendations are based on a wealth of evidence showing that cognitive remediation is effective not only in improving cognitive performance but also in improving the psychosocial functioning of people living with schizophrenia.

Keywords

Schizophrenia · Cognition · Cognitive remediation · Effectiveness · Functioning

13.1 Background

Cognitive impairment represents one of the core features of schizophrenia, with an important negative impact on several real-world functional outcomes: as such, it has become a key target in the treatment and in the rehabilitation process of schizophrenia spectrum disorders [25, 30, 46, 47, 84]. However, antipsychotic medications that are consistently effective in reducing and resolving positive symptoms do not provide substantial benefits as regards cognitive performance [18], and first-generation molecules and medications with a significant anticholinergic component might also worsen the impairment [3, 38, 39]. Medications that specifically target cognitive impairment in schizophrenia are currently being developed, and their effectiveness is being investigated in clinical trials in different phases [78, 89], but they are currently not available for use in everyday clinical settings. In this context, psychosocial interventions that target cognitive performance currently represent the most viable and effective treatment for cognitive impairment in people living with schizophrenia [6, 55, 70]. Among these, Cognitive Remediation (CR) represents the intervention with the most consistent evidence of effectiveness. In the present chapter, the features, the evidence of effectiveness, and the core elements of CR, as well as future research and clinical implications, will be reported and discussed.

13.2 What Is Cognitive Remediation?

According to the definition of the Cognitive Remediation Experts Workshop, updated in 2023, CR is "an intervention targeting cognition (attention, memory, executive function, social cognition, or metacognition) using scientific principles of learning with the goal of improving functional outcomes" [94]. This definition was first produced in 2012 and remained substantially unchanged in the following years, with the exception of the term "cognition" and its included domains replacing the previous terms of "cognitive deficits". This was due to different factors: the lack of a clear definition of cognitive deficits, the fact that most clinical trials on CR did not specifically include individuals showing cognitive deficits, and the need to highlight that even participants without prominent cognitive impairment may consistently benefit from the treatment.

Besides stating that CR is a psychosocial intervention and that its target is cognition, this definition includes two other elements of relevance: the first is that CR is based on scientific principles of learning, and the second is that the ultimate goal is the improvement of real-world psychosocial functioning. This emphasis on psychosocial functioning is of relevance, as it represent a treatment target of utmost importance in the rehabilitation and recovery process of schizophrenia not only in the perspective of researchers and clinicians but also most importantly in the perspectives of mental health service users living with schizophrenia [13, 19, 46, 57, 81, 86].

As regards the learning technique, CR is a training-based intervention, and the learning process should be provided in the most scientifically effective manner, including the principles of scaffolding, errorless learning, and massed practice [76]. This means that the learning process is based on exercises that are cognitively challenging and stimulating for the participant, without representing an impossible or excessively difficult endeavor: in this regard, the difficulty of exercises should be continuously monitored and adjusted. The exercises should also be repeated several times in order to provide an effective learning process, and the difficulty should be increased only when the previous steps have been completely acquired by the participants.

Besides these defining elements, CR interventions can take a wide variety of forms: they can be delivered individually or in a group context, include the use of a computer or pencil-and-paper exercises only, focus on different neurocognitive domains or include elements of social cognition and metacognition. The intensity of hours and sessions per week and the overall duration of the interventions also vary between different programs and protocols.

One element that is remarkably not included in the definition of CR is the fact that it should be offered to service users diagnosed with schizophrenia, but this is no oversight. In fact, the earliest iterations of CR interventions were devised for patients with traumatic brain injuries, and only in the 1990s were they being investigated in trials including participants diagnosed with schizophrenia [11, 34, 48, 71, 91]. More recently, CR interventions (or interventions based on CR principles) have shown consistent evidence of effectiveness in severe mental illness besides schizophrenia, including major depressive disorder [27, 44, 72] and bipolar disorder [9, 50, 74, 80]. As individuals living with conditions that are not schizophrenia appear to benefit from CR interventions, limiting the definition of CR to the diagnosis of schizophrenia does not appear appropriate. However, it is also true that schizophrenia spectrum disorders represent the field where CR interventions have the largest scientific basis of evidence of effectiveness.

13.3 Evidence of Effectiveness

CR currently represents the psychosocial intervention with the highest degree of recommendation in the European Psychiatric Association guidance dedicated to the treatment of cognitive impairment in schizophrenia [84]. In fact, Recommendation 2a states that "Cognitive remediation is recommended for the treatment of cognitive impairment in people living with schizophrenia" (Grade of Recommendation: A), and Recommendation 2b states that "Social cognitive training is recommended for the treatment of social cognitive deficits in people living with schizophrenia" (Grade of Recommendation: A).

Other national and international guidelines focused on the treatment of schizophrenia or on the psychiatric rehabilitation process also mention CR as a treatment for cognitive impairment [31, 41, 53].

These recommendations are based on a wealth of evidence showing that CR is effective not only in improving cognitive performance but also in improving the psychosocial functioning of people living with schizophrenia.

One of the first systematic assessments of the effectiveness of CR was conducted by the research group of Til Wykes and was published in 2011 [92]. This systematic review and meta-analysis included 40 different randomized clinical trials on the effectiveness of cognitive remediation in people living with schizophrenia and a total of 2104 participants. The meta-analyses showed a statistically significant, moderate-sized positive effect of CR on global cognition (Cohen's $d = 0.45$, $p < 0.001$) and global functioning (Cohen's $d = 0.42$, $p = 0.003$). Similar effects were observed in the different cognitive domains when considered separately, and the positive effects were significantly persistent in follow-up observations.

Since this seminal and impactful work, which in the following years became a point of reference in the context of cognition-oriented treatments and of schizophrenia rehabilitation, several other meta-analyses have explored specific aspects of CR, such as its effectiveness in patients with recent onset [64] or in inpatient settings [12], the effects of computer-assisted CR [40] or the role of integrating CR in psychiatric rehabilitation [75].

More recently, another systematic review and meta-analysis investigated the effectiveness of CR in a comprehensive manner: this research updated and expanded the results of the 2011 work, including ten more years of results [83]. This systematic review was conducted using broad inclusion criteria in order to provide a comprehensive overview of CR trials and to investigate the role of treatment- and participant-related moderators of effect: this led to the inclusion of 194 reports on 130 different studies, with a total of 8851 participants. The meta-analyses showed that CR interventions provided significant improvements in both primary outcomes: small-to-moderate positive effects were observed for global cognitive performance (Cohen's $d = 0.29, p < 0.001$) and functioning (Cohen's $d = 0.22, p < 0.001$). Similar effects were observed in the analyses exploring single cognitive domains.

Another systematic review and meta-analysis on the effects of CR for schizophrenia was conducted in a completely independent manner by another research group and published in the same year [45]. This systematic review was also conducted using a methodological approach that was similar to that of previous works, but presented more strict inclusion criteria as it included only CR intervention focused on neurocognitive performance and excluded social cognition training interventions. Interestingly, very similar results were presented, with significant positive effects of small-to-moderate effect size in both global cognitive performance and psychosocial functioning.

Other recent meta-analyses conducted with a comprehensive approach focused on specific aspects of CR trials. One systematic review and meta-analysis investigated the acceptability of CR interventions in schizophrenia by comparing drop-out rates in treatment and control arms of CR trials [85]. Including 151 studies and a total of 10,477 participants, a 16.58% dropout rate for CR programs was observed, confirming that CR has a good acceptability profile for mental health service users diagnosed with schizophrenia, in line with that of other evidence-based psychosocial interventions.

One systematic review and meta-analysis focused on the durability of effects of CR interventions and analyzed the effects reported in follow-up observation of CR trials [88]. This work included 130 reports on 67 studies with a total of 5334 participants and compared baseline and end-of follow-up observations: significant positive effects were observed in both global cognitive performance (Cohen's $d = 0.23$, $p < 0.001$) and functioning (Cohen's $d = 0.26$, $p < 0.001$), showing that the positive effects of CR interventions are durable and persist at the conclusion of follow-up observations. To further test this hypothesis, in this investigation, the results of follow-up observations were also compared to end-of-treatment observations: indeed, no significant difference was observed, confirming the stability and durability of CR positive effects.

Another systematic review and meta-analysis focused on social cognition-oriented CR and social cognition training programs, including 42 trials and 1868 participants [95]: significant positive effects were observed in different social cognition domains, and small signals of generalization to neurocognitive domains such as executive functions were also observed.

Interestingly, evidence of CR effectiveness in people with schizophrenia has also been supported by research on neurobiological correlates: CR has been shown to produce significant effects both in neuroimaging studies [59, 61] and in increasing Brain-Derived Neurotrophic Factor (BDNF) [22, 60].

It could be argued that more recent meta-analyses, such as that of Vita et al. and that of Lejeune et al., reported significant positive effects of CR interventions, but the effect sizes were considerably smaller than those reported by Wykes et al. in [92]. This could be due to the inclusion of a much larger number of studies, increasing the heterogeneity of study results and therefore of the analyses. In fact, exploring the reasons beyond the heterogeneity of study results by analyzing moderator effects revealed very interesting findings, particularly related to those that can be defined as the "core elements" (or "core ingredients") of CR effectiveness.

13.4 Core Elements and Moderators of Response

Besides the elements included in the definition of CR, the Cognitive Remediation Experts Workshop also worked to identify which were the core elements of the effectiveness of CR. As reported in the expert working group white paper, four elements were theorized as essential to obtain positive effects for people living with schizophrenia in real-world clinical contexts [10]. These elements are the repeated practice of cognitive exercises, the presence of an active and trained therapist delivering the intervention, the structured development of novel cognitive strategies, and the use of techniques to improve the transfer of cognitive gains to real-world functioning.

The repeated practice of cognitive exercises is the foundation of CR as per its definition, and this element encompasses the aforementioned scientific principles of learning that act as the basis of CR effectiveness. Exercises should be explained and tested with the participant before training, and orientation should be provided on how to navigate the computer program or how to engage with non-computerized tasks. Multiple repetition of exercises is encouraged to provide intensive training aimed at producing meaningful effects, and the difficulty of the exercises should be adjusted and gradually increased to always represent an engaging and challenging cognitive stimulus. In this regard, feedback should be provided to the participant during the training session in order to better understand the exercises and the program.

Regardless of the modality (computer or pen and paper) and setting (individual or group) of delivery, CR intervention should be delivered by a therapist who should be specifically trained for the task. The therapist should work with the participants and help him/her formulate cognitive problems and link these problems to specific goals. The therapist should also be actively engaged alongside the participant in monitoring the progress toward goals, identifying strengths and barriers, and adjusting both short- and long-term goals to optimize the results of the intervention. Moreover, the therapist can deliver effective feedback to participants during the exercise and provide praise, producing a positive reinforcement for participation and training; this praise should be directed toward the training process, for points such as staying engaged with the task and attempting new strategies, rather than on the participant's performance in the specific exercise.

The development of novel strategies represents another focal point of CR, particularly considering the aspects of cognitive rigidity and inflexibility that can be often observed as an issue alongside performance deficits in schizophrenia. In this regard, the cognitive exercises and tasks should provide opportunities to identify inefficient strategies and foster the development and monitoring of new and more effective strategies by the participant. In this regard, the presence of an active and trained therapist could be synergic, as the therapist could suggest potentially effective new strategies and monitor the participant's reaction and progress.

Finally, the CR intervention should include elements to transfer cognitive gains to real-world psychosocial functioning. In this regard, cognitive goals with clear links to personal and community functioning should be collaboratively defined with the participant. Promoting realistic expectations regarding the resolution of cognitive problems linked to specific functional goals could help foster a positive process of change and increase the real-world effectiveness of the CR intervention. An optimal way of promoting the transfer of cognitive gains into real-world functioning is the integration of the CR intervention with a structured psychiatric rehabilitation program [75]. In this context, CR could also be combined with other evidence-based psychosocial interventions, improving the effectiveness of the interventions on several different outcomes [15, 56, 84].

The relevance of these elements was not only highlighted in a theoretical manner, but was also empirically investigated in the meta-analysis by Vita et al., published in 2021 [83]. In fact, with the exception of the repeated practice of cognitive exercise that was present in almost all trials and could not represent a discriminating feature in subgroup analyses, all elements emerged as a significant moderator of CR effectiveness. A secondary analysis was conducted including only the trials on interventions featuring all four core elements: in this instance, significant positive effects were observed for both global cognitive performance (Cohen's $d = 0.40$, $p < 0.001$) and psychosocial functioning (Cohen's $d = 0.43$, $p < 0.001$), and the size of the effects increased to a moderate range, akin to those observed by Wykes et al., in [92]. These observations remained unchanged in a sub-analysis conducted, including studies featuring a high level of methodological quality. Conversely, the mode and setting of delivery of the intervention did not emerge as moderators of treatment effects.

Considering these findings, the core elements of CR effectiveness should always be included in CR programs to maximize the benefits for participants and to improve the real-world recovery process of mental health service users living with a diagnosis of schizophrenia.

Participant characteristics that have an impact on treatment effectiveness were also investigated in recent meta-analyses, as identifying individuals that would benefit in a more significant way from CR programs could represent a considerably useful information to provide this intervention in mental health and rehabilitation services in everyday clinical practice [5, 65, 68]. Fewer years of education, lower premorbid IQ, and higher symptom severity all emerged as significant moderators of positive CR treatment effect [83]. In this regard, CR represents an intervention that can provide substantial benefits even in more clinically compromised participants.

Age of participants did not emerge as a significant moderator of treatment effects in two recent meta-analyses [45, 83]. These results are in line with those of a recent dedicated research: a single-participant data meta-analysis including databases from 14 different studies and a total of 1084 participants, focused specifically on the age of participants as a potential moderator of CR response [69]. In this dedicated investigation, no significant effect of age was observed on most CR treatment outcomes, aside from a small effect of younger age on executive functioning gains. Considering these results, CR appears as an effective intervention regardless of the age of participants, and even older individuals might benefit from receiving CR treatment. However, in a longitudinal perspective, offering CR in younger, early course of the illness individuals and in early intervention services could represent an effective strategy, granting earlier cognitive and functional improvements and allowing for a better opportunity of reaching recovery and personal life goals for participants [4, 20, 24, 79].

13.5 Future Perspectives

Considering this wealth of evidence, the effectiveness of CR can be considered as well-established. However, CR programs, as well as other evidence-based psychosocial interventions, are still implemented unevenly and in a piecemeal manner in mental health services in real-world clinical settings, even in high-income countries and in contexts with high levels of available resources [82, 90, 93]. In fact, results from recent studies show that CR interventions can be effectively implemented and produce significant positive effects in clinical settings using currently available resources [51, 54], and CR programs have been successfully delivered also in contexts with very limited resources and in low-income countries such as India [33], Iran [32], Togo and Benin [16, 87].

Better understanding the barriers and facilitators to a widespread dissemination and implementation of CR programs, as well as other evidence-based psychosocial interventions for people living with schizophrenia, represents an important perspective for future research. According to available data, barriers hindering the implementation of CR programs appear to be mostly related to the limited understanding of the benefits of this type of intervention on the part of mental health professionals, of public health services, of policymaker organizations, and of regulatory entities [1, 2, 90, 94, 96]. In this regard, academic and clinical partnership and collaboration, as well as advocacy from service users, families, carers, and stakeholder organizations, can represent valuable instruments to improve implementation, as shown in recent literature [14, 66]. All in all, more effort is required to better inform and better disseminate knowledge regarding the impact and the importance of improving cognitive performance in people living with schizophrenia, and to deliver effective cognition-targeting treatments in mental health services.

Another area that could represent a field of interest for future research is the use of novel technologies to deliver and implement CR interventions.

Among these, telemedicine represents one of the recent developments that shows greater promise, as remote delivery of CR interventions could improve their dissemination and accessibility for users [36]. In fact, the acceptability profile and the effectiveness of remotely delivered CR programs have been the focus of several studies [8, 17, 21, 28, 29, 52, 77].

A systematic review and meta-analysis recently showed that remotely delivered evidence-based psychosocial interventions for schizophrenia present a good efficacy profile: in particular, remotely delivered CR interventions can provide significant improvements in cognitive and functional outcomes [7]. CR programs delivered via smartphone are also being developed, and preliminary evidence of their effectiveness has been reported [23, 43]. This represents another avenue for future research.

Virtual reality represents another technology that has recently seen novel interest in both research settings and in public opinion. In fact, it has been hypothesized that the use of virtual reality could represent an effective instrument to deliver treatments for different conditions [42, 49, 58], and evidence of the feasibility and efficacy of virtual reality psychosocial interventions for people living with schizophrenia is emerging [26, 67]. A recent systematic review and meta-analysis focused on virtual reality CR interventions; however, among the included studies, only one included a sample of participants diagnosed with schizophrenia [63].

While findings in this field cannot be considered as conclusive, virtual reality media could represent promising instruments to provide immersive and ecologically valid CR interventions. In fact, several trials are currently underway to accurately investigate the efficacy and effectiveness of virtual reality CR for people living with schizophrenia [37, 73], as well as with other severe mental health conditions [35, 62].

Future steps in this area of research include comparing virtual reality-based CR interventions to already established programs, considering efficacy, effectiveness, acceptability, participant engagement, and subjective benefits and assessing cost-effectiveness outcomes of virtual reality CR.

References

1. Altman RAE, Tan EJ, Rossell SL. Factors impacting access and engagement of cognitive remediation therapy for people with schizophrenia: a systematic review. Can J Psychiatr. 2023;68:139–51. https://doi.org/10.1177/07067437221129073.
2. Altman RAE, Reser M, Tan EJ, Rossell SL. Cognitive remediation for schizophrenia: clinician perspectives on implementation barriers and facilitators. Rehabil Psychol. 2024;69:171–83. https://doi.org/10.1037/rep0000552.
3. Ang MS, Abdul Rashid NA, Lam M, Rapisarda A, Kraus M, Keefe RSE, Lee J. The impact of medication anticholinergic burden on cognitive performance in people with schizophrenia. J Clin Psychopharmacol. 2017;37:651. https://doi.org/10.1097/JCP.0000000000000790.
4. Barlati S, de Peri L, Deste G, Fusar-Poli P, Vita A. Cognitive remediation in the early course of schizophrenia: a critical review. Curr Pharm Des. 2012;18:534–41. https://doi.org/10.2174/138161212799316091.
5. Barlati S, Deste G, Galluzzo A, Perin AP, Valsecchi P, Turrina C, Vita A. Factors associated with response and resistance to cognitive remediation in schizophrenia: a critical review. Front Pharmacol. 2019;9 https://doi.org/10.3389/fphar.2018.01542.
6. Barlati S, Nibbio G, Vita A. Evidence-based psychosocial interventions in schizophrenia: a critical review. Curr Opin Psychiatry. 2024; https://doi.org/10.1097/YCO.0000000000000925.
7. Best MW, Romanowska S, Zhou Y, Wang L, Leibovitz T, Onno KA, Jagtap S, Bowie CR. Efficacy of remotely delivered evidence-based psychosocial treatments for schizophrenia-Spectrum disorders: a series of systematic reviews and meta-analyses. Schizophr Bull. 2023;49:973–86. https://doi.org/10.1093/schbul/sbac209.
8. Biagianti B, Fisher M, Howard L, Rowlands A, Vinogradov S, Woolley J. Feasibility and preliminary efficacy of remotely delivering cognitive training to people with schizophrenia using tablets. Schizophr Res Cogn. 2017;10:7–14. https://doi.org/10.1016/j.scog.2017.07.003.
9. Bonnin CM, Reinares M, Martínez-Arán A, Balanzá-Martínez V, Sole B, Torrent C, Tabarés-Seisdedos R, García-Portilla MP, Ibáñez A, Amann BL, Arango C, Ayuso-Mateos JL, Crespo JM, González-Pinto A, Colom F, Vieta E, Group, T.C.F.R. Effects of functional remediation on neurocognitively impaired bipolar patients: enhancement of verbal memory. Psychol Med. 2016;46:291–301. https://doi.org/10.1017/S0033291715001713.
10. Bowie CR, Bell MD, Fiszdon JM, Johannesen JK, Lindenmayer J-P, McGurk SR, Medalia AA, Penadés R, Saperstein AM, Twamley EW, Ueland T, Wykes T. Cognitive remediation for schizophrenia: an expert working group white paper on core techniques. Schizophr Res. 2020;215:49–53. https://doi.org/10.1016/j.schres.2019.10.047.

11. Burda PC, Starkey TW, Dominguez F, Vera V. Computer-assisted cognitive rehabilitation of chronic psychiatric inpatients. Comput Hum Behav. 1994;10:359–68.
12. Cella M, Price T, Corboy H, Onwumere J, Shergill S, Preti A. Cognitive remediation for inpatients with psychosis: a systematic review and meta-analysis. Psychol Med. 2020;50:1062–76. https://doi.org/10.1017/S0033291720000872.
13. Correll CU, Ismail Z, McIntyre RS, Rafeyan R, Thase ME. Patient functioning, life engagement, and treatment goals in schizophrenia. J Clin Psychiatry. 2022;83:42345. https://doi.org/10.4088/JCP.LU21112AH2.
14. Dark FL, Amado I, Erlich MD, Ikezawa S. International experience of implementing cognitive remediation for people with psychotic disorders. Schizophr Bull. 2024:sbae071. https://doi.org/10.1093/schbul/sbae071.
15. Deste G, Corbo D, Nibbio G, Italia M, Dell'Ovo D, Calzavara-Pinton I, Lisoni J, Barlati S, Gasparotti R, Vita A. Impact of physical exercise alone or in combination with cognitive remediation on cognitive functions in people with schizophrenia: a qualitative critical review. Brain Sci. 2023;13:320. https://doi.org/10.3390/brainsci13020320.
16. Deste G, Kakli M, Barlati S, Nibbio G, Dossou P, Degila SL, Ceraso A, Lisoni J, Calzavara-Pinton I, Villa S, Vita A. Application of cognitive remediation in the world: new experiences from two schizophrenia rehabilitation centers in Togo and Benin. Soc Psychiatry Psychiatr Epidemiol. 2024; https://doi.org/10.1007/s00127-023-02603-z.
17. Donohoe G, Dillon R, Hargreaves A, Mothersill O, Castorina M, Furey E, Fagan AJ, Meaney JF, Fitzmaurice B, Hallahan B, McDonald C, Wykes T, Corvin A, Robertson IH. Effectiveness of a low support, remotely accessible, cognitive remediation training programme for chronic psychosis: cognitive, functional and cortical outcomes from a single blind randomised controlled trial. Psychol Med. 2018;48:751–64. https://doi.org/10.1017/S0033291717001982.
18. Feber L, Peter N, Schneider-Thoma J, Siafis S, Bighelli I, Hansen W-P, Prates Baldez D, Salanti G, Keefe RSE, Engel RR, Leucht S. Antipsychotic drugs and their effects on cognitive function: protocol for a systematic review, pairwise, and network meta-analysis. Syst Rev. 2023;12:54. https://doi.org/10.1186/s13643-023-02213-5.
19. Fiorillo A, Sartorius N. Mortality gap and physical comorbidity of people with severe mental disorders: the public health scandal. Ann General Psychiatry. 2021;20:52. https://doi.org/10.1186/s12991-021-00374-y.
20. Fisher M, Loewy R, Hardy K, Schlosser D, Vinogradov S. Cognitive interventions targeting brain plasticity in the prodromal and early phases of schizophrenia. Annu Rev Clin Psychol. 2013;9:435–63. https://doi.org/10.1146/annurev-clinpsy-032511-143134.
21. Fisher M, Loewy R, Carter C, Lee A, Ragland JD, Niendam T, Schlosser D, Pham L, Miskovich T, Vinogradov S. Neuroplasticity-based auditory training via laptop computer improves cognition in young individuals with recent onset schizophrenia. Schizophr Bull. 2015;41:250–8. https://doi.org/10.1093/schbul/sbt232.
22. Fisher M, Mellon SH, Wolkowitz O, Vinogradov S. Neuroscience-informed auditory training in schizophrenia: a final report of the effects on cognition and serum brain-derived neurotrophic factor. Schizophr Res Cogn. 2016;3:1–7. https://doi.org/10.1016/j.scog.2015.10.006.
23. Fisher M, Etter K, Murray A, Ghiasi N, LaCross K, Ramsay I, Currie A, Fitzpatrick K, Biagianti B, Schlosser D, Loewy R, Vinogradov S. The effects of remote cognitive training combined with a Mobile app intervention on psychosis: double-blind randomized controlled trial. J Med Internet Res. 2023;25:e48634. https://doi.org/10.2196/48634.
24. Frawley E, Heary C, Berry C, Cella M, Fowler D, Wykes T, Donohoe G. Participant perspectives on cognitive remediation and social recovery in early psychosis (CReSt-R): an acceptability study. Early Interv Psychiatry. 2024;18:34–41. https://doi.org/10.1111/eip.13424.
25. Galderisi S, Rucci P, Kirkpatrick B, Mucci A, Gibertoni D, Rocca P, Rossi A, Bertolino A, Strauss GP, Aguglia E, Bellomo A, Murri MB, Bucci P, Carpiniello B, Comparelli A, Cuomo A, De Berardis D, Dell'Osso L, Di Fabio F, Gelao B, Marchesi C, Monteleone P, Montemagni C, Orsenigo G, Pacitti F, Roncone R, Santonastaso P, Siracusano A, Vignapiano A, Vita A, Zeppegno P, Maj M, Italian Network for Research on Psychoses. Interplay among psychopathologic variables, personal resources, context-related factors, and real-life functioning

in individuals with schizophrenia: a network analysis. JAMA Psychiatry. 2018;75:396–404. https://doi.org/10.1001/jamapsychiatry.2017.4607.

26. Gillouin P-A, Mattatia D, Bouvet C. Virtual reality for clinical evaluation and treatment in schizophrenia: a systematic review. Psychosis. 2024;16:212–29. https://doi.org/10.1080/17522439.2023.2197030.
27. Hammar Å, Ronold EH, Rekkedal GÅ. Cognitive impairment and neurocognitive profiles in major depression-a clinical perspective. Front Psych. 2022;13:764374. https://doi.org/10.3389/fpsyt.2022.764374.
28. Hargreaves A, Dillon R, Anderson-Schmidt H, Corvin A, Fitzmaurice B, Castorina M, Robertson IH, Donohoe G. Computerised working-memory focused cognitive remediation therapy for psychosis–A preliminary study. Schizophr Res. 2015;169:135–40. https://doi.org/10.1016/j.schres.2015.09.004.
29. Harris AW, Kosic T, Xu J, Walker C, Gye W, Redoblado Hodge A. Web-based cognitive remediation improves supported employment outcomes in severe mental illness: randomized controlled trial. JMIR Ment Health. 2017;4:e30. https://doi.org/10.2196/mental.6982.
30. Harvey PD, Bosia M, Cavallaro R, Howes OD, Kahn RS, Leucht S, Müller DR, Penadés R, Vita A. Cognitive dysfunction in schizophrenia: an expert group paper on the current state of the art. Schizophr Res Cogn. 2022;29:100249. https://doi.org/10.1016/j.scog.2022.100249.
31. Hasan A, Falkai P, Wobrock T, Lieberman J, Glenthoj B, Gattaz WF, Thibaut F, Möller H-J. World Federation of Societies of Biological Psychiatry (WFSBP) guidelines for biological treatment of schizophrenia, part 1: update 2012 on the acute treatment of schizophrenia and the management of treatment resistance. World J Biol Psychiatry. 2012;13:318–78. https://doi.org/10.3109/15622975.2012.696143.
32. Hatami S, Mirsepassi Z, Sedighnia A, Tehranidoost M, Masoomi M, Sharifi V. A short course computer-assisted cognitive remediation in patients with schizophrenia spectrum disorders: a randomized clinical trial. Basic Clin Neurosci. 2021;12:551–62. https://doi.org/10.32598/bcn.2021.3112.1.
33. Hegde S. A review of Indian research on cognitive remediation for schizophrenia. Asian J Psychiatr. 2017;25:54–9. https://doi.org/10.1016/j.ajp.2016.10.001.
34. Hermanutz M, Gestrich J. Computer-assisted attention training in schizophrenics: a comparative study. Eur Arch Psychiatry Clin Neurosci. 1991;240:282–7.
35. Hernandez MEH, Michalak E, Choudhury N, Hewko M, Torres I, Menon M, Lam RW, Chakrabarty T. Co-design of a virtual reality cognitive remediation program for depression (bWell-D) with patient end users and clinicians: qualitative interview study among patients and clinicians. JMIR Serious Games. 2023;11:e43904. https://doi.org/10.2196/43904.
36. Jagtap S, Romanowska S, Leibovitz T, Onno KA, Burhan AM, Best MW. Can cognitive remediation therapy be delivered remotely? A review examining feasibility and acceptability of remote interventions. Schizophr Res Cogn. 2022;28:100238. https://doi.org/10.1016/j.scog.2022.100238.
37. Jespersen AE, Lumbye A, Vinberg M, Glenthøj L, Nordentoft M, Wæhrens EE, Knudsen GM, Makransky G, Miskowiak KW. Effect of immersive virtual reality-based cognitive remediation in patients with mood or psychosis spectrum disorders: study protocol for a randomized, controlled, double-blinded trial. Trials. 2024;25:82. https://doi.org/10.1186/s13063-024-07910-7.
38. Joshi YB. Cholinergic functioning, cognition, and anticholinergic medication burden in schizophrenia. In: Barch DM, Young JW, editors. Cognitive functioning in schizophrenia: leveraging the RDoC framework, current topics in behavioral neurosciences. Cham: Springer International Publishing; 2023. p. 393–406. https://doi.org/10.1007/7854_2022_400.
39. Joshi YB, Thomas ML, Braff DL, Green MF, Gur RC, Gur RE, Nuechterlein KH, Stone WS, Greenwood TA, Lazzeroni LC, MacDonald LR, Molina JL, Nungaray JA, Radant AD, Silverman JM, Sprock J, Sugar CA, Tsuang DW, Tsuang MT, Turetsky BI, Swerdlow NR, Light GA. Anticholinergic medication burden–associated cognitive impairment in schizophrenia. AJP. 2021;178:838–47. https://doi.org/10.1176/appi.ajp.2020.20081212.
40. Kambeitz-Ilankovic L, Betz LT, Dominke C, Haas SS, Subramaniam K, Fisher M, Vinogradov S, Koutsouleris N, Kambeitz J. Multi-outcome meta-analysis (MOMA) of cognitive

remediation in schizophrenia: revisiting the relevance of human coaching and elucidating interplay between multiple outcomes. Neurosci Biobehav Rev. 2019;107:828–45. https://doi.org/10.1016/j.neubiorev.2019.09.031.
41. Keepers GA, Fochtmann LJ, Anzia JM, Benjamin S, Lyness JM, Mojtabai R, Servis M, Walaszek A, Buckley P, Lenzenweger MF, Young AS, Degenhardt A, Hong S-H. The American Psychiatric Association practice guideline for the treatment of patients with schizophrenia. AJP. 2020;177:868–72. https://doi.org/10.1176/appi.ajp.2020.177901.
42. Kim S, Kim E. The use of virtual reality in psychiatry: a review. Soa Chongsonyon Chongsin Uihak. 2020;31:26–32. https://doi.org/10.5765/jkacap.190037.
43. Krzystanek M, Krysta K, Borkowski M, Skałacka K, Przybyło J, Pałasz A, Mucic D, Martyniak E, Waszkiewicz N. The effect of smartphone-based cognitive training on the functional/cognitive markers of schizophrenia: a one-year randomized study. J Clin Med. 2020;9:3681. https://doi.org/10.3390/jcm9113681.
44. Legemaat AM, Semkovska M, Brouwer M, Geurtsen GJ, Burger H, Denys D, Bockting CL. Effectiveness of cognitive remediation in depression: a meta-analysis. Psychol Med. 2022;52:4146–61. https://doi.org/10.1017/S0033291721001100.
45. Lejeune JA, Northrop A, Kurtz MM. A meta-analysis of cognitive remediation for schizophrenia: efficacy and the role of participant and treatment factors. Schizophr Bull. 2021;47:997–1006. https://doi.org/10.1093/schbul/sbab022.
46. Maj M, van Os J, De Hert M, Gaebel W, Galderisi S, Green MF, Guloksuz S, Harvey PD, Jones PB, Malaspina D, McGorry P, Miettunen J, Murray RM, Nuechterlein KH, Peralta V, Thornicroft G, van Winkel R, Ventura J. The clinical characterization of the patient with primary psychosis aimed at personalization of management. World Psychiatry. 2021;20:4–33. https://doi.org/10.1002/wps.20809.
47. McCutcheon RA, Keefe RSE, McGuire PK. Cognitive impairment in schizophrenia: aetiology, pathophysiology, and treatment. Mol Psychiatry. 2023:1–17. https://doi.org/10.1038/s41380-023-01949-9.
48. Medalia A, Aluma M, Tryon W, Merriam AE. Effectiveness of attention training in schizophrenia. Schizophr Bull. 1998;24:147–52.
49. Mishkind MC, Norr AM, Katz AC, Reger GM. Review of virtual reality treatment in psychiatry: evidence versus current diffusion and use. Curr Psychiatry Rep. 2017;19:80. https://doi.org/10.1007/s11920-017-0836-0.
50. Miskowiak KW, Seeberg I, Jensen MB, Balanzá-Martínez V, Del Mar Bonnin C, Bowie CR, Carvalho AF, Dols A, Douglas K, Gallagher P, Hasler G, Lafer B, Lewandowski KE, López-Jaramillo C, Martinez-Aran A, McIntyre RS, Porter RJ, Purdon SE, Schaffer A, Stokes P, Sumiyoshi T, Torres IJ, Van Rheenen TE, Yatham LN, Young AH, Kessing LV, Burdick KE, Vieta E. Randomised controlled cognition trials in remitted patients with mood disorders published between 2015 and 2021: a systematic review by the International Society for Bipolar Disorders Targeting Cognition Task Force. Bipolar Disord. 2022;24:354–74. https://doi.org/10.1111/bdi.13193.
51. Montemagni C, Del Favero E, Riccardi C, Canta L, Toye M, Zanalda E, Rocca P. Effects of cognitive remediation on cognition, metacognition, and social cognition in patients with schizophrenia. Front Psych. 2021;12:649737. https://doi.org/10.3389/fpsyt.2021.649737.
52. Moura BM, Avila A, Chendo I, Frade P, Barandas R, Vian J, Croca M, Roberto A, Almeida C, Antunes F, Linhares L, Crawford J, Ferreira C, Santos J, Abreu M, Levy P, Figueira ML, Mendes T. Facilitating the delivery of cognitive remediation in first-episode psychosis: pilot study of a home-delivered web-based intervention. J Nerv Ment Dis. 2019;207:951–7. https://doi.org/10.1097/NMD.0000000000001055.
53. National Institute for Health and Care Excellence. Overview. Rehabilitation for adults with complex psychosis. Guidance. NICE [WWW Document]. 2020. https://www.nice.org.uk/guidance/ng181. Accessed 29 July 24.
54. Nibbio G, Barlati S, Cacciani P, Corsini P, Mosca A, Ceraso A, Deste G, Vita A. Evidence-based integrated intervention in patients with schizophrenia: a pilot study of

feasibility and effectiveness in a real-world rehabilitation setting. Int J Environ Res Public Health. 2020;17:3352. https://doi.org/10.3390/ijerph17103352.

55. Nibbio G, Pinton IC, Bertoni L, Cicale A, Necchini N, Zardini D, Poddighe L, Lisoni J, Deste G, Barlati S, Vita A. Non-pharmacological interventions to improve cognitive impairment and psychosocial functioning in schizophrenia: a critical review. J Psychopathol. 2024;30 https://doi.org/10.36148/2284-0249-N455.
56. Nuechterlein KH, McEwen SC, Ventura J, Subotnik KL, Turner LR, Boucher M, Casaus LR, Distler MG, Hayata JN. Aerobic exercise enhances cognitive training effects in first-episode schizophrenia: randomized clinical trial demonstrates cognitive and functional gains. Psychol Med. 2022:1–11. https://doi.org/10.1017/S0033291722001696.
57. Onitsuka T, Hirano Y, Nakazawa T, Ichihashi K, Miura K, Inada K, Mitoma R, Yasui-Furukori N, Hashimoto R. Toward recovery in schizophrenia: current concepts, findings, and future research directions. Psychiatry Clin Neurosci. 2022;76:282–91. https://doi.org/10.1111/pcn.13342.
58. Park MJ, Kim DJ, Lee U, Na EJ, Jeon HJ. A literature overview of Virtual Reality (VR) in treatment of psychiatric disorders: recent advances and limitations. Front Psych. 2019;10 https://doi.org/10.3389/fpsyt.2019.00505.
59. Penadés R, Pujol N, Catalán R, Massana G, Rametti G, García-Rizo C, Bargalló N, Gastó C, Bernardo M, Junqué C. Brain effects of cognitive remediation therapy in schizophrenia: a structural and functional neuroimaging study. Biol Psychiatr Schizophr Context Memory. 2013;73:1015–23. https://doi.org/10.1016/j.biopsych.2013.01.017.
60. Penadés R, López-Vílchez I, Catalán R, Arias B, González-Rodríguez A, García-Rizo C, Masana G, Ruíz V, Mezquida G, Bernardo M. BDNF as a marker of response to cognitive remediation in patients with schizophrenia: a randomized and controlled trial. Schizophr Res. 2018;197:458–64. https://doi.org/10.1016/j.schres.2017.12.002.
61. Penadés R, Segura B, Inguanzo A, García-Rizo C, Catalán R, Masana G, Bernardo M, Junqué C. Cognitive remediation and brain connectivity: a resting-state fMRI study in patients with schizophrenia. Psychiatry Res Neuroimaging. 2020;303:111140. https://doi.org/10.1016/j.pscychresns.2020.111140.
62. Perra A, Galetti A, Zaccheddu R, Locci A, Piludu F, Preti A, Primavera D, Di Natale L, Nardi AE, Kurotshka PK, Cossu G, Sancassiani F, Stella G, De Lorenzo V, Zreik T, Carta MG. A recovery-oriented program for people with bipolar disorder through virtual reality-based cognitive remediation: results of a feasibility randomized clinical trial. J Clin Med. 2023a;12:2142. https://doi.org/10.3390/jcm12062142.
63. Perra A, Riccardo CL, De Lorenzo V, De Marco E, Di Natale L, Kurotschka PK, Preti A, Carta MG. Fully immersive virtual reality-based cognitive remediation for adults with psychosocial disabilities: a systematic scoping review of methods intervention gaps and meta-analysis of published effectiveness studies. Int J Environ Res Public Health. 2023b;20:1527. https://doi.org/10.3390/ijerph20021527.
64. Revell ER, Neill JC, Harte M, Khan Z, Drake RJ. A systematic review and meta-analysis of cognitive remediation in early schizophrenia. Schizophr Res. 2015;168:213–22. https://doi.org/10.1016/j.schres.2015.08.017.
65. Sampedro A, Peña J, Sánchez P, Ibarretxe-Bilbao N, Iriarte-Yoller N, Pavón C, Ojeda N. Moderators of functional improvement after integrative cognitive remediation in schizophrenia: toward a personalized treatment approach. Psychiatry Res. 2023;329:115495. https://doi.org/10.1016/j.psychres.2023.115495.
66. Saperstein AM, Bello I, Nossel I, Dixon LB, Medalia A. Implementation of cognitive health services in large systems of care: highlights from coordinated specialty care for first episode psychosis. Schizophr Bull. 2024:sbae030. https://doi.org/10.1093/schbul/sbae030.
67. Schroeder AH, Bogie BJM, Rahman TT, Thérond A, Matheson H, Guimond S. Feasibility and efficacy of virtual reality interventions to improve psychosocial functioning in psychosis: systematic review. JMIR Ment Health. 2022;9:e28502. https://doi.org/10.2196/28502.

68. Seccomandi B, Tsapekos D, Newbery K, Wykes T, Cella M. A systematic review of moderators of cognitive remediation response for people with schizophrenia. Schizophr Res Cogn. 2020;19:100160. https://doi.org/10.1016/j.scog.2019.100160.
69. Seccomandi B, Agbedjro D, Bell M, Keefe RSE, Keshavan M, Galderisi S, Fiszdon J, Mucci A, Cavallaro R, Ojeda N, Peña J, Müller D, Roder V, Wykes T, Cella M. Exploring the role of age as a moderator of cognitive remediation for people with schizophrenia. Schizophr Res. 2021;228:29–35. https://doi.org/10.1016/j.schres.2020.11.060.
70. Solmi M, Croatto G, Piva G, Rosson S, Fusar-Poli P, Rubio JM, Carvalho AF, Vieta E, Arango C, DeTore NR, Eberlin ES, Mueser KT, Correll CU. Efficacy and acceptability of psychosocial interventions in schizophrenia: systematic overview and quality appraisal of the meta-analytic evidence. Mol Psychiatry. 2023;28:354–68. https://doi.org/10.1038/s41380-022-01727-z.
71. Spaulding WD, Reed D, Sullivan M, Richardson C, Weiler M. Effects of cognitive treatment in psychiatric rehabilitation. Schizophr Bull. 1999;25:657–76. https://doi.org/10.1093/oxfordjournals.schbul.a033409.
72. Thérond A, Pezzoli P, Abbas M, Howard A, Bowie CR, Guimond S. The efficacy of cognitive remediation in depression: a systematic literature review and meta-analysis. J Affect Disord. 2021;284:238–46. https://doi.org/10.1016/j.jad.2021.02.009.
73. Thibaudeau E, Peyroux E, Franck N, Carling H, Lepage M. Navigating social cognitive impairments in schizophrenia Spectrum disorders: protocol for a pilot pre-post quasi-experimental study for remote avatar-assisted cognitive remediation therapy. JMIR Res Protoc. 2024;13:e54251. https://doi.org/10.2196/54251.
74. Torrent C, del Bonnin CM, Martínez-Arán A, Valle J, Amann BL, González-Pinto A, Crespo JM, Ibáñez Á, Garcia-Portilla MP, Tabarés-Seisdedos R, Arango C, Colom F, Solé B, Pacchiarotti I, Rosa AR, Ayuso-Mateos JL, Anaya C, Fernández P, Landín-Romero R, Alonso-Lana S, Ortiz-Gil J, Segura B, Barbeito S, Vega P, Fernández M, Ugarte A, Subirà M, Cerrillo E, Custal N, Menchón JM, Saiz-Ruiz J, Rodao JM, Isella S, Alegría A, Al-Halabi S, Bobes J, Galván G, Saiz PA, Balanzá-Martínez V, Selva G, Fuentes-Durá I, Correa P, Mayoral M, Chiclana G, Merchan-Naranjo J, Rapado-Castro M, Salamero M, Vieta E. Efficacy of functional remediation in bipolar disorder: a multicenter randomized controlled study. Am J Psychiatry. 2013;170:852–9. https://doi.org/10.1176/appi.ajp.2012.12070971.
75. van Duin D, de Winter L, Oud M, Kroon H, Veling W, van Weeghel J. The effect of rehabilitation combined with cognitive remediation on functioning in persons with severe mental illness: systematic review and meta-analysis. Psychol Med. 2019;49:1414–25. https://doi.org/10.1017/S003329171800418X.
76. van Duin D, de Winter L, Kroon H, Veling W, van Weeghel J. Associations between learning principles of cognitive remediation and cognitive and vocational outcomes in early psychosis-an exploratory analysis of RCT data. Appl Neuropsychol Adult. 2023:1–11. https://doi.org/10.1080/23279095.2023.2199935.
77. Ventura J, Wilson SA, Wood RC, Hellemann GS. Cognitive training at home in schizophrenia is feasible. Schizophr Res. 2013;143:397–8. https://doi.org/10.1016/j.schres.2012.11.033.
78. Veselinović T, Neuner I. Progress and pitfalls in developing agents to treat neurocognitive deficits associated with schizophrenia. CNS Drugs. 2022;36:819–58. https://doi.org/10.1007/s40263-022-00935-z.
79. Vidarsdottir OG, Roberts DL, Twamley EW, Gudmundsdottir B, Sigurdsson E, Magnusdottir BB. Integrative cognitive remediation for early psychosis: results from a randomized controlled trial. Psychiatry Res. 2019;273:690–8. https://doi.org/10.1016/j.psychres.2019.02.007.
80. Vieta E, Torrent C. Functional remediation: the pathway from remission to recovery in bipolar disorder. World Psychiatry. 2016;15:288–9. https://doi.org/10.1002/wps.20351.
81. Vita A, Barlati S. Recovery from schizophrenia: is it possible? Curr Opin Psychiatry. 2018;31:246–55. https://doi.org/10.1097/YCO.0000000000000407.
82. Vita A, Barlati S. The implementation of evidence-based psychiatric rehabilitation: challenges and opportunities for mental health services. Front Psychiatry. 2019;10:147.
83. Vita A, Barlati S, Ceraso A, Nibbio G, Ariu C, Deste G, Wykes T. Effectiveness, core elements, and moderators of response of cognitive remediation for schizophrenia: a systematic review

and meta-analysis of randomized clinical trials. JAMA Psychiatry. 2021;78:848–58. https://doi.org/10.1001/jamapsychiatry.2021.0620.
84. Vita A, Gaebel W, Mucci A, Sachs G, Barlati S, Giordano GM, Nibbio G, Nordentoft M, Wykes T, Galderisi S. European Psychiatric Association guidance on treatment of cognitive impairment in schizophrenia. Eur Psychiatry. 2022;65:e57. https://doi.org/10.1192/j.eurpsy.2022.2315.
85. Vita A, Barlati S, Ceraso A, Deste G, Nibbio G, Wykes T. Acceptability of cognitive remediation for schizophrenia: a systematic review and meta-analysis of randomized controlled trials. Psychol Med. 2023a;53:3661–71. https://doi.org/10.1017/S0033291722000319.
86. Vita A, Barlati S, Deste G, Nibbio G, Penn DL, Pinkham AE, McIntyre RS, Harvey PD. Life engagement in people living with schizophrenia: predictors and correlates of patient life engagement in a large sample of people living in the community. Psychol Med. 2023b;53:7943–52. https://doi.org/10.1017/S0033291723002106.
87. Vita A, Kakli M, Barlati S, Nibbio G, Dossou P, Degila SL, Villa S, Deste G. Implementation of cognitive rehabilitation interventions for schizophrenia in low-income countries: an experience from Togo and Benin. Int J Soc Psychiatry. 2023c; https://doi.org/10.1177/00207640231181526.
88. Vita A, Barlati S, Ceraso A, Nibbio G, Durante F, Facchi M, Deste G, Wykes T. Durability of effects of cognitive remediation on cognition and psychosocial functioning in schizophrenia: a systematic review and meta-analysis of randomized clinical trials. AJP. 2024a;181:520–31. https://doi.org/10.1176/appi.ajp.20230396.
89. Vita A, Nibbio G, Barlati S. Pharmacological treatment of cognitive impairment associated with schizophrenia: state of the art and future perspectives. Schizophr Bull Open. 2024b;5:sgae013. https://doi.org/10.1093/schizbullopen/sgae013.
90. Wykes T. Cognitive remediation—where are we now and what should we do next? Off J Ital Soc Psychopathol. 2018;24:57–61.
91. Wykes T, Reeder C, Corner J, Williams C, Everitt B. The effects of neurocognitive remediation on executive processing in patients with schizophrenia. Schizophr Bull. 1999;25:291–307. https://doi.org/10.1093/oxfordjournals.schbul.a033379.
92. Wykes T, Huddy V, Cellard C, McGurk SR, Czobor P. A meta-analysis of cognitive remediation for schizophrenia: methodology and effect sizes. AJP. 2011;168:472–85. https://doi.org/10.1176/appi.ajp.2010.10060855.
93. Wykes T, Stringer D, Boadu J, Tinch-Taylor R, Csipke E, Cella M, Pickles A, McCrone P, Reeder C, Birchwood M, Fowler D, Greenwood K, Johnson S, Perez J, Ritunnano R, Thompson A, Upthegrove R, Wilson J, Kenny A, Isok I, Joyce EM. Cognitive remediation works but how should we provide it? An adaptive randomized controlled trial of delivery methods using a patient nominated recovery outcome in first-episode participants. Schizophr Bull. 2023; https://doi.org/10.1093/schbul/sbac214.
94. Wykes T, Bowie CR, Cella M. Thinking about the future of cognitive remediation therapy revisited: what is left to solve before patients have access? Schizophr Bull. 2024:sbae075. https://doi.org/10.1093/schbul/sbae075.
95. Yeo H, Yoon S, Lee J, Kurtz MM, Choi K. A meta-analysis of the effects of social-cognitive training in schizophrenia: the role of treatment characteristics and study quality. Br J Clin Psychol. 2022;61:37–57. https://doi.org/10.1111/bjc.12320.
96. Zbukvic I, Bryce S, Moullin J, Allott K. The use of implementation science to close the research-to-treatment gap for cognitive impairment in psychosis. Aust N Z J Psychiatry. 2023; https://doi.org/10.1177/00048674231160987.

Part III

Negative and Cognitive Symptoms in Schizophrenia: An Integrative Perspective

Relationship Between Cognitive Impairment and Negative Symptoms in Schizophrenia

14

Antonio Melillo, Andrea Perrottelli, Armida Mucci, and Silvana Galderisi

Abstract

The chapter explores the complex relationship between cognitive impairment (CI) and negative symptoms (NS). It will first discuss the evidence and theoretical models concerning the relationship between CI and individual negative symptom domains. Second, it will address the hypothesized causal relationships among NS, CI, and real-life functioning. Finally, it will present the current methodological limitations and future research perspectives.

With respect to anhedonia, dysfunctions in episodic memory may contribute to difficulties in anticipating future or hypothetical positive scenarios, while difficulties in working memory might interfere with the mental representation of pleasant experiences. Avolition might be influenced by impairments in working memory, attention, and cognitive control, given the contribution of these cognitive domains to reward learning and goal-oriented behaviour. In relation to asociality, dysfunctions in social cognition might contribute to social motivation impairments, but new approaches, such as ecological momentary assessment, are needed to provide further insight into the complex pathways that lead to asociality. A limited cognitive resources hypothesis has been proposed to explain the strong relationship of alogia with cognitive impairment. Speech production in social situations places high demands on multiple cognitive processes, and if cognitive resources are limited, subjects might reduce their speech production. Social cognition impairment, especially dysfunctions in emotion recognition and expression, may contribute to blunted affect, with a possible link to the dysfunctions of the mirror neuron system.

The chapter discusses cognitive-behavioral models and how dysfunctional beliefs, such as defeatist performance and asocial beliefs, may mediate the

A. Melillo · A. Perrottelli · A. Mucci (✉) · S. Galderisi
University of Campania "Luigi Vanvitelli", Naples, Italy

A. Vita et al. (eds.), *Negative and Cognitive Symptoms in Schizophrenia*, Comprehensive Approach to Psychiatry 3,
https://doi.org/10.1007/978-3-032-06209-3_14

relationship between CI and NS. Finally, it addresses the relationship of NS and CI with functional outcomes and discusses the methodological limitations of the current literature, including the use of outdated assessment tools, the failure to differentiate between primary and secondary negative symptoms, and to address the potential impact of the various confounding factors, such as age, gender, and medication effects.

Keywords

Cognitive impairment · Negative symptoms · Avolition · Anhedonia · Asociality · Alogia · Blunted affect · Cognitive-behavioral models

14.1 Introduction

Negative symptoms (NS) and cognitive impairment (CI) represent core features of schizophrenia and share some characteristics. Both dimensions can be detected in subjects at-risk for the disorder, and in prodromal stages before the onset of the disease [1–3], and have a strong association with real-life functioning [4–6]. Finally, both dimensions do not respond well to current treatment options, representing unmet needs of patients' treatment [7].

The relationships between NS and CI have been studied from numerous perspectives and using different approaches, leading to valuable insights into their nature. Studies conducted in subjects with schizophrenia have shown that the severity of NS is negatively correlated with neuropsychological performance on tests assessing executive functions, verbal learning, working memory, attention, and processing speed [8–11]. However, findings are inconsistent, or weak, and indirect associations are sometimes reported [9, 12].

The current chapter will first discuss the evidence and theoretical models concerning the relationship between CI and negative symptom domains. Second, it will address the hypothesized causal relationships among NS, CI, and real-life functioning. Finally, it will present the current methodological limitations and future research perspectives.

14.2 Relationships Between Negative Symptoms and Cognitive Domains: Proposed Theoretical Models and Research Evidence

14.2.1 Anhedonia, Memory, and the Accessibility Model

Anhedonia, one of the three symptoms of the Experiential dimension of negative symptoms, is defined as a reduction of the ability to experience (consummatory anhedonia) or to anticipate (anticipatory anhedonia) pleasure [13]. As discussed in previous chapters, current evidence supports the presence of anticipatory anhedonia in individuals with schizophrenia [14, 15], while findings are still controversial for the presence of consummatory anhedonia in the same subjects [13, 14, 16–18].

Indeed, inconsistent results are found in the ability to experience in-the-moment pleasure, particularly depending on the chosen assessment method. Specifically, self-report questionnaires [14, 15], which assess non-current consummatory pleasure (report of past pleasant experiences), show greater reductions of pleasure in individuals with schizophrenia as compared to emotion-induction paradigms or ecologically momentary assessment, which assess in-the-moment and current pleasure experience [16–18]. One proposed theory is that consummatory anhedonia, for most subjects with schizophrenia, might reflect the presence of an altered emotional experience, with abnormal levels of negative emotion or concurrent positive and negative emotions during experiences that usually are perceived as pleasant [17, 19, 20]. Other studies highlight that individuals with schizophrenia might show alterations in the processes sustaining in-the-moment pleasant experiences [21, 22]. Indeed, few studies have hypothesized that, in patients with schizophrenia, the initial and immediate response to pleasant stimuli is intact. In contrast, the ability to sustain this positive response after receiving a reward is impaired [21, 22]. This decay in response to pleasant stimuli has been linked to difficulties in maintaining the mental representation of the stimulus, which might be related to working memory impairment [22]. As a matter of fact, in individuals with schizophrenia, working memory tasks have been reported to show intact initial encoding but impaired maintenance of task-relevant mental representations [23]. An fMRI study that presented stimuli with emotional content to individuals with schizophrenia reported that patients' brain activity during the presentation of these stimuli was similar to that observed in healthy controls [24]. However, during the time delay between stimulus offset and emotion ratings, individuals with schizophrenia showed decreased activity in the dorsolateral prefrontal cortex (DLPFC), an area that is thought to be critically involved in working memory processes [25]; notably, this decreased activity correlated with clinical measures of anhedonia [24].

Regarding anticipatory anhedonia, several processes are involved in the anticipation of pleasant events, including reward prediction (i.e., the ability to associate cues with potential rewards), the ability to predict how good we will feel when a positive event occurs (affective forecasting), and the capacity to experience positive emotions in response to the cues associated with future rewards (anticipatory affect) [26]. These processes involve episodic memory, as past experiences with pleasant stimuli are necessary to predict future or hypothetical scenarios and how we will feel about them. Consistent with this, simulations of future events in individuals with schizophrenia have been described as less detailed and lacking contextual information [27, 28].

The accessibility model argues that difficulties in recollecting memories of rewarding and pleasant experiences have a primary role in determining retrospective reports of reduced experience of pleasure (which might be discrepant with laboratory results), as well as in anticipating future or hypothetical scenarios, leading to consummatory and anticipatory anhedonia [22, 29, 30]. In line with this model, Strauss and Gold [29] analyzed evidence from previous studies and reported a correlation between the impairment of episodic memory and reduced referred pleasure in retrospective reports, as well as between impaired working memory and reduced anticipation of pleasure [29]. By contrast, reports of current feelings in response to positive stimuli were not significantly influenced by either working memory or episodic memory impairment [29]. Some studies have hypothesized that subjects with schizophrenia show specific and more severe impairment in recalling pleasant and rewarding stimuli, as compared to neutral stimuli [26].

A recent meta-analysis of 103 clinical studies found that episodic memory impairment, measured through different test batteries, showed a moderate association with the overall severity of NS [31]. However, an association with only a small effect size was found between memory impairment and each of the two NS dimensions (the Experiential and Expressive dimensions) when analyzed separately. Analyses applying the five-factor model were not performed due to insufficient data availability [31]. The authors concluded that, given the small effect size and the absence of specificity of the association, the evidence did not support the hypothesis that anhedonia and motivational deficits arise primarily from deficits in episodic memory [31].

In summary, current evidence does not fully support the assumption that dysfunctions in working or episodic memory can explain anhedonia in schizophrenia. However, episodic memory dysfunctions might partially contribute to impairments in the anticipation of future or hypothetical positive scenarios, and working memory impairment might impact the mental representation of pleasant experiences, factors influencing the assessment of anhedonia through clinical interviews.

14.2.2 Avolition and the Influence of Memory and Executive Functions Deficits

Avolition, included in the experiential dimension, is currently defined as reduced initiation and persistence of goal-directed activity [13]. Overall, research studies in individuals with schizophrenia have shown heterogeneous results concerning the association of clinically-rated avolition and neurocognitive impairments assessed by standard tests [9, 13, 32, 33]. Therefore, studies aimed to dissect the different cognitive processes involved in defective goal-directed behavior have examined effort-based decision-making (EBDM) and reinforcement learning constructs, offering insights into the potential involvement of motivation and reward-related cognitive processes.

In particular, EBDM paradigms have been employed to distinguish different phases of desire and initiation of goal-oriented behaviors: option generation (the generation of behavioral options, either self-generated or prompted by an environmental stimulus), option selection (based on the evaluation and prediction of different parameters and particularly reward, effort and risk), preparation for action,

action initiation and maintenance, processing of outcome and reinforcement learning [34–37].

Option generation is thought to rely on working and long-term memory, verbal fluency, and processing speed [38, 39]. Interestingly, one study found that deficits in option generation were associated with the severity of symptoms of the Experiential dimension, assessed through BNSS, and that this association was partially mediated by verbal fluency [34].

Option selection requires the computation of information regarding both the magnitude and the probability of possible positive and negative events to build mental representations of potential reward, effort, and risk associated with each action. As discussed in previous chapters, individuals with schizophrenia may show an impairment in information processing, resulting in difficulties in updating their mental representations [40]. A recent study applying a computational modelling approach to data on an EBDM task, the Effort Expenditure for Rewards Task (EEfRT), found that the subset of patients showing difficulties using information about reward magnitude and probability was characterized by greater cognitive impairment [36]. Moreover, while neurocognitive impairment was associated with difficulties in reward information processing, higher sensitivity to effort was associated with avolition scores on the SANS and motivation and pleasure subscale scores of the CAINS [36]. A study involving 40 individuals with treatment-resistant schizophrenia executing an EEfRT task confirmed that cognitive impairment was associated with difficulties in reward information processing [41]. However, the study found that individuals with schizophrenia showed reduced sensitivity to effort, which was associated with the symptoms of the Expressive dimension and not with the ones of the Experiential dimension [41].

A reduction in goal-oriented behavior has also been linked to deficits in reinforcement learning, defined as the ability to dynamically update learned associations between outcomes and the stimuli or actions associated with them [42, 43]. Studies employing reinforcement learning tasks have hypothesized a potential role of cognitive deficits, specifically of working memory and attention domains [42]. In fact, Collins et al. hypothesized that deficits in working memory might be at the core of the alterations in reward learning performance and bias individuals with schizophrenia to undervalue rewards [43]. In addition to working memory, attention is also thought to contribute to reward learning through the selective processing of stimuli that are relevant to improve task performance [44]. Indeed, individuals with schizophrenia have shown unnecessary allocation of attentional resources to irrelevant cues [45]. During probabilistic reversal learning (PRL) tasks, which are also employed in characterizing reinforcement learning processes, participants are required to rapidly update their previously learned associations of cues with potential rewards. Studies employing PRL tasks have shown that individuals with schizophrenia were unable to adaptively modulate attention, which has been associated with a reduction in the deactivation of the default mode network components [46]. These findings suggest that, similarly to working memory, impairments in attention may influence reward learning and thus cause a reduction in goal-oriented behavior.

Neuroimaging studies showed that during learning tasks and goal-oriented behavior, striatal regions are strongly connected with areas of the cognitive control network and are concomitantly activated to maintain and integrate relevant information [46, 47]. Studies focusing on schizophrenia have linked impairments in reward processing and learning to dysfunctional activity of mesolimbic structures, particularly in the bilateral striatum, thalamus, amygdala, insula, and hippocampus [47]. Specifically, subjects with schizophrenia showed a reduction in the activity of the mesolimbic circuitry in responses to prediction errors (i.e., discrepancies between expected and actual outcomes), which drives the updating of mental representations [47]. The current evidence concerning the correlation between these brain activity abnormalities during prediction tasks and negative symptoms is heterogeneous [48–50]. For instance, the meta-analysis by Yang et al. found no significant association between task-related brain activity during prediction error processing and the score of the negative subscale of the PANSS [47].

Several studies have investigated how deficits in motivation may also contribute to worsening cognitive impairment. In healthy individuals, the possibility of obtaining rewards based on performance influences their efficiency in experimental paradigms testing episodic memory [51–55], working memory [56], and attention [57–59]. In relation to episodic memory, several studies have reported that monetary incentives enhance memory formation through the facilitation of encoding-related activity in the hippocampus [52, 54, 55]. In a study by Duan and colleagues in healthy subjects, both intrinsic (based on an internal desire such as curiosity) and extrinsic motivation (linked to tangible rewards such as monetary ones) were reported to enhance memory task performance [52]. In the same study, curiosity-driven activity induced the activation of the ventral striatum and the frontal-parietal attention network, thus enhancing memory formation. Extrinsic motivation was found to deactivate parietal midline regions, and was thus hypothesized to enhance memory through the suppression of task-irrelevant processes [52]. Furthermore, employing a memory task in which the value of potential rewards was disclosed to participants only during recollection, Shigemune et al. [53] reported that reward magnitude enhanced the retrieval of episodic memories, mainly when this was more difficult [53]. Similarly, rewards have been reported to improve task performance on sustained attention tasks [57, 58].

Regarding schizophrenia, few and heterogeneous findings do not yet clarify whether affected individuals show impairments in the reward-modulated enhancements of cognitive performances, as observed in healthy individuals [60–62]. Furthermore, studies that found a significant deficit in reward-enhanced cognitive performance showed inconsistent and often non-significant associations of these alterations with avolition or overall NS severity, although these inconsistencies may partly be explained by the use of first-generation instruments for the assessment of NS, which do not provide adequate evaluations of the symptoms related to the Experiential dimension [60–62].

In summary, findings from both healthy individuals and individuals with schizophrenia show that reward-related and cognition-related systems are strictly connected and involved in goal-oriented behavior. Bi-directional interactions between dysfunctional cognition-related and reward-related networks may exert synergistic effects and amplify both cognitive and motivational deficits [63]. Given the contributions of working memory, attention, and cognitive control functions to reward processing and goal-oriented behavior, the impairment of these cognitive functions may contribute to avolition [64]. The heterogeneity of findings concerning the association of abnormalities in these processes and the severity of avolition might be due to the employment of first-generation scales for NS assessment in most of the studies, as well as to the intrinsic heterogeneity of negative symptoms (e.g., none of the studies attempted to distinguish secondary and primary negative symptoms).

14.2.3 Asociality, Social Motivation, and Social Cognition

Asociality is currently defined as a reduction in social motivation [13], and factorial analyses confirm that this symptom is strictly related to avolition and anhedonia and is conceptualized as part of the Experiential dimension [65]. Different studies have investigated social reward learning processes (employing social reward tasks such as the Social Incentive Delay task) and reported that individuals with schizophrenia present a significant reduction of social reward anticipation [66, 67], and social reward sensitivity [68, 69], and an increased sensitivity to effort, as well as deficits of effort-reward integration [70] and social reward learning [71]. The impairment of these processes is involved in the so-called social anhedonia (as measured using the Chapman Scale for Social Anhedonia] and might lead to reduced social motivation. However, the relationships between social anhedonia and asociality ratings are not always assessed in the same studies.

Different studies have reported that social and non-social reward anticipation (e.g., in monetary incentive tasks] are supported by similar neural circuits and involve the deactivation of the default mode network and the activation of the ventral striatum, insula and areas involved in cognitive control functions, such as the anterior cingulate cortex, to modulate and shift attention and to maintain and integrate information [72]. However, other neuroimaging results suggest that the processing of social reward is also supported by cerebral areas thought to be involved specifically in social-cognitive processes (e.g., the temporoparietal junction, the dorsomedial prefrontal cortex, the precuneus, and the superior temporal gyrus) [73–75].

As for non-social reward processes, a dysfunction in neurocognition-related circuits may impact social reward processing and learning [76]. Consistent with this hypothesis, cognitive remediation programs, focusing mainly on working memory, have been reported to improve sensitivity to social rewards [76].

It should be noted that social motivation is inherently more complex than non-social reward processing [77] and involves the elaboration of verbal and non-verbal social cues and social cognition skills, such as the interpretation of others' intentions and dispositions (theory of mind; ToM) and recognition of facial affect. Social reward tasks require efficient social cognition functioning. In fact, behavioral, fMRI, and electrophysiological studies reported that individuals with schizophrenia show less sensitivity and altered neurophysiological responses to social rewards as compared to non-social rewards [68, 78, 79].

However, results have not been univocal, and studies employing social reward processing models reported mixed results on whether social cognition impairments may account for deficits in social reward anticipation and learning [80–82]. Similarly, correlation analyses of asociality and social cognition [83–86] have provided inconsistent results. It should be noted, however, that this stream of literature has particularly suffered from methodological limitations, as first-generation scales for the assessment of NS typically measure asociality in behavioral terms (i.e., quantity and quality of social interactions) [13], which are strongly influenced by several confounding factors, including contextual factors (e.g., lack of social opportunities and socio-economic resources) and depend on the recollection of past experiences or the production of hypothetical scenarios of social situations. However, recent network analyses showed that social cognition and NS are not directly connected [87, 88], even when employing new generation scales to assess NS [87].

In addition, the relationship between asociality and social cognition may be complex and bidirectional: an impairment in social cognition may compromise the attainment and perception of reward signals during social interactions; at the same time, a decrease in motivation to participate in social activities, particularly during developmental stages, may result in social isolation and hence poor social cognition development [64]. Animal models suggest that social isolation may also influence other cognitive domains, as socially isolated rodents show impairments in attention, visual and spatial learning, and memory [89]. Importantly, deficits in social cognition are likely to make social interactions in everyday life less rewarding, increasing the probability of punishment (social rejection) and reinforcing social avoidance tendencies and beliefs [70, 90].

In summary, social and non-social motivations may rely on only partially overlapping networks, and social cognitive deficits are likely to contribute to social motivation impairments. New approaches, such as ecological momentary assessment and longitudinal studies, may provide further insight into the complex pathways that lead to asociality, clarifying the role of neurocognition, social cognition, and contextual factors.

14.2.4 Alogia, Memory, Verbal Fluency, and the Cognitive Resource Model

Language comprehension and production, particularly in social contexts, are among the most complex brain functions [91], which require the coordination of multiple neural processes and the involvement of several cognitive domains, including attention for word recognition [92, 93], access to episodic and semantic memory [94], working memory to maintain the relevant information [95] and social cognition to process information (e.g., social cues) related to the ongoing interaction.

Given this multifaceted involvement of higher cognitive functions, alogia, defined as a reduction in speech production and one of the two symptoms of the

Expressive dimension, is likely to be influenced by the availability of cognitive resources [96, 97].

Higher severity of alogia has been consistently found to be associated with worse general neurocognitive functioning and impaired social cognition [65, 98]. Concerning individual neurocognitive domains, alogia has been associated with deficits in semantic memory and verbal fluency, as assessed through category fluency tasks. In particular, in these tasks, individuals with schizophrenia showed increased latency, which could represent a manifestation of impaired controlled retrieval (i.e., activation and selection of information from semantic memory) [99] and a reduced number of total words retrieved [100, 101], reflecting impairments in semantic memory. Furthermore, patients with clinically significant levels of alogia showed less coherent and more disorganized semantic memory, as compared to patients with lower levels of alogia [100]. However, a recent study applying a machine learning (ML) processing of natural language recorded during speaking tasks found no correlation between cognitive functioning and either clinically rated or ML-predicted alogia scores [102].

In relation to working memory, studies employing dual-task paradigms, in which participants are asked to speak while simultaneously performing a working memory task, demonstrated that individuals with schizophrenia show a greater reduction in speech production than healthy controls [103]. This finding was replicated in transdiagnostic samples [104]. Furthermore, in cross-sectional analyses that included individuals at the first psychotic episode, the severity of alogia, as assessed by the global rating on the SANS subscale, was associated with working memory impairment [105]. However, this finding was not replicated in another study [106], which considered only the poverty of speech item of the SANS alogia subscale in line with current conceptualizations.

Additionally, speech production in individuals with schizophrenia is characterized by a variety of linguistic abnormalities [91], such as reduced lexical diversity [107], which are currently conceptualized as language-related manifestations of formal thought disorders [91]. Relevant to our subject, linguistic abnormalities in individuals with schizophrenia have been associated with impairments in various cognitive domains (working memory, semantic memory, and verbal fluency) [108], as well as with clinician-rated alogia [107]. In addition, individuals with schizophrenia show difficulties in speech comprehension, which have been associated with impairments in executive functions, working, and semantic memory [91, 109] and may contribute to reduced speech production.

Concerning social cognition, different studies have proposed an association between alogia and several aspects of social cognition. In particular, alogia and the Expressive dimension have been associated with deficits in mentalization in both chronic [85] and early-course schizophrenia [110, 111]. In another study, alogia in individuals at the first episode of schizophrenia correlated with social perception, mentalization, and overall social cognition deficits [112]. More recently, a study employing a mediation model reported that verbal memory indirectly affected the Expressive dimension, through emotion recognition and ToM [113].

To summarize, despite some inconsistencies in the current literature, alogia has been reported to be associated with deficits in working and semantic memory and social cognition. Similar associations were found for difficulties in speech comprehension and syntax processing, which may contribute to the reductions in speech production observed in clinical and naturalistic settings. Some of the studies, however, used clinical ratings of alogia that included disorganization aspects (e.g., the alogia subscale of SANS, including poverty of content of speech, blocking, and increased latency of response), which might be related to cognitive functioning more than alogia as presently conceptualized (reduced speech production and spontaneous elaboration). The association with overall neurocognitive and social cognition functioning is a robust finding replicated with adequate alogia assessment. However, more studies are needed to establish the specific neurocognitive and social cognitive domains associated with alogia as presently conceptualized.

14.2.5 Blunted Affect, Social Cognition, the Active Inference Model, and the Mirror-neuron System

The decrease in emotional expression and reactivity, and particularly the reduction in facial expressivity, vocal prosody, and gestures, conceptualized as blunted affect (part of the Expressive dimension), is currently thought to be caused by deficits in either social cognition or motor activity, rather than by a reduction in the internal experience of emotion [13]. Emotional expression and reactivity, which can be either facial, vocal, or gestural, arise in response to non-social emotional stimuli and to a wide range of social stimuli. Therefore, these functions are influenced by the processing of these stimuli through neurocognitive and social cognition functions. In line with this notion, cerebral areas involved in social cognition are active during the production of facial expressions [114, 115].

In relation to individual aspects of social cognition, blunted affect severity has been associated particularly with lower performance in emotion recognition tasks [84, 116, 117], which could lead to an impairment in emotion expressivity. Furthermore, an association with ToM deficits has also been reported [85]. However, there are conflicting results in the literature, as blunted affect, even when assessed with second-generation scales, did not show associations with global neurocognitive functioning or social cognition in other studies [65]. A recent meta-analysis found no evidence of a specific association of ToM impairment with either blunted affect or alogia [83].

The active social inference model was proposed to provide an explanatory framework for the effects of social cognition impairments in producing deficits in affective behaviors [118]. Drawing from predictive coding models of brain function, the model proposes that blunted affect arises from dysfunctions within social prediction circuits, particularly from imprecise predictions of the social consequences of the expression of emotions. Indeed, expressing facial emotions, similarly to any action, is preceded by the formulation of predictions of its social consequences (e.g., the predicted reaction of the interlocutor to a smile). The interaction of faulty predictions with perceptual information leads to social prediction error signaling. Consequently, the network adjusts future behavior strategies towards inaction (i.e., blunting of affect to minimize prediction error signaling). In this model, bottom-up feedbacks deriving from the interlocutor's reaction are interpreted through aberrant social cognition processes and contributes to the faulty top-down predictions [118].

Other studies investigating the association between social cognition and NS focused on the mirror neuron system (MNS). The MNS is not only involved in motor action, but has a primary role in supporting social cognition [119]; in fact, it contributes to processing the actions of others, understanding their intentions [120], responding to their facial expressions, and imitating them [121]. A fMRI study found a significant negative correlation between the severity of blunted affect and brain activity recorded during imitation of emotional expression in the premotor cortex, motor cortex, and inferior parietal lobule, all included within the MNS [122]. The MNS is thought to be involved in the development of the so-called self-other correspondence, which is considered ontogenetically as the foundation of mature social cognition and ToM [123]. In line with the neurodevelopmental hypothesis of schizophrenia, research on MNS in affected individuals has reported several alterations in this neural system [119, 124], which have been correlated with deficits in social cognition [125] and in ToM [124]. Given the role of MNS in motor function, social interactions, and social cognition, it was hypothesized that this system plays a role in the pathogenesis of expressive deficits and particularly blunted affect [122]. However, the association between MNS and negative symptoms, and more specifically blunted affect, is not fully understood, as a recent meta-analysis retrieved four studies reporting inconsistent results, with two of them reporting associations between impaired MNS and the anhedonia subscale [119, 126], or the anhedonia-asociality domain and particularly the "Ability to Feel Intimacy and Closeness" item score [127, 128]. Further research is needed to understand the potential role of MNS in blunted affect since, in the last years, the evidence supporting the MNS model has been inconsistent [129], and the scientific interest in the model has declined [130].

In summary, while blunted affect does not arise from a lowered inner experience of emotions, it may depend on faulty processing of social stimuli, social interactions, and contexts. In line with this hypothesis, impaired social cognition may contribute to blunted affect, and a dysfunction of the MNS may impair both social cognitive functioning and the motor execution of emotion expression. Current evidence does not fully support these hypotheses, and further research is needed, particularly in drug-free subjects and in early stages of the disorder.

14.3 Cognitive-Behavioral Models and the Mediating Role of Dysfunctional Beliefs

Several studies have highlighted that individuals with schizophrenia maintain a variety of dysfunctional beliefs about the self, the environment, and the future, which influence their behavior and the interpretation of events [131]. These beliefs have been associated with NS [131]. For instance, defeatist performance beliefs (i.e., negative assumptions concerning the ability to execute goal-oriented tasks) have been found to predict both NS severity and real-life functioning in individuals with schizophrenia [132]. Similarly, asocial beliefs (e.g., "I'm better by myself") have been associated with reduced social motivation and social functioning in individuals with schizophrenia [131, 133]. Relevant to our topic, cognitive-behavioral models have provided a framework to explain the role of dysfunctional beliefs within the context of the association between NS and CI. Beck and colleagues first proposed that CI, by contributing to the unsuccessful execution of goal-oriented behavior, causes and progressively reinforces over time defeatist beliefs, thus discouraging individuals from engaging in future activities [134]. According to this model, NS may represent the behavioral result of this pathway, reflecting a general pattern of disengagement and withdrawal from the environment [135]. In line with this hypothesis, Beck et al. reported that defeatist beliefs statistically mediated the relationship of neurocognitive impairment with both NS and functioning [134]. The finding has been more recently replicated [136]. Additionally, defeatist beliefs have been found to mediate the relationship between NS and willingness to exert effort in EBDM tasks [137–139]. However, in these studies, the relationship between the investigated parameters and the neurocognitive performances of participants was not reported [138].

In relation to social cognition and asociality, a study reported that impairments in ToM, and particularly difficulties in inferring emotions, predicted NS in individuals with schizophrenia and with low self-esteem beliefs [133]. Asocial beliefs could also mediate the relationship between social skills and social functioning. In fact, a study reported that the efficacy of cognitive-behavioral social skills training on experiential negative symptoms and functioning was mediated by its effect on defeatist and asocial beliefs [140].

The mediating role of dysfunctional beliefs in the relationship between neurocognition and NS symptoms was not confirmed in other studies [141]. For instance, in a study by Ventura et al., both NS and neurocognitive impairment were found to mediate the relationship between functioning and self-efficacy, which, in turn, did not mediate the relationship between neurocognition and NS [141]. In the structural equation model by Gonzalez-Blanch et al. [142], social relatedness (i.e., beliefs concerning social support and loneliness) partially mediated the impact of social cognition and NS on social functioning. However, the effect of social cognition impairment on NS severity was not mediated by social relatedness, and both social cognition and NS independently influenced social relatedness [142].

Importantly, the role of defeatist beliefs has also been interpreted in light of the models exposed in previous paragraphs. First, difficulties in either accessing the memories of past pleasant stimuli or in learning from them may concur to bias individuals with schizophrenia to report low pleasant experiences [31, 143]. Second, impaired reward processing and learning may impair the ability to update negative beliefs on the basis of positive/rewarding experiences, while intact learning from negative experiences may reinforce dysfunctional beliefs, resulting in a reduction of goal-oriented behavior [144]. Similarly, impaired social reward learning may lead to a failure to update beliefs of trust towards others [71, 80, 145] and about one's social ability [146] based on positive experiences, while intact learning from negative outcomes may reinforce mistrust beliefs and social avoidance, resulting in social withdrawal [146]. In summary, dysfunctional beliefs may contribute to the influence of CI on the development and severity of NS. Refining these models may provide a deeper understanding of the pathways linking NS and CI.

14.4 Negative Symptoms, Cognitive Impairment, and Their Relation with Functional Outcome Measures

In recent years, several studies have investigated the relationship between functional outcome with NS and CI. While both NS and CI are reported to have strong associations with real-life functioning in people with schizophrenia [4, 5, 147–151], the complex interplay between these parameters is still to be fully understood. Specifically, it is unclear whether their respective influence on functional outcomes are independent of each other. Several studies indicate that CI and NS may be associated with different aspects of functioning. In a longitudinal study, neurocognition has been found to directly and indirectly impact two important domains of functioning, work skills and everyday life skills [4, 151]. The impact was mediated by several variables, particularly service engagement, functional capacity, internalized stigma (through resilience), and social cognition, which, in turn, showed a direct association with functioning [4]. In the same study, these influences were independent of NS, which had a weaker association with real-life functioning: in particular, avolition was associated both directly and indirectly (through service engagement, resilience, and internalized stigma) with interpersonal relationships and work skills domains [87]. At follow-up, avolition was again directly associated with interpersonal functioning, and this association was more substantial than that of neurocognition, which was only indirect through social cognition, and comparable to the association of social cognition with the same domain of functioning [151]. Furthermore, at follow-up, avolition did not present a direct or indirect association with everyday life or work skills [151]. These data highlight the independent contribution of CI and negative symptoms, particularly the experiential negative symptoms, to different domains of real-life functioning.

Other studies found that NS outweighed CI as a predictor of real-life functioning [152] or mediated the impact of CI on functioning [153]. In a large European multicentric study, the Experiential dimension mediated the relationship between processing speed and personal and social functioning [154]. Furthermore, in agreement with the study by Galderisi et al. [4], a measure of social cognition (emotion recognition) partially mediated the impact of neurocognition on the real-life functioning domain of "socially useful activities" [154]. The studies in which negative symptoms outweighed cognitive dysfunction as determinants of real-life functioning did not examine the different domains of real-life functioning and measures of social cognition or included subjects with deficit schizophrenia (characterized by primary and persistent negative symptoms) or treatment resistance, in which the relative contribution of negative symptoms to real-life functioning might be higher than in more heterogeneous samples. In a 20-year longitudinal study, neurocognition, mainly processing speed (one of the only two cognitive domains assessed), partially mediated the impact of NS on occupational functioning over time [155]. In the same study, cross-sectional analysis at baseline showed that NS mediated the impact of neurocognition on functioning [155]. The study did not assess avolition, but only the expressive dimension of negative symptoms (poverty of speech and blunted affect), and the only assessed neurocognitive domains included processing speed and general knowledge.

In summary, the complex interplay between NS, CI, and real-life functioning has been examined with heterogeneous instruments, not always including the negative symptoms or neurocognitive domains with the highest impact on real-life functioning (i.e., avolition or working memory). Most studies did not include a measure of social cognition, which mediates the impact of neurocognition on social functioning and has an independent effect on the same domain of real-life functioning. The few studies including the above variables [4, 151, 154] showed that NS and social cognition might mediate, at least partially, the impact of neurocognition on some domains of real-life functioning. Still, they did not address the presence of bi-directional relationships, which might be present through the pathophysiological mechanisms described in the present chapter, but also through their interaction with factors that are not related to the disease, but to the individual's personal resources and their socio-economic context.

14.5 Methodological Issues, Potential Confounding Factors, and Future Directions in Research

The complexity and the inconsistencies of some of the findings on the associations between NS and CI presented in this chapter must be addressed, and some methodological limitations must be considered. First, in clinical models, several important, potentially confounding factors must be considered, including the heterogeneity of clinical presentations. This aspect can vary greatly between individuals in terms of both severity and pattern; subgroups of patients can show relatively intact levels of cognitive functioning [156, 157] or low or transient levels of NS [158]. Other confounding factors include differentiating primary and secondary NS and the influence of age, gender, education, premorbid functioning, medication effects, institutionalization, positive and affective symptoms, and poor physical health [159, 160].

Second, the assessment methods can have a severe impact on the results. In relation to NS, the predominant use of first-generation scales (i.e., scales developed before the NIMH-Negative Symptom Consensus Development Conference) [161] creates a potential overlap of negative symptoms with both functioning and cognitive deficits [148, 162, 163]. Similarly, experimental tasks for the investigation of mechanisms underlying NS, such as reinforcement learning tasks, require the involvement of neurocognitive functions. In fact, variations in task parameters (e.g., the number of stimuli in a set) produce differences in the involvement of cognitive domains, influencing the inferences on negative symptoms and reinforcement learning processes [42, 43]. Methodological heterogeneity concerning the assessment of neurocognitive domains may limit the interpretation of findings as well. For instance, while complex working memory capacity is known as a crucial domain for both speech comprehension and production, more widely used tests mainly assess short-term working memory, which does not consistently predict language-related functioning [164, 165]. More generally, the primary methods of assessing NS and cognitive functioning in clinical samples differ: while NS are typically measured using interview-based rating scales that focus on behavior and, in more recent scales, on inner experience as reported by the individual, cognitive deficits are measured using batteries that are performance-based. Finally, previous research has predominantly studied both NS and cognitive functioning as singular, unitary dimensions [65], so reports on specific associations between NS and individual cognitive deficits are often not addressed [31].

In recent years, novel approaches promise to contribute to our understanding of the pathophysiological mechanisms of NS and CI in schizophrenia. The refinement of digital phenotyping approaches, such as passive data acquisition and ecological momentary assessment [166–168] may improve assessments by offering high-resolution real-world data on the interplay between individuals' behavior and its environmental surroundings. About NS, and particularly the Experiential dimension, these technologies hold promise to potentially overcome the risk of bias due to poor insight and memory recall deficits [169] and to contribute to disentangling their influence on clinician-rated assessments [170, 171]. In relation to cognition, smartphones have been utilized to acquire both active (e.g., digital cognitive tasks) and passive (e.g., novel digital markers of cognition, such as the number of phone screen touches) data acquisition strategies [172]. In addition, machine learning technologies for facial affect production, prosody, and alogia [102] can quantitatively assess expressive deficits and acquire information at higher temporal resolution [173]. Additionally, while they were initially applied to laboratory-based studies of simulated and/or non-interactive behaviors, more recent studies have employed semi-naturalistic settings to study interpersonal interactions [174].

Overall, despite the existing digital divide across countries and between individuals with serious mental illnesses and the healthy population, the potential outreach of these strategies is remarkably more extensive than that of expert assessment in clinical settings [169]. However, these approaches have often produced results poorly associated with clinicians' assessments, and further studies with larger samples and control arms collecting relevant normative data will be needed to fill current knowledge gaps.

References

1. Millman ZB, Roemer C, Vargas T, Schiffman J, Mittal VA, Gold JM. Neuropsychological performance among individuals at clinical high-risk for psychosis vs putatively low-risk peers with other psychopathology: a systematic review and meta-analysis. Schizophr Bull. 2022;48(5):999–1010.
2. Devoe DJ, Braun A, Seredynski T, Addington J. Negative symptoms and functioning in youth at risk of psychosis: a systematic review and meta-analysis. Harv Rev Psychiatry. 2020;28(6):341–55.
3. Salazar de Pablo G, Catalan A, Vaquerizo Serrano J, Pedruzo B, Alameda L, Sandroni V, et al. Negative symptoms in children and adolescents with early-onset psychosis and at clinical high-risk for psychosis: systematic review and meta-analysis. Br J Psychiatry. 2023;223(1):282–94.
4. Galderisi S, Rossi A, Rocca P, Bertolino A, Mucci A, Bucci P, et al. The influence of illness-related variables, personal resources and context-related factors on real-life functioning of people with schizophrenia. World Psychiatry. 2014;13(3):275–87.
5. Galderisi S, Rucci P, Mucci A, Rossi A, Rocca P, Bertolino A, et al. The interplay among psychopathology, personal resources, context-related factors and real-life functioning in schizophrenia: stability in relationships after 4 years and differences in network structure between recovered and non-recovered patients. World Psychiatry. 2020;19(1):81–91.
6. Shmukler AB, Gurovich IY, Agius M, Zaytseva Y. Long-term trajectories of cognitive deficits in schizophrenia: a critical overview. Eur Psychiatry. 2015;30(8):1002–10.
7. Galderisi S. Negative symptoms of schizophrenia: trying to answer unanswered research questions. Psychiatry Res. 2023;320:115043.
8. Giordano GM, Caporusso E, Pezzella P, Galderisi S. Updated perspectives on the clinical significance of negative symptoms in patients with schizophrenia. Expert Rev Neurother. 2022;22(7):541–55.
9. Au-Yeung C, Penney D, Rae J, Carling H, Lassman L, Lepage M. The relationship between negative symptoms and MATRICS neurocognitive domains: a meta-analysis and systematic review. Prog Neuro-Psychopharmacol Biol Psychiatry. 2023;127:110833.
10. Hegde S, Thirthalli J, Rao SL, Raguram A, Philip M, Gangadhar BN. Cognitive deficits and its relation with psychopathology and global functioning in first episode schizophrenia. Asian J Psychiatr. 2013;6(6):537–43.
11. Mørch-Johnsen L, Smelror RE, Andreou D, Barth C, Johannessen C, Wedervang-Resell K, Wortinger LA, et al. Negative symptom domains are associated with verbal learning in adolescents with early onset psychosis. Front Psych. 2022;12:825681.
12. Harvey PD, Koren D, Reichenberg A, Bowie CR. Negative symptoms and cognitive deficits: what is the nature of their relationship? Schizophr Bull. 2006;32(2):250–8.
13. Marder SR, Galderisi S. The current conceptualization of negative symptoms in schizophrenia. World Psychiatry. 2017;16(1):14–24.
14. Visser KF, Chapman HC, Ruiz I, Raugh IM, Strauss GP. A meta-analysis of self-reported anticipatory and consummatory pleasure in the schizophrenia-spectrum. J Psychiatr Res. 2020;121:68–81.
15. Hallford DJ, Sharma MK. Anticipatory pleasure for future experiences in schizophrenia spectrum disorders and major depression: a systematic review and meta-analysis. Br J Clin Psychol. 2019;58(4):357–83.
16. Riehle M, Straková A, Lincoln TM. Emotional experience of people with schizophrenia and people at risk for psychosis: a meta-analysis. JAMA Psychiatry. 2024;81(1):57–66.
17. Cohen AS, Minor KS. Emotional experience in patients with schizophrenia revisited: meta-analysis of laboratory studies. Schizophr Bull. 2010;36(1):143–50.
18. Cho H, Gonzalez R, Lavaysse LM, Pence S, Fulford D, Gard DE. Do people with schizophrenia experience more negative emotion and less positive emotion in their daily lives? A meta-analysis of experience sampling studies. Schizophr Res. 2017;183:49–55.

19. Abel DB, Minor KS. Social pleasure in daily life: a meta-analysis of experience sampling studies in schizophrenia. Schizophr Res. 2023;260:56–64.
20. Strauss GP, Frost KH, Lee BG, Gold JM. The positivity offset theory of anhedonia in schizophrenia. Clin Psychol Sci. 2017;5(2):226–38.
21. Strauss GP, Cohen AS. A Transdiagnostic review of negative symptom phenomenology and etiology. Schizophr Bull. 2017;43(4):712–9.
22. Kring AM, Germans Gard M, Gard DE. Emotion deficits in schizophrenia: timing matters. J Abnorm Psychol. 2011;120(1):79–87.
23. Gotra MY, Keedy SK, Hill SK. Interactive effects of maintenance decay and interference on working memory updating in schizophrenia. Schizophr Res. 2022;239:103–10.
24. Ursu S, Kring AM, Gard MG, Minzenberg MJ, Yoon JH, Ragland JD, et al. Prefrontal cortical deficits and impaired cognition-emotion interactions in schizophrenia. Am J Psychiatry. 2011;168(3):276–85.
25. Nejati V, Majdi R, Salehinejad MA, Nitsche MA. The role of dorsolateral and ventromedial prefrontal cortex in the processing of emotional dimensions. Sci Rep. 2021;11(1):1971.
26. Frost KH, Strauss GP. A review of anticipatory pleasure in schizophrenia. Curr Behav Neurosci Rep. 2016;3(3):232–47.
27. Cohen JD, Barch DM, Carter C, Servan-Schreiber D. Context-processing deficits in schizophrenia: converging evidence from three theoretically motivated cognitive tasks. J Abnorm Psychol. 1999;108(1):120–33.
28. Raffard S, Esposito F, Boulenger JP, Van der Linden M. Impaired ability to imagine future pleasant events is associated with apathy in schizophrenia. Psychiatry Res. 2013;209(3):393–400.
29. Strauss GP, Gold JM. A new perspective on anhedonia in schizophrenia. Am J Psychiatry. 2012;169(4):364–73.
30. Strauss GP, Cohen AS. The schizophrenia spectrum anhedonia paradox. World Psychiatry. 2018;17(2):221–2.
31. Pillny M, Krkovic K, Buck L, Lincoln TM. From memories of past experiences to present motivation? A meta-analysis on the association between episodic memory and negative symptoms in people with psychosis. Schizophr Bull. 2022;48(2):307–24.
32. Hartmann-Riemer MN, Hager OM, Kirschner M, Bischof M, Kluge A, Seifritz E, et al. The association of neurocognitive impairment with diminished expression and apathy in schizophrenia. Schizophr Res. 2015;169(1–3):427–32.
33. Melillo A, Caporusso E, Giordano GM, Giuliani L, Pezzella P, Perrottelli A, et al. Correlations between negative symptoms and cognitive deficits in individuals at first psychotic episode or at high risk of psychosis: a systematic review. J Clin Med. 2023;12(22):7095.
34. Hartmann MN, Kluge A, Kalis A, Mojzisch A, Tobler PN, Kaiser S. Apathy in schizophrenia as a deficit in the generation of options for action. J Abnorm Psychol. 2015;124(2):309–18.
35. McCarthy JM, Treadway MT, Bennett ME, Blanchard JJ. Inefficient effort allocation and negative symptoms in individuals with schizophrenia. Schizophr Res. 2016;170(2–3):278–84.
36. Cooper JA, Barch DM, Reddy LF, Horan WP, Green MF, Treadway MT. Effortful goal-directed behavior in schizophrenia: computational subtypes and associations with cognition. J Abnorm Psychol. 2019;128(7):710–22.
37. Gold JM, Waltz JA, Matveeva TM, Kasanova Z, Strauss GP, Herbener ES, et al. Negative symptoms and the failure to represent the expected reward value of actions: behavioral and computational modeling evidence. Arch Gen Psychiatry. 2012;69(2):129–38.
38. Leder J, Häusser JA, Krumm S, Germar M, Schlemmer A, Kaiser S, et al. The cognitive underpinnings of option generation in everyday life decision-making: a latent variable analysis. Cogn Sci. 2018;42(8):2562–91.
39. Husain M, Roiser JP. Neuroscience of apathy and anhedonia: a transdiagnostic approach. Nat Rev Neurosci. 2018;19(8):470–84.
40. Gold JM, Waltz JA, Prentice KJ, Morris SE, Heerey EA. Reward processing in schizophrenia: a deficit in the representation of value. Schizophr Bull. 2008;34(5):835–47.

41. Saleh Y, Jarratt-Barnham I, Petitet P, Fernandez-Egea E, Manohar SG, Husain M. Negative symptoms and cognitive impairment are associated with distinct motivational deficits in treatment resistant schizophrenia. Mol Psychiatry. 2023;28(11):4831–41.
42. Collins AG, Brown JK, Gold JM, Waltz JA, Frank MJ. Working memory contributions to reinforcement learning impairments in schizophrenia. J Neurosci. 2014;34(41):13747–56.
43. Collins AGE, Albrecht MA, Waltz JA, Gold JM, Frank MJ. Interactions among working memory, reinforcement learning, and effort in value-based choice: a new paradigm and selective deficits in schizophrenia. Biol Psychiatry. 2017;82(6):431–9.
44. Radulescu A, Niv Y, Ballard I. Holistic reinforcement learning: the role of structure and attention. Trends Cogn Sci. 2019;23(4):278–92.
45. Morris R, Griffiths O, Le Pelley ME, Weickert TW. Attention to irrelevant cues is related to positive symptoms in schizophrenia. Schizophr Bull. 2013;39(3):575–82.
46. Waltz JA, Kasanova Z, Ross TJ, Salmeron BJ, McMahon RP, Gold JM, et al. The roles of reward, default, and executive control networks in set-shifting impairments in schizophrenia. PLoS One. 2013;8(2):e57257.
47. Yang X, Song Y, Zou Y, Li Y, Zeng J. Neural correlates of prediction error in patients with schizophrenia: evidence from an fMRI meta-analysis. Cereb Cortex. 2023;34(1):bhad471.
48. Culbreth AJ, Gold JM, Cools R, Barch DM. Impaired activation in cognitive control regions predicts reversal learning in schizophrenia. Schizophr Bull. 2016;42(2):484–93.
49. Katthagen T, Kaminski J, Heinz A, Buchert R, Schlagenhauf F. Striatal dopamine and reward prediction error signaling in Unmedicated schizophrenia patients. Schizophr Bull. 2020;46(6):1535–46.
50. Ermakova AO, Knolle F, Justicia A, Bullmore ET, Jones PB, Robbins TW, et al. Abnormal reward prediction-error signalling in antipsychotic naive individuals with first-episode psychosis or clinical risk for psychosis. Neuropsychopharmacology. 2018;43(8):1691–9.
51. Nagel J, Morgan DP, Gürsoy NÇ, Sander S, Kern S, Feld GB. Memory for rewards guides retrieval. Commun Psychol. 2024;2(1):31.
52. Duan H, Fernández G, van Dongen E, Kohn N. The effect of intrinsic and extrinsic motivation on memory formation: insight from behavioral and imaging study. Brain Struct Funct. 2020;225(5):1561–74.
53. Shigemune Y, Tsukiura T, Nouchi R, Kambara T, Kawashima R. Neural mechanisms underlying the reward-related enhancement of motivation when remembering episodic memories with high difficulty. Hum Brain Mapp. 2017;38(7):3428–43.
54. Miendlarzewska EA, Bavelier D, Schwartz S. Influence of reward motivation on human declarative memory. Neurosci Biobehav Rev. 2016;61:156–76.
55. Adcock RA, Thangavel A, Whitfield-Gabrieli S, Knutson B, Gabrieli JD. Reward-motivated learning: mesolimbic activation precedes memory formation. Neuron. 2006;50(3):507–17.
56. Yee DM, Crawford JL, Lamichhane B, Braver TS. Dorsal anterior cingulate cortex encodes the integrated incentive motivational value of cognitive task performance. J Neurosci. 2021;41(16):3707–20.
57. Esterman M, Poole V, Liu G, DeGutis J. Modulating reward induces differential neurocognitive approaches to sustained attention. Cereb Cortex. 2017;27(8):4022–32.
58. Esterman M, Reagan A, Liu G, Turner C, DeGutis J. Reward reveals dissociable aspects of sustained attention. J Exp Psychol Gen. 2014;143(6):2287–95.
59. Weinstein AM. Reward, motivation and brain imaging in human healthy participants – a narrative review. Front Behav Neurosci. 2023;17:1123733.
60. Cho YT, Lam NH, Starc M, Santamauro N, Savic A, Diehl CK, et al. Effects of reward on spatial working memory in schizophrenia. J Abnorm Psychol. 2018;127(7):695–709.
61. Hager OM, Kirschner M, Bischof M, Hartmann-Riemer MN, Kluge A, Seifritz E, et al. Reward-dependent modulation of working memory is associated with negative symptoms in schizophrenia. Schizophr Res. 2015;168(1–2):238–44.
62. Thornton AE, Boudreau VG, Griffiths SY, Woodward TS, Fawkes-Kirby T, Honer WG. The impact of monetary reward on memory in schizophrenia spectrum disorder. Neuropsychology. 2007;21(5):631–45.

63. Robison AJ, Thakkar KN, Diwadkar VA. Cognition and reward circuits in schizophrenia: synergistic, not separate. Biol Psychiatry. 2020;87(3):204–14.
64. Bègue I, Kaiser S, Kirschner M. Pathophysiology of negative symptom dimensions of schizophrenia—current developments and implications for treatment. Neurosci Biobehav Rev. 2020;116:74–88.
65. Giordano GM, Sanmarchi F, Mucci A, Rucci P, Brando F, Caporusso E, et al. External validation of the five domains of negative symptoms: focus on cognition, functional capacity, and real-world functioning. Eur Psychiatry. 2024;67(1):e3.
66. Engel M, Fritzsche A, Lincoln TM. Anticipation and experience of emotions in patients with schizophrenia and negative symptoms. An experimental study in a social context. Schizophr Res. 2016;170(1):191–7.
67. Zhang R-t, Yang Z-y, Wang Y-m, Wang Y, Yang T-X, Cheung EF, et al. Affective forecasting in individuals with social anhedonia: the role of social components in anticipated emotion, prospection and neural activation. Schizophr Res. 2020;215:322–9.
68. Lee J, Jimenez AM, Reavis EA, Horan WP, Wynn JK, Green MF. Reduced neural sensitivity to social vs nonsocial reward in schizophrenia. Schizophr Bull. 2019;45(3):620–8.
69. Chevallier C, Tonge N, Safra L, Kahn D, Kohls G, Miller J, et al. Measuring social motivation using signal detection and reward responsiveness. PLoS One. 2016;11(12):e0167024.
70. Catalano LT, Green MF. Social motivation in schizophrenia: what's effort got to do with it? Schizophr Bull. 2023;49(5):1127–37.
71. Campellone TR, Truong B, Gard D, Schlosser DA. Social motivation in people with recent-onset schizophrenia spectrum disorders. J Psychiatr Res. 2018;99:96–103.
72. Martins D, Rademacher L, Gabay AS, Taylor R, Richey JA, Smith DV, et al. Mapping social reward and punishment processing in the human brain: a voxel-based meta-analysis of neuroimaging findings using the social incentive delay task. Neurosci Biobehav Rev. 2021;122:1–17.
73. Goerlich KS, Votinov M, Lammertz SE, Winkler L, Spreckelmeyer KN, Habel U, et al. Effects of alexithymia and empathy on the neural processing of social and monetary rewards. Brain Struct Funct. 2017;222:2235–50.
74. Barman A, Richter S, Soch J, Deibele A, Richter A, Assmann A, et al. Gender-specific modulation of neural mechanisms underlying social reward processing by Autism Quotient. Soc Cogn Affect Neurosci. 2015;10(11):1537–47.
75. Spreckelmeyer KN, Rademacher L, Paulus FM, Gründer G. Neural activation during anticipation of opposite-sex and same-sex faces in heterosexual men and women. NeuroImage. 2013;66:223–31.
76. Li X, Xiao Y-H, Zou L-Q, Li H-H, Yang Z-Y, Shi H-S, et al. The effects of working memory training on enhancing hedonic processing to affective rewards in individuals with high social anhedonia. Psychiatry Res. 2016;245:482–90.
77. Fulford D, Campellone T, Gard DE. Social motivation in schizophrenia: how research on basic reward processes informs and limits our understanding. Clin Psychol Rev. 2018;63:12–24.
78. Catalano LT, Wynn JK, Green MF, Gold JM. Reduced neural activity when anticipating social versus nonsocial rewards in schizophrenia: preliminary evidence from an ERP study. Schizophr Res. 2022;246:7–16.
79. Catalano LT, Green MF, Wynn JK, Lee J. People with schizophrenia do not show the normal benefits of social versus nonsocial attentional cues. Neuropsychology. 2020;34(6):620–8.
80. Campellone TR, Kring AM. Anticipated pleasure for positive and negative social interaction outcomes in schizophrenia. Psychiatry Res. 2018;259:203–9.
81. Merchant JT, Barch DM, Ermel JA, Moran EK, Butler PD. Differential deficits in social versus monetary reinforcement learning in schizophrenia: associations with facial emotion recognition. J Psychopathol Clin Sci. 2024;133(1):37–47.
82. Catalano LT, Heerey EA, Gold JM. The valuation of social rewards in schizophrenia. J Abnorm Psychol. 2018;127(6):602–11.

83. Thibaudeau E, Rae J, Raucher-Chéné D, Bougeard A, Lepage M. Disentangling the relationships between the clinical symptoms of schizophrenia Spectrum disorders and theory of mind: a meta-analysis. Schizophr Bull. 2023;49(2):255–74.
84. Zierhut M, Böge K, Bergmann N, Hahne I, Braun A, Kraft J, et al. The relationship between the recognition of basic emotions and negative symptoms in individuals with schizophrenia spectrum disorders – an exploratory study. Front Psych. 2022;13:865226.
85. Andrzejewska M, Wójciak P, Domowicz K, Rybakowski J. Emotion recognition and theory of mind in chronic schizophrenia: association with negative symptoms. Arch Psychiatry Psychother. 2017;19(4):7–12.
86. Silver H, Shlomo N. Perception of facial emotions in chronic schizophrenia does not correlate with negative symptoms but correlates with cognitive and motor dysfunction. Schizophr Res. 2001;52(3):265–73.
87. Galderisi S, Rucci P, Kirkpatrick B, Mucci A, Gibertoni D, Rocca P, et al. Interplay among psychopathologic variables, personal resources, context-related factors, and real-life functioning in individuals with schizophrenia: a network analysis. JAMA Psychiatry. 2018;75(4):396–404.
88. Hajdúk M, Abplanalp SJ, Jimenez AM, Fisher M, Haut KM, Hooker CI, et al. Linking social motivation, general motivation, and social cognition to interpersonal functioning in schizophrenia: insights from exploratory graph analysis. Eur Arch Psychiatry Clin Neurosci. 2023; https://doi.org/10.1007/s00406-023-01733-4.
89. Powell SB, Swerdlow NR. The relevance of animal models of social isolation and social motivation for understanding schizophrenia: review and future directions. Schizophr Bull. 2023;49(5):1112–26.
90. Patterson TL, Moscona S, McKibbin CL, Davidson K, Jeste DV. Social skills performance assessment among older patients with schizophrenia. Schizophr Res. 2001;48(2–3):351–60.
91. Ehlen F, Montag C, Leopold K, Heinz A. Linguistic findings in persons with schizophrenia-a review of the current literature. Front Psychol. 2023;14:1287706.
92. Eckert MA, Menon V, Walczak A, Ahlstrom J, Denslow S, Horwitz A, et al. At the heart of the ventral attention system: the right anterior insula. Hum Brain Mapp. 2009;30(8):2530–41.
93. Kenneth I, Vaden J, Kuchinsky SE, Cute SL, Ahlstrom JB, Dubno JR, Eckert MA. The Cingulo-Opercular network provides word-recognition benefit. J Neurosci. 2013;33(48):18979–86.
94. Geranmayeh F, Wise RJS, Mehta A, Leech R. Overlapping networks engaged during spoken language production and its cognitive control. J Neurosci. 2014;34(26):8728–40.
95. Yumba WK. Cognitive processing speed, working memory, and the intelligibility of hearing aid-processed speech in persons with hearing impairment. Front Psychol. 2017;8:1308.
96. Mattys SL, Barden K, Samuel AG. Extrinsic cognitive load impairs low-level speech perception. Psychon Bull Rev. 2014;21(3):748–54.
97. Cohen AS, McGovern JE, Dinzeo TJ, Covington MA. Speech deficits in serious mental illness: a cognitive resource issue? Schizophr Res. 2014;160(1–3):173–9.
98. Kaiser S, Lyne J, Agartz I, Clarke M, Mørch-Johnsen L, Faerden A. Individual negative symptoms and domains – relevance for assessment, pathomechanisms and treatment. Schizophr Res. 2017;186:39–45.
99. Docherty AR, Berenbaum H, Kerns JG. Alogia and formal thought disorder: differential patterns of verbal fluency task performance. J Psychiatr Res. 2011;45(10):1352–7.
100. Sumiyoshi C, Sumiyoshi T, Nohara S, Yamashita I, Matsui M, Kurachi M, et al. Disorganization of semantic memory underlies alogia in schizophrenia: an analysis of verbal fluency performance in Japanese subjects. Schizophr Res. 2005;74(1):91–100.
101. Fervaha G, Takeuchi H, Foussias G, Agid O, Remington G. Using poverty of speech as a case study to explore the overlap between negative symptoms and cognitive dysfunction. Schizophr Res. 2016;176(2–3):411–6.
102. Cohen AS, Cox CR, Le TP, Cowan T, Masucci MD, Strauss GP, et al. Using machine learning of computerized vocal expression to measure blunted vocal affect and alogia. NPJ Schizophr. 2020;6(1):26.

103. Melinder MR, Barch DM. The influence of a working memory load manipulation on language production in schizophrenia. Schizophr Bull. 2003;29(3):473–85.
104. Le TP, Najolia GM, Minor KS, Cohen AS. The effect of limited cognitive resources on communication disturbances in serious mental illness. Psychiatry Res. 2017;248:98–104.
105. Trampush JW, Lencz T, DeRosse P, John M, Gallego JA, Petrides G, et al. Relationship of cognition to clinical response in first-episode schizophrenia Spectrum disorders. Schizophr Bull. 2015;41(6):1237–47.
106. Wong SCY, Chang WC, Hui CLM, Chan SKW, Lee EHM, Suen YN, et al. Relationship of subjective quality of life with symptomatology, neurocognition and psychosocial functioning in first-episode psychosis: a structural equation modelling approach. Eur Arch Psychiatry Clin Neurosci. 2021;271(8):1561–9.
107. Compton MT, Ku BS, Covington MA, Metzger C, Hogoboom A. Lexical diversity and other linguistic measures in schizophrenia: associations with negative symptoms and neurocognitive performance. J Nerv Ment Dis. 2023;211(8):613–20.
108. Bora E, Yalincetin B, Akdede BB, Alptekin K. Neurocognitive and linguistic correlates of positive and negative formal thought disorder: a meta-analysis. Schizophr Res. 2019;209:2–11.
109. di San B, Pietro C, Barbieri E, Marelli M, de Girolamo G, Luzzatti C. Processing argument structure and syntactic complexity in people with schizophrenia Spectrum disorders. J Commun Disord. 2022;96:106182.
110. Ventura J, Ered A, Gretchen-Doorly D, Subotnik KL, Horan WP, Hellemann GS, et al. Theory of mind in the early course of schizophrenia: stability, symptom and neurocognitive correlates, and relationship with functioning. Psychol Med. 2015;45(10):2031–43.
111. Ditlevsen JV, Simonsen A, Bliksted VF. Predicting mentalizing deficits in first-episode schizophrenia from different subdomains of negative symptoms. Schizophr Res. 2020;215:439–41.
112. Bliksted V, Videbech P, Fagerlund B, Frith C. The effect of positive symptoms on social cognition in first-episode schizophrenia is modified by the presence of negative symptoms. Neuropsychology. 2017;31(2):209–19.
113. Raucher-Chéné D, Thibaudeau E, Sauvé G, Lavigne KM, Lepage M. Understanding others as a mediator between verbal memory and negative symptoms in schizophrenia-spectrum disorder. J Psychiatr Res. 2021;143:429–35.
114. Schilbach L, Eickhoff SB, Mojzisch A, Vogeley K. What's in a smile? Neural correlates of facial embodiment during social interaction. Soc Neurosci. 2008;3(1):37–50.
115. Hennenlotter A, Schroeder U, Erhard P, Castrop F, Haslinger B, Stoecker D, et al. A common neural basis for receptive and expressive communication of pleasant facial affect. NeuroImage. 2005;26(2):581–91.
116. Gur RE, Kohler CG, Ragland JD, Siegel SJ, Lesko K, Bilker WB, et al. Flat affect in schizophrenia: relation to emotion processing and neurocognitive measures. Schizophr Bull. 2006;32(2):279–87.
117. Lepage M, Sergerie K, Benoit A, Czechowska Y, Dickie E, Armony J. Emotional face processing and flat affect in schizophrenia: functional and structural neural correlates. Psychol Med. 2011;41(9):1833–44.
118. Jeganathan J, Breakspear M. An active inference perspective on the negative symptoms of schizophrenia. Lancet Psychiatry. 2021;8(8):732–8.
119. Valizadeh A, Mbwogge M, Rasouli Yazdi A, Hedayati Amlashi N, Haadi A, Shayestefar M, et al. The mirror mechanism in schizophrenia: a systematic review and qualitative meta-analysis. Front Psych. 2022;13:884828.
120. Bonini L, Rozzi S, Serventi FU, Simone L, Ferrari PF, Fogassi L. Ventral premotor and inferior parietal cortices make distinct contribution to action organization and intention understanding. Cereb Cortex. 2010;20(6):1372–85.
121. Ferrari PF, Gerbella M, Coudé G, Rozzi S. Two different mirror neuron networks: the sensorimotor (hand) and limbic (face) pathways. Neuroscience. 2017;358:300–15.
122. Lee JS, Chun JW, Yoon SY, Park H-J, Kim J-J. Involvement of the mirror neuron system in blunted affect in schizophrenia. Schizophr Res. 2014;152(1):268–74.

123. Marshall PJ, Meltzoff AN. Neural mirroring mechanisms and imitation in human infants. Philos Trans R Soc Lond Ser B Biol Sci. 2014;369(1644):20130620.
124. Mehta UM, Thirthalli J, Aneelraj D, Jadhav P, Gangadhar BN, Keshavan MS. Mirror neuron dysfunction in schizophrenia and its functional implications: a systematic review. Schizophr Res. 2014;160(1):9–19.
125. Thakkar KN, Peterman JS, Park S. Altered brain activation during action imitation and observation in schizophrenia: a translational approach to investigating social dysfunction in schizophrenia. Am J Psychiatry. 2014;171(5):539–48.
126. Singh F, Pineda J, Cadenhead KS. Association of impaired EEG mu wave suppression, negative symptoms and social functioning in biological motion processing in first episode of psychosis. Schizophr Res. 2011;130(1–3):182–6.
127. Saito Y, Kubicki M, Koerte I, Otsuka T, Rathi Y, Pasternak O, et al. Impaired white matter connectivity between regions containing mirror neurons, and relationship to negative symptoms and social cognition, in patients with first-episode schizophrenia. Brain Imaging Behav. 2018;12(1):229–37.
128. ElShahawi HH, Sakr HM, Hashim MA, Mohamed HH, Abdeen MS. Social cognition correlation to white matter integrity alteration in mirror neurons of schizophrenic patients: DTI study. Neurol Psychiatry Brain Res. 2020;38:65–73.
129. Hickok G. Eight problems for the mirror neuron theory of action understanding in monkeys and humans. J Cogn Neurosci. 2009;21(7):1229–43.
130. Heyes C, Catmur C. What happened to Mirror neurons? Perspect Psychol Sci. 2022;17(1):153–68.
131. Abram SV, Weittenhiller LP, Bertrand CE, McQuaid JR, Mathalon DH, Ford JM, et al. Psychological dimensions relevant to motivation and pleasure in schizophrenia. Front Behav Neurosci. 2022;16:827260.
132. Campellone TR, Sanchez AH, Kring AM. Defeatist performance beliefs, negative symptoms, and functional outcome in schizophrenia: a meta-analytic review. Schizophr Bull. 2016;42(6):1343–52.
133. Lincoln TM, Mehl S, Kesting ML, Rief W. Negative symptoms and social cognition: identifying targets for psychological interventions. Schizophr Bull. 2011;37(Suppl 2):S23–32.
134. Grant PM, Beck AT. Defeatist beliefs as a mediator of cognitive impairment, negative symptoms, and functioning in schizophrenia. Schizophr Bull. 2009;35(4):798–806.
135. Rector NA, Beck AT, Stolar N. The negative symptoms of schizophrenia: a cognitive perspective. Can J Psychiatr. 2005;50(5):247–57.
136. Quinlan T, Roesch S, Granholm E. The role of dysfunctional attitudes in models of negative symptoms and functioning in schizophrenia. Schizophr Res. 2014;157(1–3):182–9.
137. Luther L, Westbrook A, Ayawvi G, Ruiz I, Raugh IM, Chu AOK, et al. The role of defeatist performance beliefs on cognitive effort-cost decision-making in schizophrenia. Schizophr Res. 2023;261:216–24.
138. Reddy LF, Horan WP, Barch DM, Buchanan RW, Gold JM, Marder SR, et al. Understanding the association between negative symptoms and performance on effort-based decision-making tasks: the importance of defeatist performance beliefs. Schizophr Bull. 2018;44(6):1217–26.
139. Granholm E, Ruiz I, Gallegos-Rodriguez Y, Holden J, Link PC. Pupillary responses as a biomarker of diminished effort associated with defeatist attitudes and negative symptoms in schizophrenia. Biol Psychiatry. 2016;80(8):581–8.
140. Granholm E, Holden J, Worley M. Improvement in negative symptoms and functioning in cognitive-behavioral social skills training for schizophrenia: mediation by defeatist performance attitudes and asocial beliefs. Schizophr Bull. 2018;44(3):653–61.
141. Ventura J, Subotnik KL, Ered A, Gretchen-Doorly D, Hellemann GS, Vaskinn A, et al. The relationship of attitudinal beliefs to negative symptoms, neurocognition, and daily functioning in recent-onset schizophrenia. Schizophr Bull. 2014;40(6):1308–18.
142. González-Blanch C, Medrano LA, Bendall S, D'Alfonso S, Cagliarini D, McEnery C, et al. The role of social relatedness and self-beliefs in social functioning in first-episode

psychosis: are we overestimating the contribution of illness-related factors? Eur Psychiatry. 2020;63(1):e92.
143. Yang Y, Yang ZY, Zou YM, Shi HS, Wang Y, Xie DJ, et al. Low-pleasure beliefs in patients with schizophrenia and individuals with social anhedonia. Schizophr Res. 2018;201:137–44.
144. Hu HX, Jiang SY, Shan HD, Chu MY, Lv QY, Yi ZH, et al. Negative belief-updating bias for positive daily life events in individuals with schizophrenia and social anhedonia. Cogn Neuropsychiatry. 2022;27(4):237–54.
145. Fett A-K, Shergill S, Korver-Nieberg N, Yakub F, Gromann P, Krabbendam L. Learning to trust: trust and attachment in early psychosis. Psychol Med. 2016;46(7):1437–47.
146. Heerey EA. Learning from social rewards predicts individual differences in self-reported social ability. J Exp Psychol Gen. 2014;143(1):332–9.
147. Galderisi S, Mucci A, Dollfus S, Nordentoft M, Falkai P, Kaiser S, et al. EPA guidance on assessment of negative symptoms in schizophrenia. Eur Psychiatry. 2021;64(1):e23.
148. Glenthøj LB, Kristensen TD, Wenneberg C, Hjorthøj C, Nordentoft M. Experiential negative symptoms are more predictive of real-life functional outcome than expressive negative symptoms in clinical high-risk states. Schizophr Res. 2020;218:151–6.
149. Halverson TF, Orleans-Pobee M, Merritt C, Sheeran P, Fett AK, Penn DL. Pathways to functional outcomes in schizophrenia spectrum disorders: meta-analysis of social cognitive and neurocognitive predictors. Neurosci Biobehav Rev. 2019;105:212–9.
150. Giuliani L, Giordano GM, Bucci P, Pezzella P, Brando F, Galderisi S. Improving knowledge on pathways to functional outcome in schizophrenia: Main results from the Italian network for research on psychoses. Front Psych. 2021;12:791117.
151. Mucci A, Galderisi S, Gibertoni D, Rossi A, Rocca P, Bertolino A, et al. Factors associated with real-life functioning in persons with schizophrenia in a 4-year follow-up study of the Italian network for research on psychoses. JAMA Psychiatry. 2021;78(5):550–9.
152. Galderisi S, Bucci P, Mucci A, Kirkpatrick B, Pini S, Rossi A, et al. Categorical and dimensional approaches to negative symptoms of schizophrenia: focus on long-term stability and functional outcome. Schizophr Res. 2013;147(1):157–62.
153. Ventura J, Hellemann GS, Thames AD, Koellner V, Nuechterlein KH. Symptoms as mediators of the relationship between neurocognition and functional outcome in schizophrenia: a meta-analysis. Schizophr Res. 2009;113(2–3):189–99.
154. Giordano GM, Pezzella P, Mucci A, Austin SF, Erfurth A, Glenthøj B, et al. Negative symptoms and social cognition as mediators of the relationship between neurocognition and functional outcome in schizophrenia. Front Psych. 2024;15:1333711.
155. Luther L, Suor JH, Rosen C, Jobe TH, Faull RN, Harrow M. Clarifying the direction of impact of negative symptoms and neurocognition on prospective work functioning in psychosis: a 20-year longitudinal study. Schizophr Res. 2020;220:232–9.
156. Kam CTK, Fung VSC, Chang WC, Hui CLM, Chan SKW, Lee EHM, et al. Cognitive subgroups and the relationships with symptoms, psychosocial functioning and quality of life in first-episode non-affective psychosis: a cluster-analysis approach. Front Psych. 2023;14:1203655.
157. Carruthers SP, Van Rheenen TE, Gurvich C, Sumner PJ, Rossell SL. Characterising the structure of cognitive heterogeneity in schizophrenia spectrum disorders. A systematic review and narrative synthesis. Neurosci Biobehav Rev. 2019;107:252–78.
158. Fekih-Romdhane F, Hajje R, Haddad C, Hallit S, Azar J. Exploring negative symptoms heterogeneity in patients diagnosed with schizophrenia and schizoaffective disorder using cluster analysis. BMC Psychiatry. 2023;23(1):595.
159. Hagi K, Nosaka T, Dickinson D, Lindenmayer JP, Lee J, Friedman J, Boyer L, et al. Association between cardiovascular risk factors and cognitive impairment in people with schizophrenia: a systematic review and meta-analysis. JAMA Psychiatry. 2021;78(5):510–8.
160. Monteleone P, Cascino G, Monteleone AM, Rocca P, Rossi A, Bertolino A, et al. Prevalence of antipsychotic-induced extrapyramidal symptoms and their association with neurocognition and social cognition in outpatients with schizophrenia in the "real-life". Prog Neuro-Psychopharmacol Biol Psychiatry. 2021;109:110250.

161. Kirkpatrick B, Fenton WS, Carpenter WT Jr, Marder SR. The NIMH-MATRICS consensus statement on negative symptoms. Schizophr Bull. 2006;32(2):214–9.
162. Galderisi S, Mucci A, Buchanan RW, Arango C. Negative symptoms of schizophrenia: new developments and unanswered research questions. Lancet Psychiatry. 2018;5(8):664–77.
163. Mucci A, Vignapiano A, Bitter I, Austin SF, Delouche C, Dollfus S, et al. A large European, multicenter, multinational validation study of the Brief Negative Symptom Scale. Eur Neuropsychopharmacol. 2019;29(8):947–59.
164. Rönnberg J, Lunner T, Zekveld A, Sörqvist P, Danielsson H, Lyxell B, et al. The Ease of Language Understanding (ELU) model: theoretical, empirical, and clinical advances. Front Syst Neurosci. 2013;7:31.
165. Unsworth N, Engle RW. On the division of short-term and working memory: an examination of simple and complex span and their relation to higher order abilities. Psychol Bull. 2007;133(6):1038–66.
166. Nebeker C, Bartlett Ellis RJ, Torous J. Development of a decision-making checklist tool to support technology selection in digital health research. Transl Behav Med. 2020;10(4):1004–15.
167. Torous J, Camacho E, Myrick K, editors. Equitable and informed digital mental health: Skills, evaluation, and integration of apps into care. TMS Proceedings 2021; 2021: PubPub.
168. Torous J, Andersson G, Bertagnoli A, Christensen H, Cuijpers P, Firth J, et al. Towards a consensus around standards for smartphone apps and digital mental health. World Psychiatry. 2019;18(1):97–8.
169. Galderisi S, Mucci A. A new approach to negative symptoms of schizophrenia. Eur Neuropsychopharmacol. 2023;75:62–4.
170. Granholm E, Ben-Zeev D, Fulford D, Swendsen J. Ecological Momentary Assessment of social functioning in schizophrenia: impact of performance appraisals and affect on social interactions. Schizophr Res. 2013;145(1–3):120–4.
171. Gard DE, Sanchez AH, Cooper K, Fisher M, Garrett C, Vinogradov S. Do people witschizophrenia have difficulty anticipating pleasure, engaging in effortful behavior, or both? J Abnorm Psychol. 2014;123(4):771–82.
172. Guimond S, Keshavan MS, Torous JB. Towards remote digital phenotyping of cognition in schizophrenia. Schizophr Res. 2019;208:36–8.
173. Gupta T, Haase CM, Strauss GP, Cohen AS, Ricard JR, Mittal VA. Alterations in facial expressions of emotion: determining the promise of ultrathin slicing approaches and comparing human and automated coding methods in psychosis risk. Emotion. 2022;22(4):714–24.
174. Shore D, Robertson O, Lafit G, Parkinson B. Facial regulation during dyadic interaction: interpersonal effects on cooperation. Affect Sci. 2023;4(3):506–16.

15 Negative Symptoms, Cognition, and Recovery in Schizophrenia

Bernardo Carpiniello, Federica Pinna, Pasquale Paribello, and Mirko Manchia

Abstract

Two parallel visions of recovery may be found in literature, one adhered to by clinicians, usually defined as "clinical recovery," the other based on the personal, subjective experience of people living with schizophrenia, usually defined as "personal recovery" or "subjective recovery." Both are considered as relevant targets of treatment in schizophrenia and related disorders. The present chapter reviews how much both negative and cognitive symptoms may affect clinical recovery, through their influence on both sustained remission and personal/social functioning, the two main dimensions of clinical recovery. Personal recovery mostly refers to subjective experiences of optimism, empowerment, interpersonal support, peer support, and stigma reduction. Although symptom remission and functioning are not considered necessary prerequisites of personal recovery, several studies seem to show some contribution of clinical dimensions, with particular reference to negative symptoms and social cognition.

Keywords

Clinical recovery · Personal recovery · Negative symptoms · Cognitive symptoms · Neurocognitive symptoms · Social cognition

B. Carpiniello (✉)
Department of Medical Sciences and Public Health, University of Cagliari, Cagliari, Italy

F. Pinna · P. Paribello · M. Manchia
Department of Medical Sciences and Public Health, University of Cagliari and Psychiatric Unit, University Hospital Trust, Cagliari, Italy

A. Vita et al. (eds.), *Negative and Cognitive Symptoms in Schizophrenia*, Comprehensive Approach to Psychiatry 3,
https://doi.org/10.1007/978-3-032-06209-3_15

15.1 Introduction

As is well known, two clear-cut parallel visions of recovery may be found in literature, one adhered to by clinicians, usually defined as "clinical recovery," the other based on the personal, subjective experience of people living with schizophrenia, usually defined as "personal recovery" or "subjective recovery" [1]. Both clinicians and mental health service users consider them as a valuable and worthwhile target of treatment, starting from their respective points of view. In any case, both of these targets should be considered nowadays as fundamental in mental health services organizations and activities. Unfortunately, the substantial differences in conceptualizing recovery by clinicians and service users make it difficult to reconcile their opposite ways of evaluating it, also considering that studies about correlations between the "objective" measures of clinical recovery, namely symptomatology and functioning, and the evaluation of the "subjective" experiences which are typical of personal recovery have reported quite inconsistent findings [2]. Considering that symptomatology and functioning are two fundamental dimensions of clinical recovery and the specific role of "negative" and "cognitive" symptoms on the latter, they will be, of course, the main focus of this chapter, although they may have a not irrelevant role in personal recovery too, as we try to demonstrate.

15.2 Determinants of Clinical Recovery

Clinical recovery, mainly focused on in the present chapter, otherwise named by some authors as "objective recovery," has been conceived in different ways: as a "unidimensional" construct, generally indicating a more or less "sustained remission," as a "bi-dimensional" construct, including both clinical remission and functional remission or as a "multidimensional" construct, including other dimensions beyond symptom remission and functioning, such as quality of life or subjective well-being [1]. Despite the lack of a clear consensus regarding how many dimensions should be considered, data from the literature show that the by far most extensively shared definition of clinical recovery is the attainment of both symptom and functional remission according to a set of specific criteria [3]. So defined, clinical recovery is considered the optimal target of therapeutical and rehabilitative interventions [4], a seemingly attainable target emerging from meta-analyses and systematic reviews. Indeed, Jääskeläinen et al. [5] reported a median proportion of cases meeting recovery criteria of 13.5%, with a range between 8.1% and 20.0%. Warner [6] found a rate of 20.4% of fully recovered among multiple episode patients and a similar rate among first episode subjects (20.7%). Huxley et al. [7] reported a 37.5% rate of "complete" recovery in "multiple episode" patients and 57.1% among first-episode patients, while Molstrom et al. [8] found that 24.2% of patients could be considered in recovery, in their meta-analysis concerning only long-term (at least 20 yrs) follow-up studies. Beyond this wide variability of data, partly due to methodological differences of the studies, it seems clear that a more optimistic vision of Schizophrenia outcome emerges, different from the substantial negative views of the past. On the other hand, a clear difficulty in getting and maintaining clinical recovery for the majority of people affected by schizophrenia still emerges from the available data. The reasons why clinical recovery may be quite difficult to obtain are related to a series of factors.

First of all, even though symptomatic remission is one of the essential prerequisites of clinical recovery, being free from relevant symptoms is not enough. Indeed, a recent meta-analysis regarding the correlations between psychopathology (i.e., symptoms) and functioning in Schizophrenia [9] shows that all psychopathological subdomains seem to be negatively correlated with measures of social functioning, i.e., the higher the levels of symptoms, the lower the levels of functioning. However, looking at correlation indices, the proportion of shared variance between symptoms and functioning is no more than 27% (Table 15.1), which means an important but not exclusive role of symptomatology in explaining schizophrenia-related disability. Such evidence is supported by other studies, for instance, an interventional prospective study evaluating the correlation between clinical outcome measures and personal and social performance functioning among patients with schizophrenia, showing that even though significant positive correlations could be found between scores at symptoms and functioning rating scales, 45% of the variation of scores at PSP, the scale adopted in this study, could not be explained by other clinical evaluation measures [10]. In other words, these findings indicate again that the level of impairment in functioning is largely independent of symptomatology. Moreover, being symptomatically "remitted" does not necessarily imply a satisfying everyday life functioning, given that only 30% of patients displaying symptomatic remission were also found to be functionally remitted in a sample of chronic subjects affected by schizophrenia treated in a community setting and cross-sectionally evaluated at baseline in the context of a prospective naturalistic follow-up study [11]. These pieces of evidence are confirmed by a seminal paper about remission in schizophrenia, where prospective studies assessing the proportion of patients in remission having a good functional level showed that only 30–38% of remitted patients displayed adequate functioning at follow-ups [12]. Thus, symptom dimensions of schizophrenia have an important but not unique role in determining disability, considering that the majority of subjects who are symptomatically remitted are not functionally remitted.

Secondly, the stability of symptomatic remission, an essential variable for clinical recovery, appears quite difficult to maintain, as shown by a longitudinal study of

Table 15.1 Correlation between psychopathological dimensions and functioning in schizophrenia

Symptom dimensions	Correlation	95% Confidence intervals	Level of significance
General psychopathology	$r = -0.52$	−0.60; −0.43	$p < 0.01$
Negative symptoms	$r = -0.52$	−0.61; −0.42	$p < 0.01$
Disorganized Symptoms	$r = -0.35$	−0.54; −0.14	$p < 0.01$
Positive symptoms	$r = -0.32$	−0.39; −0.25	$p < 0.01$
Depressive symptoms	$r = -0.23$	−0.33; −0.11	$p < 0.01$

From: Handest et al. [9], modified

a cohort of community-dwelling outpatients with schizophrenia, which demonstrates how the proportion of remitted patients who had a sustained remission progressively decreased during the 2 years follow up, from 86% at 6 months to only 45% at 24 months [13].

We should also consider that the other major pillar of clinical recovery, the real-life functioning, is multidetermined, as clearly shown by a series of studies of the Italian Network on Psychoses, demonstrating that not only a series of psychopathologic factors, in particular negative symptoms and cognitive (neurocognitive and social cognitive) impairments but also other factors as disorganization, depression, autistic symptoms, together with a series of personal and environmental variables [14–18].

Independent of how symptoms affect directly or indirectly recovery, in a lifelong illness such as schizophrenia, symptoms that emerge in late adolescence/early adulthood and often tend to persist throughout the rest of the patient's life, such as negative symptoms, especially primary negative symptoms, and cognitive impairments, should be considered of outstanding importance for outcome [19].

15.3 Negative Symptoms and Clinical Recovery

15.3.1 Negative Symptoms as Predictors of Recovery

Examining data from prospective follow-up studies is probably the best way to evaluate the role of negative symptoms in relation to clinical recovery, given that the latter has been included more and more frequently as an outcome measure in these studies, with particular reference to those concerning "first episode" patients.

In a naturalistic 2-year follow-up study of a cohort consisting of 1175 first-episode patients, 345 (29.4%) recovered at the end of follow-up; multiple logistic regression revealed lower baseline PANSS negative scores as the only symptom emerging as a significant predictor of clinical recovery, together with female gender, being married, younger age, higher education, shorter DUP, and early treatment response at month 3 [20].

The Danish Opus study, one of the longest follow-up studies of first-episode patients, found that in a cohort of 394 participants only 42 subjects (14%) met the criteria for clinical recovery at the follow-up; according to a multivariable binary logistic regression analysis, baseline predictors accounted for 22% of the variance of full recovery and only lower severity of negative symptoms and earlier age at diagnosis predicted better rates of recovery after 10 years, leading the authors to conclude that "... negative symptoms could play a central role in the process of recovery from schizophrenia..." and that "a challenge for clinicians and researchers is to understand the mechanisms behind negative symptoms and develop interventions that can prevent or ameliorate these symptoms in order to promote recovery" [21].

A study examined predictors for clinical recovery in 375 patients with early-stage schizophrenia spectrum disorders followed up to 6.5 years, using as a definition of full recovery having the score of the Positive and Negative Syndrome Scale (PANSS) 8-item ≤2 and adequate functional recovery for at least 1-year; evaluating both total and acute patients, the study found a higher level of negative symptoms of the PANSS, higher duration of untreated psychosis (DUP) and non-professional job were identified in patients with early-stage SSD as independent predictors of unfavourable recovery through a stepwise Cox regression [22].

A cohort study of 255 patients with non-affective psychosis at their first episode examined subjects 5 years after first diagnosis and treatment, defining recovery as working or studying, having a GAF-function score of 60 or above, having remission of negative and psychotic symptoms, and not living in a supported housing facility or being hospitalized during the last 2 years before the five-year follow-up interview; according to these criteria, 40 (15.7%) patients were found to be recovered; recovery was predicted by female sex, higher age, pre-morbid social adaptation, growing up with both parents and low levels of negative symptoms at baseline as a unique clinical predictor [23].

A prospective follow-up study included patients meeting the criteria for a first episode of a schizophrenia spectrum or affective psychosis who were followed up for 2 years with a series of established predictors, verbal memory, and length of positive and negative symptom remission which were regressed on functioning at 1 ($n = 208$) and 2 ($n = 159$) years; regressions demonstrated that length of combined positive and negative symptom remission explained the most variance in functioning at 1 and 2 years, with verbal memory contributed only slightly to such outcome; length of remission of negative symptoms was a stronger predictor of functioning than remission of positive symptoms at 1 year, but length of positive symptom remission also made a large contribution at 2 years, [24]. These results highlighted the importance of achieving and maintaining remission of both negative and positive symptoms for longer periods in patients with a first episode of a psychotic disorder.

The Schizophrenia Outpatients Health Outcomes (SOHO) study, considered the largest naturalistic prospective cohort study in patients affected by schizophrenia under antipsychotic treatment [25], used a quite stringent definition of recovery, including not only long-lasting symptomatic and functional remission but also an adequate quality of life for a minimum of 24 months and until the 36-month visit; on these bases, the study found only 4% of subjects recovered among the 6642 patients evaluated during the 3 year follow-up period; logistic regression analysis showed that higher negative symptom severity, together with higher BMI and lack of effectiveness as the reason for change of medication at baseline, were the only factors associated with a lower likelihood of achieving recovery.

15.3.2 How Negative Symptoms Affect Recovery

15.3.2.1 Negative Symptoms and Sustained Symptom Remission

Two sets of evidence explain why negative symptoms predict clinical recovery. The first one is that lower levels of negative symptoms are directly associated with a higher probability of symptomatic remission. In the previously cited study regarding sustained remission, considered as the persistence of remission status of a cohort of 112 patients with schizophrenia or schizoaffective disorder in the six-month antecedent each one of the evaluations, regression analysis identified a lower level of positive and negative symptoms and cognitive deficits, together with a higher level of antipsychotic dosages expressed in chlorpromazine equivalents at baseline, as predictors of duration and stability of symptomatic status at the 2 years endpoint [13]. In a 3 years follow-up study of 92 Chinese patients with first-episode schizophrenia spectrum disorder, at the end of the follow only 4.6% of patients achieved sustained symptomatic remission; the latter was associated with both fewer 1-year positive and negative symptoms and better premorbid adjustment, better baseline occupational status and better 1-year verbal memory; a logistic regression showed that verbal memory impairment, premorbid functioning, and negative symptom severity independently predicted remission status [26]. Persistent negative symptoms were associated with a lower probability of remission in the European First Episode Schizophrenia Trial (EUFEST) study, where 6.7% of a sample of 345 first-episode patients presented unconfounded persistent negative symptoms at 1 year, together with more positive and disorganization symptoms [27]. A large cohort of patients in the early stage of schizophrenia, schizophreniform or schizoaffective disorder, was studied in the context of the OPTiMiSE trial regarding symptomatic remission, attrition rate and psychosocial functioning, assessed in subjects with short-term (4 weeks) persistent, unconfounded negative symptoms (PNS) and in those with non-persistent ones (N-PNS); this study shows that 11% of subjects had PNS unconfounded at baseline and 7.9% had PNS unconfounded at both baseline and end of 4-week treatment; with PNS subjects showing lower remission and higher attrition rates at the end of all treatment phases [28].

15.3.2.2 Negative Symptoms and Functioning

Besides being associated with a lower probability of symptomatic remission, the presence of negative symptoms affects negative functioning [29]. Indeed, with their two subdomains motivation (i.e., avolition/apathy and asociality) and diminished expression (i.e., poverty of speech and affective flattening), negative symptoms are considered among the best predictors of functional outcomes in schizophrenia, both through a direct role and an indirect effect through cognition, with a severe interference with real-life functioning, particularly when they are primary and persistent [30]. The OPTiMiSE trial shows that psychosocial functioning was comparable in PNS and N-PNS subjects at baseline but was significantly worse in the former group after 4-weeks [28]. A follow-up study of patients diagnosed as having deficit schizophrenia (DS) compared to patients affected by Non-Deficit schizophrenia (NDS) patients demonstrated that DS patients showed more severe negative symptoms and greater social dysfunction than NDS ones [31]. In a multicenter prospective cohort study regarding 921 patients enrolled in a cross-sectional study who were contacted after 4 years for reassessment, a series of baseline variables were directly associated with aspects of real-life functioning at follow-up; among these, avolition emerged as significantly associated with the interpersonal relationship domain [17].

Considering the currently used two-dimensional model of negative symptoms, where avolition, anhedonia, and asociality constitute the Experiential Deficit domain, and blunted affect and alogia the Expressive Deficit domain [30], some studies addressed the question of the impact of these negative domains on everyday life functioning, showing that the Experiential domain, in particular avolition, has a more evident predictive value regarding functional outcome respect to the expressive domain [17, 18, 32, 33]. In particular, motivation, which may be considered a particular aspect of avolition, appears to be related to the functional outcome [34–36].

15.4 Cognition and Clinical Recovery

15.4.1 Cognition as a Predictor of Recovery

A series of longitudinal studies, including first-episode ones, clearly show how cognitive domains (neurocognitive and social cognition) predict clinical recovery in schizophrenia. A study of 118 subjects in their first episode of schizophrenia or schizoaffective disorder ($N = 118$) at a 5-year follow-up found that 13.7% of subjects met full clinical recovery criteria (clinical remission plus adequate social functioning for 2 years or more) and that better cognitive functioning at stabilization, together with a shorter duration of psychosis at study entry and a more evident cerebral asymmetry [37] were associated with full recovery. A 2-year follow-up study of 255 individuals at their first episode of psychosis found that early-onset subjects with low premorbid IQ (as a measure of general cognitive functioning) showed persistent negative symptoms, lower functioning, and a lower recovery likelihood [38]. In a one-year observational follow-up study of 1010 outpatients with schizophrenia, a subgroup of 452 patients in symptomatic remission was followed for 1 year, showing at baseline that 103 (22.8%) of these patients fulfilled the recovery definition (having both symptomatic and functional remission), and that among the 338 patients maintaining symptomatic remission after1 year, the proportion of those recovered was increased to 27.1%; the study also found that better premorbid adjustment and social cognition measures correlated with recovery at baseline; moreover, after 1 year, better social cognition and other factors (lower duration of untreated psychosis, type of pharmacotherapy, positive attitudes toward medication and lower depressive symptoms) were associated to the likelihood of recovery [39].

15.4.2 How Cognition Affects Recovery

15.4.2.1 Cognition and Sustained Symptom Remission

The previously cited 2-year prospective studies of a cohort of chronic outpatients affected by schizophrenia spectrum disorders found that a lower level of cognitive deficits, together with several other factors, is one of the predictors of duration and stability of symptomatic status at the 2-year endpoint [13]. Similarly, the above-cited three-year follow-up study of patients with first-episode schizophrenia spectrum disorder found that sustained remission was associated, among several factors, with better 1-year verbal memory, which was also found as an independent predictor of the remission status at the endpoint through logistic regression analysis [26]. A one-year prospective study of 54 first-episode patients found that within the neurocognitive domain, only processing speed was associated with the 1-year remission. In contrast, the other baseline cognitive variables were not associated with remission [40].

15.4.2.2 Cognition and Functioning

Activities including the neurocognitive domain (NC) (processing speed, attention/vigilance, working memory, verbal learning and memory, visual learning and memory, reasoning and problem solving and verbal comprehension) and those belonging to the domain of social cognition (SC) (emotional processing; social perception and knowledge; theory of mind and attributional bias) are considered salient domains that should be considered in the clinical characterization of the patients with primary psychos such as schizophrenia [41]. Cognitive impairment in schizophrenia represents one of the main obstacles to clinical and functional recovery [42]. Other than being a "core feature" of schizophrenia, cognitive deficits are significantly associated with the impaired functioning of the disorder [43], as shown by a consistent series of studies. The one-year first-episode study previously cited [40] found that occupational status and maintaining life goals, as expression of functioning of patients, were predicted by processing speed at baseline, while measures of social cognition were associated with occupational status a year later; this study also shows that most of the associations were retained beyond baseline positive and affective symptom levels, but when negative symptoms were taken into account, cognition no longer predicted 1-year outcomes, a finding interpreted by the authors as a demonstration of an "overlapping" of negative symptoms with cognition. The relationship between negative symptoms and cognition concerning functional outcome was addressed by a study including 58 patients with schizophrenia, specifically devoted to evaluating the additional impact of negative symptoms, showing that global negative symptoms scores added a substantial additional variance to the one accounted for by neurocognition and functional capacity in predicting real-world functioning measures and employment status; moreover, both diminished expression and avolition-apathy predicted real-world functioning after controlling for cognition and functional capacity, with avolition-apathy also accounting for additional variance in employment status [44]. Substantially, findings from this study indicate an additional contribution of negative symptoms in predicting real-world functioning and employment outcomes after accounting for neurocognition and functional capacity. Another research group studied 131 first-episode patients, divided into two groups: those with functional deficits ($n = 97$) and those with functional recovery ($n = 34$), as emerging from a 1-year follow-up; the study showed that sustained attention together with negative symptoms emerged as significant predictors of good global functional outcome; however, while the model revealed a high accuracy (91%) in the classification of patients with functional deficits, it resulted too low (26%) in the classification of patients with global functional recovery; according the authors, this limitation in the prediction of a favourable global functional outcome highlights the need to address the role of other factors which are not commonly included in longitudinal studies of long-term outcomes in schizophrenia [45]. This knowledge gap was overcome by the Italian Network for Research on Psychoses study about determinants of real-world functioning in schizophrenia, probably the largest one conducted so far, which specifically addressed the interplay among illness-related variables, personal resources, context-related factors and real-life functioning in a large sample of patients with schizophrenia by using network analysis [15]. This cross-sectional study, including 921 subjects with schizophrenia, 379 unaffected first-degree relatives of these patients, and 780 healthy controls, demonstrated that social and non-social cognition, functional capacity, negative symptoms, resilience, and family or social incentives strongly influence a measure of global functioning. A subsequent follow up study including 618 patients from the original sample, through a series of papers demonstrated the key role of cognition, functional capacity, the experiential domain of negative symptoms, and everyday life skills in predicting functional outcome; in particular social cognition and the experiential domain of negative symptoms emerged as factors having an impact on interpersonal functioning, with non-social cognition having an impact on everyday life skills and both non-social cognition and social cognition predicting work skills; moreover, the relationships of cognitive impairment and negative symptoms was investigated using both a structural equation model (SEM) and a network analysis, showing that only the experiential domain of negative symptoms had a distinct direct effect on functioning, with network analysis demonstrating that the expressive deficit and social and non-social cognitive variables correlated to both functional capacity, and to disorganization [16, 17, 46]. A closer evaluation of the two subgroups constituted by 124 recovered patients (20.1%) and 494 (79.9%) non-recovered at the 4-year follow-up [18], functional capacity and everyday life skills were found to maintain a high betweenness and closeness in the network analysis at baseline, while psychopathological variables remained more peripheral; the network structure and connectivity of non-recovered patients were quite similar to the whole sample, but largely different from recovered subjects, which were characterized by few connections only; according to the authors, these findings suggested that "tightly coupled symptoms/dysfunctions tend to maintain each other's activation, contributing to poor outcome in schizophrenia." The contribution of cognition to functioning in schizophrenia has also been examined through four systematic reviews and/or meta-analyses. The first one [47], based upon 52 studies comprising 2692 subjects and 12 a priori, identified neurocognitive and social cognitive domains and four domains of functional outcome yielded a number of 25 significant mean correlations; overall, meta-analyses demonstrated that social cognition was more strongly associated with community functioning than neurocognition, with the strongest associations being between theory of mind and functional outcomes; anyway, the study shows also that about three-quarters of variance in outcome were left unexplained considering only cognitive factors impact on outcome. The second and more recent systematic review and meta-analysis [48] specifically evaluated predictors of functional recovery in first-episode patients considering only longitudinal studies with a minimum 12-month follow-up, based on 50 eligible studies regarding 6669 subjects, duration of untreated psychosis (DUP), most cognitive variables, and concurrent remission of positive and negative symptoms were all independently related to functional recovery; in particular, a significant association was found between overall functioning and general cognitive ability, attention, working memory, verbal fluency/language, verbal memory/learning, processing speed but not with executive functioning, nonverbal memory and learning and visuo-motor skills. The third meta-analysis [49] regards specifically one of the components of social cognition, namely the theory of mind (ToM) and different domains of functioning (social functioning, productive activities, and instrumental activities of daily living), each assessed separately for functional performance and functional outcome and for different aspects of functioning (functional performance and functional outcome) in schizophrenia; the meta-analysis includes fifty-nine studies and 4369 patients, showing a moderate association between ToM and all domains of functioning with a stronger association with productive activities compared with social functioning in terms of functional outcome; as regards to the different aspects of functioning, a stronger association was found between ToM and functional performance, compared with functional outcome, and for overall functioning and social functioning. In general, these findings show a stronger association of ToM with productive activities and functional performance. The fourth meta-analysis [50] specifically estimates the magnitude of the associations between ToM and neurocognition in 91 studies regarding 5462 subjects with schizophrenia; moderate associations emerged between ToM and each neurocognitive domain, with no significant difference between domains; considering executive functions, abstraction showed a stronger association with ToM; moreover. Several ToM task characteristics (e.g., modality of stimuli, type of mental state), emerged as significantly related to the magnitude of the associations between ToM and executive functions, visuo-spatial/problem solving, attention and episodic memory, suggesting that ToM is linked to a wide range of neurocognitive abilities in schizophrenia, and that ToM tasks act as a significant moderator of these associations.

In summary, cognitive impairment is strongly associated with impairments in functioning, with social cognition among the strongest known correlates of social functioning, probably even more strongly associated than neurocognition; moreover, social cognition appears to mediate the relationship between neurocognition and functioning and both social cognition and neurocognition likely impact real-world functioning through their negative effects on functional capacity as the ability to perform everyday living and social skills [51].

15.5 Negative Symptoms, Cognitive Impairment, and Personal Recovery

The concept of personal recovery relates to the point of view of mental health services users and prioritizes individually meaningful, personalized treatment goals as a target of interventions. Thus, personal recovery does not necessarily imply symptom remission nor attainment of normal functioning but mostly refers to subjective experiences of optimism, empowerment, interpersonal support, peer support, and stigma reduction [1]. Personal recovery encompasses various dimensions such as perceived respect, competence, empowerment, personal responsibility, sense of self-determination, hope, person-centered treatment, understanding of the recovery process, and peer support [52]. It may be conceived based on five processes: connectedness, hope and optimism about the future, identity, meaning in life, and empowerment [53]. Although symptom remission and functioning are not considered necessary prerequisites of personal recovery, several studies seem to show some contribution of clinical dimensions. A study of a total of 51 first-episode patients [54] shows significant albeit modest correlations (r = 0.38–0.51) found between personal recovery and several aspects of clinical recovery, such as negative symptoms and functioning; indeed, Personal Recovery, as measured by a specific scale (QPR-15), was significantly and negatively correlated with some negative symptoms domains of the SANS scale (affective $r = 0.414$, alogia 0.414, anhedonia 0.388) and with only one aspect of functioning at PSP (PSP social, $r = 0.512$); moreover, reductions in negative symptoms (global scores, apathy and anhedonia scores) during treatment were associated with moving towards personal recovery. A study based upon baseline data from four different studies of 971 people affected with schizophrenia-spectrum disorders, which were amalgamated for the analyses, found that affective symptoms were the most strongly associated with personal recovery, accounting for 30% of the variance in personal recovery, with negative and disorganized symptoms found related to both functional and personal recovery, and excitement symptoms only related to personal recovery, without any correlation of measures of positive symptoms with either functional or personal recovery [55]. A long-term, 20-year study of 172 first episodes of psychosis (FEP) evaluating also outcomes other than the symptomatic domain demonstrated that social cognition deficits were associated with poor personal recovery, while neurocognitive deficits such as processing speed, verbal memory deficits and visual memory impairments showed significant associations with poor outcomes on symptomatic status, social functioning, social disadvantage, and other clinical outcomes; moreover, lower cognitive global scores were significantly associated with all outcome dimensions except for drug abuse and physical status [56]. Finally, a meta-analysis investigating the relationship between clinical dimensions and personal measure of recovery in mainly chronic patients with schizophrenia spectrum disorders [57] found a significant albeit modest negative correlation between Personal recovery and overall symptom severity ($r = -0.21$, $p < 0.001$), Positive symptoms ($r = -0.14$, $p = 0.004$), Negative Symptoms ($r = -0.26$, $p < 0.001$) and with Affective symptoms at a significantly moderate correlation ($r = -0.43$, $p < 0.001$); a small but significant positive correlation was also found for the association with general functioning ($r = 0.21$ $p < 0.001$).

15.6 Concluding Remarks

Negative symptoms and cognitive impairment are certainly among the most relevant factors related to the relatively low rates of clinical recovery of people affected by schizophrenia spectrum disorders, with a more modest, albeit significant impact even on personal Recovery. These findings may have different levels of explanation. Indeed, they may be partly due to the still modest results of pharmacological treatments over negative symptoms in general and over persistent negative ones in particular [30, 58, 59], with similar lines of evidence regarding pharmacological treatments of cognitive symptoms, for which second-generation antipsychotics are recommended, mostly for their favourable cognitive profile compared to first generation antipsychotics, without no clear superiority of a single second-generation antipsychotic, with a series of new and hopefully effective drugs and other somatic interventions still in development [60]. Moreover, while now having psychosocial interventions of proven efficacy, such as social skill training for negative symptoms [61] and cognitive remediation for cognitive deficits [62], these interventions continue to be inadequately applied in common clinical settings, as some studies show [63, 64]. Hopefully, it should be expected that the timely implementation of personalized intervention plans starting from the first episodes, based upon the integration of pharmacological and psychosocial interventions, may further improve the recovery rates of people affected by schizophrenia and related disorders.

References

1. Carpiniello B, Pinna F, Manchia M, et al. Dimensions and course of clinical recovery in schizophrenia and related disorders. In: Carpiniello B, Vita A, Mencacci C, editors. Recovery and major mental disorders. Springer; 2022. p. 3–22.
2. Rossi R, Socci V, Rossi A. Personal recovery in schizophrenia: a narrative review. In: Carpiniello B, Vita A, Mencacci C, editors. Recovery and major mental disorders. Springer; 2022. p. 45–56.
3. Harvey PD, Bellack AS. Toward a terminology for functional recovery in schizophrenia: is functional remission a viable concept? Schizophr Bull. 2009;35(2):300–6. https://doi.org/10.1093/schbul/sbn171.
4. Vita A, Barlati S. Recovery from schizophrenia: is it possible? Curr Opin Psychiatry. 2018;31:246–55.
5. Jääskeläinen E, Juola P, Hirvonen N, et al. A systematic review and meta-analysis of recovery in schizophrenia. Schizophr Bull. 2013;39(6):1296–306.
6. Warner R. Recovery of schizophrenia: psychiatry and political economy. London: Routledge; 2004.
7. Huxley P, Krayer A, Poole R, et al. Schizophrenia outcomes in the 21st century: a systematic review. Brain Behav. 2021;11(6):e02172. https://doi.org/10.1002/brb3.2172.

8. Molstrom IM, Nordgaard J, Urfer-Parnas A, et al. The prognosis of schizophrenia: a systematic review and meta-analysis with meta-regression of 20-year follow-up studies. Schizophr Res. 2022;250:152–63. https://doi.org/10.1016/j.schres.2022.11.010.
9. Handest R, Molstrom IM, Henriksen MG, et al. A systematic review and meta-analysis of the association between psychopathology and social functioning in schizophrenia. Schizophr Bull. 2023;49:1470–85. https://doi.org/10.1093/schbul/sbad075.
10. Vauth R, Carpiniello B, Turczyński J, et al. Relationship between clinical outcomes measures and personal and social performance functioning in a prospective, interventional study in schizophrenia. Int J Methods Psychiatr Res. 2021;30(2):e1855. https://doi.org/10.1002/mpr.1855.
11. Pinna F, Deriu L, Lepori T, et al. Is it true remission? A study of remitted patients affected by schizophrenia and schizoaffective disorders. Psychiatry Res. 2013;210(3):739–44. https://doi.org/10.1016/j.psychres.2013.08.022.
12. Lambert M, Karow A, Leucht S, et al. Remission in schizophrenia: validity, frequency, predictors, and patients' perspective 5 years later. Dialogues Clin Neurosci. 2010;12(3):393–407. https://doi.org/10.31887/DCNS.2010.12.3/mlambert.
13. Carpiniello B, Pinna F, Manchia M, et al. Sustained symptomatic remission in schizophrenia: course and predictors from a two-year prospective study. Schizophr Res. 2022;239:34–41. https://doi.org/10.1016/j.schres.2021.11.023.
14. Giordano GM, Galderisi S, Pezzella P, et al. Determinants of clinical recovery in schizophrenia. In: Carpiniello B, Vita A, Mencacci C, editors. Recovery and major mental disorders. Springer; 2022. p. 23–43.
15. Galderisi S, Rossi A, Rocca P, et al. The influence of illness-related variables, personal resources and context-related factors on real-life functioning of people with schizophrenia. World Psychiatry. 2014;13(3):275–87. https://doi.org/10.1002/wps.20167.
16. Galderisi S, Bucci P, Kirkpatrick B, et al. Interplay among psychopathologic variables, personal resources, context-related factors, and real-life functioning in individuals with schizophrenia: a network analysis. JAMA Psychiatry. 2018;75(4):396–404. https://doi.org/10.1001/jamapsychiatry.2017.4607.
17. Mucci A, Galderisi S, Gibertoni D, et al. Factors associated with real-life functioning in persons with schizophrenia in a 4-year follow-up study of the Italian network for research on psychoses. JAMA. Psychiatry. 2021;78(5):550–9. https://doi.org/10.1001/jamapsychiatry.2020.4614.
18. Galderisi S, Bucci P, Mucci A, et al. The interplay among psychopathology, personal resources, context-related factors and real-life functioning in schizophrenia: stability in relationships after 4 years and differences in network structure between recovered and non-recovered patients. World Psychiatry. 2020;19(1):81–91. https://doi.org/10.1002/wps.2070019.
19. Tamminga CA, Buchanan RW, Gold JM. The role of negative symptoms and cognitive dysfunction in schizophrenia outcome. Int Clin Psychopharmacol. 1998;13(Suppl 3):S21–6. https://doi.org/10.1097/00004850-199803003-00004.
20. Verma S, Subramaniam M, Abdin E, et al. Symptomatic and functional remission in patients with first-episode psychosis. Acta Psychiatr Scand. 2012;126:282–9. https://doi.org/10.1111/j.1600-0447.2012.01883.x.
21. Austin SF, Mors O, Secher RG, et al. Predictors of recovery in first episode psychosis: the OPUS cohort at10 year follow-up. Schizophr Res. 2013;150:163–8. https://doi.org/10.1016/j.schres.2013.07.031.
22. Li L, Rami FZ, Lee BM, et al. Predictors of full recovery in patients with early stage schizophrenia spectrum disorders. Psychiatry Res. 2023;320:115035. https://doi.org/10.1016/j.psychres.2022.115035.
23. Albert N, Bertelsen M, Thorup A, et al. Predictors of recovery from psychosis analyses of clinical and social factors associated with recovery among patients with first-episode psychosis after 5 years. Schizophr Res. 2011;125:257–66. https://doi.org/10.1016/j.schres.2010.10.013.
24. Jordan G, Lutgens D, Joober R, et al. The relative contribution of cognition and symptomatic remission to functional outcome following treatment of a first episode of psychosis. J Clin Psychiatry. 2014;75(6):e566–72. https://doi.org/10.4088/JCP.13m08606.

25. Novick D, Haro JM, Suarez D, et al. Recovery in the outpatient setting: 36-month results from the Schizophrenia Outpatients Health Outcomes (SOHO) study. Schizophr Res. 2009;108:223–30. https://doi.org/10.1016/j.schres.2008.11.007.
26. Chang WC, Hui MCL, Yan Wong GH, et al. Symptomatic remission and cognitive impairment in first-episode schizophrenia: a prospective 3-year follow-up study. J Clin Psychiatry. 2013;74(11):e1046–53. https://doi.org/10.4088/JCP.13m08355.
27. Galderisi S, Mucci A, Bitter I, et al. Persistent negative symptoms in first episode patients with schizophrenia: results from the European first episode schizophrenia trial. Eur Neuropsychopharmacol. 2013;23(3):196–204. https://doi.org/10.1016/j.euroneuro.2012.04.019.
28. Bucci P, Mucci A, van Rossum IW, et al. Persistent negative symptoms in recent-onset psychosis: relationship to treatment response and psychosocial functioning. Eur Neuropsychopharmacol. 2020;34:76–86. https://doi.org/10.1016/j.euroneuro.2020.03.010.
29. Carpiniello B, Carta MG. Disability in schizophrenia. Intrinsic factors and prediction of psychosocial outcome. An analysis of literature. Epidemiol Psichiatr Soc. 2002;11(1):45–58. https://doi.org/10.1017/s1121189x00010149.
30. Galderisi S, Mucci A, Dollfus S, et al. EPA guidance on assessment of negative symptoms in schizophrenia. Eur Psychiatry. 2021;64(1):e23. https://doi.org/10.1192/j.eurpsy.2021.1131.
31. Galderisi S, Bucci P, Mucci A, et al. Categorical and dimensional approaches to negative symptoms of schizophrenia: focus on long-term stability and functional outcome. Schizophr Res. 2013;147(1):157–62. https://doi.org/10.1016/j.schres.2013.03.020.
32. Chang WC, Hui CL, Chan SK, et al. Impact of avolition and cognitive impairment on functional outcome in first-episode schizophrenia-spectrum disorder: a prospective one-year follow-up study. Schizophr Res. 2016;170(2–3):318–21. https://doi.org/10.1016/j.schres.2016.01.004.
33. Pillny M, Lincoln TM. Predictors of improved functioning in patients with psychosis: the role of amotivation and defeatist performance beliefs. Psychiatry Res. 2016;244:117–22. https://doi.org/10.1016/j.psychres.2016.07.018.
34. Foussias G, Mann S, Zakzanis L, et al. Prediction of longitudinal functional outcomes in schizophrenia: the impact of baseline motivational deficits. Schizophr Res. 2011;132(1):24–7. https://doi.org/10.1016/j.schres.2011.06.026.
35. Fervaha G, Foussias G, Agid O, et al. Motivational deficits in early schizophrenia: prevalent, persistent, and key determinants of functional outcome. Schizophr Res. 2015;166(1–3):9–16. https://doi.org/10.1016/j.schres.2015.04.040.
36. Fervaha G, Foussias G, Agid O, et al. Motivational and neurocognitive deficits are central to the prediction of longitudinal functional outcome in schizophrenia. Acta Psychiatr Scand. 2014;130(4):290–9. https://doi.org/10.1111/acps.12289.
37. Robinson DG, Woerner WG, McMeniman M, et al. Symptomatic and functional recovery from a first episode of schizophrenia or schizoaffective disorder. Am J Psychiatry. 2004;161:473–9. https://doi.org/10.1176/appi.ajp.161.3.473.
38. Molina-García M, David Fraguas D, del Rey-Mejías Á, et al. The role of premorbid IQ and age of onset as useful predictors of clinical, functional outcomes, and recovery of individuals with a first episode of psychosis. J Clin Med. 2021;10(11):2474. https://doi.org/10.3390/jcm10112474.
39. Bobes J, Ciudad A, Álvarez E, et al. Recovery from schizophrenia: results from a 1-year follow-up observational study of patients in symptomatic remission. Schizophr Res. 2009;115:58–66. https://doi.org/10.1016/j.schres.2009.07.003.
40. Lindgren M, Holm M, Kieseppä T, et al. Neurocognition and social cognition predicting 1-year outcomes in first-episode psychosis front. Psychiatry. 2020;11:603933. https://doi.org/10.3389/fpsyt.2020.603933.
41. Maj M, van Os J, De Hert M, et al. The clinical characterization of the patient with primary psychosis aimed at personalization of management. World Psychiatry. 2021;20(1):4–33. https://doi.org/10.1002/wps.20809.

42. Harvey PD, Bosia M, Cavallaro R, et al. Cognitive dysfunction in schizophrenia: an expert group paper on the current state of the art. Schizophr Res Cogn. 2022;22(29):100249. https://doi.org/10.1016/j.scog.2022.100249.
43. McCutcheon RA, Keefe RSE, McGuire PK. Cognitive impairment in schizophrenia: aetiology, pathophysiology, and treatment. Mol Psychiatry. 2023;28(5):1902–18. https://doi.org/10.1038/s41380-023-01949-9.
44. Yang Z, Hong Lee S, Rashid NAA, et al. Predicting real-world functioning in schizophrenia: the relative contributions of neurocognition, functional capacity, and negative symptoms. Front Psychiatry. 12:639536. https://doi.org/10.3389/fpsyt.2021.639536.
45. González-Blanch C, Perez-Iglesias R, Pardo-García G, et al. Prognostic value of cognitive functioning for global functional recovery in first-episode schizophrenia. Psychol Med. 2010;40(6):935–44. https://doi.org/10.1017/S0033291709991267.
46. Giuliani L, Giordano GM, Bucci P, et al. Improving knowledge on pathways to functional outcome in schizophrenia: main results from the Italian network for research on psychoses front. Psychiatry. 2021;12:791117. https://doi.org/10.3389/fpsyt.2021.791117.
47. Fett AKJ, Viechtbauer W, de Gracia Domingueza M, et al. The relationship between neurocognition and social cognition with functional outcomes in schizophrenia: a meta-analysis. Neurosci Biobehav Rev. 2011;35:573–88. https://doi.org/10.1016/j.neubiorev.2010.07.001.
48. Santesteban-Echarria O, Painod M, Ricea S, et al. Predictors of functional recovery in first-episode psychosis: a systematic review and meta-analysis of longitudinal studies. Clin Psychol Rev. 2017;58:59–75. https://doi.org/10.1016/j.cpr.2017.09.007.
49. Thibaudeau E, Cellard C, Turcotte M, et al. Functional impairments and theory of mind deficits in schizophrenia: a meta-analysis of the associations, meta-Analysis. Schizophr Bull. 2021;47(3):695–711. https://doi.org/10.1093/schbul/sbaa182.
50. Thibaudeau E, Achim AM, Parent C, et al. A meta-analysis of the associations between theory of mind and neurocognition in schizophrenia. Schizophr Res. 2020;216:118–28. https://doi.org/10.1016/j.schres.2019.12.017.
51. Jones MT, Harvey PD. Major neuropsychological impairments in schizophrenia patients: clinical implications. Curr Psychiatry Rep. 2020;22:59. https://doi.org/10.1007/s11920-020-01181-053.
52. Chiu MY, Ho WW, Lo WT, et al. Operationalization of the SAMHSA model of recovery: a quality of life perspective. Qual Life Res. 2010;19:1–13. https://doi.org/10.1007/s11136-009-9555-2.
53. Leamy M, Bird V, Le Boutillier C, et al. Conceptual framework for personal recovery in mental health: systematic review and narrative synthesis. Br J Psychiatry. 2011;199(6):445–52. https://doi.org/10.1192/bjp.bp.110.083733.
54. Austin SF, Hjorthøj C, Baagland H, et al. Investigating personal and clinical recovery processes in people with first episode psychosis. Early Interv Psychiatry. 2022;16:1102–11. https://doi.org/10.1111/eip.13258.
55. Best MW, Law H, Pyle M, et al. Relationships between psychiatric symptoms, functioning and personal recovery in psychosis. Schizophr Res. 2020;223:112–8. https://doi.org/10.1016/j.schres.2020.06.026.
56. Cuesta MJ, Sánchez-Torres AM, Moreno-Izco L, et al. Neurocognitive correlates of the varied domains of outcomes at 20 year follow-up of first-episode psychosis. Psychiatry Res. 2022;318:114933. https://doi.org/10.1016/j.psychres.2022.114933.
57. Van Eck RM, Burger TJ, Vellinga A, et al. The relationship between clinical and personal recovery in patients with schizophrenia spectrum disorders: a systematic review and meta-analysis. Schizophr Bull. 2018;44(3):631–42. https://doi.org/10.1093/schbul/sbx088.
58. Remington G, Foussias G, Fervaha G, et al. Treating negative symptoms in schizophrenia: an update. Curr Treat Options Psychiatry. 2016;3:133–50. https://doi.org/10.1007/s40501-016-0075-8.
59. Krause M, Zhu Y, Huhn M, Schneider-Thoma J, et al. Antipsychotic drugs for patients with schizophrenia and predominant or prominent negative symptoms: a systematic review

and meta-analysis. Eur Arch Psychiatry Clin Neurosci. 2018;268(7):625–39. https://doi.org/10.1007/s00406-018-0869-3.
60. Vita A, Gaebel W, Mucci A, et al. European Psychiatric Association guidance on treatment of cognitive impairment in schizophrenia. Eur Psychiatry. 2022;65:e57. https://doi.org/10.1192/j.eurpsy.2022.2315.
61. Turner DT, McGlanaghy E, Cuijpers P, et al. A meta-analysis of social skills training and related interventions for psychosis. Schizophr Bull. 2018;44(3):475–91. https://doi.org/10.1093/schbul/sbx146.
62. Vita A, Barlati S, Ceraso A, et al. Durability of effects of cognitive remediation on cognition and psychosocial functioning in schizophrenia: a systematic review and meta-analysis of randomized clinical trials. Am J Psychiatry. 2024;181(6):520–31. https://doi.org/10.1176/appi.ajp.20230396.
63. Ince P, Haddock G, Tai S. A systematic review of the implementation of recommended psychological interventions for schizophrenia: rates, barriers, and improvement strategies. Psychol Psychother. 2016;89(3):324–50. https://doi.org/10.1111/papt.12084.
64. Vita A, Barlati S, Porcellana M, et al. The patient journey project in Italian mental health services: results from a co-designed survey on clinical interventions and current barriers to improve the care of people living with schizophrenia. Front Psych. 2024;15:1382326. https://doi.org/10.3389/fpsyt.2024.1382326.

Methodologic Considerations on Developing Drugs for Negative Symptoms and Cognitive Impairment in Schizophrenia

16

Michael Davidson, Jay Saoud, and Remy Luthringer

Abstract

Several reviews have enumerated the plethora of drug trials aimed at improving negative symptoms and/or cognitive performance in patients affected by schizophrenia (Sabe et al. Neurosci Biobehav Rev 144: 104979, 2023). The aim of this chapter is to highlight two of the reasons for the high rate of failures in these indications and to suggest potential remedies: (a) reduce heterogeneity of patients recruited for the trials by improving sample stratification and (b) employ monotherapy rather than add-on trial design.

Keywords

Negative symptoms · Schizophrenia · Methodology · Treatment

16.1 Background

With the advent of the DSM classification, which emphasizes positive symptoms, coupled with the availability of antipsychotic drugs that interfere with dopamine (DA) neurotransmission and ameliorate positive psychotic symptoms, the attention of the lay public and the medical community focused on positive psychotic symptoms of schizophrenia. However, for most individuals who meet DSM/ICD criteria for schizophrenia, psychosis manifests only intermittently and for a subgroup of

M. Davidson (✉)
Minerva Neurosciences, Waltham, MA, USA

Basic and Clinical Sciences Nicosia University Medical School, Nicosia, Cyprus

J. Saoud · R. Luthringer
Minerva Neurosciences, Waltham, MA, USA
e-mail: jsaoud@minervaneurosciences.com; RLuthringer@minervaneurosciences.com

A. Vita et al. (eds.), *Negative and Cognitive Symptoms in Schizophrenia*,
Comprehensive Approach to Psychiatry 3,
https://doi.org/10.1007/978-3-032-06209-3_16

individuals only once or twice lifelong [2]. On the other hand, avolition, apathy, asociality, alogia, and flat affect, which cluster under the notion of negative symptoms, are persistently present in most patients, often before the first psychosis episode and remain after remission of psychosis [3].

Negative symptoms are viewed by researchers and clinicians as primary negative symptoms that are intrinsic to the schizophrenic illness and secondary negative symptoms which are a common adverse effect of most antipsychotic drugs. Like negative symptoms, aspects of cognitive functioning are impaired to a mild to moderate degree before the first manifestation of psychosis and before the formal diagnosis of schizophrenia and persist lifelong. Negative symptoms and cognitive impairment, which some evidence suggest are related [4], constitute the main reason for the poor social and vocational performance of individuals affected by schizophrenia, justifying the efforts to find an appropriate treatment.

Evidence of the intense research activity in the field are the consistent increase in related publications [1] and a search of clinicalTrials.gov, which lists most but not all trials. Using *schizophrenia* and *cognition* or *negative symptoms* as keywords revealed >500 trials mostly active or recent. The majority of the listed trials are repositioning trials in which a drug approved for a different indication is tested in patients with schizophrenia and negative symptoms and/or cognitive impairment. There exist several paths driving drug repositioning: (1) serendipitous clinical observation that the drug benefits the target symptom in schizophrenia and (2) hypothesizing that a drug already demonstrated to benefit a symptom similar to negative symptoms or cognitive impairment in another mental disorder would have the same effect in schizophrenia. This is not unreasonable, considering the overlap between DSM/ICD disorders [5] (3), translating basic science findings into clinical trials. Such serendipitous observations and hypotheses are then tested in randomized controlled trials aimed at proving or disproving the observation or the hypotheses [6].

Examples of serendipitous observations leading to marketed drugs are the classic antipsychotics and antidepressants. However, to appreciate the many potential missteps separating a serendipitous observation or a hypothesis-driven idea and a positive RCT, just imagine that before the multiple causes of dyspnea were understood, it would have been observed, serendipitously, that a compound with diuretic properties occasionally ameliorated dyspnea and the resulting poor quality of life. A rational follow-up RCT would test the diuretic in patients with dyspnea and poor quality of life. Although the observation was correct, the results of the RCT would be highly unpredictable because of the sample's biological heterogeneity.

Examples of hypotheses-driven reposition attempts for cognitive impairment in schizophrenia are trials with amphetamine-like compounds, which improve attention in ADHD [7] or modafinil, which indirectly may improve aspects of cognitive performance by improving vigilance [8]. Because of phenomenological similarities between Major Depressive Disorder (MDD) and in negative symptoms, drugs developed and approved as antidepressants were repeatedly tested for negative symptoms, to no avail [9]. Also, the observation that phencyclidine, a drug of abuse with NMDA receptor blocking properties, causes withdrawal behaviour has led to the hypothesis that NMDA receptor modulators or agonists might improve negative symptoms. However, trials with available NMDA agonists and modulators like serine or d-serine were added to antipsychotics [10], or de-novo created compounds like biopterin [11] could not replicate the initial positive results. More recently, repositioning has been based on screening of databases of electronic clinical records and repositories of compounds with the help of computerized algorithms aiming to match between characteristics of the compound(s) and of the disease [12].

However, despite these efforts, at the present time, there exit no drug benefiting negative symptoms or cognitive impairment in schizophrenia. The following are some of the hurdles delaying such efforts.

16.2 Heterogeneity of the Trial Populations

Despite considerable basic and clinical research efforts, the notion "schizophrenia" remains a fuzzy description of convenience and a framework for distinctively diverse behaviours, perceptions, and emotions, on a continuum between moderately aberrant to severely abnormal [13]. The Krepelanian notion of the unitary ilness manifestation and course with inevitable deterioration has been challenged by long-term follow up cohort studies [14], genome-wide association studies (GWAS) studies [15] and by biological abnormalities driven by subgroups of outliers rather than distinctive distributions between patients and controls.

Progress in basic neuroscience has been slow to translate into providing clinically useful biological markers for the complex behaviours and emotions manifested in schizophrenia; hence, trial participants are still selected and grouped based on partial phenomenological semblance. The general characteristics of the patients participating in trials targeting cognition and/or negative symptoms are: (1) DSM/ICD criteria for schizophrenia, (2) stable psychotic symptoms, often bellow an overall or symptom specific severity threshold overall so that the targeted symptoms can be assessed without the confounders of active psychosis and (3) cognitive impairment or negative symptoms assessed with the help of comprehensive scales, above a certain level of severity, so that there is room for improvement. Assuming strict adherence to the criteria, the resulting patient' sample remains very heterogeneous, hence reducing the likelihood of matching between a drug's pharmacologic effect and a potential biologic abnormality related to the targeted symptom. This heterogeneity might dilute and obscure a drug effect on a specific aspect of the targeted symptoms and/or in a specific subgroup of patients, even if a few participants in the study experience clinically meaningful benefits. Post-hoc analyses are supposed to reveal such specific symptoms or subgroups, and statistical control for multiple comparisons reduces the possibility that this is a chance finding. Unfortunately, too often such findings tend the be dismissed and undervalued, discouraging investigators from incorporating the post-hoc finding into the hypothesis of the next trial.

All trials, including schizophrenia trials, need to commit a priori to a detailed statistical plan including primary and secondary outcomes to prevent control of the type 1 error. This is for example the case for a new drug intended to treat acute psychotic exacerbations, where after 60 years of availability in clinical practice, the expected specific therapeutic effects and timelines are well-known. However, this might be more difficult for a condition for which no effective treatment exists, like negative symptoms or cognitive impairment. Considering all the unknown variables described above, it is hard to predict which aspects of cognition or negative symptoms will benefit by the drug even when the pharmacologic mechanism of action of the new drug is well known. A secondary rather than a primary outcome might be benefited by the experimental drug yet a strict adherence to the a priori outcome runs the risk of inflating the type 2 error. The standard solution for such an occurrence is a repeated trial in which the positive but secondary outcome becomes the primary outcome. However, in an environment with multiple unknown variables as described above, long study duration, and high costs, this might not be always feasible. Assuming that the new compound has demonstrated a definite positive signal and large safety margins, an alternative solution is to allow the new drug on the market and let clinical practice and post-marketing trials accumulate real-life data.

Specifically, cognitive performance in general, including individuals suffering from schizophrenia- involves a large number of psychometrically validated components, some reflecting immutable genetic influence further shaped by a large array of environmental variables, such as education, quality of parenting, peer interactions, and nutrition, to name a few. Depending on the genetic configuration and the interacting circumstances, each variable has a different impact on the overall cognitive performance. The precise areas of cognitive impairment, degree of severity, and percentage of cognitively impaired individuals are still disputed among investigators. It is still disputed if a specific area of cognitive function, such as processing speed, attention, or working memory, constitutes the main area of impairment or if global cognitive impairment is the more common impairment [16]. Also, the degree of cognitive impairment compared to healthy individuals ranges between 1 SD and > 2 SD, with a few patients scoring I the normal range [17]. Furthermore, mild deviations from cognitive performance population norms are present across almost all DSM categories [18]. Therefore, considering the diverse factors that affect cognitive performance, selecting trial participants based on performance below a threshold on a scale summing up very diverse cognitive functions would unavoidably result in a phenomenological and probably biological heterogeneous sample.

Like cognitive dysfunctions, negative symptoms assessed by specific scales encompass experiential and expressional manifestations, which in turn can be divided in more granular manifestations with diverse impact on the overall manifestations of negative symptoms. For example, it has been suggested that avolition is a central component of negative symptoms, which interacts with and often generates the rest of the negative symptoms, and drug-induced improvement in avolition is associated with improvement in the rest of the negative symptoms [19]. Furthermore, like cognitive impairment, avolition is present across several DSM/ICD categories consistent with the hypothesis that it might have a distinct biological underpinning regardless of the specific DSM/ICD condition [20]. It would therefore be reasonable to stratify trial participants based on the presence and severity of avolition rather than overall negative symptoms manifested in patients who meet diagnostic criteria for schizophrenia.

Despite the poor replications of any biological findings in schizophrenia, in future trials, patients can be stratified a priori or analysed separately based on brain imaging or electroencephalographic signal demonstrated in areas hypothesized to be related to the presenting symptom(s). For example, imaging findings have been associated with both negative symptoms and cognitive impairment [21]. More recently, striatal functional hypoconnectivity has been demonstrated in patients suffering from schizophrenia and negative symptoms [22]. Example of such a desirable match between a biological marker and a target would be the amyloid load demonstrated by a PET scan in Alzheimer's disease patients recruited for trials of antibodies aimed at reducing the amyloid load [23].

16.3 Trial Design

Most of the trials attempting to improve cognition and/or negative symptoms have been add-on trials, by which an experimental drug or placebo is added to antipsychotic drugs. This is done under the assumption that all patients who have had more than one psychotic exacerbation should be administered antipsychotic drugs for a long time to reduce the risk of future psychotic exacerbation. However, because DA neurotransmission is involved in the brain reward system [24], and most antipsychotic drugs interfere with DA neurotransmission, mostly by blocking the D2 subtype receptors, it is not surprising that these drugs cause secondary negative symptoms such as avolition, which is both an expression of and a core generator of negative symptoms. Furthermore, DA receptor blockade in the nigrostriatum causes Parkinson-like manifestations such as flat affect with reduced facial expression, rigidity, and slow movements, causing secondary negative symptoms, thus adding to the primary negative symptoms and enhancing the overall negative symptoms manifestations [25]. Moreover, antipsychotic drugs, by reducing motivation, might reduce eagerness to perform on cognitive tests, particularly speed-dependent tests [26]. Despite these shortcomings, the add-on design is employed under the expectation that the beneficial effects of the experimental compound on cognition or negative symptoms are sufficiently large to overcome the detrimental effect of the antipsychotics, an assumption which has not been proven in RCTs.

An alternative to the add-on design would be a targeted or intermittent treatment design by which symptomatically stable patients are randomized to monotherapy with the experimental drug targeting negative symptoms or cognitive impairment or to placebo, as recently recommended by a panel of schizophrenia trialists [27]. In case of reemergence of psychotic symptoms, an antipsychotic drug is administered and then discontinued upon stabilization of psychotic symptoms [28]. Whether the administration of the experimental drug should continue during the administration of the antipsychotic should depend on safety considerations and the potential interaction between the 2 drugs. Monotherapy with drugs targeting specific symptoms rather than continuous administration of "major tranquilizers" antipsychotics might fit better with the patients' goals, the majority of whom discontinue medication on their own.

References

1. Sabe M, Chen C, Perez N, Solmi M, Mucci A, Galderisi S, Strauss GP, Kaiser S. Thirty years of research on negative symptoms of schizophrenia: a scientometric analysis of hotspots, bursts, and research trends. Neurosci Biobehav Rev. 2023;144:104979., ISSN 0149-7634. https://doi.org/10.1016/j.neubiorev.2022.104979.
2. Carpenter WT, Kirkpatrick B. The heterogeneity of the long-term course of schizophrenia. Schizophr Bull. 1988;14(4):645–52. https://doi.org/10.1093/schbul/14.4.645.
3. Haro JM, Altamura C, Corral R, Elkis H, Evans J, Krebs MO, Zink M, Malla A, Méndez JI, Bernasconi C, Lalonde J, Nordstroem AL. Understanding the course of persistent symptoms in schizophrenia: longitudinal findings from the pattern study. Psychiatry Research. 2018;267:56–62.
4. Mellilo, et al. J Clin Med. 2023;12(22)
5. Davidson M, Gabos C. Do DSM classifications help or hinder drug development. Dialogues Clin Neurosci. 2020;22(1):73–9.
6. Torrey EF, Davis JM. Clin Schizophr Relat Psychoses. 2012;5(4):208–16. https://doi.org/10.3371/CSRP.5.4.5. PMID: 22182458
7. Barch DM, Carter CS. Amphetamine improves cognitive function in medicated individuals with schizophrenia and in healthy volunteers. Schizophr Res. 2005;77(1):43–58, ISSN 0920-9964,. https://doi.org/10.1016/j.schres.2004.12.019.
8. Ortiz-Orendain J, Covarrubias-Castillo SA, Vazquez-Alvarez AO, et al. Modafinil for people with schizophrenia or related disorders: a Cochrane review. BJPsych Adv. 2022;28(5):280–90. https://doi.org/10.1192/bja.2022.37.
9. Fusar-Poli P, Papanastasiou E, Stahl D, et al. Treatments of negative symptoms in schizophrenia: meta-analysis of 168 randomized placebo-controlled trials. Schizophr Bull. 2015;41(4):892–9.
10. Kantrowitz JT, Correll CU, Jain R, Andrew J. Cutler new developments in the treatment of schizophrenia: an expert roundtable. Int J Neuropsychopharmacol. 2023;26(5):322–30.
11. Bugarski-Kirola D, Arango C, Fava M, et al. Pimavanserin for negative symptoms of schizophrenia: results from the ADVANCE phase 2 randomised, placebo-controlled trial in North America and Europe. Lancet Psychiatry. 2021;9(1):46–58.
12. Jayatunga MKP, Ayers M, Bruens L, Jayanth D, Meier C. How successful are AI-discovered drugs in clinical trials? A first analysis and emerging lessons. Drug Discov Today. 2024;29(6):104009.
13. Guloksuz S, van Os J. The slow death of the concept of schizophrenia and the painful birth of the psychosis spectrum. Psychol Med. 2018;48(2):229–44. https://doi.org/10.1017/S0033291717001775.
14. Harrow M, Jobe TH, Faull RN. Do all schizophrenia patients need antipsychotic treatment continuously throughout their lifetime? A 20-year longitudinal study. Psych Med. 2012;42(10):2145–55.
15. Working Group of the Psychiatric Genomics Consortium. Biological insights from 108 schizophrenia-associated genetic loci Schizophrenia. Nature. 2014;511:421–7.
16. McCutcheon RA, Keefe RSE, McGuire PK. Cognitive impairment in schizophrenia: aetiology, pathophysiology, and treatment. Mol Psychiatry. 2023;28:1902–18. https://doi.org/10.1038/s41380-023-01949-9.
17. Carruthers SP, Van Rheenen TE, Karantonis JA, et al. Characterising demographic, clinical and functional features of cognitive subgroups in schizophrenia Spectrum disorders: a systematic review. Neuropsychol Rev. 2022;32:807–27. https://doi.org/10.1007/s11065-021-09525-0.

18. Weiser M, Reichenberg A, Rabinowitz J, Knobler HY, Lubin G, Yazvitzky R, Nahon D, Gur R, Davidson M. Cognitive performance of male adolescents is lower than controls across psychiatric disorders: a population-based study. Acta Psychiatr Scand. 2004;110(6):471–5.
19. Strauss GP, Bartolomeo LA, Harvey PD. Avolition as the core negative symptom in schizophrenia: relevance to pharmacological treatment development. NPJ Schizophr. 2021;7(1):16.
20. Masud CH, Husain, Roiser JP. Apathy and motivation: biological basis and drug treatment. Ann Rev Pharm Toxicol. 2024;64:313–38. (Volume publication date January 2024)
21. Shaffer JJ, Peterson MJ, McMahon MA, et al. Neural correlates of schizophrenia negative symptoms: distinct subtypes impact dissociable brain circuits. Mol Neuropsychiatry. 2015;1:191–200.
22. Howes OD, et al. New drug treatments for schizophrenia: a review of approaches to target circuit dysfunction. Biol Psychiatry. 2024;96:638.
23. Na S, Kang DW, Kim GH, Kim KW, Kim Y, Kim HJ, Park KH, Park YH, Byeon G, Suh J, Shin JH, Shim Y, Yang Y, Um YH, Oh SI, Wang SM, Yoon B, Yoon HJ, Lee SM, Lee J, Lee JS, Rhee HY, Lim JS, Jung YH, Chin J, Hong YJ, Jang H, Choi H, Choi M, Jang JW, Korean Dementia Association. The usefulness of ^{18}F-FDG PET to differentiate subtypes of dementia: the systematic review and meta-analysis. Dement Neurocogn Disord. 2024;23(1):54–66. https://doi.org/10.12779/dnd.2024.23.1.54.
24. Webber HE, Lopez-Gamundi P, Stamatovich SN, de Wit H, Wardle MC. Using pharmacological manipulations to study the role of dopamine in human reward functioning: a review of studies in healthy adults. Neurosci Biobehav Rev. 2021;120:123–58.
25. Matthias K, Aleman A, Kaiser S. Secondary negative symptoms—a review of mechanisms, assessment and treatment. Schizophr Res. 2017;186:29–38.
26. Haddad C, Salameh P, Sacre H, et al. Effects of antipsychotic and anticholinergic medications on cognition in chronic patients with schizophrenia. BMC Psychiatry. 2023;23:61. https://doi.org/10.1186/s12888-023-04552-y.
27. Marder SR, Davidson M, Zaragoza S, et al. Issues and perspectives in designing clinical trials for negative symptoms in schizophrenia: consensus statements. Schizophr Bull Open. 2020;1(1):sgz001.
28. Davidson M, Carpenter WT Jr. Targeted treatment of schizophrenia symptoms as they manifest, or continuous treatment to reduce the risk of psychosis recurrence. Schizophr Bull. 2024;50(1):14–21. https://doi.org/10.1093/schbul/sbad145.

17 Physical Exercise to Improve Negative Symptoms, Cognitive Deficits, and Levels of Daily-Life Functioning in Schizophrenia Spectrum Disorders: Current Evidence and Treatment Recommendations

Lukas Roell, Isabel Maurus, Andrea Schmitt, and Peter Falkai

Abstract

In recent years, physical exercise therapies have been proposed as an add-on treatment to achieve additional improvements in people with schizophrenia spectrum disorders (SSD). As indicated by numerous large-scale meta-analyses, physical exercise treatments in SSD lead to improvements in positive, negative, and total symptom severity, as well as cognitive impairment and daily

L. Roell
Department of Psychiatry and Psychotherapy, LMU University Hospital, LMU Munich, Munich, Germany

NeuroImaging Core Unit Munich (NICUM), LMU University Hospital, LMU Munich, Munich, Germany

Max Planck Institute of Psychiatry, Munich, Germany
e-mail: Lukas.Roell@med.uni-muenchen.de

I. Maurus
Department of Psychiatry and Psychotherapy, LMU University Hospital, LMU Munich, Munich, Germany
e-mail: Isabel.Maurus@med.uni-muenchen.de

A. Schmitt
Department of Psychiatry and Psychotherapy, LMU University Hospital, LMU Munich, Munich, Germany

Max Planck Institute of Psychiatry, Munich, Germany

Laboratory of Neuroscience (LIM27), Institute of Psychiatry, University of Sao Paulo, São Paulo, Brazil
e-mail: Andrea.Schmitt@med.uni-muenchen.de

A. Vita et al. (eds.), *Negative and Cognitive Symptoms in Schizophrenia*, Comprehensive Approach to Psychiatry 3,
https://doi.org/10.1007/978-3-032-06209-3_17

functioning, whereas the impact on physical health is only minimal. This chapter summarizes the current evidence of physical exercise therapies applied in people with SSD and derives practical recommendations for establishing such therapies in the clinical setting.

Keywords

Schizophrenia · Exercise · Sport · Physical activity · Negative symptoms · Cognition · Functioning · Physical health

In the fifth edition of the Diagnostic and Statistical Manual of Mental Disorders (DSM-5), schizophrenia spectrum disorders (SSD) are characterized by a variety of different clinical symptoms. These include positive symptoms, such as delusions, hallucinations, disorganized speech, and disorganized or catatonic behavior, and negative symptoms, which consist of anhedonia (restricted ability to experience pleasure), avolition (lack of motivation), blunted affect (reduced emotional expression), alogia (poverty of speech), and asociality. Additionally, cognitive deteriorations and associated impairments in social and occupational functioning serve as indicators for diagnosis [1].

Apart from the psychiatric domain, SSD is accompanied by several severe physical comorbidities. People with SSD more frequently show unhealthy lifestyle habits, such as physical inactivity [2], unhealthy diet [3], and the use of nicotine, alcohol, and other substances [4–6]. Accordingly, the prevalence of cardiovascular diseases [7–11], overweight [12–14], type 2 diabetes [15–17], and metabolic syndrome [18] is significantly higher in people with SSD compared to the general population. As a result, current large-scale evidence reveals that general body and organ health are severely disturbed in SSD, even more so than in other psychiatric conditions [19]. Consequently, the life expectancy of people with SSD ranges from 61.1 to 71.3 years, representing an average reduction of 11.2 to 17.8 years compared to the general population [20, 21].

Current treatment strategies encompass different approaches, ranging from tailored pharmacotherapy with antipsychotics to cognitive behavioral therapy and cognitive remediation. In cases of medication resistance, the use of non-invasive brain stimulation techniques such as electroconvulsive therapy or repetitive transcranial magnetic stimulation is recommended [22]. The efficacy of antipsychotics on

P. Falkai (✉)
Department of Psychiatry and Psychotherapy, LMU University Hospital, LMU Munich, Munich, Germany

Max Planck Institute of Psychiatry, Munich, Germany

German Center for Mental Health (DZPG), partner site Munich/Augsburg, Munich, Augsburg, Germany
e-mail: Peter.Falkai@med.uni-muenchen.de

positive symptoms has been demonstrated in several large-scale meta-analyses [23, 24], as has cognitive remediation to treat cognitive deficits and negative symptoms [25–27] and cognitive behavioral therapy to cover positive and negative symptoms and daily-life functioning [28]. Interventions, including brain stimulation techniques, also yield the potential to ameliorate negative symptoms [29] and cognitive impairments [30]. Despite the promising extension of applicable treatment approaches, many patients experience chronic impairments and encounter an unfavorable disease course, particularly with regard to social and occupational functioning [31]. Thus, there is an urgent need for additional therapeutic interventions to improve the long-term prognosis for SSD. Reducing negative symptoms, cognitive deficits, impaired daily life functioning, and antipsychotic side effects is crucial to achieve such an aim. At this point, physical exercise therapies could play an important role in leveraging the present treatment of SSD.

17.1 Effects of Physical Exercise on Negative Symptoms in SSD

In recent years, various types of physical exercise have emerged as a promising adjunctive treatment in SSD [32]. As demonstrated by multiple meta-analyses and by a recent large-scale multicenter randomized-controlled trial (RCT) in people with SSD, physical exercise treatments such as aerobic or anaerobic endurance programs, resistance trainings or mind-body interventions in additional to the usual antipsychotic treatment lead to further improvements in total symptoms [33–39] and positive symptom severity [33–36, 39–41] with small to moderate effect sizes (total symptoms: Hedge's $g = 0.16$ to 0.72; positive symptoms: Hedge's $g = 0.17$ to 0.54). These additional enhancements to already established antipsychotic medication effects highlight the potential of exercise therapies to improve schizophrenia treatment.

Importantly, the positive impact of physical exercise in SSD is not restricted to total and positive symptoms. Ten meta-analyses published between 2015 and 2024 examined the effects of different types of exercise treatments on negative symptom severity [33–38, 40–43]. The included exercise therapies comprised aerobic endurance training, anaerobic exercise programs, resistance training, or mind-body interventions such as yoga lessons. The study population among all meta-analyses was diverse, ranging from post-acute and clinically stable outpatients without prominent positive symptoms to more acute psychotic inpatients. Across all types of exercise treatments and all patient groups, physical exercise interventions compared to active or passive control conditions led to improvements in negative symptom severity with effect sizes ranging from small to medium (Hedge's $g = 0.30$ to 0.65) [33–38, 40–43]. Hence, the current state of research supports the general efficacy of physical exercise interventions to improve negative symptoms in SSD.

However, there remain several important aspects to clarify in order to gain a more comprehensive conclusion on the efficacy of physical exercise programs on negative symptom severity:

First, statistically significant effects of a treatment such as physical exercise on a certain outcome, like negative symptoms, are not necessarily clinically meaningful [44]. In order to evaluate the clinical relevance of an observed effect, it is essential to consider the respective effect sizes and to link the treatment-induced changes in the outcome of interest to potentially co-occurring changes in outcomes covering the overall clinical status of the patients [45]. Additionally, the interpretation of the clinical relevance of treatment effects depends on the underlying study population. For instance, strong beneficial effects of physical exercise on negative symptom severity may be less likely in already stable patient populations with only residual negative symptom severity at baseline. With respect to commonly used rating scales to cover negative symptoms in SSD, an average reduction of the respective subscale of the Positive and Negative Syndrome Scale (PANSS) [46] by 24% has been linked to a minimal clinical improvement as assessed by the Clinical Global Impression (CGI) scale [47] in patients with predominant negative symptoms [48], while a reduction of 13 to 21% in the Scale for the Assessment of Negative Symptoms (SANS) [49] is considered to reflect a minimal clinical improvement [50]. Regarding the effects of physical exercise in SSD, the average percentage reductions in negative symptom severity across all original studies have not been systematically accumulated in the respective meta-analyses, but current preliminary estimates hover between 18 and 51% reduction of the negative subscale of the PANSS and between 30 to 41% reduction of the SANS [33, 38]. Consequently, the obtained beneficial effects of physical exercise on negative symptom severity are likely to be clinically meaningful, although there remains a small uncertainty due to a lack of systematic evaluation of the clinical relevance of the meta-analytic effects.

Second, a differentiation between primary and secondary negative symptoms, as well as a focus on patients with persistent or at least predominant or prominent negative symptoms, is essential to evaluate if therapies such as physical exercise interventions can be considered to be effective [51]. As the underlying meta-analyses do not systematically assess the effects of physical exercise only in patients with persistent, predominant, or prominent negative symptoms, an accurate estimate of exercise effects on negative symptoms remains elusive. However, in contrast to antipsychotic treatment, the beneficial effects of physical exercise on negative symptoms across all meta-analyses are larger and more stable compared to positive symptoms [33–38, 40–43]. Furthermore, Sabe et al. [42] in their meta-analysis found no correlation between improvements in positive and negative symptoms. The authors further consider the positive effects of physical exercise on negative symptom severity to be of "potential clinical relevance", as the underlying study samples mostly consist of patients with chronic negative symptoms. Nevertheless, there is an urgent need to systematically investigate the effects of physical exercise only in patients with persistent, predominant, or prominent negative symptoms.

Third, various types of physical exercise may have a different impact on negative symptom severity, but the results are inconclusive between meta-analyses. Ziebart et al. [36] report that aerobic exercise but not yoga interventions improve negative symptoms, whereas Vogel et al. [43] obtained medium effects for mind-body exercise programs such as yoga, but only small effects of aerobic exercise. Sabe et al.

[42] identified aerobic exercise interventions to be superior to non-aerobic exercise treatments, while Kim et al. [40] suggest combined treatments of aerobic and resistance training to have the largest effect sizes, followed by only aerobic exercise programs. Considering recent developments toward precision psychiatry, these partially inconsistent group-level findings may indicate that the effectiveness of a specific type of exercise program depends on the underlying study population. In other words, there is no physical exercise intervention that generally outperforms the other, but there may exist a specific optimal treatment for each individual or at least for certain patient subgroups. However, given the lack of comprehensive machine learning approaches in this area, it is not yet possible to assign individual patients or subgroups to their particular optimal exercise treatment in advance. Hence, further multicenter large-scale exercise trials with harmonized multimodal assessments are warranted.

Finally, it is essential to clarify the dose–response relationship in the context of physical exercise treatments. Respective recommendations vary between meta-analyses. Firth et al. [33] conclude that at least 90 minutes of exercise with moderate-to-vigorous intensity are required to achieve optimal benefit in negative symptoms, while Sabe et al. [42] suggest at least 60 minutes per week, split into two sessions over a minimum of six months. Kim et al. [40] emphasize the importance of a total duration of the intervention of at least 20 weeks to achieve optimal efficacy. Synthesizing these findings, it is recommended to engage patients with SSD in at least 60 to 90 minutes of moderate-to-vigorous intensity physical exercise per week for a minimum duration of six months to optimize the improvement of negative symptom severity.

In sum, current evidence supports the efficacy of physical exercise treatments in improving negative symptoms in SSD. Observed effects are likely to be clinically meaningful, but further investigations are required to better distinguish between primary and secondary negative symptoms. Furthermore, future large-scale trials are needed to develop personalized exercise treatments for the individual patient, but generally, at least 60 to 90 minutes of moderate-to-vigorous exercise per week are recommended to optimize treatment effects on negative symptoms in SSD.

17.2 Effects of Physical Exercise on Cognitive Deficits in SSD

Besides the effects of physical exercise on negative symptoms, three meta-analyses and one large-scale multicenter randomized-controlled trial published between 2017 and 2023 examined the effects of exercise treatments on cognitive deficits in people with SSD [39, 52–54]. These studies mostly incorporated aerobic endurance training. The study population among all meta-analyses was diverse with regard to symptom patterns and general disorder severity. In general, physical exercise interventions improved cognitive performance in people with schizophrenia with small effect sizes (Hedge's $g = 0.21$ to 0.33) compared to active or passive control conditions [39, 52–54]. Thus, the current state of research confirms the general efficacy of physical exercise interventions in improving cognitive functioning in people with SSD.

Analogously to the effects of physical exercise on negative symptoms in SSD, several essential aspects need to be clarified to evaluate the efficacy of physical exercise therapies on cognitive functioning in more detail.

First, the clinical relevance of the pro-cognitive effects of physical exercise in SSD remains to be determined. The small effect sizes mentioned above are comparable to those observed in multiple large-scale meta-analyses of cognitive remediation treatments [25, 26, 55]. These cognitive improvements after cognitive remediation therapy elicit ameliorations in global functioning [26], supporting the clinical relevance of small pro-cognitive effects. However, a systematic evaluation of clinical relevance requires additional assessments such as self-ratings by the patient and external ratings by persons close to the patient to evaluate if cognitive improvements transfer into daily life [56]. Thus, given the comparable effect sizes of physical exercise and cognitive remediation, the exercise-induced improvements in cognitive functioning in people with SSD are likely to be clinically meaningful, but future studies should additionally examine clinical relevance using self- and external ratings.

Furthermore, in case of cognitive deficits, it is essential to differentiate between several subdomains of cognitive functioning, as the respective effect sizes differ. For instance, Firth et al. [52] demonstrate domain-specific improvements in working memory, attention, and social cognition, whereas no significant effects are obtained for processing speed, verbal and visual learning, or reasoning and problem solving. In contrast, Xu and colleagues [54] report pronounced effects on verbal learning and memory, as well as reasoning and problem solving, but do not find significant effects on working memory, attention, and social cognition. Shimada et al. [53] reveal that physical exercise interventions lead to specific ameliorations of attention, working memory, and verbal learning. Despite consistent positive findings for global cognition, the results regarding specific cognitive subdomains differ between meta-analyses and thus need further evaluation in future exercise trials.

A comprehensive comparison of different types of exercise treatments remains challenging, as most meta-analyses specifically focus on aerobic exercise programs. However, Firth et al. [52] conclude that aerobic exercise may be more efficient to treat cognitive deficits in SSD than mind-body exercise such as yoga, because it has been shown to positively impact a broader range of cognitive subdomains. Particularly, yoga appears to improve only long-term memory [34], whereas aerobic exercise leads to ameliorations in multiple cognitive domains such as working memory, attentional processes, and social cognition [52]. Importantly, a current large-scale multicenter randomized-controlled trial in SSD demonstrated that both aerobic exercise and resistance training have positive effects on global cognitive functioning [39]. To conclude, although the evidence is most robust for aerobic exercise treatments, we cannot derive decisive conclusions on its superiority regarding cognitive effects compared to other types of exercise, yet. Therefore, future studies should systematically compare the pro-cognitive effects of different exercise programs, while aiming for precision psychiatry approaches to determine if there exists a specific exercise treatment for certain patient subgroups to achieve cognitive improvements.

Lastly, the current state of research provides first insights into the dose-response relationship in the context of physical exercise interventions to treat cognitive deficits in SSD. Shimada et al. [53] propose at least 90 minutes of aerobic exercise per week over a period of at least twelve weeks to achieve the highest efficacies. Moreover, professional supervision and group settings increase the effect sizes [52–54].

In sum, current evidence reveals that physical exercise treatments can improve cognitive functioning in SSD. These effects are likely to be clinically meaningful, but further studies including subjective self- and external ratings are warranted to elucidate the clinical relevance of pro-cognitive effects of physical exercise therapy. Furthermore, more large-scale trials are required to develop individualized exercise treatments. In general, at least 90 minutes of supervised group exercise per week at a moderate intensity are recommended to optimize treatment effects on cognitive functioning in SSD.

17.3 Additional Effects of Physical Exercise on Global Functioning in SSD

Three meta-analyses published between 2016 and 2023 studied the effects of exercise treatments on global functioning in people with SSD [34, 35, 57]. These studies included a broad range of different exercise programs. The study population was heterogeneous with regard to symptom patterns and general disorder severity. In general, physical exercise interventions improved global functioning with small effect sizes (Hedge's $g = 0.32$ to 0.40) compared to active or passive control conditions [34, 35, 57]. Consequently, physical exercise interventions are effective in improving global functioning in people with SSD.

Again, these effects have to be evaluated with regard to their clinical relevance, the efficacy of different types of physical exercise therapies, and the dose-response relationship.

Most underlying studies used the Global Assessment of Functioning (GAF) scale to cover global functioning [58]. On average, an increase of 4.5 points in the GAF has been found to reflect a minimal clinical improvement as measured by the CGI [59]. Importantly, the abovementioned meta-analyses do not report the average change in the GAF after exercise therapy, complicating straightforward inferences about the clinical relevance of the effects reported. However, a recent large-scale multicenter randomized-controlled exercise trial found that 30 percent of the patients achieved an improvement in the GAF of at least ten points [39], reflecting a clear, clinically meaningful improvement [60]. Therefore, at least a subgroup of patients shows a strong clinical amelioration in response to physical exercise treatment. Further large-scale examinations should be conducted to evaluate more comprehensively the clinical relevance of exercise effects on global functioning.

When comparing different types of physical exercise treatments in people with SSD, Fernandez-Abascal and colleagues [35] demonstrate the largest effect sizes for aerobic exercise interventions at moderate to vigorous intensity, followed by mind-body exercise treatments, and low-intensity aerobic exercise programs. This is in line with the meta-analysis by Korman et al. [57], who also propose the superiority of aerobic exercise treatments in ameliorating global functioning. Consequently, at the group-level, aerobic exercise therapies in SSD appear to be most beneficial with regard to improvements in global functioning, but other types of physical exercise interventions, such as resistance training, have not been studied sufficiently in this context. Therefore, future large-scale trials should aim to determine if different clinical subgroups of patients reveal particular benefits in global functioning after certain exercise treatments.

In accordance with the recommendations for the treatment of cognitive treatments, the authors of current meta-analyses conclude that at least 30 minutes per day, three times per week over a period of twelve weeks are required to achieve improvements in global functioning [34, 35].

To conclude, current evidence demonstrates that physical exercise treatments lead to ameliorations of global functioning in people with SSD. Despite being likely, more evidence is required to explore the clinical relevance of these effects, as is the case for the identification of individualized exercise therapies designed for specific subgroups of patients. In general, at least 90 minutes of supervised group exercise per week at a moderate intensity are recommended to optimize treatment effects on global functioning in SSD.

17.4 Additional Effects of Physical Exercise on Physical Health in SSD

Between 2015 and 2024, six meta-analyses investigated the effects of physical exercise treatments on various physical health outcomes in people with SSD, such as body mass index (BMI), cardiorespiratory fitness, and typical laboratory blood markers [33, 35, 37, 38, 61, 62]. These studies considered different types of exercise programs. The study population was heterogeneous. Four meta-analyses examined the effects of physical exercise therapies on BMI and revealed heterogeneous results. Two meta-analyses report significant reductions after physical exercise treatment compared to active or passive control conditions [35, 62], whereas two others do not obtain significant effects [33, 38]. Accordingly, three meta-analyses reveal improvements in cardiorespiratory fitness induced by physical exercise treatments [33, 61, 62], whereas another recent meta-analysis did not find a significant effect [37]. Two current meta-analyses explore the impact of physical exercise interventions on typical blood markers such as glucose, cholesterol, low-density lipoproteins, high-density lipoproteins, or triglycerides. Apart from beneficial effects on triglycerides observed by Fernandez-Abascal et al. [35], these meta-analyses do not find significant effects on these blood markers [35, 37]. Correspondingly, results of a recent large multicenter randomized-controlled trial do not demonstrate beneficial effects on any physical health outcome of either aerobic exercise treatment or resistance training [39]. Taken together, physical exercise treatments may yield the potential to improve physical health in people with SSD, especially cardiorespiratory fitness, but current evidence on the group level is not robust.

Therefore, these findings require further examination with regard to the varying efficacy between different types of exercise, the identification of particularly responsive subgroups of patients, and the dose-response relationships.

With regard to improvements in cardiorespiratory fitness, current evidence is most consistent for aerobic exercise interventions [33, 61, 62]. Considering the impact of physical exercise therapies on physical health in general, Fernandez-Abascal et al. [35] highlight that combined therapies consisting of exercise and psychotherapy are most effective. Accordingly, other authors [33, 39] conclude that physical exercise treatments should be combined with nutrition therapies to achieve reliable ameliorations of physical health in people with SSD, as diet modifications have already been proven to be effective in populations with severe mental illnesses [63]. Therefore, future research in this area should systematically evaluate if combined therapies, comprising physical exercise, psychotherapy, and nutritional treatments, lead to more robust and consistent ameliorations of physical health in SSD. Moreover, as already described for negative symptoms, cognition, and global functioning, there is a need to identify patient characteristics that facilitate improvements in physical health after a certain combined treatment approach, aiming at providing personalized exercise therapies that address physical health in SSD.

Concerning the required dosage of physical exercise to address physical health in SSD, no clear recommendations can be made based on the current state of research [33, 35, 37, 38, 61, 62]. However, given that the overall effects of exercise on physical health are less consistent compared to the impact on psychiatric outcomes, it appears plausible that dosages of the abovementioned combined therapies need to be higher than 90 minutes per week for twelve weeks.

17.5 Mechanisms Explaining Beneficial Effects of Physical Exercise

While the evidence in favor of physical exercise treatments to achieve additional health benefits in people with SSD is promising, only preliminary findings of the underlying mechanisms are available to date.

Current studies in people with SSD have been mainly focused on exercise-induced neurobiological mechanisms that may drive clinical ameliorations. In particular, physical exercise is said to facilitate the expression of growth factors such as the brain-derived neurotrophic factor, which in turn stimulates synaptogenesis, synaptic transmission, and dendritic spine density [64–66]. Moreover, exercise can lead to increased expression of genes involved in cell growth, synaptic trafficking, or signal transduction, leading to general improvements in synaptic plasticity [64]. Likewise, physical exercise treatments are expected to affect several core neurotransmitter systems, including the serotonergic, dopaminergic, glutamatergic, acetylcholinergic, or norepinephrinergic pathways, which are essential for efficient neural information processing [64]. However, these insights into neural mechanisms on the microlevel have been mostly derived from animal studies. In humans, preliminary evidence indicates that higher aerobic fitness, as an indicator of regular involvement in physical exercise of patients with SSD, is linked to both larger hippocampal structure and widespread functional connectivity patterns, which are partly associated with symptom severity [67–69]. In the context of physical exercise interventions in people with SSD, first findings indicate that especially aerobic exercise may lead to increases in hippocampal subfield volumes [70] and changes in functional connectivity within certain brain networks [71], but the clinical implications of these effects remain to be determined.

In addition to the neurobiological level, ameliorations of physical health after aerobic exercise treatment [61] may modulate neurobiological adaptations and resulting clinical outcomes. Impaired physical and metabolic health are crucial in schizophrenia [19] and thus can lead to widespread neural deteriorations in both brain structure and function [72–75]. Improving physical and metabolic health through physical exercise treatments may thus yield the potential to stimulate regenerative neural processes and associated clinical outcomes. However, empirical evidence is scarce and further physical exercise studies are warranted to explore the interaction between physical and neural mechanisms that explain clinical improvements.

To conclude, the underlying neurobiological and physical mechanisms of beneficial exercise effects on clinical outcomes in SSD remain largely unclear, although preliminary evidence provides a first potential mechanistic target. Elucidating such mechanisms is essential to gain a better understanding of the pathophysiology of SSD and can provide a basis for the development of targeted treatments, such as brain stimulation approaches that specifically address these mechanisms.

17.6 Feasibility, Adherence, and Safety

As outlined previously, physical exercise interventions induce additional benefits in negative symptom severity, cognitive performance, and global functioning in people with SSD. These advantageous effects are expected to hold clinical significance, although this has not been definitively confirmed yet. Given the remaining uncertainty of the clinical relevance of exercise effects on negative symptoms, cognition, and global functioning, it is important to comprehensively assess the feasibility, adherence, and safety of physical exercise interventions in SSD. A well-founded evaluation of the positive impact of physical exercise in SSD and the required resources and potential risks of such interventions is essential to derive final recommendations.

The feasibility of physical exercise therapies in people with SSD has been demonstrated through multiple successful studies conducted in the past two decades and summarized within the abovementioned meta-analyses [33–36, 39–41, 52–54, 57, 62]. Many of these meta-analyses included both trials in inpatient and outpatient settings that were partially embedded in the clinical routine setting without any additional research funding. To implement a regular physical exercise program in the clinical setting, at least one broadly educated, professional sport coach with good interpersonal skills is required, as is one exercise room with sufficient space for at least three patients and respective materials (ergometers, treadmills, sport mats, elastic thera bands, etc.). Hence, the resources needed are manageable, supporting the feasibility of physical exercise programs in daily clinical care.

Further, current evidence demonstrates that the overall adherence rates across all types of physical exercise interventions and all study settings vary between 36 to 87% [33, 36, 38, 39, 76]. Importantly, adherence is improved substantially in professionally supervised group exercise settings, reaching stable rates of 78% [33, 76]. Moreover, adherence rates have been found to be higher in inpatient than in outpatient settings, supporting their implementation in daily clinical care [76]. Consequently, physical exercise interventions are well-accepted by people with SSD if they are supervised by a professional and if they are performed in a group setting.

With respect to the safety of physical exercise interventions, no relevant concerns have been reported. Documented adverse events in the context of respective exercise trials have not been linked to the interventions themselves [36, 39]. Furthermore, physical exercise interventions do not increase the risk for relapse or rehospitalization [36]. However, it is essential to design the particular exercise program according to the needs and physical prerequisites of the individual patient to avoid injuries and to ensure safety. Therefore, respective physical examinations and exercise readiness checks prior to the start of the treatment are recommended [36].

In sum, besides their beneficial impact, physical exercise interventions in people with SSD can be considered feasible, well-tolerated, and safe.

17.7 Summary and Recommendations

Physical exercise treatments in people with SSD lead to improvements in negative symptom severity, cognitive performance, and global functioning, while stable beneficial effects on physical health require combined therapies of physical exercise, psychotherapy, and diet modifications. Ameliorations in negative symptoms, cognitive deficits, and global functioning are likely to be clinically meaningful, but further investigations are required to confirm this notion. However, given that physical exercise treatments in people with SSD are feasible, well-accepted, and safe, their systematic application in clinical routine care is strongly recommended. In particular, current large-scale evidence proposes the clinical implementation of individually adjusted, supervised group exercise programs at moderate intensity with a weekly duration of at least 90 minutes split into two to three sessions per week over three months.

References

1. American Psychiatric Association. Diagnostic and statistical manual of mental disorders: DSM-5™. Washington, DC: American Psychiatric Publishing; 2013.
2. Vancampfort D, Rosenbaum S, Schuch F, Ward PB, Richards J, Mugisha J, et al. Cardiorespiratory fitness in severe mental illness: a systematic review and meta-analysis. Sports Med. 2017;47(2):343–52. https://doi.org/10.1007/s40279-016-0574-1.
3. Dipasquale S, Pariante CM, Dazzan P, Aguglia E, McGuire P, Mondelli V. The dietary pattern of patients with schizophrenia: a systematic review. J Psychiatr Res. 2013;47(2):197–207. https://doi.org/10.1016/j.jpsychires.2012.10.005.
4. Hartz SM, Pato CN, Medeiros H, Cavazos-Rehg P, Sobell JL, Knowles JA, et al. Comorbidity of severe psychotic disorders with measures of substance use. JAMA Psychiatry. 2014;71(3):248–54. https://doi.org/10.1001/jamapsychiatry.2013.3726.
5. Ohi K, Shimada T, Kuwata A, Kataoka Y, Okubo H, Kimura K, et al. Smoking rates and number of cigarettes smoked per day in schizophrenia: a large cohort meta-analysis in a Japanese population. Int J Neuropsychopharmacol. 2019;22(1):19–27. https://doi.org/10.1093/ijnp/pyy061.
6. Šagud M, Vuksan-Ćusa B, Jakšić N, Mihaljević-Peleš A, Rojnić Kuzman M, Pivac N. Smoking in schizophrenia: an updated review. Psychiatr Danub. 2018;30(4):216–23.
7. De Hert M, Detraux J, Vancampfort D. The intriguing relationship between coronary heart disease and mental disorders. Dialogues Clin Neurosci. 2018;20(1):31–40. https://doi.org/10.31887/DCNS.2018.20.1/mdehert.
8. Fan Z, Wu Y, Shen J, Ji T, Zhan R. Schizophrenia and the risk of cardiovascular diseases: a meta-analysis of thirteen cohort studies. J Psychiatr Res. 2013;47(11):1549–56. https://doi.org/10.1016/j.jpsychires.2013.07.011.
9. Foguet-Boreu Q, Fernandez San Martin MI, Flores Mateo G, Zabaleta Del Olmo E, Ayerbe García-Morzon L, Perez-Piñar López M, et al. Cardiovascular risk assessment in patients with a severe mental illness: a systematic review and meta-analysis. BMC Psychiatry. 2016;16:141. https://doi.org/10.1186/s12888-016-0833-6.
10. Ringen PA, Engh JA, Birkenaes AB, Dieset I, Andreassen OA. Increased mortality in schizophrenia due to cardiovascular disease - a non-systematic review of epidemiology, possible causes, and interventions. Front Psych. 2014;5:137. https://doi.org/10.3389/fpsyt.2014.00137.
11. von Hausswolff-Juhlin Y, Bjartveit M, Lindström E, Jones P. Schizophrenia and physical health problems. Acta Psychiatr Scand. 2009;438:15–21. https://doi.org/10.1111/j.1600-0447.2008.01309.x.
12. Manu P, Dima L, Shulman M, Vancampfort D, De Hert M, Correll CU. Weight gain and obesity in schizophrenia: epidemiology, pathobiology, and management. Acta Psychiatr Scand. 2015;132(2):97–108. https://doi.org/10.1111/acps.12445.
13. Rege S. Antipsychotic induced weight gain in schizophrenia:mechanisms and management. Aust N Z J Psychiatry. 2008;42(5):369–81. https://doi.org/10.1080/00048670801961123.
14. Tek C, Kucukgoncu S, Guloksuz S, Woods SW, Srihari VH, Annamalai A. Antipsychotic-induced weight gain in first-episode psychosis patients: a meta-analysis of differential effects of antipsychotic medications. Early Interv Psychiatry. 2016;10(3):193–202. https://doi.org/10.1111/eip.12251.
15. Holt RI. The prevention of diabetes and cardiovascular disease in people with schizophrenia. Acta Psychiatr Scand. 2015;132(2):86–96. https://doi.org/10.1111/acps.12443.
16. Stubbs B, Vancampfort D, De Hert M, Mitchell AJ. The prevalence and predictors of type two diabetes mellitus in people with schizophrenia: a systematic review and comparative meta-analysis. Acta Psychiatr Scand. 2015;132(2):144–57. https://doi.org/10.1111/acps.12439.
17. Vancampfort D, Correll CU, Galling B, Probst M, De Hert M, Ward PB, et al. Diabetes mellitus in people with schizophrenia, bipolar disorder and major depressive disorder: a systematic review and large scale meta-analysis. World Psychiatry. 2016;15(2):166–74. https://doi.org/10.1002/wps.20309.

18. Mitchell AJ, Vancampfort D, Sweers K, van Winkel R, Yu W, De Hert M. Prevalence of metabolic syndrome and metabolic abnormalities in schizophrenia and related disorders--a systematic review and meta-analysis. Schizophr Bull. 2013;39(2):306–18. https://doi.org/10.1093/schbul/sbr148.
19. Tian YE, Di Biase MA, Mosley PE, Lupton MK, Xia Y, Fripp J, et al. Evaluation of brain-body health in individuals with common neuropsychiatric disorders. JAMA Psychiatry. 2023;80(6):567–76. https://doi.org/10.1001/jamapsychiatry.2023.0791.
20. Hjorthøj C, Stürup AE, McGrath JJ, Nordentoft M. Years of potential life lost and life expectancy in schizophrenia: a systematic review and meta-analysis. Lancet Psychiatry. 2017;4(4):295–301. https://doi.org/10.1016/s2215-0366(17)30078-0.
21. Laursen TM, Plana-Ripoll O, Andersen PK, McGrath JJ, Toender A, Nordentoft M, et al. Cause-specific life years lost among persons diagnosed with schizophrenia: is it getting better or worse? Schizophr Res. 2019;206:284–90. https://doi.org/10.1016/j.schres.2018.11.003.
22. Gaebel W, Hasan A, Falkai P. S3-Leitlinie Schizophrenie. Berlin: Springer; 2019.
23. Huhn M, Nikolakopoulou A, Schneider-Thoma J, Krause M, Samara M, Peter N, et al. Comparative efficacy and tolerability of 32 oral antipsychotics for the acute treatment of adults with multi-episode schizophrenia: a systematic review and network meta-analysis. Lancet. 2019;394(10202):939–51. https://doi.org/10.1016/s0140-6736(19)31135-3.
24. Krause M, Zhu Y, Huhn M, Schneider-Thoma J, Bighelli I, Nikolakopoulou A, et al. Antipsychotic drugs for patients with schizophrenia and predominant or prominent negative symptoms: a systematic review and meta-analysis. Eur Arch Psychiatry Clin Neurosci. 2018;268(7):625–39. https://doi.org/10.1007/s00406-018-0869-3.
25. Vita A, Barlati S, Ceraso A, Nibbio G, Ariu C, Deste G, et al. Effectiveness, core elements, and moderators of response of cognitive remediation for schizophrenia: a systematic review and meta-analysis of randomized clinical trials. JAMA Psychiatry. 2021;78(8):848–58. https://doi.org/10.1001/jamapsychiatry.2021.0620.
26. Kambeitz-Ilankovic L, Betz LT, Dominke C, Haas SS, Subramaniam K, Fisher M, et al. Multi-outcome meta-analysis (MOMA) of cognitive remediation in schizophrenia: revisiting the relevance of human coaching and elucidating interplay between multiple outcomes. Neurosci Biobehav Rev. 2019;107:828–45. https://doi.org/10.1016/j.neubiorev.2019.09.031.
27. Cella M, Preti A, Edwards C, Dow T, Wykes T. Cognitive remediation for negative symptoms of schizophrenia: a network meta-analysis. Clin Psychol Rev. 2017;52:43–51. https://doi.org/10.1016/j.cpr.2016.11.009.
28. Solmi M, Croatto G, Piva G, Rosson S, Fusar-Poli P, Rubio JM, et al. Efficacy and acceptability of psychosocial interventions in schizophrenia: systematic overview and quality appraisal of the meta-analytic evidence. Mol Psychiatry. 2023;28(1):354–68. https://doi.org/10.1038/s41380-022-01727-z.
29. Tseng PT, Zeng BS, Hung CM, Liang CS, Stubbs B, Carvalho AF, et al. Assessment of non-invasive brain stimulation interventions for negative symptoms of schizophrenia: a systematic review and network meta-analysis. JAMA Psychiatry. 2022;79(8):770–9. https://doi.org/10.1001/jamapsychiatry.2022.1513.
30. Begemann MJ, Brand BA, Ćurčić-Blake B, Aleman A, Sommer IE. Efficacy of non-invasive brain stimulation on cognitive functioning in brain disorders: a meta-analysis. Psychol Med. 2020;50(15):2465–86. https://doi.org/10.1017/s0033291720003670.
31. Jauhar S, Johnstone M, McKenna PJ. Schizophrenia. Lancet. 2022;399(10323):473–86. https://doi.org/10.1016/s0140-6736(21)01730-x.
32. Schmitt A, Reich-Erkelenz D, Hasan A, Falkai P. Aerobic exercise in mental disorders: from basic mechanisms to treatment recommendations. Eur Arch Psychiatry Clin Neurosci. 2019;269(5):483–4. https://doi.org/10.1007/s00406-019-01037-6.
33. Firth J, Cotter J, Elliott R, French P, Yung AR. A systematic review and meta-analysis of exercise interventions in schizophrenia patients. Psychol Med. 2015;45(7):1343–61. https://doi.org/10.1017/s0033291714003110.

34. Dauwan M, Begemann MJ, Heringa SM, Sommer IE. Exercise improves clinical symptoms, quality of life, global functioning, and depression in schizophrenia: a systematic review and meta-analysis. Schizophr Bull. 2016;42(3):588–99. https://doi.org/10.1093/schbul/sbv164.
35. Fernández-Abascal B, Suárez-Pinilla P, Cobo-Corrales C, Crespo-Facorro B, Suárez-Pinilla M. In- and outpatient lifestyle interventions on diet and exercise and their effect on physical and psychological health: a systematic review and meta-analysis of randomised controlled trials in patients with schizophrenia spectrum disorders and first episode of psychosis. Neurosci Biobehav Rev. 2021;125:535–68. https://doi.org/10.1016/j.neubiorev.2021.01.005.
36. Ziebart C, Bobos P, MacDermid JC, Furtado R, Sobczak DJ, Doering M. The efficacy and safety of exercise and physical activity on psychosis: a systematic review and meta-analysis. Front Psych. 2022;13:807140. https://doi.org/10.3389/fpsyt.2022.807140.
37. Gallardo-Gómez D, Noetel M, Álvarez-Barbosa F, Alfonso-Rosa RM, Ramos-Munell J, Del Pozo Cruz B, et al. Exercise to treat psychopathology and other clinical outcomes in schizophrenia: a systematic review and meta-analysis. Eur Psychiatry. 2023;66(1):e40. https://doi.org/10.1192/j.eurpsy.2023.24.
38. Guo J, Liu K, Liao Y, Qin Y, Yue W. Efficacy and feasibility of aerobic exercise interventions as an adjunctive treatment for patients with schizophrenia: a meta-analysis. Schizophrenia (Heidelb). 2024;10(1):2. https://doi.org/10.1038/s41537-023-00426-0.
39. Maurus I, Roell L, Lembeck M, Papazova I, Greska D, Muenz S, et al. Exercise as an add-on treatment in individuals with schizophrenia: results from a large multicenter randomized controlled trial. Psychiatry Res. 2023;328:115480. https://doi.org/10.1016/j.psychres.2023.115480.
40. Kim M, Lee Y, Kang H. Effects of exercise on positive symptoms, negative symptoms, and depression in patients with schizophrenia: a systematic review and meta-analysis. Int J Environ Res Public Health. 2023;20(4) https://doi.org/10.3390/ijerph20043719.
41. Wei GX, Yang L, Imm K, Loprinzi PD, Smith L, Zhang X, et al. Effects of mind-body exercises on schizophrenia: a systematic review with meta-analysis. Front Psych. 2020;11:819. https://doi.org/10.3389/fpsyt.2020.00819.
42. Sabe M, Kaiser S, Sentissi O. Physical exercise for negative symptoms of schizophrenia: systematic review of randomized controlled trials and meta-analysis. Gen Hosp Psychiatry. 2020;62:13–20. https://doi.org/10.1016/j.genhosppsych.2019.11.002.
43. Vogel JS, van der Gaag M, Slofstra C, Knegtering H, Bruins J, Castelein S. The effect of mind-body and aerobic exercise on negative symptoms in schizophrenia: a meta-analysis. Psychiatry Res. 2019;279:295–305. https://doi.org/10.1016/j.psychres.2019.03.012.
44. Fusar-Poli P, Papanastasiou E, Stahl D, Rocchetti M, Carpenter W, Shergill S, et al. Treatments of negative symptoms in schizophrenia: meta-analysis of 168 randomized placebo-controlled trials. Schizophr Bull. 2015;41(4):892–9. https://doi.org/10.1093/schbul/sbu170.
45. Leucht S, Kane JM, Etschel E, Kissling W, Hamann J, Engel RR. Linking the PANSS, BPRS, and CGI: clinical implications. Neuropsychopharmacology. 2006;31(10):2318–25. https://doi.org/10.1038/sj.npp.1301147.
46. Kay SR, Fiszbein A, Opler LA. The positive and negative syndrome scale (PANSS) for schizophrenia. Schizophr Bull. 1987;13(2):261–76. https://doi.org/10.1093/schbul/13.2.261.
47. Busner J, Targum SD. The clinical global impressions scale: applying a research tool in clinical practice. Psychiatry (Edgmont). 2007;4(7):28–37.
48. Leucht S, Barabássy Á, Laszlovszky I, Szatmári B, Acsai K, Szalai E, et al. Linking PANSS negative symptom scores with the Clinical Global Impressions Scale: understanding negative symptom scores in schizophrenia. Neuropsychopharmacology. 2019;44(9):1589–96. https://doi.org/10.1038/s41386-019-0363-2.
49. Andreasen NC. The scale for the assessment of negative symptoms (SANS): conceptual and theoretical foundations. Br J Psychiatry. 1989;155(S7):49–52. https://doi.org/10.1192/S0007125000291496.
50. Levine SZ, Leucht S. Identifying clinically meaningful symptom response cut-off values on the SANS in predominant negative symptoms. Schizophr Res. 2013;145(1–3):125–7. https://doi.org/10.1016/j.schres.2012.12.032.

51. Galderisi S, Mucci A, Dollfus S, Nordentoft M, Falkai P, Kaiser S, et al. EPA guidance on assessment of negative symptoms in schizophrenia. Eur Psychiatry. 2021;64(1):e23. https://doi.org/10.1192/j.eurpsy.2021.11.
52. Firth J, Stubbs B, Rosenbaum S, Vancampfort D, Malchow B, Schuch F, et al. Aerobic exercise improves cognitive functioning in people with schizophrenia: a systematic review and meta-analysis. Schizophr Bull. 2017;43(3):546–56. https://doi.org/10.1093/schbul/sbw115.
53. Shimada T, Ito S, Makabe A, Yamanushi A, Takenaka A, Kawano K, et al. Aerobic exercise and cognitive functioning in schizophrenia: an updated systematic review and meta-analysis. Psychiatry Res. 2022;314:114656. https://doi.org/10.1016/j.psychres.2022.114656.
54. Xu Y, Cai Z, Fang C, Zheng J, Shan J, Yang Y. Impact of aerobic exercise on cognitive function in patients with schizophrenia during daily care: a meta-analysis. Psychiatry Res. 2022;312:114560. https://doi.org/10.1016/j.psychres.2022.114560.
55. Lejeune JA, Northrop A, Kurtz MM. A meta-analysis of cognitive remediation for schizophrenia: efficacy and the role of participant and treatment factors. Schizophr Bull. 2021;47(4):997–1006. https://doi.org/10.1093/schbul/sbab022.
56. Keefe RS, Davis VG, Spagnola NB, Hilt D, Dgetluck N, Ruse S, et al. Reliability, validity and treatment sensitivity of the Schizophrenia Cognition Rating Scale. Eur Neuropsychopharmacol. 2015;25(2):176–84. https://doi.org/10.1016/j.euroneuro.2014.06.009.
57. Korman N, Stanton R, Vecchio A, Chapman J, Parker S, Martland R, et al. The effect of exercise on global, social, daily living and occupational functioning in people living with schizophrenia: a systematic review and meta-analysis. Schizophr Res. 2023;256:98–111. https://doi.org/10.1016/j.schres.2023.04.012.
58. Endicott J, Spitzer RL, Fleiss JL, Cohen J. The global assessment scale. A procedure for measuring overall severity of psychiatric disturbance. Arch Gen Psychiatry. 1976;33(6):766–71. https://doi.org/10.1001/archpsyc.1976.01770060086012.
59. Samara MT, Engel RR, Millier A, Kandenwein J, Toumi M, Leucht S. Equipercentile linking of scales measuring functioning and symptoms: examining the GAF, SOFAS, CGI-S, and PANSS. Eur Neuropsychopharmacol. 2014;24(11):1767–72. https://doi.org/10.1016/j.euroneuro.2014.08.009.
60. Falkai P, Maurus I, Schmitt A, Malchow B, Schneider-Axmann T, Röll L, et al. Improvement in daily functioning after aerobic exercise training in schizophrenia is sustained after exercise cessation. Eur Arch Psychiatry Clin Neurosci. 2021;271(7):1201–3. https://doi.org/10.1007/s00406-021-01282-8.
61. Vancampfort D, Rosenbaum S, Ward PB, Stubbs B. Exercise improves cardiorespiratory fitness in people with schizophrenia: a systematic review and meta-analysis. Schizophr Res. 2015;169(1–3):453–7. https://doi.org/10.1016/j.schres.2015.09.029.
62. Bredin SSD, Kaufman KL, Chow MI, Lang DJ, Wu N, Kim DD, et al. Effects of aerobic, resistance, and combined exercise training on psychiatric symptom severity and related health measures in adults living with schizophrenia: a systematic review and meta-analysis. Front Cardiovasc Med. 2021;8:753117. https://doi.org/10.3389/fcvm.2021.753117.
63. Daumit GL, Dickerson FB, Wang NY, Dalcin A, Jerome GJ, Anderson CA, et al. A behavioral weight-loss intervention in persons with serious mental illness. N Engl J Med. 2013;368(17):1594–602. https://doi.org/10.1056/NEJMoa1214530.
64. Maurus I, Hasan A, Röh A, Takahashi S, Rauchmann B, Keeser D, et al. Neurobiological effects of aerobic exercise, with a focus on patients with schizophrenia. Eur Arch Psychiatry Clin Neurosci. 2019;269(5):499–515. https://doi.org/10.1007/s00406-019-01025-w.
65. Firth J, Cotter J, Carney R, Yung AR. The pro-cognitive mechanisms of physical exercise in people with schizophrenia. Br J Pharmacol. 2017;174(19):3161–72. https://doi.org/10.1111/bph.13772.
66. Liu PZ, Nusslock R. Exercise-mediated neurogenesis in the hippocampus via BDNF. Front Neurosci. 2018;12:52. https://doi.org/10.3389/fnins.2018.00052.
67. Roell L, Maurus I, Keeser D, Karali T, Papazov B, Hasan A, et al. Association between aerobic fitness and the functional connectome in patients with schizophrenia. Eur Arch Psychiatry Clin Neurosci. 2022;272(7):1253–72. https://doi.org/10.1007/s00406-022-01411-x.

68. Maurus I, Roell L, Keeser D, Papazov B, Papazova I, Lembeck M, et al. Fitness is positively associated with hippocampal formation subfield volumes in schizophrenia: a multiparametric magnetic resonance imaging study. Transl Psychiatry. 2022;12(1):388. https://doi.org/10.1038/s41398-022-02155-x.
69. Maurus I, Röll L, Keeser D, Karali T, Papazov B, Hasan A, et al. Associations between aerobic fitness, negative symptoms, cognitive deficits and brain structure in schizophrenia-a cross-sectional study. Schizophr. 2022;8(1):63. https://doi.org/10.1038/s41537-022-00269-1.
70. Roell L, Fischer T, Maurus I. Effects of aerobic exercise on hippocampal formation volume in people with schizophrenia–a systematic review and meta-analysis with original data from a randomized-controlled trial. OSF Preprints 2023. https://doi.org/10.31219/osf.io/y2phs.
71. Roell L, Keeser D, Papazov B, Lembeck M, Papazova I, Greska D, et al. Effects of exercise on structural and functional brain patterns in schizophrenia-data from a multicenter randomized-controlled study. Schizophr Bull. 2024;50(1):145–56. https://doi.org/10.1093/schbul/sbad113.
72. Li L, Yu H, Zhong M, Liu S, Wei W, Meng Y, et al. Gray matter volume alterations in subjects with overweight and obesity: evidence from a voxel-based meta-analysis. Front Psych. 2022;13:955741. https://doi.org/10.3389/fpsyt.2022.955741.
73. Herrmann MJ, Tesar AK, Beier J, Berg M, Warrings B. Grey matter alterations in obesity: a meta-analysis of whole-brain studies. Obes Rev. 2019;20(3):464–71. https://doi.org/10.1111/obr.12799.
74. Syan SK, McIntyre-Wood C, Minuzzi L, Hall G, McCabe RE, MacKillop J. Dysregulated resting state functional connectivity and obesity: a systematic review. Neurosci Biobehav Rev. 2021;131:270–92. https://doi.org/10.1016/j.neubiorev.2021.08.019.
75. Parsons N, Steward T, Clohesy R, Almgren H, Duehlmeyer L. A systematic review of resting-state functional connectivity in obesity: refining current neurobiological frameworks and methodological considerations moving forward. Rev Endocr Metab Disord. 2022;23(4):861–79. https://doi.org/10.1007/s11154-021-09665-x.
76. Vancampfort D, Rosenbaum S, Schuch FB, Ward PB, Probst M, Stubbs B. Prevalence and predictors of treatment dropout from physical activity interventions in schizophrenia: a meta-analysis. Gen Hosp Psychiatry. 2016;39:15–23. https://doi.org/10.1016/j.genhosppsych.2015.11.008.

Cognitive Impairment and Negative Symptoms in Schizophrenia: A Summary and Future Research Perspectives

18

Antonio Vita, Bernardo Carpiniello, and Silvana Galderisi

Abstract

Negative symptoms are common and pervasive in people living with schizophrenia. These symptoms significantly impact an individual's life and are strong predictors of functional outcomes. However, they are often underrecognized and undertreated compared to positive symptoms. Addressing negative symptoms requires a multimodal approach.

Cognitive impairment represents a core feature of schizophrenia, is present at any stage of the disease and in subjects at high risk for psychosis, and also has a significant impact on patients' psychosocial functioning. Also in this case, treatment of cognitive symptoms requires a multimodal approach.

However, the relationships between cognitive and negative dimensions of schizophrenia are still poorly addressed and understood. Both negative and cognitive symptoms may be in part secondary to the other dimension, and treating each of them could be beneficial also on the other. A core part of these dimensions remains however independent and may represent a nuclear feature of the disease process in the lifespan of patients living with schizophrenia.

A. Vita (✉)
Department of Clinical and Experimental Sciences, University of Brescia, Brescia, Italy

Department of Mental Health and Addiction Services, ASST Spedali Civili of Brescia, Brescia, Italy
e-mail: antonio.vita@unibs.it

B. Carpiniello
Department of Medical Sciences and Public Health, University of Cagliari, Cagliari, Italy

S. Galderisi
Department of Mental and Physical Health, University of Campania "Luigi Vanvitelli", Naples, Italy

A. Vita et al. (eds.), *Negative and Cognitive Symptoms in Schizophrenia*,
Comprehensive Approach to Psychiatry 3,
https://doi.org/10.1007/978-3-032-06209-3_18

Keywords

Schizophrenia · Negative symptoms · Cognitive symptoms

18.1 Cognitive Impairment in Schizophrenia

Cognitive impairment represents one of the core components of schizophrenia [69, 109, 113].

A global cognitive performance score that is one standard deviation or more below the mean observed in the general population can be considered as a measure of significant cognitive impairment, and this can be found in more than 80% of assessed individuals diagnosed with schizophrenia [74, 112]. Recent studies have also investigated this issue in Asian and African countries and observed that the prevalence of cognitive impairment in people living with schizophrenia is quite consistent across the world [26, 68, 123].

The role of cognitive impairment as an essential feature of schizophrenia is further confirmed by the fact that it can be observed in first-episode subjects and even in the earliest phases of the disorder, as it often predates the onset of psychotic symptoms, as attested by the results of studies assessing individuals at high risk for psychosis [19, 44, 57, 103]. A poor cognitive performance has also been observed on several tests in non-affected first-degree relatives of individuals living with schizophrenia [122, 162]. However, the importance of cognitive impairment is related not only to its frequency, but also, and most importantly, to its impact on the lives of people with schizophrenia. In fact, cognitive impairment represents one of the main determinants of real-world functional impairment in people living with schizophrenia, with a strong negative impact on community functioning, personal, social, and work-related outcomes [22, 39, 64, 75, 167]. Moreover, cognitive impairment in people with schizophrenia has also been shown to be related to worse quality of life [3, 94, 108, 185] and worse life engagement [31, 182]. For these reasons, cognitive impairment has become an issue of crucial importance not only in research and clinical settings, but also among mental health service users [23, 29, 46, 149]. As such, it is becoming increasingly relevant to the identification of treatment targets and to the definition of current research goals [74, 109, 113, 184].

In the following paragraphs, we summarize the most recent evidence on research and clinical aspects of cognitive impairment and highlight future perspectives in both fields.

18.2 Assessment of Cognitive Impairment in Schizophrenia

Assessment of cognitive performance represents the first step in the clinical management of cognitive impairment associated with schizophrenia.

For this reason, the Section of the European Psychiatric Association focusing on Schizophrenia recently produced a guidance document dedicated specifically to the assessment of cognitive impairment [185], aiming at providing information and evidence-based guidance on the best approach to the issue in research and clinical settings.

Both neurocognition and social cognition show significant impairment in people living with schizophrenia [69, 142], and the need to assess both areas of cognition has the highest grade of recommendation (Grade A). Furthermore, several domains of neurocognition are involved: in particular, the domains of speed of processing, attention/vigilance, working memory, verbal learning and memory, visuospatial learning and memory, and reasoning and problem solving/executive functions, i.e., the six domains recognized by the National Institute of Mental Health (NIMH) Measurement and Treatment Research to Improve Cognition in Schizophrenia (MATRICS), should all be separately assessed to have the most detailed evaluation of neurocognitive performance [129]. These domains should be carefully assessed in all phases of the disorder as well as in clinically high-risk subjects (Grade A recommendation).

Similarly, the domains of social cognition identified by the NIMH Social Cognition Psychometric Evaluation (SCOPE), i.e., emotional processing, theory of mind/mentalization, attributional style/bias and social perception, should be assessed [143] (again, Grade A recommendation).

The assessment should be carried out as soon as possible in the clinical course and recovery process of individuals with schizophrenia to assess the baseline cognitive impairment and evaluate cognitive strengths and weaknesses of the subject, allowing for an in-depth assessment in the absence of potential confounding factors (e.g., effects of chronicity and medication), and representing the ideal starting point to plan and devise a rehabilitation program. Moreover, the assessment of cognitive impairment should not be considered a once-in-a-lifetime evaluation: to monitor potential changes across time and measure the effectiveness of treatments, the assessment should be repeated at different times, particularly when a novel pharmacological or psychosocial treatment is provided to the subject [185].

To perform the assessment, several different validated instruments are available. The assessment tool providing the most detailed evaluation is the MATRICS Consensus Cognitive Battery (MCCB), which evaluates all six MATRICS domains and includes a brief assessment of social cognitive performance [92, 130]. However, this battery requires 50–60 min to be completed. An alternative, which is also largely used and well-validated, includes five MATRICS domains, provides a less detailed assessment, and requires 30–40 min to complete, is the Brief Assessment of Cognition in Schizophrenia (BACS) [90]. The Screen for Cognitive Impairment in Psychiatry (SCIP) is another validated assessment tool: this instrument requires 15 min to be completed, and while it cannot be considered a full assessment of cognitive impairment, it covers four MATRICS domains and can be used as an effective screening tool in clinical practice and for the inclusion of participants into research projects [147]. In order to assess social cognition performance, the most effective and validated tools are the Bell Lysaker Emotion Recognition task (BLERT) for emotion processing, the Hinting Task for theory of mind/mentalization and The Awareness of Social Inference Test (TASIT) that measures both; for attributional style/bias and social perception several tests are available, but none of them shows optimal reliability and psychometric characteristics [142, 144].

As regards the assessment of cognition in schizophrenia, an important future research perspective is represented by the need to further understand and define the distinction between primary and secondary cognitive impairment [187]. Cognitive impairment in schizophrenia appears to be composed of two elements: primary cognitive impairment, which stems from the neurobiological alterations that are responsible for the core psychopathological manifestations of the disorder, and secondary cognitive impairment, which is the consequence of several source issues that are frequent in people living with schizophrenia. Sources of secondary cognitive impairment include autistic symptoms [1, 40, 183], depressive symptoms [27, 88, 119], inadequate pharmacological treatment [42, 54, 85, 86, 184], metabolic syndrome [18, 70, 71], substance abuse [20, 56, 107, 146, 166], social deprivation [170], and sleep disorders [25, 102, 110]. Better understanding the differences, the correlates, and the cognitive profiles related to primary and secondary cognitive impairment and introducing this distinction in a systematic manner in research and clinical settings could provide valuable knowledge, with important repercussions on the clinical management and treatment of cognitive impairment in schizophrenia [187].

It has been argued that cognitive performance assessment tools require extensive and dedicated staff training, and even the tests that have a good balance between accuracy and administration time may be too lengthy for effective use in the everyday clinical practice of real-world treatment and rehabilitation services, leading to significant cost-effectiveness and sustainability issues [132]. This is a significant problem, and current research is aiming to provide feasible solutions. On the one hand, the development of novel assessment tools that have superior accuracy and an administration time balance could represent a valid approach. On the other hand, current and future technologies may provide viable and effective solutions. For instance, the use of telemedicine could allow the remote administration of different cognitive assessment tools: this is being explored for the MCCB tests [150], and a tablet version of the BACS (BAC App) has shown high levels of feasibility and reliability compared to the pen-and-paper version of the test [8] and could allow remote testing and even self-administration [7]. However, the accuracy and the cost-effectiveness of these approaches require further dedicated study.

Another issue that is increasingly gaining attention in current research and represents an important future perspective is the role of subjective cognitive impairment.

Subjective cognitive impairment appears to represent an important mediator of functional capacity and real-world psychosocial functioning in people living with schizophrenia (Pezzella et al., 2024), and interview-based measures of cognitive impairment such as the Cognitive Assessment Interview (CAI) or the Schizophrenia Cognition Rating Scale (SCoRS) [174], which provide more insight into the subjective experience of the individual, are recommended as co-primary measures of cognitive impairment in clinical setting and trials [185]. However, the issue of patient reported outcomes is gaining increasing attention not only in a research perspective but also for regulatory agencies and at a policymaking level [12, 153], and the development of patient reported outcomes for cognitive impairment [125, 171, 189] represents a focus for future research. Other important future research perspectives include the implementation of novel technologies in the assessment of cognitive impairment, with new developments such as digital assessment and Ecological Momentary Assessment [120].

18.3 Treatment of Cognitive Impairment in Schizophrenia

Currently available medications that are routinely used in the treatment of schizophrenia, such as antipsychotic medication, do not appear to provide significant benefits as regards cognitive impairment, and no molecule is currently approved or recommended for the management of cognitive impairment in schizophrenia [11, 66, 91, 148, 184]. However, providing an appropriate pharmacological treatment could provide significant benefits to cognitive performance, particularly from the perspective of avoiding secondary cognitive impairment [184, 187]. Second-generation antipsychotics should be preferred to first-generation ones for their favorable cognitive profile [9, 36, 128], with more recently developed medications providing preliminary evidence of better cognitive outcomes [30, 52, 134, 138].

Anticholinergic burden should be kept to the lowest possible level, avoiding anticholinergic medications and considering the anticholinergic component of antipsychotic medications [6, 42, 43, 85, 86, 93, 135, 160, 177]. Treatment with benzodiazepines should also be kept to a minimum [54, 158]. All these recommendations are included in the European Psychiatric Association guidance dedicated specifically to the treatment of cognitive impairment in schizophrenia [184].

Some molecules have shown preliminary evidence of positive effects on cognitive performance in people living with schizophrenia [176, 188]: acetylcholinesterase inhibitors, such as memantine [76, 81, 155] and galantamine [98], and anti-inflammatory molecules [28, 80, 137, 165], particularly N-Acetylcysteine [193] and Minocycline [24, 28, 195], could provide some benefit in terms of cognitive outcomes.

Several novel molecules are under study and could provide substantial benefits in the treatment of several core dimensions of schizophrenia, including cognitive impairment: iclepertin [53, 117, 118, 173], luvadaxistat [55, 133] and xanomeline-trospium [89, 154] are currently being investigated in undergoing Phase 2 and Phase 3 studies, and their potential role as cognitive enhancers in people living with schizophrenia will be confirmed or confuted by the results of these studies.

Considering the limited benefits of currently available pharmacological treatment on cognitive outcomes, psychosocial interventions represent the most effective instrument to address cognitive impairment in people living with schizophrenia [10, 51, 127, 164].

Cognitive remediation (CR) is recommended with the highest degree of recommendation in the European Psychiatry Association guidance [184], as several large and recent systematic reviews and meta-analyses have confirmed its effectiveness on cognitive and functional outcomes [87, 104, 180, 192], its good acceptability profile [179], and the durability of its positive effects in schizophrenia patients [181]. Core elements of effectiveness are represented by the presence of an active and trained therapist delivering the intervention, the structured repetition of exercises, the development of novel cognitive strategies, and the implementation of techniques to transfer cognitive gains into the real world [21, 180], while the age of participants does not limit the positive effects of the intervention [159, 180].

However, the dissemination and use of CR in rehabilitation settings and mental health services remain uneven also in contexts with high levels of available resources [178, 190, 191], even if recent evidence shows that limited available resources do not represent a true barrier to CR use [116, 126]: in fact, effective CR programs have been recently implemented also in low-income countries such as India [78], Iran [77], Togo and Benin [38, 186].

On the other hand, the limited understanding of the benefits and effectiveness of CR by part of mental health professionals, public health services, policymaker organizations, and regulatory entities represents a consistent barrier to CR implementation [4, 5, 190, 191, 194]. In this perspective, academic and clinical partnership and collaboration, as well as advocacy from service users, families, carers, and stakeholders' organizations, could represent effective strategies to improve the widespread dissemination of CR intervention in real-world clinical contexts [34, 152].

Physical Exercise (PE) based interventions (or physical activity interventions), particularly aerobic exercise, represent another category of effective interventions to improve not only metabolic and health-related outcomes in people living with schizophrenia [59, 61, 63, 91, 156], but also core dimensions of the disorder such as positive and negative symptoms [45, 47, 151], cognitive impairment [35, 45, 48] and, consequently, psychosocial functioning [99]. PE interventions, like CR interventions, appear to provide greater benefits when delivered by an active and specifically trained therapist, when delivered in a group context, and when the program includes an intensity of at least 90 min of exercise per week and a total duration of at least 12 weeks [161].

Moreover, CR and PE can be combined into integrated interventions, providing superior benefits compared to each intervention alone [32, 37, 131]. Telemedicine could represent a valid instrument not only in the assessment of cognitive impairment, but also to improve its treatment, as it could allow the remote delivery of evidence-based psychosocial interventions such as CR [82]. In fact, the effectiveness and the acceptability of remotely delivered evidence-based psychosocial interventions have been the focus of several primary studies [15, 41, 50, 72, 73, 121, 175], and recent meta-analytic evidence attests that remotely delivered CR interventions can provide significant improvements in both cognitive and functional outcomes [14]. App-based and smartphone-based CR interventions are also being developed and evaluated, providing preliminary evidence of effectiveness [49, 101]. This type of approach could improve the accessibility and the dissemination of CR interventions, but more evidence regarding its effectiveness compared to already established CR programs and its cost-effectiveness is currently required.

Virtual reality represents another technology that could be used to deliver treatments for several different mental health conditions [97, 115, 139], and systematic reviews attest that virtual reality-based interventions are feasible in people living with schizophrenia [67, 157]. Recent meta-analytic evidence reports preliminary evidence suggesting that virtual reality CR interventions appear to produce improvements in cognitive impairment [141], and different large and high-quality trials are currently investigating the effectiveness of virtual reality-mediated CR intervention in schizophrenia [83, 169] and in other severe mental illnesses [79, 140]. Future research should compare virtual reality-based CR intervention to currently available CR programs, considering also measures of engagement, accessibility, and cost-effectiveness.

Non-invasive brain stimulation (NIBS) techniques, such as transcranial direct current stimulation (tDCS) and transcranial magnetic stimulation (TMS), represent another option that requires further evidence as a treatment for cognitive impairment [13, 163].

TMS [84] and tDCS [105, 106, 124, 168] have both shown evidence of efficacy on cognitive outcomes in people with schizophrenia, particularly in the working memory domain. However, limited evidence has been provided regarding the impact of these approaches on global cognitive performance and, most importantly of its effectiveness in improving real-life functional outcomes: from this perspective, more research is currently needed to assess the role of NIBS in the treatment of cognitive impairment in schizophrenia. It should also be noted that NIBS could be combined with evidence-based interventions, such as CR, into structured programs [145]: this approach represents another interesting perspective for future studies.

18.4 Negative Symptoms of Schizophrenia

Negative symptoms have traditionally been regarded as a core aspect of schizophrenia, significantly impacting the lives of persons suffering from the disorder, and posing challenges to healthcare systems globally. They often precede the first acute psychotic episode or appear early in the course of the illness, and exhibit high stability over time, with a huge impact on functioning, especially in the presence of poor premorbid adjustment. However, they may also emerge later in the course of the disorder [60, 62, 65].

Despite their acknowledged importance, these symptoms have been largely neglected and understudied, partly due to their marginal relevance for diagnosing the disorder and partly due to the belief that they cannot be assessed reliably and cannot be treated. As they are not required for the diagnosis of schizophrenia in either ICD-11 or DSM-5, they are neither regularly assessed by clinicians nor regularly addressed by treatment plans. The documented impact of these symptoms on real-life of people suffering from schizophrenia and of their informal caregivers has been largely documented [64, 75], urging the scientific and clinical community to regard negative symptoms as an important unmet need in the treatment of schizophrenia and, as a consequence multiply the efforts to improve their evaluation methods, increase their dissemination and hopefully come to a good degree of agreement in their detection and evaluation. These steps will not only improve our clinical practice but will likely provide a more adequate basis for advancing our knowledge on their etiopathological mechanisms.

18.5 Assessment of Negative Symptoms in Schizophrenia

Progress in the definition and classification of negative symptoms, implementation and dissemination of new assessment tools, and strong evidence of their profound influence on functional outcomes and response to treatment, have fueled renewed interest in relevant research and clinical practice. A consensus on which symptoms should be considered negative symptoms and how they should be defined was reached in 2006 [95]. At that time, it began clear that largely used instruments for the assessment of negative symptoms had several limitations: they focused mostly on behavioral aspects, in addition to items relevant to the evaluation of cognition, they included items relevant to disorganization or psychosocial functioning (thus leading to a high degree of overlap between measures of negative symptoms and psychosocial functioning) and, ignoring persons' internal experience, they did not help to differentiate between primary and secondary negative symptoms. The introduction of assessment tools aimed to overcome such limitations has been an important step forward, especially for training purposes and research. The second-generation rating scales for the assessment of negative symptoms of schizophrenia, i.e., the Brief Negative Symptom Scale (BNSS, [96]) and the Clinical Assessment Interview for Negative Symptoms (CAINS, [100]), have certainly represented an important step forward, and in fact are recommended.

However, the uptake of the new instruments has been unsatisfactory, their psychometric validation is still in progress, and they are not yet recommended by regulatory agencies. As a result, in clinical settings, negative symptoms remain either non-assessed or assessed by using traditional instruments, particularly the PANSS; regulatory agencies have not changed their requirements, and clinical trials for treatments targeting negative symptoms still rely upon old definitions and assessment tools, and centres recruiting patients for clinical trials often lack a specific expertise in the correct assessment of negative symptoms. It is not surprising that breakthrough discoveries leading to the introduction of game changers in the treatment of negative symptoms are still to come.

On the positive side, there is no longer a discussion that negative symptoms do represent an important unmet need in the treatment of persons with schizophrenia, and investments in training and research are needed to advance the field. Important guidance papers for the assessment and treatment of these psychopathological dimensions have been implemented, such as the ones produced by the Schizophrenia section of the European Psychiatric Association [59, 61], including, among its members, renowned experts in the field of negative symptoms. These guidance papers can contribute to improve the standards in training, clinical practice, and research, while highlighting the many areas still in need of further investigation and consensus. To provide an example, the endless debate on the opportunity to use, in clinical trials and neurobiological research, the total score of two main factors in which consented upon negative symptoms tend to cluster (i.e., the Experiential and the Expressive factor) in the light of more recent evidence supporting the high degree of independence of the five domains (Avolition, Asociality, Anhedonia, Alogia and Blunted affect) [2]. In addition to the need to favor a shift towards second-generation rating scales in clinical trials, the need to learn more from the person's internal experience of negative symptoms by using self-assessment instruments should also be emphasized, as done in this book by Dollfus et al. [114].

Important innovation in the field can be expected from the implementation of digital measures for negative symptoms [33], that currently are in an early stage of development and require input by multiple stakeholders to improve data protection and sustainability of available apps, to collect normative data, and provide validated measures of negative symptoms in real-life.

18.6 Neurobiological Research on Negative Symptoms

Research on neurobiological substrates of negative symptoms has recently focused on the two factors and, as highlighted in the chapter by Pallis et al. [136]. As summarized by the authors, the main hypothesis emerging from a recent metanalysis supports an involvement of striatal regions in the pathogenesis of negative symptoms: a hypoactivation of the ventral striatum during the anticipation of a reward seems to stem from the use of first generation antipsychotics, and therefore mainly accounting for negative symptoms secondary to a blockade of the D2 receptors, while a reduced activation in the dorsal striatum during similar tasks has been linked to apathy, affecting avolition possibly more than anhedonia. Dysfunctional effort-based decision making has also been studied as a potential mechanism underlying the amotivational psychopathology [111]. Worth noting are also the studies investigating the relationships between cognition and motivation, the main unmet needs in the current treatment of persons with schizophrenia. Further mechanisms explored as potential substrates of the Experiential factor of negative symptoms are summarized in Pallis et al.'s chapter [136], though none of them seem to be fully supported by current evidence. The Expressive domain of negative symptoms, for its close relationships with different cognitive abilities (e.g., insufficiency of cognitive resources or motor system dysfunctions), is still awaiting a fully developed cognitive model to substantiate neuroimaging findings relevant to it.

Future research on the pathophysiology of negative symptoms using structural, functional, and molecular neuroimaging, as well as genetics, might benefit from a systematic uptake of the most recent assessment tools reflecting current conceptualizations and classifications of negative symptoms, and of large cohorts enabling the use of five instead of two negative symptoms domains.

18.7 Treatment of Negative Symptoms

To date, no drug has received an indication for the treatment of negative symptoms from regulatory agencies.

Treatment with first-generation antipsychotics in patients with schizophrenia presenting negative symptoms should always be avoided [61]. In addition, as emphasized in Bitter and Czobor's chapter and described in Blasio et al.'s [17], the treatment of negative symptoms should start with the distinction between primary and secondary negative symptoms. In fact, providing an effective treatment for secondary negative symptoms is possible, while for primary negative symptoms, current tools might lead to an improvement, but cannot be regarded as an effective treatment.

Recommendations for the treatment of negative symptoms secondary to positive symptoms, depression, or extrapyramidal symptoms can be found in the Guidance paper of the European Psychiatric Association focusing on the treatment of negative symptoms [61], as well as in Bitter and Czobor [16]. The latter contribution also addressed the treatment and prevention of negative symptoms secondary to or associated with diabetes mellitus in schizophrenia and highlights the need for combining pharmacological and psychosocial treatments to properly address the complex relationship.

It is well known that the introduction in the clinical practice of second-generation antipsychotics created great expectations about their potential to treat negative symptoms of schizophrenia. A meta-analysis found that second- but not first-generation antipsychotics had a significant effect on negative symptoms, regarded by the authors as mainly due to the effect of second-generation antipsychotics on secondary negative symptoms [58]. The relatively higher affinity for the 5-HT2 (serotonin) receptor than for the D2 (dopamine) receptor might minimize the adverse effects associated with D2-receptor blockade and explain the finding [58]. However, even a statistically significant effect did not reach the threshold for being regarded as clinically meaningful.

Unfortunately, up to now, the research in the field has not yet led to the discovery of effective pharmacological treatments for primary and/or persistent negative symptoms. As described in Bitter & Czobor [16] several factors may account for the failure, including the difficult to explain high placebo response in randomized controlled trials aimed to demonstrate superiority of new drugs versus placebo, the frequent use of a total score for negative symptoms on first generation assessment tools (deemed inadequate by the scientific community but still required by the regulatory agencies) and the failure to distinguish primary/persistent from secondary/transient negative symptoms. New mechanisms of action, beyond D2 antagonism/partial agonism, have been explored, as either adjunctive or monotherapy, but results have been disappointing: Pimavanserin, a 5-HT2A receptor inverse agonist/antagonist, Roluperidone, 5-HT2A and σ2 receptor antagonist, Luvadaxistat, a D-amino acid oxidase (DAAO) inhibitor, are recent examples of unfulfilled expectations. Stimulants of the N-methyl-D-aspartate receptor, modulators of the metabotropic-glutamate receptor, or of the cholinergic receptors, represent new mechanisms awaiting evidence relevant to the treatment of schizophrenia negative symptoms [172].

Available evidence suggests a potential role of non-invasive brain stimulation techniques, in particular the stimulation of the left dorsolateral prefrontal cortex (DLPFC) by using the transcranial magnetic stimulation or the transcranial direct-current stimulation appears promising, but in need of further studies focusing on primary/persistent negative symptoms, and on the integration with pharmacological and psychosocial treatments.

In light of the limited benefits of currently available pharmacological treatment on negative symptoms, as also reviewed in Pillny et al.'s chapter on psychological interventions [136], cognitive behavioral therapy has shown a small/moderate impact; however, the failure to examine its impact on the different domains of negative symptoms might have obscured the impact on the avolition domain. Social skills training programs and cognitive remediation have proven effective for addressing negative symptoms. Innovative methodologies, including the use of virtual reality and ecological momentary assessment, should be tested further in people with negative symptoms, hoping for a greater involvement in the therapeutic programs. Further evidence is needed to provide recommendations relevant to primary/persistent negative symptoms, to the different domains of negative symptoms, and to define the specific characteristics of the intervention to target individual needs, thus improving integration and personalization of the intervention.

References

1. Abu-Akel A, Wood SJ, Upthegrove R, Chisholm K, Lin A, Hansen PC, Gillespie SM, Apperly IA, Montag C. Psychosocial functioning in the balance between autism and psychosis: evidence from three populations. Mol Psychiatry. 2022;27:2976–84. https://doi.org/10.1038/s41380-022-01543-5.
2. Ahmed AO, Kirkpatrick B, Galderisi S, Mucci A, Rossi A, Bertolino A, Rocca P, Maj M, Kaiser S, Bischof M, Hartmann-Riemer MN, Kirschner M, Schneider K, Garcia-Portilla MP, Mane A, Bernardo M, Fernandez-Egea E, Jiefeng C, Jing Y, Shuping T, Gold JM, Allen DN, Strauss GP. Cross-cultural validation of the 5-factor structure of negative symptoms in schizophrenia. Schizophr Bull. 2019;45(2):305–14. https://doi.org/10.1093/schbul/sby050.
3. Alptekin K, Akvardar Y, Akdede BBK, Dumlu K, Işık D, Pirinçci F, Yahssin S, Kitiş A. Is quality of life associated with cognitive impairment in schizophrenia? Prog Neuro-Psychopharmacol Biol Psychiatry. 2005;29:239–44. https://doi.org/10.1016/j.pnpbp.2004.11.006.
4. Altman RAE, Reser M, Tan EJ, Rossell SL. Cognitive remediation for schizophrenia: clinician perspectives on implementation barriers and facilitators. Rehabil Psychol. 2024;69:171–83. https://doi.org/10.1037/rep0000552.
5. Altman RAE, Tan EJ, Rossell SL. Factors impacting access and engagement of cognitive remediation therapy for people with schizophrenia: a systematic review. Can J Psychiatr. 2023;68:139–51. https://doi.org/10.1177/07067437221129073.
6. Ang MS, Abdul Rashid NA, Lam M, Rapisarda A, Kraus M, Keefe RSE, Lee J. The impact of medication anticholinergic burden on cognitive performance in people with schizophrenia. J Clin Psychopharmacol. 2017;37:651. https://doi.org/10.1097/JCP.0000000000000790.
7. Atkins AS, Kraus MS, Welch M, Yuan Z, Stevens H, Welsh-Bohmer KA, Keefe RSE. Remote self-administration of digital cognitive tests using the brief assessment of cognition: feasibility,

reliability, and sensitivity to subjective cognitive decline. Front Psych. 2022;13:910896. https://doi.org/10.3389/fpsyt.2022.910896.
8. Atkins AS, Tseng T, Vaughan A, Twamley EW, Harvey P, Patterson T, Narasimhan M, Keefe RSE. Validation of the tablet-administered brief assessment of cognition (BAC app). Schizophr Res. 2017;181:100–6. https://doi.org/10.1016/j.schres.2016.10.010.
9. Baldez DP, Biazus TB, Rabelo-da-Ponte FD, Nogaro GP, Martins DS, Kunz M, Czepielewski LS. The effect of antipsychotics on the cognitive performance of individuals with psychotic disorders: network meta-analyses of randomized controlled trials. Neurosci Biobehav Rev. 2021;126:265–75. https://doi.org/10.1016/j.neubiorev.2021.03.028.
10. Barlati S, Nibbio G, Vita A. Evidence-based psychosocial interventions in schizophrenia: a critical review. Curr Opin Psychiatry. 2024;37:131–9. https://doi.org/10.1097/YCO.0000000000000925.
11. Barnes TR, Drake R, Paton C, Cooper SJ, Deakin B, Ferrier IN, Gregory CJ, Haddad PM, Howes OD, Jones I, Joyce EM, Lewis S, Lingford-Hughes A, MacCabe JH, Owens DC, Patel MX, Sinclair JM, Stone JM, Talbot PS, Upthegrove R, Wieck A, Yung AR. Evidence-based guidelines for the pharmacological treatment of schizophrenia: updated recommendations from the British Association for Psychopharmacology. J Psychopharmacol. 2020;34:3–78. https://doi.org/10.1177/0269881119889296.
12. Basch E. Patient-reported outcomes—harnessing patients' voices to improve clinical care. N Engl J Med. 2017;376:105–8. https://doi.org/10.1056/NEJMp1611252.
13. Begemann MJ, Brand BA, Ćurčić-Blake B, Aleman A, Sommer IE. Efficacy of non-invasive brain stimulation on cognitive functioning in brain disorders: a meta-analysis. Psychol Med. 2020;50:2465–86. https://doi.org/10.1017/S0033291720003670.
14. Best MW, Romanowska S, Zhou Y, Wang L, Leibovitz T, Onno KA, Jagtap S, Bowie CR. Efficacy of remotely delivered evidence-based psychosocial treatments for schizophrenia-Spectrum disorders: a series of systematic reviews and meta-analyses. Schizophr Bull. 2023;49:973–86. https://doi.org/10.1093/schbul/sbac209.
15. Biagianti B, Fisher M, Howard L, Rowlands A, Vinogradov S, Woolley J. Feasibility and preliminary efficacy of remotely delivering cognitive training to people with schizophrenia using tablets. Schizophr Res Cogn. 2017;10:7–14. https://doi.org/10.1016/j.scog.2017.07.003.
16. Bitter I, Czobor P. Negative symptoms in schizophrenia: the role of pharmacotherapy (this book)
17. Blasio P, Giordano GM., Pezzella P, Mucci A, Galderisi S. Negative symptoms in schizophrenia: definition, classification and course (this book).
18. Bora E, Akdede BB, Alptekin K. The relationship between cognitive impairment in schizophrenia and metabolic syndrome: a systematic review and meta-analysis. Psychol Med. 2017;47:1030–40. https://doi.org/10.1017/S0033291716003366.
19. Bora E, Lin A, Wood SJ, Yung AR, McGorry PD, Pantelis C. Cognitive deficits in youth with familial and clinical high risk to psychosis: a systematic review and meta-analysis. Acta Psychiatr Scand. 2014;130:1–15. https://doi.org/10.1111/acps.12261.
20. Bourque J, Potvin S. Cannabis and cognitive functioning: from acute to residual effects, from randomized controlled trials to prospective designs. Front Psychiatry. 2021;12
21. Bowie CR, Bell MD, Fiszdon JM, Johannesen JK, Lindenmayer J-P, McGurk SR, Medalia AA, Penadés R, Saperstein AM, Twamley EW, Ueland T, Wykes T. Cognitive remediation for schizophrenia: an expert working group white paper on core techniques. Schizophr Res. 2020;215:49–53. https://doi.org/10.1016/j.schres.2019.10.047.
22. Bowie CR, Leung WW, Reichenberg A, McClure MM, Patterson TL, Heaton RK, Harvey PD. Predicting schizophrenia patients' real-world behavior with specific neuropsychological and functional capacity measures. Biol Psychiatry. 2008;63:505–11. https://doi.org/10.1016/j.biopsych.2007.05.022.
23. Bridges JFP, Slawik L, Schmeding A, Reimer J, Naber D, Kuhnigk O. A test of concordance between patient and psychiatrist valuations of multiple treatment goals for schizophrenia. Health Expect. 2013;16:164–76. https://doi.org/10.1111/j.1369-7625.2011.00704.x.

24. Çakici N, van Beveren NJM, Judge-Hundal G, Koola MM, Sommer IEC. An update on the efficacy of anti-inflammatory agents for patients with schizophrenia: a meta-analysis. Psychol Med. 2019;49:2307–19. https://doi.org/10.1017/S0033291719001995.
25. Carruthers SP, Brunetti G, Rossell SL. Sleep disturbances and cognitive impairment in schizophrenia spectrum disorders: a systematic review and narrative synthesis. Sleep Med. 2021;84:8–19. https://doi.org/10.1016/j.sleep.2021.05.011.
26. Charernboon T, Chompookard P. Detecting cognitive impairment in patients with schizophrenia with the Addenbrooke's cognitive examination. Asian J Psychiatr. 2019;40:19–22. https://doi.org/10.1016/j.ajp.2019.01.006.
27. Chen XJ, Wang DM, Zhou HX, Zhu RR, Tian Y, Du YX, Chen JJ, Chen DC, Wang L, Zhang XY. Association of depressive symptoms with cognitive impairment in patients with never-treated first-episode schizophrenia: analysis of the depression in schizophrenia in China (DISC) study. Gen Hosp Psychiatry. 2021;71:108–13. https://doi.org/10.1016/j.genhosppsych.2021.04.010.
28. Cho M, Lee TY, Kwak YB, Yoon YB, Kim M, Kwon JS. Adjunctive use of anti-inflammatory drugs for schizophrenia: a meta-analytic investigation of randomized controlled trials. Aust N Z J Psychiatry. 2019;53:742–59. https://doi.org/10.1177/0004867419835028.
29. Contreras NA, Lee S, Tan EJ, Castle DJ, Rossell SL. How is cognitive remediation training perceived by people with schizophrenia? A qualitative study examining personal experiences. J Ment Health. 2016;25:260–6. https://doi.org/10.3109/09638237.2016.1167856.
30. Corponi F, Fabbri C, Bitter I, Montgomery S, Vieta E, Kasper S, Pallanti S, Serretti A. Novel antipsychotics specificity profile: a clinically oriented review of lurasidone, brexpiprazole, cariprazine and lumateperone. Eur Neuropsychopharmacol. 2019;29:971–85. https://doi.org/10.1016/j.euroneuro.2019.06.008.
31. Correll CU, Ismail Z, McIntyre RS, Rafeyan R, Thase ME. Patient functioning, life engagement, and treatment goals in schizophrenia. J Clin Psychiatry. 2022;83:42345. https://doi.org/10.4088/JCP.LU21112AH2.
32. Dai Y, Ding H, Lu X, Wu X, Xu C, Jiang T, Ming L, Xia Z, Song C, Shen H, Hao W, Huang S. CCRT and aerobic exercise: a randomised controlled study of processing speed, cognitive flexibility, and serum BDNF expression in schizophrenia. Schizophrenia (Heidelb). 2022;8:84. https://doi.org/10.1038/s41537-022-00297-x.
33. Daniel DG, Cohen AS, Velligan D, Harvey PD, Alphs L, Davidson M, Potter W, Kott A, Schooler N, Brodie CR, Moore RC, Lindenmeyer P, Marder SR. Remote assessment of negative symptoms of schizophrenia. Schizophr Bull Open. 2023;4(1):sgad001 https://doi.org/10.1093/schizbullopen/sgad001.
34. Dark FL, Amado I, Erlich MD, Ikezawa S. International experience of implementing cognitive remediation for people with psychotic disorders. Schizophr Bull. 2024;sbae071 https://doi.org/10.1093/schbul/sbae071.
35. Dauwan M, Begemann MJH, Heringa SM, Sommer IE. Exercise improves clinical symptoms, quality of life, global functioning, and depression in schizophrenia: a systematic review and meta-analysis. Schizophr Bull. 2016;42:588–99. https://doi.org/10.1093/schbul/sbv164.
36. Désaméricq G, Schurhoff F, Meary A, Szöke A, Macquin-Mavier I, Bachoud-Lévi AC, Maison P. Long-term neurocognitive effects of antipsychotics in schizophrenia: a network meta-analysis. Eur J Clin Pharmacol. 2014;70:127–34. https://doi.org/10.1007/s00228-013-1600-y.
37. Deste G, Corbo D, Nibbio G, Italia M, Dell'Ovo D, Calzavara-Pinton I, Lisoni J, Barlati S, Gasparotti R, Vita A. Impact of physical exercise alone or in combination with cognitive remediation on cognitive functions in people with schizophrenia: a qualitative critical review. Brain Sci. 2023;13:320. https://doi.org/10.3390/brainsci13020320.
38. Deste G, Kakli M, Barlati S, Nibbio G, Dossou P, Degila SL, Ceraso A, Lisoni J, Calzavara-Pinton I, Villa S, Vita A. Application of cognitive remediation in the world: new experiences from two schizophrenia rehabilitation centers in Togo and Benin. Soc Psychiatry Psychiatr Epidemiol. 2024; https://doi.org/10.1007/s00127-023-02603-z.

39. Deste G, Vita A, Nibbio G, Penn DL, Pinkham AE, Harvey PD. Autistic symptoms and social cognition predict real-world outcomes in patients with schizophrenia. Front Psych. 2020a;11:524. https://doi.org/10.3389/fpsyt.2020.00524.
40. Deste G, Vita A, Penn DL, Pinkham AE, Nibbio G, Harvey PD. Autistic symptoms predict social cognitive performance in patients with schizophrenia. Schizophr Res. 2020b;215:113–9. https://doi.org/10.1016/j.schres.2019.11.008.
41. Donohoe G, Dillon R, Hargreaves A, Mothersill O, Castorina M, Furey E, Fagan AJ, Meaney JF, Fitzmaurice B, Hallahan B, McDonald C, Wykes T, Corvin A, Robertson IH. Effectiveness of a low support, remotely accessible, cognitive remediation training programme for chronic psychosis: cognitive, functional and cortical outcomes from a single blind randomised controlled trial. Psychol Med. 2018;48:751–64. https://doi.org/10.1017/S0033291717001982.
42. Eum S, Hill SK, Alliey-Rodriguez N, Stevenson JM, Rubin LH, Lee AM, Mills LJ, Reilly JL, Lencer R, Keedy SK, Ivleva E, Keefe RSE, Pearlson GD, Clementz BA, Tamminga CA, Keshavan MS, Gershon ES, Sweeney JA, Bishop JR. Genome-wide association study accounting for anticholinergic burden to examine cognitive dysfunction in psychotic disorders. Neuropsychopharmacology. 2021;46(10):1802–10. https://doi.org/10.1038/s41386-021-01057-8.
43. Eum S, Hill SK, Rubin LH, Carnahan RM, Reilly JL, Ivleva EI, Keedy SK, Tamminga CA, Pearlson GD, Clementz BA, Gershon ES, Keshavan MS, Keefe RSE, Sweeney JA, Bishop JR. Cognitive burden of anticholinergic medications in psychotic disorders. Schizophr Res. 2017;190:129–35. https://doi.org/10.1016/j.schres.2017.03.034.
44. Fatouros-Bergman H, Cervenka S, Flyckt L, Edman G, Farde L. Meta-analysis of cognitive performance in drug-naïve patients with schizophrenia. Schizophr Res. 2014;158:156–62. https://doi.org/10.1016/j.schres.2014.06.034.
45. Fernández-Abascal B, Suárez-Pinilla P, Cobo-Corrales C, Crespo-Facorro B, Suárez-Pinilla M. In- and outpatient lifestyle interventions on diet and exercise and their effect on physical and psychological health: a systematic review and meta-analysis of randomised controlled trials in patients with schizophrenia spectrum disorders and first episode of psychosis. Neurosci Biobehav Rev. 2021;125:535–68. https://doi.org/10.1016/j.neubiorev.2021.01.005.
46. Fiorillo A, Barlati S, Bellomo A, Corrivetti G, Nicolò G, Sampogna G, Stanga V, Veltro F, Maina G, Vita A. The role of shared decision-making in improving adherence to pharmacological treatments in patients with schizophrenia: a clinical review. Ann General Psychiatry. 2020;19:43. https://doi.org/10.1186/s12991-020-00293-4.
47. Firth J, Cotter J, Elliott R, French P, Yung AR. A systematic review and meta-analysis of exercise interventions in schizophrenia patients. Psychol Med. 2015;45:1343–61. https://doi.org/10.1017/S0033291714003110.
48. Firth J, Stubbs B, Rosenbaum S, Vancampfort D, Malchow B, Schuch F, Elliott R, Nuechterlein KH, Yung AR. Aerobic exercise improves cognitive functioning in people with schizophrenia: a systematic review and meta-analysis. Schizophr Bull. 2017;43:546–56. https://doi.org/10.1093/schbul/sbw115.
49. Fisher M, Etter K, Murray A, Ghiasi N, LaCross K, Ramsay I, Currie A, Fitzpatrick K, Biagianti B, Schlosser D, Loewy R, Vinogradov S. The effects of remote cognitive training combined with a Mobile app intervention on psychosis: double-blind randomized controlled trial. J Med Internet Res. 2023;25:e48634. https://doi.org/10.2196/48634.
50. Fisher M, Loewy R, Carter C, Lee A, Ragland JD, Niendam T, Schlosser D, Pham L, Miskovich T, Vinogradov S. Neuroplasticity-based auditory training via laptop computer improves cognition in young individuals with recent onset schizophrenia. Schizophr Bull. 2015;41:250–8. https://doi.org/10.1093/schbul/sbt232.
51. Fitapelli B, Lindenmayer J-P. Advances in cognitive remediation training in schizophrenia: a review. Brain Sci. 2022;12:129. https://doi.org/10.3390/brainsci12020129.
52. Fleischhacker W, Galderisi S, Laszlovszky I, Szatmári B, Barabássy Á, Acsai K, Szalai E, Harsányi J, Earley W, Patel M, Németh G. The efficacy of cariprazine in negative symptoms of schizophrenia: post hoc analyses of PANSS individual items and PANSS-derived factors. Eur Psychiatry. 2019;58:1–9. https://doi.org/10.1016/j.eurpsy.2019.01.015.

53. Fleischhacker WW, Podhorna J, Gröschl M, Hake S, Zhao Y, Huang S, Keefe RSE, Desch M, Brenner R, Walling DP, Mantero-Atienza E, Nakagome K, Pollentier S. Efficacy and safety of the novel glycine transporter inhibitor BI 425809 once daily in patients with schizophrenia: a double-blind, randomised, placebo-controlled phase 2 study. Lancet Psychiatry. 2021;8:191–201. https://doi.org/10.1016/S2215-0366(20)30513-7.
54. Fond G, Berna F, Boyer L, Godin O, Brunel L, Andrianarisoa M, Aouizerate B, Capdevielle D, Chereau I, Danion JM, Dubertret C, Dubreucq J, Faget C, Gabayet F, Le Gloahec T, Llorca PM, Mallet J, Misdrahi D, Rey R, Richieri R, Passerieux C, Portalier C, Roux P, Vehier A, Yazbek H, Schürhoff F, Bulzacka E, Andrianarisoa M, Aouizerate B, Berna F, Blanc O, Brunel L, Bulzacka E, Capdevielle D, Chéreau-Boudet I, Chesnoy-Servanin G, Danion J-M, D'Amato T, Deloge A, Delorme C, Denizot H, Dorey J-M, Dubertret C, Dubreucq J, Faget C, Fluttaz C, Fond G, Fonteneau S, Gabayet F, Giraud-Baro E, Hardy-Baylé M-C, Lacelle D, Lançon C, Laouamri H, Leboyer M, Gloahec L, Tifenn LS, Y., Llorca P-M, Mallet, Jasmina, Metairie, E, Misdrahi, David, Offerlin-Meyer I, Passerieux, Christine, Peri, P, Pires, S, Portalier, Céline, Rey, Romain, Roman, C, Sebilleau, M, Schandrin, A, Schurhoff, F, Tessier, A, Tronche, A-M, Urbach, M, Vaillant, F, Vehier, Aurélie, Vidailhet, P, Vilà, E, Yazbek, Hanan, Zinetti-Bertschy, A, The FACE-SZ (FondaMental Academic Centers of Expertise for Schizophrenia) Group. Benzodiazepine long-term administration is associated with impaired attention/working memory in schizophrenia: results from the national multicentre FACE-SZ data set. Eur Arch Psychiatry Clin Neurosci. 2018;268:17–26. https://doi.org/10.1007/s00406-017-0787-9.
55. Fradley R, Goetghebeur P, Miller D, Burley R, Almond S, Gruart I, Massó A, Delgado, García JM, Zhu B, Howley E, Neill JC, Grayson B, Gaskin P, Carlton M, Gray I, Serrats J, Davies CH. Luvadaxistat: a novel potent and selective D-amino acid oxidase inhibitor improves cognitive and social deficits in rodent models for schizophrenia. Neurochem Res. 2023;48:3027–41. https://doi.org/10.1007/s11064-023-03956-2.
56. Frazer KM, Richards Q, Keith DR. The long-term effects of cocaine use on cognitive functioning: a systematic critical review. Behav Brain Res. 2018;348:241–62. https://doi.org/10.1016/j.bbr.2018.04.005.
57. Fusar-Poli P, Deste G, Smieskova R, Barlati S, Yung AR, Howes O, Stieglitz R-D, Vita A, McGuire P, Borgwardt S. Cognitive functioning in prodromal psychosis: a meta-analysis. Arch Gen Psychiatry. 2012;69:562–71. https://doi.org/10.1001/archgenpsychiatry.2011.1592.
58. Fusar-Poli P, Papanastasiou E, Stahl D, Rocchetti M, Carpenter W, Shergill S, McGuire P. Treatments of negative symptoms in schizophrenia: meta-analysis of 168 randomized placebo-controlled trials. Schizophr Bull. 2015;41(4):892–9. https://doi.org/10.1093/schbul/sbu170.
59. Galderisi S, De Hert M, Del Prato S, Fagiolini A, Gorwood P, Leucht S, Maggioni AP, Mucci A, Arango C. Identification and management of cardiometabolic risk in subjects with schizophrenia spectrum disorders: a Delphi expert consensus study. Eur Psychiatry. 2021a;64:e7. https://doi.org/10.1192/j.eurpsy.2020.115.
60. Galderisi S, Kaiser S. The pathophysiology of negative symptoms of schizophrenia: main hypotheses and open challenges. Br J Psychiatry. 2023;223(1):298–300. https://doi.org/10.1192/bjp.2023.63.
61. Galderisi S, Kaiser S, Bitter I, Nordentoft M, Mucci A, Sabé M, Giordano GM, Nielsen MØ, Glenthøj LB, Pezzella P, Falkai P, Dollfus S, Gaebel W. EPA guidance on treatment of negative symptoms in schizophrenia. Eur Psychiatry. 2021b;64(1):e21. https://doi.org/10.1192/j.eurpsy.2021.13.
62. Galderisi S, Mucci A, Buchanan RW, Arango C. Negative symptoms of schizophrenia: new developments and unanswered research questions. Lancet Psychiatry. 2018a;5(8):664–77. https://doi.org/10.1016/S2215-0366(18)30050-6.
63. Galderisi S, Mucci A, Dollfus S, Nordentoft M, Falkai P, Kaiser S, Giordano GM, Vandevelde A, Nielsen MØ, Glenthøj LB, Sabé M, Pezzella P, Bitter I, Gaebel W. EPA guidance on assessment of negative symptoms in schizophrenia. Eur Psychiatry. 2021c;64(1):e23. https://doi.org/10.1192/j.eurpsy.2021.11.

64. Galderisi S, Rossi A, Rocca P, Bertolino A, Mucci A, Bucci P, Rucci P, Gibertoni D, Aguglia E, Amore M, Bellomo A, Biondi M, Brugnoli R, Dell'Osso L, De Ronchi D, Di Emidio G, Di Giannantonio M, Fagiolini A, Marchesi C, Monteleone P, Oldani L, Pinna F, Roncone R, Sacchetti E, Santonastaso P, Siracusano A, Vita A, Zeppegno P, Maj M, Italian Network For Research on Psychoses. The influence of illness-related variables, personal resources and context-related factors on real-life functioning of people with schizophrenia. World Psychiatry. 2014;13(3):275–87. https://doi.org/10.1002/wps.20167.
65. Galderisi S, Rucci P, Kirkpatrick B, Mucci A, Gibertoni D, Rocca P, Rossi A, Bertolino A, Strauss GP, Aguglia E, Bellomo A, Murri MB, Bucci P, Carpiniello B, Comparelli A, Cuomo A, De Berardis D, Dell'Osso L, Di Fabio F, Gelao B, Marchesi C, Monteleone P, Montemagni C, Orsenigo G, Pacitti F, Roncone R, Santonastaso P, Siracusano A, Vignapiano A, Vita A, Zeppegno P, Maj M, Italian Network for Research on Psychoses. Interplay among psychopathologic variables, personal resources, context-related factors, and real-life functioning in individuals with schizophrenia: a network analysis. JAMA Psychiatry. 2018b;75:396–404. https://doi.org/10.1001/jamapsychiatry.2017.4607.
66. Galletly C, Castle D, Dark F, Humberstone V, Jablensky A, Killackey E, Kulkarni J, McGorry P, Nielssen O, Tran N. Royal Australian and New Zealand College of Psychiatrists clinical practice guidelines for the management of schizophrenia and related disorders. Aust NZ J Psychiatry. 2016;50:410–72.
67. Gillouin P-A, Mattatia D, Bouvet C. Virtual reality for clinical evaluation and treatment in schizophrenia: a systematic review. Psychosis. 2024;16:212–29. https://doi.org/10.1080/17522439.2023.2197030.
68. Goonathilake P, Ediriweera D, Ruban R, Isuru A. Prevalence and correlates of cognitive impairment in schizophrenia: a cross-sectional study from a teaching hospital southern Sri Lanka. BMC Psychiatry. 2022;22:716. https://doi.org/10.1186/s12888-022-04368-2.
69. Green MF, Horan WP, Lee J. Nonsocial and social cognition in schizophrenia: current evidence and future directions. World Psychiatry. 2019;18:146–61. https://doi.org/10.1002/wps.20624.
70. Grover SRP, Sahoo S, Gopal S, Nehra R, Ganesh A, Raghavan V, Sankaranarayan A. Relationship of metabolic syndrome and neurocognitive deficits in patients with schizophrenia. Psychiatry Res. 2019;278:56–64. https://doi.org/10.1016/j.psychres.2019.05.023.
71. Hagi K, Nosaka T, Dickinson D, Lindenmayer JP, Lee J, Friedman J, Boyer L, Han M, Abdul-Rashid NA, Correll CU. Association between cardiovascular risk factors and cognitive impairment in people with schizophrenia: a systematic review and meta-analysis. JAMA Psychiatry. 2021;78:510–8. https://doi.org/10.1001/jamapsychiatry.2021.0015.
72. Hargreaves, A., Dillon, R., Anderson-Schmidt, H., Corvin, A., Fitzmaurice, B., Castorina, M., Robertson, I.H., Donohoe, G., 2015. Computerised working-memory focused cognitive remediation therapy for psychosis—a preliminary study. Schizophr Res 169, 135–140. doi:https://doi.org/10.1016/j.schres.2015.09.004.
73. Harris AW, Kosic T, Xu J, Walker C, Gye W, Redoblado Hodge A. Web-based cognitive remediation improves supported employment outcomes in severe mental illness: randomized controlled trial. JMIR Ment Health. 2017;4:e30. https://doi.org/10.2196/mental.6982.
74. Harvey PD, Bosia M, Cavallaro R, Howes OD, Kahn RS, Leucht S, Müller DR, Penadés R, Vita A. Cognitive dysfunction in schizophrenia: an expert group paper on the current state of the art. Schizophr Res Cogn. 2022;29:100249. https://doi.org/10.1016/j.scog.2022.100249.
75. Harvey PD, Strassnig M. Predicting the severity of everyday functional disability in people with schizophrenia: cognitive deficits, functional capacity, symptoms, and health status. World Psychiatry. 2012;11:73–9. https://doi.org/10.1016/j.wpsyc.2012.05.004.
76. Hassanpour F, Zarghami M, Mouodi S, Moosazadeh M, Barzegar F, Bagheri M, Hendouei N. Adjunctive memantine treatment of schizophrenia: a double-blind, randomized placebo-controlled study. J Clin Psychopharmacol. 2019;39:634–8. https://doi.org/10.1097/JCP.0000000000001115.
77. Hatami S, Mirsepassi Z, Sedighnia A, Tehranidoost M, Masoomi M, Sharifi V. A short course computer-assisted cognitive remediation in patients with schizophrenia Spectrum disorders:

a randomized clinical trial. Basic Clin Neurosci. 2021;12:551–62. https://doi.org/10.32598/bcn.2021.3112.1.

78. Hegde S. A review of Indian research on cognitive remediation for schizophrenia. Asian J Psychiatr. 2017;25:54–9. https://doi.org/10.1016/j.ajp.2016.10.001.
79. Hernandez MEH, Michalak E, Choudhury N, Hewko M, Torres I, Menon M, Lam RW, Chakrabarty T. Co-design of a virtual reality cognitive remediation program for depression (bWell-D) with patient end users and clinicians: qualitative interview study among patients and clinicians. JMIR Serious Games. 2023;11:e43904. https://doi.org/10.2196/43904.
80. Hong J, Bang M. Anti-inflammatory strategies for schizophrenia: a review of evidence for therapeutic applications and drug repurposing. Clin Psychopharmacol Neurosci. 2020;18:10–24. https://doi.org/10.9758/cpn.2020.18.1.10.
81. Hosseini SA, Sepehrmanesh Z, Gilasi H, Ghoraishi FS. The effect of memantine versus folic acid on cognitive impairment in patients with schizophrenia: a randomized clinical trial. Iran J Psychiatry. 2023;18:258–65. https://doi.org/10.18502/ijps.v18i3.13002.
82. Jagtap S, Romanowska S, Leibovitz T, Onno KA, Burhan AM, Best MW. Can cognitive remediation therapy be delivered remotely? A review examining feasibility and acceptability of remote interventions. Schizophrenia Research: Cognition. 2022;28:100238. https://doi.org/10.1016/j.scog.2022.100238.
83. Jespersen AE, Lumbye A, Vinberg M, Glenthøj L, Nordentoft M, Wæhrens EE, Knudsen GM, Makransky G, Miskowiak KW. Effect of immersive virtual reality-based cognitive remediation in patients with mood or psychosis spectrum disorders: study protocol for a randomized, controlled, double-blinded trial. Trials. 2024;25:82. https://doi.org/10.1186/s13063-024-07910-7.
84. Jiang Y, Guo Z, Xing G, He L, Peng H, Du F, McClure MA, Mu Q. Effects of high-frequency transcranial magnetic stimulation for cognitive deficit in schizophrenia: a meta-analysis. Front Psych. 2019;10:135. https://doi.org/10.3389/fpsyt.2019.00135.
85. Joshi YB. Cholinergic functioning, cognition, and anticholinergic medication burden in schizophrenia. In: Barch DM, Young JW, editors. Cognitive functioning in schizophrenia: leveraging the RDoC framework, current topics in behavioral neurosciences. Cham: Springer International Publishing; 2023. p. 393–406. https://doi.org/10.1007/7854_2022_400.
86. Joshi YB, Thomas ML, Braff DL, Green MF, Gur RC, Gur RE, Nuechterlein KH, Stone WS, Greenwood TA, Lazzeroni LC, MacDonald LR, Molina JL, Nungaray JA, Radant AD, Silverman JM, Sprock J, Sugar CA, Tsuang DW, Tsuang MT, Turetsky BI, Swerdlow NR, Light GA. Anticholinergic Medication Burden–Associated Cognitive Impairment in Schizophrenia. AJP. 2021;178:838–47. https://doi.org/10.1176/appi.ajp.2020.20081212.
87. Kambeitz-Ilankovic L, Betz LT, Dominke C, Haas SS, Subramaniam K, Fisher M, Vinogradov S, Koutsouleris N, Kambeitz J. Multi-outcome meta-analysis (MOMA) of cognitive remediation in schizophrenia: revisiting the relevance of human coaching and elucidating interplay between multiple outcomes. Neurosci Biobehav Rev. 2019;107:828–45. https://doi.org/10.1016/j.neubiorev.2019.09.031.
88. Kanchanatawan B, Thika S, Anderson G, Galecki P, Maes M. Affective symptoms in schizophrenia are strongly associated with neurocognitive deficits indicating disorders in executive functions, visual memory, attention and social cognition. Prog Neuro-Psychopharmacol Biol Psychiatry. 2018;80:168–76. https://doi.org/10.1016/j.pnpbp.2017.06.031.
89. Kaul I, Sawchak S, Correll CU, Kakar R, Breier A, Zhu H, Miller AC, Paul SM, Brannan SK. Efficacy and safety of the muscarinic receptor agonist KarXT (xanomeline–trospium) in schizophrenia (EMERGENT-2) in the USA: results from a randomised, double-blind, placebo-controlled, flexible-dose phase 3 trial. Lancet. 2024;403:160–70. https://doi.org/10.1016/S0140-6736(23)02190-6.
90. Keefe RSE, Goldberg TE, Harvey PD, Gold JM, Poe MP, Coughenour L. The brief assessment of cognition in schizophrenia: reliability, sensitivity, and comparison with a standard neurocognitive battery. Schizophr Res. 2004;68:283–97. https://doi.org/10.1016/j.schres.2003.09.011.

91. Keepers GA, Fochtmann LJ, Anzia JM, Benjamin S, Lyness JM, Mojtabai R, Servis M, Walaszek A, Buckley P, Lenzenweger MF, Young AS, Degenhardt A, Hong S-H. The American Psychiatric Association practice guideline for the treatment of patients with schizophrenia. AJP. 2020;177:868–72. https://doi.org/10.1176/appi.ajp.2020.177901.
92. Kern RS, Nuechterlein KH, Green MF, Baade LE, Fenton WS, Gold JM, Keefe RSE, Mesholam-Gately R, Mintz J, Seidman LJ, Stover E, Marder SR. The MATRICS consensus cognitive battery, part 2: co-norming and standardization. Am J Psychiatry. 2008;165:214–20. https://doi.org/10.1176/appi.ajp.2007.07010043.
93. Khan WU, Ghazala Z, Brooks HJ, Subramaniam P, Mulsant BH, Kumar S, Voineskos AN, Blumberger DM, Kern RS, Rajji TK. The impact of anticholinergic burden on functional capacity in persons with schizophrenia across the adult life span. Schizophr Bull. 2021;47:249–57. https://doi.org/10.1093/schbul/sbaa093.
94. Kharawala S, Hastedt C, Podhorna J, Shukla H, Kappelhoff B, Harvey PD. The relationship between cognition and functioning in schizophrenia: a semi-systematic review. Schizophrenia Research: Cognition. 2022;27:100217. https://doi.org/10.1016/j.scog.2021.100217.
95. Kirkpatrick B, Fenton WS, Carpenter WT Jr, Marder SR. The NIMH-MATRICS consensus statement on negative symptoms. Schizophr Bull. 2006;32(2):214–9. https://doi.org/10.1093/schbul/sbj053.
96. Kirkpatrick B, Strauss GP, Nguyen L, Fischer BA, Daniel DG, Cienfuegos A, Marder SR. The brief negative symptom scale: psychometric properties. Schizophr Bull. 2011;37(2):300–5. https://doi.org/10.1093/schbul/sbq059.
97. Kim S, Kim E. The use of virtual reality in psychiatry: a review. Soa Chongsonyon Chongsin Uihak. 2020;31:26–32. https://doi.org/10.5765/jkacap.190037.
98. Koola MM, Looney SW, Hong H, Pillai A, Hou W. Meta-analysis of randomized controlled trials of galantamine in schizophrenia: significant cognitive enhancement. Psychiatry Res. 2020;291:113285. https://doi.org/10.1016/j.psychres.2020.113285.
99. Korman N, Stanton R, Vecchio A, Chapman J, Parker S, Martland R, Siskind D, Firth J. The effect of exercise on global, social, daily living and occupational functioning in people living with schizophrenia: a systematic review and meta-analysis. Schizophr Res. 2023;256:98–111. https://doi.org/10.1016/j.schres.2023.04.012.
100. Kring AM, Gur RE, Blanchard JJ, Horan WP, Reise SP. The clinical assessment interview for negative symptoms (CAINS): final development and validation. Am J Psychiatry. 2013;170(2):165–72. https://doi.org/10.1176/appi.ajp.2012.12010109.
101. Krzystanek M, Krysta K, Borkowski M, Skałacka K, Przybyło J, Pałasz A, Mucic D, Martyniak E, Waszkiewicz N. The effect of smartphone-based cognitive training on the functional/cognitive markers of schizophrenia: a one-year randomized study. J Clin Med. 2020;9:3681. https://doi.org/10.3390/jcm9113681.
102. Laskemoen JF, Büchmann C, Barrett EA, Collier-Høegh M, Haatveit B, Vedal TJ, Ueland T, Melle I, Aas M, Simonsen C. Do sleep disturbances contribute to cognitive impairments in schizophrenia spectrum and bipolar disorders? Eur Arch Psychiatry Clin Neurosci. 2020;270:749–59. https://doi.org/10.1007/s00406-019-01075-0.
103. Lee TY, Hong SB, Shin NY, Kwon JS. Social cognitive functioning in prodromal psychosis: a meta-analysis. Schizophr Res. 2015;164:28–34. https://doi.org/10.1016/j.schres.2015.02.008.
104. Lejeune JA, Northrop A, Kurtz MM. A meta-analysis of cognitive remediation for schizophrenia: efficacy and the role of participant and treatment factors. Schizophr Bull. 2021;47:997–1006. https://doi.org/10.1093/schbul/sbab022.
105. Lisoni J, Baldacci G, Nibbio G, Zucchetti A, Butti Lemmi Gigli E, Savorelli A, Facchi M, Miotto P, Deste G, Barlati S, Vita A. Effects of bilateral, bipolar-nonbalanced, frontal transcranial direct current stimulation (tDCS) on negative symptoms and neurocognition in a sample of patients living with schizophrenia: results of a randomized double-blind sham-controlled trial. J Psychiatr Res. 2022;155:430–42. https://doi.org/10.1016/j.jpsychires.2022.09.011.
106. Liu Y, Gu N, Cao X, Zhu Y, Wang J, Smith RC, Li C. Effects of transcranial electrical stimulation on working memory in patients with schizophrenia: a systematic review and meta-analysis. Psychiatry Res. 2021;296:113656. https://doi.org/10.1016/j.psychres.2020.113656.

107. Lovell ME, Akhurst J, Padgett C, Garry MI, Matthews A. Cognitive outcomes associated with long-term, regular, recreational cannabis use in adults: a meta-analysis. Exp Clin Psychopharmacol. 2020;28:471–94. https://doi.org/10.1037/pha0000326.
108. Maat A, Fett A-K, Derks E. Social cognition and quality of life in schizophrenia. Schizophr Res. 2012;137:212–8. https://doi.org/10.1016/j.schres.2012.02.017.
109. Maj M, Stein DJ, Parker G, Zimmerman M, Fava GA, De Hert M, Demyttenaere K, McIntyre RS, Widiger T, Wittchen H-U. The clinical characterization of the adult patient with depression aimed at personalization of management. World Psychiatry. 2020;19:269–93. https://doi.org/10.1002/wps.20771.
110. Mayeli A, LaGoy AD, Smagula SF, Wilson JD, Zarbo C, Rocchetti M, Starace F, Zamparini M, Casiraghi L, Calza S, Rota M, D'Agostino A, de Girolamo G, DiAPAson Consortium, Ferrarelli, F. Shared and distinct abnormalities in sleep-wake patterns and their relationship with the negative symptoms of Schizophrenia Spectrum disorder patients. Mol Psychiatry. 2023;28:2049–57. https://doi.org/10.1038/s41380-023-02050-x.
111. Marder SR, Galderisi S. The current conceptualization of negative symptoms in schizophrenia. World Psychiatry. 2017;16(1):14–24. https://doi.org/10.1002/wps.20385.
112. McCleery A, Nuechterlein KH. Cognitive impairment in psychotic illness: prevalence, profile of impairment, developmental course, and treatment considerations. Dialogues Clin Neurosci. 2019;21:239–48. https://doi.org/10.31887/DCNS.2019.21.3/amccleery.
113. McCutcheon RA, Keefe RSE, McGuire PK. Cognitive impairment in schizophrenia: aetiology, pathophysiology, and treatment. Mol Psychiatry. 2023;28:1902–18. https://doi.org/10.1038/s41380-023-01949-9.
114. Métivier L, Dollfus S. The assessment of negative symptoms in schizophrenia. (this book).
115. Mishkind MC, Norr AM, Katz AC, Reger GM. Review of virtual reality treatment in psychiatry: evidence versus current diffusion and use. Curr Psychiatry Rep. 2017;19:80. https://doi.org/10.1007/s11920-017-0836-0.
116. Montemagni C, Del Favero E, Riccardi C, Canta L, Toye M, Zanalda E, Rocca P. Effects of cognitive remediation on cognition, metacognition, and social cognition in patients with schizophrenia. Front Psych. 2021;12:649737. https://doi.org/10.3389/fpsyt.2021.649737.
117. Moschetti V, Desch M, Goetz S, Liesenfeld K-H, Rosenbrock H, Kammerer K-P, Wunderlich G, Wind S. Safety, tolerability and pharmacokinetics of Oral BI 425809, a glycine transporter 1 inhibitor, in healthy male volunteers: a partially randomised, single-blind, placebo-controlled, first-in-human study. Eur J Drug Metab Pharmacokinet. 2018a;43:239–49. https://doi.org/10.1007/s13318-017-0440-z.
118. Moschetti V, Schlecker C, Wind S, Goetz S, Schmitt H, Schultz A, Liesenfeld K-H, Wunderlich G, Desch M. Multiple rising doses of oral BI 425809, a GlyT1 inhibitor, in Young and elderly healthy volunteers: a randomised, double-blind, phase I study investigating safety and pharmacokinetics. Clin Drug Investig. 2018b;38:737–50. https://doi.org/10.1007/s40261-018-0660-2.
119. Möser C, Krieg JC, Zihl J, Lautenbacher S. Attention and memory deficits in schizophrenia: the role of symptoms of depression. Cogn Behav Neurol. 2006;19:150–6.
120. Mote J, Fulford D. Ecological momentary assessment of everyday social experiences of people with schizophrenia: a systematic review. Schizophr Res. 2020;216:56–68. https://doi.org/10.1016/j.schres.2019.10.021.
121. Moura BM, Avila A, Chendo I, Frade P, Barandas R, Vian J, Croca M, Roberto A, Almeida C, Antunes F, Linhares L, Crawford J, Ferreira C, Santos J, Abreu M, Levy P, Figueira ML, Mendes T. Facilitating the delivery of cognitive remediation in first-episode psychosis: pilot study of a home-delivered web-based intervention. J Nerv Ment Dis. 2019;207:951–7. https://doi.org/10.1097/NMD.0000000000001055.
122. Mucci A, Galderisi S, Green MF, Nuechterlein K, Rucci P, Gibertoni D, Rossi A, Rocca P, Bertolino A, Bucci P, Hellemann G, Spisto M, Palumbo D, Aguglia E, Amodeo G, Amore M, Bellomo A, Brugnoli R, Carpiniello B, Dell'Osso L, Di Fabio F, di Giannantonio M, Di Lorenzo G, Marchesi C, Monteleone P, Montemagni C, Oldani L, Romano R, Roncone R, Stratta P, Tenconi E, Vita A, Zeppegno P, Maj M, Italian Network for Research on Psychoses.

Familial aggregation of MATRICS Consensus Cognitive Battery scores in a large sample of outpatients with schizophrenia and their unaffected relatives. Psychol Med. 2018;48:1359–66. https://doi.org/10.1017/S0033291717002902.
123. Mwesiga EK, Robbins R, Akena D, Koen N, Nakku J, Nakasujja N, Stein DJ. Prevalence, profile and associations of cognitive impairment in Ugandan first-episode psychosis patients. Schizophr Res Cogn. 2022;28:100234. https://doi.org/10.1016/j.scog.2021.100234.
124. Narita Z, Stickley A, DeVylder J, Yokoi Y, Inagawa T, Yamada Y, Maruo K, Koyanagi A, Oh H, Sawa A, Sumiyoshi T. Effect of multi-session prefrontal transcranial direct current stimulation on cognition in schizophrenia: a systematic review and meta-analysis. Schizophr Res. 2020;216:367–73. https://doi.org/10.1016/j.schres.2019.11.011.
125. Nasrallah H. The cognition self-assessment rating scale for patients with schizophrenia. Curr Psychiatr Ther. 2023;22:30–4.
126. Nibbio G, Barlati S, Cacciani P, Corsini P, Mosca A, Ceraso A, Deste G, Vita A. Evidence-based integrated intervention in patients with schizophrenia: a pilot study of feasibility and effectiveness in a real-world rehabilitation setting. Int J Environ Res Public Health. 2020;17:3352. https://doi.org/10.3390/ijerph17103352.
127. Nibbio G, Pinton IC, Bertoni L, Cicale A, Necchini N, Zardini D, Poddighe L, Lisoni J, Deste G, Barlati S, Vita A. Non-pharmacological interventions to improve cognitive impairment and psychosocial functioning in schizophrenia: a critical review. J Psychopathol. 2024;30 https://doi.org/10.36148/2284-0249-N455.
128. Nielsen RE, Levander S, Kjaersdam Telléus G, Jensen SOW, Østergaard Christensen T, Leucht S. Second-generation antipsychotic effect on cognition in patients with schizophrenia—a meta-analysis of randomized clinical trials. Acta Psychiatr Scand. 2015;131:185–96. https://doi.org/10.1111/acps.12374.
129. Nuechterlein KH, Green M. MATRICS consensus cognitive battery. Los Angeles: Manual. MATRICS Assessment Inc.; 2006.
130. Nuechterlein KH, Green MF, Kern RS, Baade LE, Barch DM, Cohen JD, Essock S, Fenton WS, Frese FJ, Gold JM, Goldberg T, Heaton RK, Keefe RSE, Kraemer H, Mesholam-Gately R, Seidman LJ, Stover E, Weinberger DR, Young AS, Zalcman S, Marder SR. The MATRICS consensus cognitive battery, part 1: test selection, reliability, and validity. AJP. 2008;165:203–13. https://doi.org/10.1176/appi.ajp.2007.07010042.
131. Nuechterlein KH, McEwen SC, Ventura J, Subotnik KL, Turner LR, Boucher M, Casaus LR, Distler MG, Hayata JN. Aerobic exercise enhances cognitive training effects in first-episode schizophrenia: randomized clinical trial demonstrates cognitive and functional gains. Psychol Med. 2022;1–11 https://doi.org/10.1017/S0033291722001696.
132. Nuechterlein KH, Nasrallah H, Velligan D. Measuring cognitive impairments associated with schizophrenia in clinical practice: overview of current challenges and future opportunities. Schizophr Bull. 2024;sbae051 https://doi.org/10.1093/schbul/sbae051.
133. O'Donnell P, Dong C, Murthy V, Asgharnejad M, Du X, Summerfelt A, Lu H, Xu L, Wendland JR, Dunayevich E, Buhl DL, Litman R, Hetrick WP, Hong LE, Rosen LB. The D-amino acid oxidase inhibitor luvadaxistat improves mismatch negativity in patients with schizophrenia in a randomized trial. Neuropsychopharmacol. 2023;48:1052–9. https://doi.org/10.1038/s41386-023-01560-0.
134. Olivola M, Bassetti N, Parente S, Arienti V, Civardi SC, Topa PA, Brondino N. Cognitive effects of Lurasidone and Cariprazine: a mini systematic review. Curr Neuropharmacol. 2023;21:2431–46. https://doi.org/10.2174/1570159X21666230727140843.
135. O'Reilly K, O'Connell P, Donohoe G, Coyle C, O'Sullivan D, Azvee Z, Maddock C, Sharma K, Sadi H, McMahon M, Kennedy HG. Anticholinergic burden in schizophrenia and ability to benefit from psychosocial treatment programmes: a 3-year prospective cohort study. Psychol Med. 2016;46:3199–211. https://doi.org/10.1017/S0033291716002154.
136. Pallis K, Kirschner M, Kaiser S, Bègue I. Negative symptoms in schizophrenia: neurobiological bases (this book).
137. Pandurangi AK, Buckley PF. Inflammation, antipsychotic drugs, and evidence for effectiveness of anti-inflammatory agents in schizophrenia. In: Khandaker GM, Meyer U,

Jones PB, editors. Neuroinflammation and schizophrenia, current topics in behavioral neurosciences. Cham: Springer International Publishing; 2020. p. 227–44. https://doi.org/10.1007/7854_2019_91.
138. Parikh NB, Robinson DM, Clayton AH. Clinical role of brexpiprazole in depression and schizophrenia. TCRM. 2017;13:299–306. https://doi.org/10.2147/TCRM.S94060.
139. Park MJ, Kim DJ, Lee U, Na EJ, Jeon HJ. A literature overview of virtual reality (VR) in treatment of psychiatric disorders: recent advances and limitations. Front Psych. 2019;10 https://doi.org/10.3389/fpsyt.2019.00505.
140. Perra A, Galetti A, Zaccheddu R, Locci A, Piludu F, Preti A, Primavera D, Di Natale L, Nardi AE, Kurotshka PK, Cossu G, Sancassiani F, Stella G, De Lorenzo V, Zreik T, Carta MG. A recovery-oriented program for people with bipolar disorder through virtual reality-based cognitive remediation: results of a feasibility randomized clinical trial. J Clin Med. 2023a;12:2142. https://doi.org/10.3390/jcm12062142.
141. Perra A, Riccardo CL, De Lorenzo V, De Marco E, Di Natale L, Kurotschka PK, Preti A, Carta MG. Fully immersive virtual reality-based cognitive remediation for adults with psychosocial disabilities: a systematic scoping review of methods intervention gaps and meta-analysis of published effectiveness studies. Int J Environ Res Public Health. 2023b;20:1527. https://doi.org/10.3390/ijerph20021527.
142. Pinkham AE, Harvey PD, Penn DL. Social cognition psychometric evaluation: results of the final validation study. Schizophr Bull. 2018;44:737–48. https://doi.org/10.1093/schbul/sbx117.
143. Pinkham AE, Penn DL, Green MF, Buck B, Healey K, Harvey PD. The social cognition psychometric evaluation study: results of the expert survey and RAND panel. Schizophr Bull. 2014;40:813–23. https://doi.org/10.1093/schbul/sbt081.
144. Pinkham AE, Penn DL, Green MF, Harvey PD. Social cognition psychometric evaluation: results of the initial psychometric study. Schizophr Bull. 2016;42:494–504. https://doi.org/10.1093/schbul/sbv056.
145. Poppe A, Ritter FD, Bais L, Pustejovsky JE, van Tol M-J, Ćurčić-Blake B, Pijnenborg GH, van Der Meer L. The efficacy of combining cognitive training and noninvasive brain stimulation: a transdiagnostic systematic review and meta-analysis. Psychol Bull. 2023;
146. Potvin S, Stavro K, Rizkallah É, Pelletier J. Cocaine and cognition: a systematic quantitative review. J Addict Med. 2014;8:368. https://doi.org/10.1097/ADM.0000000000000066.
147. Purdon S. SCIP Manual, 2005
148. Remington G, Addington D, Honer W, Ismail Z, Raedler T, Teehan M. Guidelines for the pharmacotherapy of schizophrenia in adults. Can J Psychiatr. 2017;62:604–16. https://doi.org/10.1177/0706743717720448.
149. Rose D, Wykes T, Farrier D, Doran A-M, Sporle T, Bogner D. What do clients think of cognitive remediation therapy?: a consumer-led investigation of satisfaction and side effects. Am J Psychiatr Rehabil. 2008;11:181–204.
150. Russell MT, Funsch KM, Springfield CR, Ackerman RA, Depp CA, Harvey PD, Moore RC, Pinkham AE. Validity of remote administration of the MATRICS consensus cognitive battery for individuals with severe mental illness. Schizophr Res Cogn. 2021;27:100226. https://doi.org/10.1016/j.scog.2021.100226.
151. Sabe M, Kaiser S, Sentissi O. Physical exercise for negative symptoms of schizophrenia: systematic review of randomized controlled trials and meta-analysis. Gen Hosp Psychiatry. 2020;62:13–20. https://doi.org/10.1016/j.genhosppsych.2019.11.002.
152. Saperstein AM, Bello I, Nossel I, Dixon LB, Medalia A. Implementation of cognitive health Services in Large Systems of care: highlights from coordinated specialty care for first episode psychosis. Schizophr Bull. 2024;sbae030 https://doi.org/10.1093/schbul/sbae030.
153. Sartorius N. Patient-reported outcomes in psychiatry. Dialogues Clin Neurosci. 2014;16:123–4. https://doi.org/10.31887/DCNS.2014.16.2/nsartorius.
154. Sauder C, Allen LA, Baker E, Miller AC, Paul SM, Brannan SK. Effectiveness of KarXT (xanomeline-trospium) for cognitive impairment in schizophrenia: post hoc analyses from a

randomised, double-blind, placebo-controlled phase 2 study. Transl Psychiatry. 2022;12:1–8. https://doi.org/10.1038/s41398-022-02254-9.
155. Schaefer M, Sarkar S, Theophil I, Leopold K, Heinz A, Gallinat J. Acute and long-term Memantine add-on treatment to risperidone improves cognitive dysfunction in patients with acute and chronic schizophrenia. Pharmacopsychiatry. 2020;53:21–9. https://doi.org/10.1055/a-0970-9310.
156. Schmitt A, Maurus I, Rossner MJ, Röh A, Lembeck M, von Wilmsdorff M, Takahashi S, Rauchmann B, Keeser D, Hasan A, Malchow B, Falkai P. Effects of aerobic exercise on metabolic syndrome, cardiorespiratory fitness, and symptoms in schizophrenia include decreased mortality. Front Psych. 2018;9:690. https://doi.org/10.3389/fpsyt.2018.00690.
157. Schroeder AH, Bogie BJM, Rahman TT, Thérond A, Matheson H, Guimond S. Feasibility and efficacy of virtual reality interventions to improve psychosocial functioning in psychosis: systematic review. JMIR Ment Health. 2022;9:e28502. https://doi.org/10.2196/28502.
158. Schuster J-P, Hoertel N, Gunten A von, Seigneurie A-S, Limosin F, Group, C.S. Benzodiazepine use among older adults with schizophrenia spectrum disorder: prevalence and associated factors in a multicenter study. Int Psychogeriatr. 2020;32:441–51. https://doi.org/10.1017/S1041610219000358.
159. Seccomandi B, Agbedjro D, Bell M, Keefe RSE, Keshavan M, Galderisi S, Fiszdon J, Mucci A, Cavallaro R, Ojeda N, Peña J, Müller D, Roder V, Wykes T, Cella M. Exploring the role of age as a moderator of cognitive remediation for people with schizophrenia. Schizophr Res. 2021;228:29–35. https://doi.org/10.1016/j.schres.2020.11.060.
160. Selvaggi P, Fazio L, Toro VD, Mucci A, Rocca P, Martinotti G, Cascino G, Siracusano A, Zeppegno P, Pergola G, Bertolino A, Blasi G, Galderisi S, Italian Network for Research on Psychosis. Effect of anticholinergic burden on brain activity during working memory and real-world functioning in patients with schizophrenia. Schizophr Res. 2023;260:76–84. https://doi.org/10.1016/j.schres.2023.08.015.
161. Shimada T, Ito S, Makabe A, Yamanushi A, Takenaka A, Kawano K, Kobayashi M. Aerobic exercise and cognitive functioning in schizophrenia: an updated systematic review and meta-analysis. Psychiatry Res. 2022;314:114656. https://doi.org/10.1016/j.psychres.2022.114656.
162. Sitskoorn MM, Aleman A, Ebisch SJH, Appels MCM, Kahn RS. Cognitive deficits in relatives of patients with schizophrenia: a meta-analysis. Schizophr Res. 2004;71:285–95. https://doi.org/10.1016/j.schres.2004.03.007.
163. Sloan NP, Byrne LK, Enticott PG, Lum JAG. Non-invasive brain stimulation does not improve working memory in schizophrenia: a meta-analysis of randomised controlled trials. Neuropsychol Rev. 2021;31:115–38. https://doi.org/10.1007/s11065-020-09454-4.
164. Solmi M, Croatto G, Piva G, Rosson S, Fusar-Poli P, Rubio JM, Carvalho AF, Vieta E, Arango C, DeTore NR, Eberlin ES, Mueser KT, Correll CU. Efficacy and acceptability of psychosocial interventions in schizophrenia: systematic overview and quality appraisal of the meta-analytic evidence. Mol Psychiatry. 2023;28:354–68. https://doi.org/10.1038/s41380-022-01727-z.
165. Sommer IE, van Westrhenen R, Begemann MJH, de Witte LD, Leucht S, Kahn RS. Efficacy of anti-inflammatory agents to improve symptoms in patients with schizophrenia: an update. Schizophr Bull. 2014;40:181–91. https://doi.org/10.1093/schbul/sbt139.
166. Spronk DB, van Wel JHP, Ramaekers JG, Verkes RJ. Characterizing the cognitive effects of cocaine: a comprehensive review. Neurosci Biobehav Rev. 2013;37:1838–59. https://doi.org/10.1016/j.neubiorev.2013.07.003.
167. Strassnig M, Bowie C, Pinkham AE, Penn D, Twamley EW, Patterson TL, Harvey PD. Which levels of cognitive impairments and negative symptoms are related to functional deficits in schizophrenia? J Psychiatr Res. 2018;104:124–9. https://doi.org/10.1016/j.jpsychires.2018.06.018.
168. Sun Y, Song H, Liu H, Mao F, Sun X, Cao F. Occupational stress, mental health, and self-efficacy among community mental health workers: a cross-sectional study during COVID-19 pandemic. Int J Soc Psychiatry. 2021;67:737–46. https://doi.org/10.1177/0020764020972131.

169. Thibaudeau E, Peyroux E, Franck N, Carling H, Lepage M. Navigating social cognitive impairments in schizophrenia Spectrum disorders: protocol for a pilot pre-post quasi-experimental study for remote avatar-assisted cognitive remediation therapy. JMIR Research Protocols. 2024;13:e54251. https://doi.org/10.2196/54251.
170. Thomas EC, Snethen G, Salzer MS. Community participation factors and poor neurocognitive functioning among persons with schizophrenia. Am J Orthopsychiatry. 2020;90:90–7. https://doi.org/10.1037/ort0000399.
171. Treichler EBH, Thomas ML, Bismark AW, Hochberger WC, Tarasenko M, Nungaray J, Cardoso L, Joshi YB, Zhang W, Sprock J, Swerdlow N, Cohen AN, Light GA. Divergence of subjective and performance-based cognitive gains following cognitive training in schizophrenia. Schizophr Res. 2019;210:215–20. https://doi.org/10.1016/j.schres.2018.12.034.
172. Tsapakis EM, Treiber M, Mitkani C, Drakaki Z, Cholevas A, Spanaki C, Fountoulakis KN. Pharmacological treatments of negative symptoms in schizophrenia-an update. J Clin Med. 2024;13(18):5637. https://doi.org/10.3390/jcm13185637.
173. Tsuda Y, Ugai H, Wunderlich G, Shin J-G. Pharmacokinetics of single doses of BI 425809 in healthy Chinese and Japanese subjects: a randomized study. Clin Ther. 2019;41:961–71. https://doi.org/10.1016/j.clinthera.2019.03.014.
174. Ventura J, Reise SP, Keefe RSE, Baade LE, Gold JM, Green MF, Kern RS, Mesholam-Gately R, Nuechterlein KH, Seidman LJ, Bilder RM. The cognitive assessment interview (CAI): development and validation of an empirically derived, brief interview-based measure of cognition. Schizophr Res. 2010;121:24–31. https://doi.org/10.1016/j.schres.2010.04.016.
175. Ventura J, Wilson SA, Wood RC, Hellemann GS. Cognitive training at home in schizophrenia is feasible. Schizophr Res. 2013;143:397–8. https://doi.org/10.1016/j.schres.2012.11.033.
176. Veselinović T, Neuner I. Progress and pitfalls in developing agents to treat neurocognitive deficits associated with schizophrenia. CNS Drugs. 2022;36:819–58. https://doi.org/10.1007/s40263-022-00935-z.
177. Vinogradov S, Fisher M, Warm H, Holland C, Kirshner MA, Pollock BG. The cognitive cost of anticholinergic burden: decreased response to cognitive training in schizophrenia. AJP. 2009;166:1055–62. https://doi.org/10.1176/appi.ajp.2009.09010017.
178. Vita A, Barlati S. The implementation of evidence-based psychiatric rehabilitation: challenges and opportunities for mental health services. Frontiers. Psychiatry. 2019;10
179. Vita A, Barlati S, Ceraso A, Deste G, Nibbio G, Wykes T. Acceptability of cognitive remediation for schizophrenia: a systematic review and meta-analysis of randomized controlled trials. Psychol Med. 2023a;53:3661–71. https://doi.org/10.1017/S0033291722000319.
180. Vita A, Barlati S, Ceraso A, Nibbio G, Ariu C, Deste G, Wykes T. Effectiveness, Core elements, and moderators of response of cognitive remediation for schizophrenia: a systematic review and meta-analysis of randomized clinical trials. JAMA Psychiatry. 2021;78:848–58. https://doi.org/10.1001/jamapsychiatry.2021.0620.
181. Vita A, Barlati S, Ceraso A, Nibbio G, Durante F, Facchi M, Deste G, Wykes T. Durability of effects of cognitive remediation on cognition and psychosocial functioning in schizophrenia: a systematic review and meta-analysis of randomized clinical trials. AJP. 2024a;181:520–31. https://doi.org/10.1176/appi.ajp.20230396.
182. Vita A, Barlati S, Deste G, Nibbio G, Penn DL, Pinkham AE, McIntyre RS, Harvey PD. Life engagement in people living with schizophrenia: predictors and correlates of patient life engagement in a large sample of people living in the community. Psychol Med. 2023b;53:7943–52. https://doi.org/10.1017/S0033291723002106.
183. Vita A, Barlati S, Deste G, Rocca P, Rossi A, Bertolino A, Aguglia E, Amore M, Bellomo A, Biondi M, Carpiniello B, Collantoni E, Cuomo A, D'Ambrosio E, Osso L, dell, di Giannantonio M, Giordano GM, Marchesi C, Monteleone P, Montemagni C, Oldani L, Pompili M, Roncone R, Rossi R, Siracusano A, Zeppegno P, Nibbio G, Galderisi S, Maj M, Psychoses, I.N. for R. The influence of autistic symptoms on social and non-social cognition and on real-life functioning in people with schizophrenia: evidence from the Italian network for research on psychoses multicenter study. Eur Psychiatry. 2020;63 https://doi.org/10.1192/j.eurpsy.2020.99.

184. Vita A, Gaebel W, Mucci A, Sachs G, Barlati S, Giordano GM, Nibbio G, Nordentoft M, Wykes T, Galderisi S. European psychiatric association guidance on treatment of cognitive impairment in schizophrenia. Eur Psychiatry. 2022a;65:e57. https://doi.org/10.1192/j.eurpsy.2022.2315.
185. Vita A, Gaebel W, Mucci A, Sachs G, Erfurth A, Barlati S, Zanca F, Giordano GM, Birkedal Glenthøj L, Nordentoft M, Galderisi S. European psychiatric association guidance on assessment of cognitive impairment in schizophrenia. Eur Psychiatry. 2022b;65:e58. https://doi.org/10.1192/j.eurpsy.2022.2316.
186. Vita A, Kakli M, Barlati S, Nibbio G, Dossou P, Degila SL, Villa S, Deste G. Implementation of cognitive rehabilitation interventions for schizophrenia in low-income countries: an experience from Togo and Benin. Int J Soc Psychiatry. 2023c;00207640231181526 https://doi.org/10.1177/00207640231181526.
187. Vita A, Nibbio G, Barlati S. Conceptualization and characterization of "primary" and "secondary" cognitive impairment in schizophrenia. Psychiatry Res. 2024b;340:116126. https://doi.org/10.1016/j.psychres.2024.116126.
188. Vita A, Nibbio G, Barlati S. Pharmacological treatment of cognitive impairment associated with schizophrenia: state of the art and future perspectives. Schizophrenia Bulletin Open. 2024c;5:sgae013. https://doi.org/10.1093/schizbullopen/sgae013.
189. Welch LC, Trudeau JJ, Silverstein SM, Sand M, Henderson DC, Rosen RC. Initial development of a patient-reported outcome measure of experience with cognitive impairment associated with schizophrenia. Patient Related Outcome Measures. 2017;8:71–81. https://doi.org/10.2147/PROM.S123266.
190. Wykes T. Cognitive remediation—where are we now and what should we do next? Off J Italian Soc Psychopathol. 2018;24:57–61.
191. Wykes T, Bowie CR, Cella M. Thinking about the future of cognitive remediation therapy revisited: what is left to solve before patients have access? Schizophr Bull. 2024;sbae075 https://doi.org/10.1093/schbul/sbae075.
192. Yeo H, Yoon S, Lee J, Kurtz MM, Choi K. A meta-analysis of the effects of social-cognitive training in schizophrenia: the role of treatment characteristics and study quality. Br J Clin Psychol. 2022;61:37–57. https://doi.org/10.1111/bjc.12320.
193. Yolland CO, Hanratty D, Neill E, Rossell SL, Berk M, Dean OM, Castle DJ, Tan EJ, Phillipou A, Harris AW, Barreiros AR, Hansen A, Siskind D. Meta-analysis of randomised controlled trials with N-acetylcysteine in the treatment of schizophrenia. Aust N Z J Psychiatry. 2020;54:453–66. https://doi.org/10.1177/0004867419893439.
194. Zbukvic I, Bryce S, Moullin J, Allott K. The use of implementation science to close the research-to-treatment gap for cognitive impairment in psychosis. Aust N Z J Psychiatry. 2023;00048674231160987 https://doi.org/10.1177/00048674231160987.
195. Zhang L, Zheng H, Wu R, Kosten TR, Zhang X-Y, Zhao J. The effect of minocycline on amelioration of cognitive deficits and pro-inflammatory cytokines levels in patients with schizophrenia. Schizophr Res. 2019;212:92–8. https://doi.org/10.1016/j.schres.2019.08.005.

MIX
Papier aus verantwortungsvollen Quellen
Paper from responsible sources
FSC® C105338

If you have any concerns about our products,
you can contact us on
ProductSafety@springernature.com

In case Publisher is established outside the EU,
the EU authorized representative is:
Springer Nature Customer Service Center GmbH
Europaplatz 3, 69115 Heidelberg, Germany

Printed by Libri Plureos GmbH
in Hamburg, Germany